AF683528

COGNITIVE STRUCTURES IN SCIENTIFIC INQUIRY

POZNAŃ STUDIES IN THE PHILOSOPHY OF THE SCIENCES AND THE HUMANITIES

VOLUME 84

EDITORS

Jerzy Brzeziński
Andrzej Klawiter
Piotr Kwieciński (assistant editor)
Krzysztof Łastowski
Leszek Nowak (editor-in-chief)
Izabella Nowakowa
Katarzyna Paprzycka (managing editor)
Marcin Paprzycki
Piotr Przybysz (assistant editor)
Michael J. Shaffer (Wilmington)

ADVISORY COMMITTEE

Joseph Agassi (Tel-Aviv)
Étienne Balibar (Paris)
Wolfgang Balzer (München)
Mario Bunge (Montreal)
Nancy Cartwright (London)
Robert S. Cohen (Boston)
Francesco Coniglione (Catania)
Andrzej Falkiewicz (Wrocław)
Dagfinn Føllesdal (Oslo)
Bert Hamminga (Tilburg)
Jaakko Hintikka (Boston)
Jacek J. Jadacki (Warszawa)
Jerzy Kmita (Poznań)
Leon Koj (Lublin)
Władysław Krajewski (Warszawa)
Theo A.F. Kuipers (Groningen)
Witold Marciszewski (Warszawa)
Ilkka Niiniluoto (Helsinki)
Günter Patzig (Göttingen)
Jerzy Perzanowski (Toruń)
Marian Przełęcki (Warszawa)
Jan Such (Poznań)
Max Urchs (Konstanz)
Jan Woleński (Kraków)
Ryszard Wójcicki (Warszawa)

Poznań Studies in the Philosophy of the Sciences and the Humanities
is partly sponsored by Adam Mickiewicz University

Address: dr Katarzyna Paprzycka . Instytut Filozofii . SWPS . ul. Chodakowska 19/31
03-815 Warszawa . Poland . fax: +48(0)22-517-9703 . E-mail: drp@swps edu.pl
http://main.amu.edu.pl/~pozn-stu/

MONOGRAPHS-IN-DEBATE

EDITORIAL COMMITTEE

Jerzy Brzeziński
Jacek Juliusz Jadacki
Andrzej Klawiter
Krzysztof Łastowski
Leszek Nowak
Marcin Paprzycki
Piotr Przybysz
Jan Woleński

Katarzyna Paprzycka (editor-in-chief)
Instytut Filozofii · Szkoła Wyższa Psychologii Społecznej
ul. Chodakowska 19/31 · 03-815 Warszawa · Poland
Katarzyna.Paprzycka@swps.edu.pl

Monographs-in-Debate is a new subseries of the Poznań *Studies in the Philosophy of the Sciences and the Humanities* book series. It publishes monographs that deal with the general area of interest of the *Poznań Studies* series. The special nature of the *Monographs-in-Debate* volumes arises from the recognition of the dialectical nature of philosophical work. Each volume contains a monograph, followed by peer commentaries and the author's replies.

POZNAŃ STUDIES IN THE PHILOSOPHY OF THE SCIENCES AND THE HUMANITIES, VOLUME 84
MONOGRAPHS-IN-DEBATE

Cognitive Structures in Scientific Inquiry

Essays in Debate with Theo Kuipers
Volume 2

Edited by
Roberto Festa, Atocha Aliseda
and Jeanne Peijnenburg

Rodopi

Amsterdam - New York, NY 2005

The paper on which this book is printed meets the requirements of "ISO 9706:1994, Information and documentation - Paper for documents - Requirements for permanence".

ISSN 0303-8157
ISBN: 90-420-1648-5 (Bound)
©Editions Rodopi B.V., Amsterdam - New York, NY 2005
Printed in The Netherlands

CONTENTS

THEORIES AND STRUCTURES

SCIENCE AND ETHICS

Poznań Studies in the Philosophy of the Sciences and the Humanities, vol. 83

CONTENTS OF THE COMPANION VOLUME

CONFIRMATION, EMPIRICAL PROGRESS, AND TRUTH APPROXIMATION

Essays in Debate with theo Kuipers, Vol. 1

INTRODUCTION

The present volume, *Cognitive Structures in Scientific Inquiry*, is the sequel of *Confirmation, Empirical Progress, and Truth Approximation*: together, they form Volume 1 and 2, respectively, of *Essays in Debate with Theo Kuipers*. The subdivision of the latter into two volumes, which can be read independently from each other, is motivated by the fact that they deal with two different, although closely related, clusters of topics and issues.[1]

In the present introduction we will describe the nature of *Essays* and the structure of this volume (Section 1).[2] Then we will summarize the contents of the relevant book of Kuipers, see below, and the seventeen contributions appearing in this volume (Section 2). Finally, we will consider some 'metaphilosophical problems' raised by the philosophical approach developed by Kuipers in the works debated in these *Essays*, with special reference to Kuipers' views about the nature and the role of cognitive structures in scientific inquiry (Section 3).[3]

---

[1] *Essays in Debate with Theo Kuipers* is the second (two volume) book of the new sub-series *Monographs-in-Debate* (MiD) of *Poznań Studies in the Philosophy of the Sciences and the Humanities*. It has been preceded by the volume devoted to the discussion of Evandro Agazzi, *Right, Wrong, and Science: The Ethical Dimensions of the Techno-Scientific Enterprise* (edited by Craig Dilworth), published in 2004. MiD continues in a systematic form those volumes of *Poznań Studies* that were devoted to important publications in philosophy. Such volumes include books about the philosophical works of Mario Bunge (vol. 18, 1990), Jonathan Cohen (vol. 21, 1991), Agnes Heller (vol. 37, 1994), Jerzy Kmita (vol. 47, 1996), Ernest Gellner (vol. 48, 1996), and Jaakko Hintikka (vol. 51, 1997).

[2] Section 1 is identical – apart from some minor adaptations – to the corresponding section of the introduction to the companion volume.

[3] Section 3 might be read together with the corresponding section of the Introduction to Volume 1, where further metaphilosophical implications of Kuipers' approach to philosophy of science are examined.

In: R. Festa, A. Aliseda and J. Peijnenburg (eds.), *Cognitive Structures in Scientific Inquiry* (*Poznań Studies in the Philosophy of the Sciences and the Humanities,* vol. 84), pp. 11-20. Amsterdam/New York, NY: Rodopi, 2005.

1. Nature and Structure of the Volume

Essays in Debate with Theo Kuipers deals with the content of two books written by Theo Kuipers and published by Kluwer Academic Publishers in the Synthese Library, viz. *From Instrumentalism to Constructive Realism. On Some Relations Between Confirmation, Empirical Progress, and Truth Approximation* (2000) and *Structures in Science. Heuristic Patterns Based on Cognitive Structures: An Advanced Textbook in Neo-Classical Philosophy of Science* (2001). We will refer to these books as ICR and SiS. The two volumes of *Essays* are devoted to a critical discussion of ICR and SiS, respectively, where the division of the work is mirrored by their titles: *Confirmation, Empirical Progress and Truth Approximation* is almost identical to the subtitle of ICR, while *Cognitive Structures in Scientific Inquiry* is essentially a condensed version of the subtitle of SiS. However, the reader should not be surprised to find frequent references also to many other papers by Kuipers: indeed, both ICR and SiS are synthetic monographs, putting much of the earlier work of Theo Kuipers in a revised form together and adding new, formerly missing links.

Essays is intended to provide an occasion for significant dialogue on – and better understanding of – issues and positions involved in ICR and SiS. Each of the thirty-four commentaries included in the *Essays* is directly related to one of the topics dealt with in ICR or SiS: all commentators explicitly discuss Kuipers' approach to the relevant topic and, in many cases, presents her or his own approach. The reader will notice that *Essays* scarcely constitutes a traditional Liber Amicorum. Instead, it provides a genuine and lively debate on theses defended in ICR and SiS. Many articles are critical and polemical, as ought to be the case in any significant debate; see, for instance, the papers of Patrick Maher and Sjoerd Zwart (in Volume 1) and those of Eric R. Scerri and Arno Wouters (in Volume 2). In the spirit of this significant dialogue, each contribution is followed by a substantial reply by Kuipers himself.

The present volume consists of three parts: (1) a synopsis of the target monograph by Kuipers; (2) seventeen commentaries on the monograph (or a related paper), each followed by a reply from Kuipers; (3) a bibliography of Kuipers' work.

Since ICR and SiS are closely related, it is not surprising that the present SiS-related volume includes several references to ICR; moreover, especially in Kuipers' replies, there are references to contributions included in the ICR-related Volume 1 of *Essays*. For a better understanding of these references, a terse Table of Contents of ICR (including only the titles of the parts and the chapters) is enclosed in Appendix 2 of Kuipers' synopsis of SiS, and the complete Table of Contents of Volume 1 of *Essays* is included immediately after the Table of Contents of this Volume.

2. Contents of the Volume

SiS is at the same time a synthetic monograph and an advanced textbook that explicates, updates, and integrates the best insights of the traditional philosophy of science – as practised within the logical-empiricist tradition – and of its main critics, starting from Popper, Kuhn and Lakatos. For this reason Kuipers' approach to philosophy of science is described by the author as "neo-classical."

A leading role in this approach is played by the idea that philosophy of science may and should become a meta-science with heuristic use-value. More specifically, a basic purpose of neo-classical philosophy of science is the identification of "structures in scientific cognition," i.e., of *cognitive structures* which may also play a heuristic role, by providing heuristic patterns for actual scientific research.[4]

The issues extensively treated in SiS include the analysis of various kinds of research programs and various ways of explaining and reducing laws and concepts. Moreover, SiS explicates the notion of empirical progress and summarizes its relation to confirmation and truth approximation (as presented in detail in ICR). Finally, it pays special attention to research programs aiming at the design of a product or process, computational philosophy of science, the structuralist approach to theories, and research ethics.

While in SiS the above issues are treated in great detail, Kuipers' synopsis (this volume) only highlights the main topics, the final emphasis being on design research and research ethics.

The seventeen commentaries included in the present volume discuss the positions and outcomes defended in SiS, as well as their ramifications and implications. For the convenience of the reader, the commentaries have been divided into five groups. The four papers of the first group discuss Kuipers' views on different TYPES OF RESEARCH AND RESEARCH PROGRAMS (SiS, Part I, Ch. 1, and Part V). The six papers within the second group deal with different aspects of Kuipers' account of several TYPES OF EXPLANATION (SiS, Parts 2 and 3). Several issues related to (Kuipers' critical analysis of) COMPUTATIONAL APPROACHES to philosophy of science (SiS, Part VI, Ch. 11) are debated in the four papers within the third group. The two papers within the fourth group, THEORIES AND STRUCTURES, are about Kuipers' version of

[4] Incidentally, the heuristic role of cognitive structures for scientific research is responsible for the subtitle, "Heuristic Patterns Based on Cognitive Structures," of SiS.

the structuralist approach to theories (SiS, Part VI, Ch. 12). Finally, under the label SCIENCE AND ETHICS, the last group consists of one paper discussing the issue of research ethics (SiS, Part VI, Ch. 13).

2.1. *Types of Research and Research Programs*

This group includes four papers due to David Atkinson, Thomas Nickles, Maarten Franssen and Jean Paul Van Bendegem.

Atkinson challenges Kuipers' typology of pure and hybrid research programs, based on four basic kinds of program (SiS, Ch. 1), by raising the question as to whether a recent development of fundamental physics, namely string theory, could be accommodated by one of them, or whether it should be classified in a new, fifth kind of research program.

Nickles argues for the central importance of reduction in philosophy of science, especially when considered in the perspective of problem reduction, and discusses various kinds of problem reduction and similar relations, illustrating them, *inter alia*, in terms of the black body problem and early quantization problems. Two central claims of his paper are the following: (1) problem reduction is important in its own right and does not "reduce" to theory reduction; and (2) problem reduction is generally more important than theory reduction for methodology as the "control theory" of inquiry.

Franssen argues that Kuipers' set-theoretic approach to design research programs (SiS, especially Ch. 1 and 10), through its conception of properties as "atomic," cannot do justice to the fact that most properties that matter in design problems come in degrees. This approach offers no help with the difficult problem of evaluating different design concepts or prototypes when multiple features or properties, each giving rise to a comparative ordering of the concepts or prototypes, have to be taken into account. The author argues that this problem, rather surprisingly, is isomorphic to the well-known problem of social choice related to Arrow's theorem.

Sharing Kuipers' view that there are important similarities between mathematics and empirical sciences, *Van Bendegem* looks for a more unified treatment of these two areas of scientific research. After arguing that, as a consequence of the popular view that mathematics is basically about producing formal proofs, these similarities are entirely lost sight of, he introduces the notion of a mathematical argument as a more liberalized version of the notion of mathematical proof. Moreover, he shows that the way in which these arguments build up our support of mathematical statements is quite similar to the way it is done in the empirical sciences.

2.2. *Types of Explanation*

The six papers in this group are due to Erik Weber and Helena de Preester, Eric R. Scerri, Jeanne Peijnenburg, Anne Ruth Mackor, Arno Wouters, and Adam Grobler and Andrzej Wiśniewski.

Weber and *De Preester* propose an extension of Kuipers' account of explanation of laws (SiS, Ch. 3), by introducing an additional type of explanation, namely functional explanation. More precisely, they argue that micro-explanations of laws can have two formats: they work either by aggregation and transformation (as Kuipers suggests) or by means of function ascriptions (a possibility neglected by Kuipers). The authors compare both types from an epistemic point of view: which information is needed to construct the explanation? Moreover, from a means-end perspective: do both types serve the same purposes, and are they equally good?

Dealing with the very difficult problem of reduction in chemistry, *Scerri* criticizes the attempt by Hettema and Kuipers (2000) to formalize the periodic table and to reduce the periodic system to atomic physics. In particular he disputes their identification of a naïve periodic table with tables having a constant periodicity of eight elements. More generally, he criticizes their views on the different conceptions of the atom by chemists and physicists, by showing that the structuralist reconstruction of a naïve and a refined version of the periodic table is in many historical respects problematic.

Peijnenburg compares Kuipers' model of action explanation with that of Anscombe and with models in the post-Anscombian tradition. She points out that the difference with post-Anscombian writers concerns the so-called intentional statements: while the models of these writers contain no intentional statement, Kuipers' own model includes no less than two intentional statements. However, such statements can be reduced to beliefs and desires; hence, contrary to appearances, Kuipers' model is not immune to the criticisms based on the call for intentional statements, since this is a call for intentions that are in fact *irreducible* to beliefs and desires.

Mackor shares with Kuipers a naturalist view of the relation between the natural and social sciences, and asks whether his analysis of different scientific levels and the relations between them (SiS, Ch. 3, 4 and 6) applies to the relation between these two scientific areas as well. In particular, she deals with some features of the social sciences that seem to cause trouble for any reductionist model, starting from the question about how we ascribe mental states to other agents. Among other things, she analyzes the implications that the recent "simulation theory" might have for the social sciences, and the trouble it might cause for the naturalist view.

Wouters evaluates Kuipers' account of functional explanation in biology. His analysis is based on the elaboration of a real example in the life sciences, regarding the explanation of why electric fishes swim backwards. According to the author, Kuipers' account fails to do justice to the main insights provided by the example explanation; hence, he proposes a refinement and an extension of this account which is consistent with Kuipers' idea that function attributions are established by means of a process of hypothetico-deductive reasoning guided by a heuristic principle.

Grobler and *Wiśniewski* claim that Kuipers' approach to explanation opens the possibility for a further refinement of his own refined HD method for the evaluation of theories. One severe problem for the (refined) HD method is theory-ladenness: in fact, since experimental results are theory-laden, the comparative evaluation of alternative hypotheses is always relative to background knowledge. The authors claim that this difficulty can be avoided by supplementing HD considerations with the principle of inference to the best explanation and sketch a program for doing this which exploits some similarities between Kuipers' account of explanation and Lipton's.

2.3. *Computational Approaches*

The authors of the four papers in this group – who use computational tools to build up their claims, partly based on Kuipers' work – are Jaap Kamps, Alexander P. M. van den Bosch, Paul Thagard and Gerard A. W. Vreeswijk.

Kamps discusses the problems that background knowledge may cause for the formalization of scientific theories and shows how some of these problems can be addressed in the context of the computational representation of scientific theories. In particular, he argues that implicit background knowledge is something we have to excavate, by using computational means, and considers the possibility that making background knowledge explicit may be viewed as a form of truth approximation.

Van den Bosch explores some issues on the frontier between the topics of computational approaches, the structuralist view of theories and design research. Indeed, he explores structuralism as a way to model theories from scientific practice by computational means. As a case study he analyzes a theory about the dynamics of the basal ganglia: he explicates the structure of the basal ganglia theory, how it explains Parkinson's disease and how it leads to treatments. He also indicates the way in which computational means can be used in drug design research.

Thagard discusses Kuipers' recent account of beauty and truth (Kuipers 2002). It challenges Kuipers' psychological account of how scientists come to appreciate beautiful theories, as well as his attempt to justify the use of aesthetic criteria on the basis of a "meta-induction." After advancing some general and specific reserves about Kuipers's naturalistic-cum-formal inductive account of the relation between beauty and truth, the author proposes an alternative psychological/philosophical account based on emotional coherence.

Thagard's theory of explanatory coherence (TEC) is a conceptual and computational framework that is used to show how new scientific theories can be judged to be superior to previous ones. Kuipers (SiS, Ch. 11) criticizes TEC as a model that does not faithfully reflect scientific practice. *Vreeswijk* presents an alternative criticism of TEC, by explaining the machinery behind TEC, and trying to indicate where TEC falls short and where it can be improved. The main constructive purpose of the author is to design a new connectionist method that claims to evaluate theories in a way that improves on TEC.

2.4. *Theories and Structures*

This group consists of a contribution by Emma Ruttkamp and one by Robert Causey. They are mainly constructive papers, providing accounts of phenomena or concepts not touched in Kuipers' account of this issue.

Ruttkamp deals with several topics from ICR and SiS, i.e., with the testing and confirmation of empirical theories (ICR, Ch. 5 and 6 and SiS, Ch. 7 and 8), scientific realism (ICR, Ch. 1 and 13), and the structure of scientific theories (SiS, Ch. 12). More specifically, her main aim is to defend a kind of realism, called model-theoretic realism that, among other things, can make sense of the problem of overdetermination of theories by empirical data, using the machinery of non-monotonic reasoning as proposed for knowledge representation in artificial intelligence.

In philosophy of science, 'structure' can refer to different uses. *Causey* intends this term in its ontological-cum-epistemological sense, referring to the nature of certain kinds of objects in the real world. After characterizing the concept of structure, by introducing the notions of "bonds" and of "structured wholes," he uses such notions to present a detailed analysis of microreductive explanations – or microreductions – which aim to explain the laws governing a structured object in terms of laws about its parts, plus a description of its structure. The author argues that his analysis applies from physics to the social sciences, and illustrates the latter by a hypothetical robotic social structure.

2.5. *Science and Ethics*

This heading only refers to the paper by *Henk Zandvoort*, dealing with the ethics of science as described by Merton and as actually practiced by scientists (see SiS, Ch. 13). The author gives reasons as to why scientists should not be permitted to proceed on the basis of Merton's "communism" norm, that scientific knowledge and its dissemination are unconditionally good. He also claims that, in contrast to Merton's norms, research ethics should take account of generally recognized ethical principles, notably those of restricted liberty and responsibility, and concludes by considering the possibility of abandoning certain themes of research so long as the present failure on the part of politics to control the hazards of science and technology persists.

3. Structures vs. Structure

When we compare Kuipers' *Structures in Science* with the book from which it derives its title, Nagel's *The Structure of Science*, the differences are clear. One of the differences that immediately catches the eye is evident in the title: where Nagel has the singular 'Structure', Kuipers uses the plural 'Structures'. It is scarcely an exaggeration to say that this tiny grammatical change reflects forty years of research in philosophy of science. While the old ideal of a unified science still attracts Nagel, it has no appeal whatsoever for Kuipers. Like the majority of his colleagues today, Kuipers reverts to the Aristotelian conviction that science is essentially manifold: there are several different sciences, with several different objects of study, using several indispensably different methods. As a result, the cognitive structures that underlie different scientific inquiries are multiple as well.

Kuipers even goes a step further, as can also be seen from his replies. Not only are the cognitive structures plural in character, they are flexible and they can, in the course of scientific research, change considerably or even cease to exist. For this reason any dogmatism is altogether misconceived: no cognitive structure, whether in the field of research programs, explanations, concept formation, or of interlevel and interfield research, can ever function as a fixed mould into which future findings must be forced. Structures do not act as procrustian beds for which new ideas and discoveries must be made suitable; on the contrary, it is quite possible that future research is best reconstructed along lines that are very different from the products and processes that we have encountered in science so far. At their best, structures might serve as heuristic means, somewhat as systems theory, that forerunner of modern cognitive studies, inspired the development of many areas of present-day research (*cf.* SiS, xiii). Thus the subtitle of SiS, *Heuristic Patterns Based on Cognitive Structures,* differs considerably from the subtitle of Nagel's book, *Problems in the Logic of Scientific Explanation*, which betrays a concentration on nomological explanations in science.

Kuipers' lenient and openminded attitude toward multiform structures in several domains should not be confused with endorsement of a Feyerabendian "anything goes." It is enough to read one or two pages of Kuipers' extensive work to convince oneself of the fact that he cannot be accused of having any relativistic or postmodernistic proclivities. "The core of the program of cognitive studies of science," Kuipers declares, "is the idea that there are systems underlying knowledge and knowledge production, and hence that theory formation about them is possible in principle" (SiS, p. x; *cf.* the synopsis of SiS in the present volume). Apart from the fact that Kuipers is here mainly thinking of quasi-empirical theory formation, the foregoing quotation echoes Nagel's non-relativistic view that philosophy of science "is primarily an examination of logical patterns exhibited in the organization of scientific knowledge" (Nagel 1961, p. viii).

Being a classical and positivistic philosopher of science, Nagel is known for his precise and rather exact style. However, Kuipers is in a sense even more rigorous than is Nagel. In one of the reviews that appeared of SiS, Kuipers is described as "the most analytical amongst the analytical philosophers in the Netherlands," on the grounds that "his work is characterized by a mathematical precision; central terms are always explicitly defined; and whenever a statement can be proved, he will not confine himself to a plausibility argument" (Douven 2002, p. 310, our translation). It is precisely in this method, rather than in its themes or subjects, that Kuipers' book as it were outstrips that of Nagel in analyticity.

With *The Structure of Science* Nagel aimed to write a book for a wider audience than that of professional students of philosophy. Hence he avoided symbolic notations and abstained from presenting his analyses in a formalized way. SiS, on the other hand, is an advanced textbook in neo-classical philosophy of science, as its second subtitle makes clear. It does not eschew the use of logical and mathematical formulas, and much use is made of schemes, figures, tables, matrices, and abbreviations (on the relations in Kuipers' work between logic and philosophy of science, see Section 3 of the introduction to Volume 1). All this is in accordance with Kuipers' standards of clarity and precision, compared with which Nagel's otherwise exemplarily book may well be deemed "not sufficiently analytical," as Kuipers states in his Foreword (SiS, p. xiii). The formal apparatus gives SiS a very systematized and organised appearance, at least for the professional reader. In this sense SiS not only *deals* with all sorts of more or less complicated structures, but is also highly structured *itself* – incidentally, a property that it shares with its companion volume *From Instrumentalism to Constructive Realism*.

Kuipers has made it repeatedly clear that he did not write his book solely for philosophers or philosophers of science; he hopes his readers will include scientists too. Indeed, his main motivation for abandoning the singular 'structure' in favor of 'structures' is that the former cannot do justice to the diversity of scientific practice. The area of discussion in the present volume about themes and positions in SiS is therefore not an exclusively philosophical bailiwick. Many of the seventeen contributions have been written either by scientists or by philosophers with a degree in one of the sciences; and all contributors, in one way or another, sympathize with Kuipers' desire to augment philosophy's use-value for science. These facts constitute another step forward in regaining the self-confidence that philosophy of science, as Kuipers rightly diagnoses (see the first page of his synopsis in the present volume), seems to have lost in our post-positivistic era.

ACKNOWLEDGMENTS

The publication of this two volume book of *Essays in Debate with Theo Kuipers* would not have been possible without the support and the cooperation of a number of persons and institutions, whose role and contribution we gratefully acknowledge.

First of all, we wish to express our gratitude to Leszek Nowak who kindly invited us, in his function as editor-in-chief of *Poznań Studies*, to design and edit this book for the new subseries *Monographs-in-Debate*. We are indebted also to Kluwer Academic Publishers, for making available the copies of ICR and SiS for all contributors, and to the Faculty of Philosophy of the University of Groningen, for providing financial support. Finally, we would like to thank four persons who played different but equally crucial roles in the production of the *Essays*, i.e., Hauke de Vries for his unflagging technical assistance, David Atkinson for grammatical corrections and stylistic improvements, Ian Priestnall (Paragraph Services) for his work of linguistic correction, and Lieke Hendriks for her work on the camera ready version.

Roberto Festa, *Trieste* (Italy)
Atocha Aliseda, *Mexico City* (Mexico)
Jeanne Peijnenburg, *Groningen* (The Netherlands)

REFERENCES

Douven, I. (2002). Review of T. Kuipers, *Structures in Science. Dutch Journal of Philosophy* (Algemeen Nederlands Tijdschrift voor Wijsbegeerte) **94** (4), 310-312 (in Dutch).

Hettema, H. and T.A.F. Kuipers (2000). The Formalisation of the Periodic Table. In: W. Balzer, J. Sneed, and C.U. Moulines (eds.), *Structuralist Knowledge Representation: Paradigmatic Examples* (*Poznań Studies in the Philosophy of the Sciences and the Humanities*, vol. 75), pp. 285-305. Amsterdam/Atlanta: Rodopi.

Kuipers, T.A.F. (2002). Beauty, a Road to The Truth. *Synthese* **131** (3), 291-328.

Nagel, E. (1961). *The Structure of Science. Problems in the Logic of Scientific Explanation.* London: Routledge & Kegan Paul. First paperpack edition: 1979.

Theo A.F. Kuipers

STRUCTURES IN SCIENTIFIC COGNITION

A SYNOPSIS OF

STRUCTURES IN SCIENCE. HEURISTIC PATTERNS BASED ON COGNITIVE STRUCTURES. AN ADVANCED TEXTBOOK IN NEO-CLASSICAL PHILOSOPHY OF SCIENCE (2001)

CONTENTS

Theo A.F. Kuipers

STRUCTURES IN SCIENTIFIC COGNITION

A SYNOPSIS OF *STRUCTURES IN SCIENCE. HEURISTIC PATTERNS BASED ON COGNITIVE STRUCTURES. AN ADVANCED TEXTBOOK IN NEO-CLASSICAL PHILOSOPHY OF SCIENCE (2001)*

ABSTRACT. The philosophy of science has lost its self-confidence. *Structures in Science* (2001) is an advanced textbook that explicates, updates and integrates the best insights of logical empiricism and its main critics. This "neo-classical approach" aims at providing heuristic patterns for research.

The book introduces four ideal types of research programs (descriptive, explanatory, design and explicative) and reanimates the distinction between observational laws and proper theories without assuming a theory-free language. It explicates various patterns of explanation by subsumption and specification as well as structures in reductive and other types of interlevel research. Its threefold analysis of theory evaluation leads to new characterizations of confirmation, empirical progress, and truth approximation. What emerges are partial analogies between progress in nomological research, presented in detail in *From Instrumentalism to Constructive Realism* (2000) and progress in explicative and design research. Finally, special chapters are devoted to design research programs, computational philosophy of science, the structuralist approach to theories, and research ethics.

The present synopsis of *Structures in Science* highlights the main topics, the final emphasis being on design research and research ethics.

Introduction

Although there is an abundance of highly specialized monographs, learned collections and general introductions to the philosophy of science, only a few synthetic monographs and advanced textbooks have appeared in the last 25 years. The philosophy of science seems to have lost its self-confidence. The main reason for such a loss is that the traditional analytical, logical-empiricist approaches to the philosophy of science had to make a number of concessions, especially in response to the work of Popper, Kuhn and Lakatos. With *Structures in Science* (SiS) I have intended to present both a synthetic monograph and an advanced textbook that accommodates and integrates the insights of these philosophers, in what I like to call a neo-classical approach.

In: R. Festa, A. Aliseda and J. Peijnenburg (eds.), *Cognitive Structures in Scientific Inquiry* (*Poznań Studies in the Philosophy of the Sciences and the Humanities,* vol. 84), pp. 23-92. Amsterdam/New York, NY: Rodopi, 2005.

Here I would like to use the plausible terminology of "idealization and concretization" in defending the use of the term "neo-classical approach." Many introductory textbooks like to speak about alternative or postpositivist approaches in philosophy. For example, Bechtel (1988) does so in his introduction to philosophy of science, which, apart from that, I find a very useful book, witness the role it plays in SiS. In my view, however, I show in SiS that it is possible to concretize the, surely very idealized, logical-empiricist point of departure, by using some of the main insights of Popper, Kuhn and Lakatos, that is, by taking some crucial logical, methodological and historical factors into account.[1]

The resulting monograph-cum-advanced-textbook provides a detailed discussion of a number of the main topics in the philosophy of science and is aimed at advanced students in philosophy and in the natural, technological, social, behavioral, cognitive and neuro-sciences. It distinguishes various kinds of research programs and various ways of explaining and reducing laws and concepts. It explicates the notion of empirical progress and summarizes its relation to confirmation and truth approximation (as presented in detail in my *From Instrumentalism to Constructive Realism*, ICR). Finally, it pays special attention to research programs aiming at the design of a product or process, computational philosophy of science, the structuralist approach to theories, and research ethics.

The present synopsis of SiS highlights some of the main topics in a necessarily selective way, sometimes by merely giving a summary of a chapter. However, it essentially follows the division of SiS into 6 parts and 13 chapters, here resulting in 6 parts and 13 sections. In view of its selective character, it may be useful from time to time to consult the complete table of contents, including section titles, which is reproduced in Appendix 1.[2] Appendix 2 presents the outline table of contents of the companion volume *From Instrumentalism to Constructive Realism* (ICR, 2000), to which references will occasionally be made. Appendix 3 gives a list of acronyms.

There are two related factors about which SiS formulates a firm position. First, the philosophy of science, which generally bears only a slight resemblance to an empirical science, should in my opinion become a genuine empirical science, or, more precisely, an empirical *meta-science*, with its own meta-theories. Second, philosophers of science usually are rather skeptical about the use-value of their results, despite the fact that these results are frequently presented in a rather prescriptive style. In my view the philosophy

---

[1] A very readable recent introduction to the philosophy of science (Ladyman 2002) seems to have been written from the same perspective.

[2] SiS, like ICR, is based on many of my publications, starting from 1982. The foreword to SiS (p. xiii) mentions those 11 papers, the text of which has partially been used in writing SiS.

of science should explicitly aim at use-values, not by strengthening its prescriptive claims, but by aiming at various kinds of *heuristic use-value*.

In SiS I have tried to show that the philosophy of science may become a meta-science with a heuristic use-value. More specifically, making an analogy with the so-called social studies of science, one may conceive the book as the product of various cognitive studies of science, looking for cognitive structures that underlie scientific inquiry and that generate multi-purpose heuristic patterns.

It is important to note that 'cognitive' in 'cognitive studies/structures/patterns' does not refer to the notion of "cognitive" used in mainstream cognitive science, where it primarily refers to the "internal/psychological processes" of knowledge formation and processing. My use is broader. Science is a kind of cognition, scientific cognition, and cognitive science needs to study scientific knowledge formation, representation and processing as well. Therefore, it seems apt to call the relevant studies "cognitive studies of science" and the findings "cognitive structures" or "structures in scientific cognition."

Cognitive studies of science may be characterized more specifically as follows: quasi-empirical studies of cognitive aspects of scientific knowledge, including its methods and its development. Empirical studies may of course be of a descriptive and explanatory nature. The qualification 'quasi-' (empirical) is used to leave room for normative problems and aims. In consequence, one or more heuristic-normative points of view frequently guide scientific inquiry. For instance, the intention to apply patterns of successful cases to areas of research that have so far not shown much progress.

The core of the program of cognitive studies of science is the idea that there are systems underlying knowledge and knowledge production, and hence that theory formation about them is possible in principle. It is also plausible that such theorizing concerns part and aspect systems and hence different patterns, which may or may not be easily interrelated and harmonized.

More or less concrete and realistic patterns have been found on the basis of case studies or formal analysis. Several cognitive structures receive extensive attention in SiS. These findings are partly restatements of views found elsewhere in the literature, partly elaborations of ideas of others, and partly results of my own research. A selection of them is presented in this synopsis.

Returning to the use-value, cognitive structures always involve informative patterns, which seem useful in one way or another. It is instructive to distinguish at least the following five kinds of possible use-value:

(a) cognitive structures may provide the "null hypothesis of ideal courses of events," which can play a guiding role in the social studies of science; they may at least raise the question of whether certain patterns are favored over others in various psychological, cultural, sociological or economic circumstances, but stronger kinds of social or external influences are also conceivable;

(b) they may clarify or even solve classical problems belonging to abstract philosophy of science; e.g., ICR based explications of the correspondence theory of truth and of dialectical concepts on its truth approximation account;

(c) they may be useful as didactic instruments for writing advanced textbooks, leading to better understanding and remembrance; for forceful general pleas for use, and illustrations of the use of insights from philosophy of science in science teaching, see Duschl (1990) and Matthews (1994);

(d) they may play a heuristic role in research policy and even in science policy, e.g., by stimulating applications of successful patterns of research in stagnating areas of research;

(e) last but not least, they may play a heuristic role in actual research, not only of the standard nature, but also in the expanding field of computational research; in the last respect cognitive structures at least provide a basis for generating new means for a further development of computational philosophy of science.

Let me elaborate a bit on the latter use-value, the heuristic role for research. It is this value that is responsible for the subtitle "Heuristic Patterns Based on Cognitive Structures" of SiS, will be worked out a little further. Traditional philosophy of science has frequently been accused of studying ready-made science, i.e., the end products of science, rather than the process that led to those products. Although it is certainly true that this can give very distorted pictures of science in the making, it is also true that structures recognizable in end products, which were apparently successful, may be used as heuristic means for new but, in one way or another, similar research: they may provide models of what one may be looking for. Or, to use the phrase of one of the main forerunners of cognitive psychology and cognitive science, Otto Selz (1924), these structures may provide the "schematic anticipation" ('schematische Antizipation') of the solution to a scientific problem. Of course, there is no guarantee that solutions to new problems should resemble solutions to old ones. For that reason it is important to stress that cognitive structures cannot provide prescriptive models for new research. At most, they can play the role of heuristic means, which may be rather a lot.

The main title of SiS, viz., "Structures in Science," alludes of course to Nagel's *The Structure of Science* (1961). That book is in some respects outdated, though much less than is generally assumed, and I make explicit and implicit use of it in SiS. Besides Nagel, a number of other authors should be mentioned as my main intellectual heroes: Carnap, Hempel, Hintikka, Popper, Lakatos, Suppes, and Sneed. In SiS I aim to show that their work can be reconciled, or more precisely, that a synthesis of some of their best insights is possible. In making this attempt I have profited a lot by freely using the work of former Ph.D. students, notably, Alexander van den Bosch, Roberto Festa, Lex Guichard, Bert Hamminga, Hinne Hettema, Maarten Janssen, Rick Looijen, Anne Ruth Mackor, Rein Vos, Henk Zandvoort, and Sjoerd Zwart, and other scholars, notably, Balzer, Bechtel, Burton, Causey, Darden, Kim, Millikan, Niiniluoto, Nowak, Panhuysen, and Thagard.

My resulting perspective on the present-day philosophy of science as a discipline becomes quite clear in SiS. However, I should mention Ilkka Niiniluoto's *Critical Scientific Realism* (1999); as far as I know, the most learned recent exposition of some of the main themes in the philosophy of science in the form of an advanced debate-book, that is, a critical exposition and assessment of the recent literature, including his own major contribution, viz. *Truthlikeness* of 1987. Despite our major differences regarding the topic of truth approximation, I would like to express my affinity to, in particular, his rare type of constructive-critical attitude in the philosophy of science.

I conclude this introduction by quoting Atkinson's bracketed remark "it is still the case that, for many scientists, 'philosophizing' is put on a par with daydreaming or sloppy reasoning" (Atkinson, this volume). Weinberg (1993) and Wolpert (1992), for example, exemplify this view on philosophy (of science) in an eloquent way. By writing SiS I have intended to show that neo-classical philosophy of science need not be daydreaming nor sloppy reasoning and can provide heuristics means for scientific research and education.

I. Units of Scientific Knowledge and Knowledge Acquisition

In the first part the emphasis is on patterns in scientific research programs, research strategies, observational laws, and theories.

Section 1 presents the more or less generally accepted view, introduced by Kuhn and Lakatos, that the development of scientific research takes place by means of encompassing cognitive units, called research programs. I distinguish four kinds of programs: descriptive, explanatory, design, and explicative. Finally, I present one major strategy for the internal development of programs, viz. idealization and concretization.

The main structural and developmental features of programs will now be indicated. SiS Ch. 1 describes them in some detail, using Dalton's atomic theory program to illustrate them. It also addresses the strategic lessons that may be drawn so far. They involve the value of programmatic research as such and the interaction between programs as a result of competition or co-operation. Finally, the chapter addresses a second strategy for the internal development of programs, in particular, viz. by focussing on interesting theorems.

In Section 2 it is argued that the theory-relative distinction between observational laws and proper theories, called the law distinction, is one of the main dynamic factors in the development of explanatory programs and in the interaction between descriptive and explanatory programs. After having indicated some structural features of proper theories, I close with a brief presentation of the main epistemological positions involved in observational and theoretical knowledge claims of increasing strength.

SiS Ch. 2, moreover, elaborates the theory-relative explication of theoretical and observation terms on which the law distinction is based. The analysis suggests an epistemological hierarchy of scientific knowledge and a disentanglement of the so-called theory-ladenness of observations. An appendix shows how a similar explication of the main points can be obtained by starting from the so-called empirical basis; this possibility makes it even more surprising that Popper did not pay attention to the law distinction.

1. *Research Programs and Research Strategies*

At the beginning of the 21st century, the more or less generally accepted view, introduced by Kuhn and Lakatos, is that the development of scientific research takes place by means of encompassing cognitive units, following Lakatos here called research programs. In my view we should analytically distinguish the goals and methods of four kinds of research and corresponding research programs: descriptive, explanatory, design, and explicative.

Types of Research Programs

Traditionally, philosophers of science have paid almost exclusive attention to descriptive and explanatory research. In my view, however, actual scientific research in the second half of the 20th century has been at least as much design research. Sometimes in the empirical sciences, and frequently in mathematics and philosophy this takes the form of conceptual design research, also called explicative research. As suggested, the resulting four types of research occur in the form of research programs of which the following characterizations can be given.

Descriptive research programs are meant to describe a certain domain of phenomena, primarily in terms of individual facts or in terms of general observable facts. Descriptive programs may be fundamentally based on experiments, in which case it is plausible to speak of experimental programs. A famous example is Boyle's search for a relation between the pressure and volume of a gas, followed by Charles, Gay-Lussac and others with their quest for the relation with temperature. The investigation by Durkheim of what he called the social facts about suicide is another typical example. Finally, a recent example is the Human Genome Project, more or less completed in 2000, aiming at the true description of the (almost unique) composition of the 23 human chromosomes as pairwise sequences of the four bases C, T, A, and G, that is, the typical vocabulary of DNA.

Descriptive research takes place by more or less selective (experimentation and successive) observation, and the resulting facts are couched in so-called observation terms. These observation terms are not given by the natural world, but form the specific spectacles through which the researcher in that program is looking. At the start of a descriptive program there usually is only some core vocabulary. For the rest it is not altogether clear which further observation terms are to be considered as relevant and how certain observation terms are to be precisely interpreted. Additional terms are selected and shaped in the course of the development of the program. It should also be stressed that, at least as a rule, observation and hence observation terms are, and remain, loaded with theoretical presuppositions that are considered to belong to the unproblematic so-called background knowledge.

Explanatory research programs have another aim. They are directed at the explanation and further prediction of the observable individual and general facts in a certain domain of phenomena. Hence, an explanatory program has a (quasi-) deductive nature and is always built on an underlying descriptive program. For this reason explanatory programs are frequently developed along with underlying descriptive programs, in which case the two types of program can be distinguished only analytically. The kinetic theory of gases, on the one hand, and Durkheim's anomy theory, on the other, provide paradigm cases of explanatory programs built on the previously mentioned descriptive programs. The primary objective of the kinetic gas theory program was the explanation and detailed prediction of the precise relation between pressure, volume and temperature. To illustrate this fact I confine myself to one representative of the many researchers conducting this type of research: Van der Waals. Similarly, Dalton's program (elaborated in SiS, Section 1.1.7) tried to explain chemical regularities on the basis of the atomic constitution of matter. Durkheim tried to explain the social facts about suicide with his anomy theory. It is important to realize that several explanatory programs may arise on the basis of the same descriptive program. They may be competitive, but need not be.

The most important tools used by explanatory programs are so-called theoretical terms, denoting fundamentally new concepts, which have not yet been firmly established as observation terms, either inside or outside the program. Of course, the terms as such may have been used before to refer to a related concept. Examples of theoretical concepts are the concept of force in

Newtonian mechanics, Chomsky's concept of the deep structure of languages, and the concept of utility in rational choice or utility theory. The new terms may refer to theoretical properties, relations and functions, as suggested by the examples, but also to newly postulated entities, such as atoms and genes. If an explanatory program introduces theoretical terms, it may also be called a theoretical program. If it does not, which certainly is possible, it belongs to the explanatory subtype of observational programs, to be distinguished from the descriptive subtype.

Design research programs involve the design and actual construction of certain products. Some examples are: programs directed at the construction of new materials, the production of new medical drugs, the improvement of breeding methods of plants, the design of training programs for certain types of handicaps, and the design of so-called expert systems. As the examples illustrate, the products of design programs need not be products in a strict sense but may also be processes, or their improvement. The product targeted by a design program has to satisfy certain previously chosen demands; these demands are of course derived from the intended use of the product being developed. Since design programs often use knowledge obtained in descriptive and explanatory programs, the design process will only be considered to belong to *scientific* research if it is not fully based on existing knowledge and techniques. That is, new theories have to be developed or new experiments have to be performed in order for a design program to be scientific in nature.

Explicative research programs are directed at concept explication, i.e., the formal construction of a simple, precise and useful concept that is, in addition, similar to a given informal concept. For example, the concepts of "logical consequence" and "probability" have given rise to very successful explicative programs in the borderland between philosophy and mathematics. Another important explicative program in philosophy of science deals with explicating the concepts of "truthlikeness" and "truth approximation."

The strategy of concept explication starts by deriving from the intuitive concept to be explicated and, if relevant, from empirical findings, conditions of adequacy that the explicated concept will have to satisfy, and evident examples and counter-examples that the explicated concept has to include or exclude. Explication may go further than the explication of intuitive concepts, it may also aim at the explication of intuitive judgments, i.e., intuitions, including their justification, demystification or even undermining. A main example in ICR concerns the intuition about the functionality of choosing empirically more successful theories in order to enhance truth approximation. The strategy of "intuition explication" is a plausible extension of that involving concept explication.

The four types of research programs have structural and developmental features, partly in common, partly characteristic of the type. Here they will only be briefly indicated. The most important common structural features of programs are, ideally, that they have 1) a *domain* of existing or not yet existing phenomena, 2) the *goal* of solving a certain problem about it, be it the finding of its true description or its true theory, or the construction of an intended product or concept, 3) a *core idea*, that is, principles couched in a certain vocabulary, about how to solve the problem, and 4) some additional ideas, *heuristics*, suggesting how to save the core idea against *prima facie* failures to solve the problem.

As common developmental features we may discern an internal and external phase. The internal phase may be decomposable in an exploratory and an evaluative phase. In the external, or application, phase the focus may be either some technological or societal problem or a problem raised by another research program. Of course, external aspirations may already be important in the internal phase. As soon as another program provides an important focus of a certain program it is plausible to speak of guide and supply programs, respectively, a very useful distinction introduced by Henk Zandvoort.

Idealization and Concretization

Important strategic lessons that may be drawn from the research program structure and development of scientific research and its differentiation involve the value of programmatic research as such and the productive interaction between programs, of a competitive or co-operative nature. Moreover, important strategies for the internal development of programs can be discerned, in particular, the strategy of idealization and concretization, articulated by Leszek Nowak, and the related strategy of focusing on an interesting theorem, articulated by Bert Hamminga. The former will be presented here.

Idealization and concretization form a strategy for the internal development of a research program, particularly for generating a succession of improving specific theories. To be sure, this strategy can also be used without assuming the boundaries of a research program, the only claim is that it is frequently used within a program. Idealization is frequently applied in empirical scientific practice as an unavoidable step in theory formation. This is certainly true in the natural sciences; in the human sciences and also in philosophy the necessity of explicit idealization is not yet generally accepted.

Surprisingly enough, on closer inspection Marx developed his ideas in *Das Kapital* rather systematically according to the method of idealization and successive concretization, as Nowak (1974, 1980) has pointed out; in particular he shows how Part I and Part III in their succession can be seen as illustrations of what Marx used to call "rising from the abstract to the concrete." Another Polish philosopher, Krajewski (1977), freely following Nowak, has also contributed importantly to the growing awareness of the systematic role of what he calls "idealization and factualization" (I&C from now on).

Although idealization-and-concretization also occurs in qualitative theorizing, it is primarily explicated for quantitative theorizing, in particular for the succession of specific theories within a research program. The general idea is that it is frequently possible to make an ordering in the degree of importance or relevance of all the factors that influence the value of a certain quantity G, which may even lead to a division of primary and secondary factors. Starting from an ordering of all (or only the secondary) factors f_0, f_1,...(f_m), in the n-th stage of concretization factors f_0 up to f_n have been accounted for, while the remaining factors are still neglected, leading to the typical I&C formulation of the n-th specific theory:

if $f_0 \neq 0, f_1 \neq 0, \ldots, f_n \neq 0$ and $f_{(n+1)} = 0, f_{(n+2)} = 0, \ldots$
Then $G = G_n(f_0, f_1, \ldots, f_n)$.

In the 0-th stage there is maximum idealization and when all factors have been concretized, maximum concretization has been achieved. Note that, although any given functional representation of a factor is allotted the value 0 on a formally arbitrary basis, the neglect of a certain factor is empirically speaking usually not arbitrary, in which case the functional representation can be chosen in accordance with this.

The transition of the ideal gas law to the Law of Van der Waals is a paradigm case. This transition can be represented in a stepwise decomposition, of which the crucial formulas include:

(1) $P = RT/V$
(2) $P = RT/V - a/V^2$ (or, alternatively, $P = RT/(V-b)$)
(3) $P = RT/(V-b) - a/V^2$ (or the standard form: $(P+a/V^2)(V-b)=RT$)

where P, V, T and R indicate pressure, volume, temperature and the ideal gas constant, respectively, and a and b refer respectively to specific gas constants related to mutual attraction between the molecules and the volume of the molecules. The book series *Poznań Studies in the Philosophy of the Sciences and the Humanities* includes many other examples of I&C.

I&C can be used to structure theories in their research stage as well as in textbooks. Although it seems very natural to do so, it is very surprising that it is seldom explicitly done. However, in general expositions about what one has been doing or how one should do it, there is frequently reference to I&C. A specific reason for the relative neglect of I&C in the social sciences may be the great social pressure to avoid strong idealizations: fear of being accused of distorting reality too much seems to be rampant.

The abovementioned paradigm example raises an interesting question concerning explanations: is it possible to (re) construct the explanation of a concretized law as a concretization of the explanation of the (more) idealized law? Chapter 10 of ICR deals with this question in some detail. Another question, at least as important, is whether and in what sense the I&C strategy is functional for truth approximation in the empirical sciences. A detailed positive answer is given in ICR (Section 10.4).

As already mentioned, the I&C heuristic can also be used in the qualitative theorizing that often occurs in mathematics and philosophy. The ordered textbook presentation, first of propositional logic and then predicate logic provides a famous example. SiS occasionally applies the I&C heuristic in the presentation and development of qualitative meta-theories, e.g., the HD method for the separate evaluation of theories in Chapter 7, the set-theoretic approach to design research in Chapter 10 and the structuralist approach to theories in Chapter 12. Finally, the theory of truth approximation, developed in ICR, is in many respects an example of the I&C-heuristic (see also Kuipers forthcoming).

2. *Observational Laws and Proper Theories*

In the empirical sciences the informal distinction between observational laws and proper theories plays a crucial role. This "law distinction" is first presented in some detail. After a brief indication of the structure of theories, in terms of the distinction between epistemologically and ontologically stratified theories, I will present the leading epistemological positions: instrumentalism, constructive empiricism, referential realism and theory realism.

The Law Distinction

Ernest Nagel has stressed the distinction between experimental laws and proper theories, where the latter aim to explain the former by introducing theoretical terms. This "law distinction" is one of the main dynamic factors in the development of explanatory programs and in the interaction between descriptive and explanatory programs. Since there do not seem to be theory-free or theory-neutral observation terms, the law distinction has to be explicated on the basis of a theory-relative explication of theoretical and observation terms. The ideal gas law turns out to be an instructive example, viz., *prima facie* it is a proper theory according to the law distinction, but on closer inspection (presented in an appendix in SiS, based on Kuipers 1982) it is an observational law, because its purported theoretical terms, viz. temperature and the gas constant, can be explicitly defined by using existence and uniqueness conditions that are provided by laws that are unproblematically observational. Hence, a proper theory is a theory with theoretical terms which cannot be explicitly defined in observational terms and laws, that is, these terms are laden with the theory itself. On the other hand, genuine observational laws are improper theories in the sense that they do not use terms that are laden with the law itself. This analysis also suggests a disentanglement of the so-called theory-ladenness of observations. In particular, an observation may not only be laden by a theory, namely when its description uses crucial terms of the theory, but even if unladen by it, observation may nevertheless be relevant to a theory, and even guided by it. Mendeleev's theory, the periodic table of the chemical elements, provides a beautiful example of this.

Proper theories arise from the two-level distinction between observation and theoretical terms, as opposed to observational laws and theories, which only use observation terms, by definition. The resulting two-level distinction between observational laws and proper theories gives rise to a short-term dynamics in the development of scientific knowledge in terms of the explanation and prediction of observational laws by theories. To improve this, theories are revised piecemeal, usually within the boundaries of a research program. The long-term dynamics is generated by the transformation of proper theories into observation theories, by accepting them as true, and giving rise to a multi-level distinction according to which proper theories may not only explain or predict a lower level observational law, but also be presupposed by a higher level one. This description of the long-term dynamics typically has a theory-realist flavor. However, other epistemological positions (see below) have their own way of describing such dynamics.

The distinction between theoretical and observational terms is an epistemological one, leading to the stratification of a theory in terms of theoretical and observational statements. However, besides epistemological stratification there is ontological stratification: they frequently go together, but are essentially independent. A theory is said to be ontologically stratified when there are two or more kinds of entities involved and when entities of one of these kinds are components of entities of the other kind. It is then plausible to speak of a lower, micro-level and a higher, macro-level. In this case some principles of the theory concern only the micro-entities, and their properties and relations, and are called *micro- or internal principles*, whereas others connect the different kinds of entities, and their properties and relations, and are called *bridge principles*. The example of the atomic theory provides a nice example of an ontologically as well as (along the same lines) epistemologically stratified theory.

Epistemological Positions

The core of the ongoing instrumentalism-realism debate concerns the nature of theoretical terms and of proper theories using such terms, or rather the attitude one should have toward them. *Prima facie*, the most important epistemological positions in that debate are certainly instrumentalism, constructive empiricism, referential realism and theory realism. They can be characterized and ordered according to the ways in which they answer a number of leading questions, where every subsequent question presupposes the affirmative answer to the previous one. For completeness, I start with two preliminary questions that get a positive answer from the major positions, but a negative one in idealist and extremely relativist postmodern circles:

Question 0: Does a natural world that is independent of human beings exist?
No: ontological idealism; Yes: ontological realism.

Question 1: Can we claim to possess true claims to knowledge about the natural world?
No: epistemological relativism; Yes: epistemological realism.

Question 2: Can we claim to possess true claims to knowledge about the natural world beyond what is observable?
No: empiricism: instrumentalism or constructive empiricism; Yes: scientific realism.

Question 3: Can we claim to possess true claims to knowledge about the natural world beyond (what is observable and) reference claims concerning theoretical terms?
No: entity or, more generally, referential realism; Yes: theory realism.

Question 4: Does there exist a correct or ideal conceptualization of the natural world?
No: constructive realism; Yes: essentialist realism.

Note first that "empiricism" has two variants. They split on the subquestion whether reference of theoretical terms and truth values of theoretical statements even have to be formally denounced, notably as category mistakes by instrumentalists, or not, as constructive empiricism concedes. The splitting of "theory realism" at the end of this question-and-answer game into "constructive realism" and "essentialist realism" suggests that we now have five main positions: instrumentalism, constructive empiricism, and referential, constructive and essentialist realism. The following scheme, starting with Question 2, presents their relation in brief.

Q2: true claims about the natural world beyond the observable?	⇒ no	empiricism – instrumentalism – constructive empiricism
yes ⇓ scientific realism		
Q3: beyond reference?	⇒ no	referential realism ⇒ entity realism
yes ⇓ theory realism		
Q4: ideal conceptualization?	⇒ no	constructive realism
yes ⇓ essentialist realism		

The main epistemological positions

Important refinements are obtained when the Questions 2-4 are considered from four perspectives on theories. On the one hand, theories supposedly deal only with "the actual world" or primarily with "the nomic world," that is, with what is possible in the natural world. On the other hand, one may only be interested in whether theories are true or false, or primarily in whether they approach "the truth," regarding the world of interest. It should be stressed that 'the truth' is always to be understood in a domain-and-vocabulary relative way. Hence, no language-independent metaphysical or essentialist notion of "the truth" is assumed. The four perspectives imply that all (non-relativistic) epistemological positions have an "actual world version" and a "nomic world version" and that they may be restricted to "true-or-false" claims, or emphasize "truth approximation claims." In both cases it is plausible to distinguish between observational, referential, and theoretical claims and corresponding inductions, that is, the acceptance of such claims as true. Instrumentalists, in parallel, speak of theories as "reliable-or-unreliable" derivation instruments or as "approaching the best derivation instrument." All four perspectives occur in particular in their realist versions, but they also make sense in adapted form in most of the other epistemological positions.

From Instrumentalism to Constructive Realism (ICR) is a study of confirmation, empirical progress and truth approximation, and their relations. With the emphasis on their nomic interpretation, the five main epistemological positions are further characterized and compared in the light of the results of that study, leading to the following conclusions. There are good reasons for the instrumentalist to become a constructive empiricist; in turn, in order to give deeper explanations of success differences, the constructive empiricist is forced to become a referential realist; in turn, there are good reasons for the referential realist to become a theory realist. The theory realist has good reasons to indulge in constructive realism, since there is no reason to assume that there are essences in the world. As a result, the way leads to constructive realism and amounts to a pragmatic argument for this position, where the good reasons mainly deal with the short-term and the long-term dynamics generated by the nature of, and the relations between, confirmation, empirical progress and truth approximation.

The suggested hierarchy of guidelines or heuristics corresponding to the epistemological positions is, of course, not to be taken in any dogmatic sense. That is, when one is unable to successfully use the constructive realist heuristic, one should not stick to it, but try weaker heuristics: first the referential realist, then the constructive empiricist, and finally the instrumentalist heuristic. For, as with other kinds of heuristics, although not everything goes all the time, *pace* Feyerabend, everything goes sometimes. Moreover, after using a weaker heuristic, a stronger heuristic may become applicable at a later stage: "reculer pour mieux sauter."

Besides epistemological conclusions, there are some general methodological lessons to be drawn. There are good reasons for all positions not to use the falsificationist but the instrumentalist or "evaluation(ist)" methodology. That is, the selection of theories should exclusively be guided by empirical (and perhaps nonempirical) success, even if the better theory has already been falsified. For this common methodology, directed at the separate and comparative evaluation of theories, see Sections 7 and 8 below.

II. Patterns of Explanation and Description

Guided by descriptive and explanatory programs, scientists generate observational laws, proper theories, and explanations of laws by theories as the products of their efforts. Although the standard model of nomological explanation or explanation by subsumption is a very important one, not all explanations fit into it. Another model, called explanation by specification, is needed to do justice to the richness of explanation in the sciences.

Section 3 presents a decomposition model for the explanation of laws by subsumption under theories, and illustrates this with the explanation of the ideal gas law by the kinetic theory. Moreover, it is indicated that there occur in the literature three different reasons for speaking of the reduction of a law.

SiS Ch. 3 presents three more examples in detail, viz., the explanation of Galileo's law of free fall, Mendel's law of interbreeding, and Olson's law about collective goods. Moreover, it is shown that the indicated diagnosis about speaking of reduction suggests a systematic explication of several distinctions made in the literature. Finally, the model of reductive and nonreductive explanations of laws and these distinctions are illustrated by briefly indicating a large number of examples.

There seem to be many explanations that do not fit into the subsumption scheme. In Section 4 it is shown in some detail that, in particular, intentional explanations satisfy a rather different general pattern, called explanation by specification. The resulting two general models enable the distinction of three styles of description and explanation.

SiS Ch. 4 argues in detail that, contrary to the claims of Nagel and Hempel, a subsumption reconstruction of intentional explanation of actions and functional explanations of biological traits, if possible at all, does not do justice to scientific practice. It also shows that functional explanation in biology and a certain type of causal explanation both fit the general pattern of explanation by specification, despite some important fundamental differences, which are easy to indicate in terms of the model. It concludes with some speculations about the existence of other subtypes of the general pattern of explanation by specification.

3. *Explanation and Reduction of Laws*

The standard view on the explanation of individual facts, such as events and states of affairs, and of general facts, that is, observational laws, is that of (deductive-)nomological explanation or explanation by *subsumption* under a law or theory. In contrast to most elementary textbooks, I lay a strong emphasis on the explanation and *reduction* of laws.

The Ideal Gas Law

I start with the decomposition of the explanation of the ideal gas law (IGL) by the kinetic theory of gases (KTG) into three steps. IGL states that for a mole (a standard amount) of gas in a container the product of volume V and (macroscopic) pressure P is proportional to the empirical absolute temperature T, such that the proportionality constant is the same for all gases; viz., the ideal gas constant R. Hence,

IGL: $PV = RT$

According to KTG an isolated amount of gas consists of molecules which move and collide with each other and with the container wall in accordance with Newton's laws of motion. In the *application* step these laws are applied to one molecule colliding with the wall. In this step we use the auxiliary hypothesis that the collision is elastic. The result is the "individual law" which states that the momentum exchange q equals $2mv_w$ (m: mass; v_w: velocity in the direction of the wall):

$q = 2mv_w$

The second step, the *aggregation* step, is an ingenious aggregation, using some statistical auxiliary hypotheses, of the momentum exchange of Avogadro's standard number N of molecules N in a mole of gas, leading to the "aggregated law" which states that the product of the resulting kinetic pressure p on the wall and the volume V is equal to $(2/3)Nu$:

$$pV = (2/3)\, N\, u$$

where u indicates the mean kinetic energy, i.e., the mean value of $(1/2)mv^2$ (v: the total velocity of a molecule).

In the third and last step, the *transformation* step or, more specifically, the *identification* step, two identity hypotheses are introduced:

$$p = P \qquad u = (3/2)(R/N)T$$

It is easy to see that they enable us to derive IGL immediately. Interestingly enough, the second, quantitative identification can essentially be reduced to the following qualitative identification: being in the same thermal state (the basic qualitative concept for the notion of empirical absolute temperature) is identical to having the same mean kinetic energy (see Kuipers 1982).

It is clear that KTG is stratified, both ontologically and epistemologically. The aggregation step amounts to a jump to a higher ontological level, and the identification step amounts to the replacement of theoretical terms by observation terms.

Five-Step Model

The analysis of a number of successful examples, notably the explanation of the ideal gas law, Galileo's law of free fall, the ideal gas law, Mendel's law of interbreeding, and Olson's law about collective goods, results in a general decomposition model according to which, in the explanation of laws by theories, there occur one or more applications of the following five well distinguished steps.

Five-step model:

"Theory X explains law L" if

empirical condition: there are good reasons for accepting theory X and the required auxiliary hypotheses $A1$, …, $A5$ as approximately true.

formal condition: there are auxiliary mutually consistent hypotheses $A1,\ldots,A5$ such that L can be derived strictly or approximately from X using one or more times one or more of five steps, here schematically represented for a paradigmatic case in which all five steps occur precisely one time in the indicated order:

Step	From	To
(1) application	*X*	*A*1
(2) aggregation	*L*1	*A*2
(3) identification	*L*2	*A*3
(4) correlation	*L*3	*A*4
(5) approximation	*L*4	*A*5 (- - -)
		*L*5 = *L*

The empirical condition is in a sense necessary for a 'genuine' explanation, but it will only be assumed when I say so explicitly. The empirical condition hides a world of problems, in particular problems of idealization and circularity, confirmation and (approximate) truth.

The five steps can be characterized as follows.

(1) *Application*: *X* is tailored to the kind of object, system or situation that *L* is, implicitly or explicitly, about. This type of modeling usually requires a lot of auxiliary hypotheses, in particular specification hypotheses concerning variables occurring in *X*. Together with *X*, this leads to the formulation of some law, which may, in the face of an aggregation step, be called an "individual law."

(2) *Aggregation*: the total effect of the individual law for many objects is calculated by a suitable addition, or composition (or synthesis) if more than one type of individual law is involved. The required auxiliary hypotheses are primarily of a statistical nature, but "macroscopic constraints" may also be involved. The resulting law is called an "aggregated law."

(3) *Identification*: the aggregated law is transformed with the aid of one or more identity hypotheses, ontologically identifying (object and predicate) terms of *L* with terms (definable in terms) of *X*. The resulting law need not yet be completely stated in terms of *L*.

(4) *Correlation*: the resulting law is transformed with the aid of some causal hypotheses, correlating terms of *L* with (primary or defined) terms of *X*. Such (causal) correlations may, but need not be, of a statistical nature.

(5) *Approximation* (*or re-idealization*): the deductively derived law is now simplified by some mathematical or logical approximation device, in particular counterfactual idealization, and justified by an auxiliary hypothesis stating implicitly or explicitly, for instance, that some relevant term is relatively small. In this way, the last step is a final opportunity to get rid of terms not occurring in *L*, viz., terms disappearing when they assume some extreme value.

The ideal gas law example uses the first three steps. The other three examples mentioned can be decomposed as follows. The explanation of

Galileo's law of free fall by Newton's theory of gravity requires, after the application step, an approximation step, viz. the initial height of the falling object is negligible relative to the radius of the earth. The explanation of the laws of interbreeding by Mendel's theory typically requires, after the application step, a (deterministic) causal correlation step, viz. correlating genetic factor combinations with observable characteristics. Finally, the explanation of Olson's law by utility theory formally resembles the ideal gas law explanation. However, here the application step and subsequent aggregation step are followed by a (statistical) causal correlation step, viz. the smaller the degree of participation in a group the smaller the chance of realization of a collective good.

The Reduction of Laws

So much for the general decomposition model. On the basis of the literature, starting with Nagel's seminal Chapter 11 of his *The Structure of Science*, it can be argued that three of the five steps are apparently the reason for speaking of the reduction of a law:

> "Theory *X* reduces law *L*" or "*L* can be reduced to/by *X*"
> if and only if
>
> *systematic condition*: *X* explains *L* in accordance with the five-step model , with at least one of the steps being aggregation (2) (provided it is non-trivial), identification (3) or approximation (5),
>
> *temporal condition*: *L* has been established before *X*.

This diagnosis suggests a systematic explication of several distinctions between reductions made in the literature: heterogeneous versus homogeneous reduction (with or without identification), corrective versus deductive reduction (with or without approximation), and micro-reduction versus other types of reduction (with or without aggregation). It should be stressed that all kinds of reduction of a law discussed above are non-eliminative in the sense that the law undergoes at most some correction, namely in the case of an approximative step. The phrase "eliminative reduction," as used by philosophers of (science and) mind usually refers to cases where a theory is replaced by a totally different one, which is certainly a non-standard use for many natural scientists. Hence, for our purposes, reduction of laws in all its variety is non-eliminative.

4. *Explanation and Description by Specification*

Although Section 3 primarily deals with the explanation of laws by subsumption under a theory, it can easily be extended to a similar subsumption explanation of individual events. However, there seem to be many explanations that do not fit into the subsumption scheme. In particular, contrary to the claims of Nagel and Hempel, a subsumption reconstruction of intentional explanation of actions and functional explanations of biological traits, if it is possible at all, does not do justice to scientific practice (SiS, 4.1.1 and 4.2.1). Such explanations satisfy another general pattern, called explanation by specification, in terms of which it is easy to indicate the fundamental differences between intentional and functional explanation and to nevertheless reconstruct the relevant thought processes of scientists in a similar way. The general scheme of explanation by specification also fits certain types of causal explanation, viz., explanations of individual events that select "the cause" out of the set of causal factors. In the three subtypes, scientists are guided by a corresponding searchlight principle, viz., intentionality, functionality, and (specific) causality, respectively. Although the results obtained in this way are typically called explanations (by specification), the programs in the context of which they are generated are usually called descriptive programs, because explanation by a certain type of specification automatically leads to a corresponding type of description, in particular classification. Together with explanation by subsumption this enables the distinction of three styles of description and explanation: intentional, functional and causal (or structural).

Intentional Explanation

For the intentional explanation of actions the following meaning analysis is crucial. According to a first meaning postulate, a *specific intentional statement* of the form:

> *x* performed action *y* with the intention of approaching goal *z*

is supposed to be decomposable into the conjunction of

> action component: *x* performed action *y*
> desire component: *x* desired goal *z*
> belief component: *x* believed *y* to be *useful* to approach *z*

In this meaning postulate it is assumed that goal *z* is always different from the *internal goal* of action *y*, that is, the goal of action *y* according to the description used for it. For example, the internal goal of "opening the door" is "having the door open." Notice that the belief component does not require that the action is believed to be necessary for achieving the goal. The postulate is a first approximation in the sense that the three components do not exhaust the meaning of the complex statement. For instance, some time conditions are plausible, in particular to exclude hindsight "rationalizations": the desire and the belief may not "start" later than the action. More importantly, it is defensible to add a *causal* component stating that the belief and desire component were causally effective in the sense that the combination of the two mental states played a significant causal role in the formation of the plan to perform the action.

The second meaning postulate interprets an *unspecific intentional statement* of the form:

> x performed action y intentionally

as

> there is a goal γ such that x performed action y with the intention of approaching goal γ

Note that this postulate entails that a specific intentional statement logically implies the corresponding unspecific one, by so-called existential generalization.

The crucial claim now is that searching for an intentional explanation for a certain action, performed by somebody, amounts to searching for a true "intentional specification" of the unspecific intentional hypothesis that is generated by the following methodological principle:

> *principle of intentionality*:
> if x performs action y then x performs action y intentionally

that is, if someone performs (or has performed) an action, he will do (have done) that intentionally. It is important to note that the notion of "intentionality" is here not used in the broad philosophical sense of "aboutness," but in the restricted straightforward sense of consciously aiming at a certain goal.

As in many other methodological principles, the principle of intentionality should be conceived as a principle that fundamentally leaves room for exceptions. Such exceptions can only improperly be called falsifications. That is, the research is guided by the idea that the qualification "intentional" is adequate, but the research may fail to substantiate this claim. Although it is impossible to have conclusive reasons for inferring that an action was not intentionally performed, the researcher may provisionally come to this conclusion, that is, for the time being. Such methodological principles are essentially heuristic variants of default rules of inference in the sense of non-monotonic logic; as long as one has no good reason to the contrary, a particular instance of a certain type is supposed to have a feature that most of the instances of that type have.

The foregoing meaning analysis enables a stepwise reconstruction of the thought process governing an intentional explanation, starting with the why-question and, if successful, leading to one true "why-answer," i.e. an intentional specification and, as a rule, a number of new, related why- and how-questions.

A simple example can illustrate the main ideas. Noticing that "Jane has opened the window," the question may arise "Why did Jane open the window?", assuming that she did it intentionally, i.e., assuming "Jane opened the window intentionally." One possible explanation by specification is that "Jane opened the window in order to let the room cool down." This hypothesis has to be tested by testing its (extra) meaning components: "Jane wished that the room would cool down" and "Jane believed that opening the window would probably cool down the room." Now suppose that it is mid summer and that Jane has standard common sense, then the belief component may be assumed to be false. Another hypothesis is that "Jane wanted to let out her tame canary." It may well be that we can obtain good reasons to conclude that its test implications are verified, and hence obtain a true why-answer. If so, we may first conclude, by existential generalization, that she did it intentionally, as we thought from the start. Then we may jump to new related why- and how-questions, such as "Why did she want to let out her tame canary?". This why-question not only brings us to the highly similar nature of the explanation of goals in terms of further goals, but also to an illustration of a multiple why-answer: she may want to please the canary and she may in addition hope for social contact.

In SiS it is shown that a similar analysis can be given of the explanation of goals and choices between different actions and goals. It is also shown that, apart from the specific meaning postulates, a formally similar analysis of functional explanations of biological traits can be given. Moreover, although causal explanations of events are seldom mentioned as examples of a type of explanation that cannot be seen as a subsumption explanation, there is a certain type of causal explanation for which the subsumption interpretation is, though formally possible, not the most adequate explication. I am especially thinking of causal explanation in the context of e.g., everyday life, medicine, jurisdiction, insurance, technology, etc. In many cases one does not hesitate to call one of the causal factors *the cause*, whereas the subsumption view is essentially neutral with respect to the causal factors. As shown in SiS, the general scheme of explanation by specification provides again a better framework for modeling such explanations, with its own specific meaning postulates, of course.

Styles of Description and Explanation

Explanation by a certain type of specification automatically leads to a corresponding type of description, in particular classification. Organs performing the same function, though in quite different ways, are primarily classified according to that function. Actions are classified according to an internal goal that may be ascribed to the relevant behavior. Hence, in general we may distinguish a *functional style* and an *intentional style* of description and explanation. Besides these we may distinguish the *causal style*, that is, causal description and explanation in the general sense of identifying causal factors as well as specific causal description and explanation. Hence, in the causal style I combine explanation by subsumption under causal laws and explanation by causal specification, along with corresponding types of description. Moreover, all styles are also supposed to include relevant conceptual and ontological claims, such as identity and part-whole relations. In the case of the causal style, this might be a reason for speaking instead of the *structural* style. Notice that, roughly speaking, the styles correspond to Dennett's intentional, design and physical stance (Dennett 1987).

The scope of the three styles of description and explanation is much broader and less exclusive than has been suggested so far. To begin with the latter point in terms of explanation by subsumption and by specification: it is important to note that, although there are many facts which can be explained only by subsumption and not by specification, this is not due to some incompatibility between these two general types of explanation. In my view the situation is as follows. In certain contexts of scientific interest we happen to focus on certain types of explanation. Besides interest in explanation by subsumption in many cases, one is frequently also or only interested in explanation and description by specification. For example, put in our terms, biologists are often primarily interested in explanation and classification by functional specification (frequently called "functional biology"), historians in explanation and classification by intentional specification, and insurance companies in explanation and classification by causal specification. By way of a slogan one might say: although explanation by subsumption may in principle always be possible, it is certainly not always interesting; on the other hand, explanation by (some relevant) specification is certainly not always possible, but, as long as the answer is not well known, it is always interesting when it is possible. What holds for explanation, holds *eo ipso* for the corresponding kind of description, i.e., classification, viz., classification according to structure, function or intention. Hence it holds for the styles of description and explanation in general. In particular, explanatory-cum-descriptive research programs may be governed partly or even completely by a style, more precisely, by a core heuristic-methodological principle, e.g. the functionality principle.

Subsequent to this line of thought I would like to draw attention to the scope of explanation by specification. In particular, the question is whether there are other subtypes of explanation by specification. More specifically, my explication raises the question of whether there are "clear and distinct" subtypes of explanation by specification between intentional explanation on the one hand and functional explanation in biology on the other. To verify this conjecture, further research may lead to proposals for the corresponding meaning postulates (keys) for these subtypes. In psychology one might think of a key for explanation by (specification of) unconscious motives. In sociology and anthropology there may also be specific keys for functional explanation, which are not strictly biological. To be sure, the latter disciplines use such explanations (for a survey, see Pettit 1996). However, as long as no convincing proposals have been formulated for a key, I believe the alternative to be the case: there is merely a diffuse continuous path from biological functional explanation to intentional explanation.

III. Structures in Interlevel and Interfield Research

In this part I explore some new patterns, notably concerning reduction and correlation of concepts, after which all patterns distinguished so far are applied in SiS to mind-body research. However, the latter application will only be summarized here.

The very idea of concept reduction requires a distinction between ontological identities and causal correlations, which is intuitively clear but difficult to explicate. Inspired by Causey's and Kim's work, three kinds of type-type concept reduction can be distinguished: singular, multiple and quasi-reduction. Special attention is paid to the radical and moderate reductionistic and holistic research strategies. Moderate or mixed strategies, in particular by means of co-operating research programs, recognize, first, that for the reduction of concepts and laws there should be something to reduce, and, second, that reduction of concepts and laws has by no means to imply their elimination.

In Section 3 I dealt with the reduction of laws by theories. In this part reduction is the dominant focus of attention. Although reduction is the mainstream point of view in many sciences, including for example "behavioral and cognitive neurosciences," it is not a popular point of view in the philosophy of the cognitive sciences in general and the philosophy of mind in particular. I would like to conclude this introduction with Kim's *cri de coeur* (2000, p. 89) of this even much more general but regretful situation:

> Expressions like 'reduction', 'reductionism', 'reductionist theory', and 'reductionist explanation' have become pejoratives not only in philosophy, on both sides of the Atlantic, but also in the general intellectual culture of today. They have become common epithets thrown at one's critical targets to tarnish them with intellectual naivete and

> backwardness. To call someone 'a reductionist', in high-culture press if not in serious philosophy, goes beyond mere criticism or expression of doctrinal disagreement; it is to put a person down, to heap scorn on him and his work. We used to read about 'bourgeois reductionism' in left-wing press; we now regularly encounter charges of 'biological reductionism', 'sociological reductionism', 'economic reductionism', and the like, in the writings about culture, race, gender, and social class. If you want to be politically correct in philosophical matters, you would not dare come anywhere near reductionism, nor a reductionist. It is interesting to note that philosophers who are engaged in what clearly seem like reductionist projects would not call themselves reductionists or advertise their work as reductionist programs.

In a note, Kim mentions work of Jerry Fodor and Fred Dretske, in which they speak of "naturalization," for that term "doesn't offend, it seems, in the way 'reduction' does." Be this as it may, with Kim and Bickle (1998), I like to continue calling the subject by its natural name.

5. *Reduction and Correlation of Concepts*

In the philosophy of science literature produced since 1960, emphasis has fallen on the reduction of laws, notwithstanding the fact that something like concept reduction frequently appears as a necessary intermediate link. As in the case of the reduction of laws, the notion of general identities, as opposed to merely causal correlations, plays an important role in the reduction of concepts. The distinction between identities and correlations is crucial for the three main variants of reduction of concepts (singular, multiple, and quasi) and the two main variants of correlation of concepts (singular and multiple) that can be distinguished. The various perspectives on reduction lead to a number of research strategies, from radical reductionistic to radical holistic ones, along with the mixed strategies in between.

Leaning heavily on Causey (1977), SiS Ch. 5 starts by explaining and defending the distinction between identities and correlations as a working hypothesis. It then analyzes in great detail the three crucial variant forms of concept reduction, using elementary physical examples. The two possible types of correlation of concepts result from substituting correlations for identities. The chapter also pays much attention to the relation between reduction of laws and concepts and the importance of multiple reduction, the latter by strongly relativizing the so-called problem of multiple realizability.

Degrees and Kinds of Reduction and Correlation of Concepts

Stimulated by Kim (1996), some general patterns have been uncovered by analyzing two, relatively simple, examples in detail, viz. "water" and "temperature." The technically somewhat complicated character of the patterns shows that the standard presentations of (kinds of) type-type reduction are rather naive, not to speak of weaker kinds of concept reduction.

Here I confine myself to a summary of the degrees and kinds of reduction and correlation of concepts that can be distinguished. I start with three degrees of concept reduction. I will use the general terminology of types and their tokens. The tokens of a type are instantiations or (ontological) realizations of the latter and constitute it together.

Singular type-type (or one-one) *reductions* are perfect (non-approximate) concept reductions of the highest, third, degree: a type of a higher level description of an aggregate, that is a macro-type, is ontologically identified with *one* type of a lower level, that is a micro-type. It amounts to the claim that the set of micro-tokens (of the same aggregate!) that ontologically realize the macro-type coincides with the set of micro-tokens that constitute the micro-type by definition. Example: roughly speaking, the reduction of "water" in terms of aggregates of H_2O molecules and 'temperature' in terms of mean kinetic energy.

Multiple (or one-many) *reductions* are perfect concept reductions of the second degree: a macro-type is ontologically identified with a union of (usually disjunct) micro-types. It amounts to the claim that the set of micro-tokens that ontologically realize the macro-type coincides with the union of the sets of tokens that constitute the relevant micro-types by definition. Example: the reduction of "uranium" in terms of pure and mixed aggregates of (molecules of) its three isotopic atoms and, on closer inspection, "temperature" in terms of mean kinetic energy of translation of the molecules in the case of gases and to mean kinetic energy of vibration of the molecules in the case of solids.

Quasi (type-type or one-one) *reductions* are perfect concept reductions of the first degree: a macro-type is ontologically identified with the union of all micro-tokens that ontologically realize the macro-type. Example: the (quasi-) reduction of a color to its corresponding wavelength interval. As long as there is no natural partition of wavelengths on that level, a genuine reduction will not be possible.

Perfect reductions of the second and third degree have approximate versions, that is, cases in which the relevant sets do not coincide perfectly but only approximately.

Since 'quasi-correlation' does not make sense, there are only two degrees of correlation of concepts, both with approximate versions.

Singular type-type (or one-one) *correlations* are perfect concept correlations of the third degree: a macro-type is causally correlated with *one* micro-type. It amounts to the claim that the set of lower level (or micro-) tokens (now not necessarily of the same aggregate!) that causally realize the macro-type coincides with the set of micro-tokens that constitute the micro-type by definition. Example: the causal correlation between a recessive phenotype and its corresponding homozygote genotype.

Multiple (or one-many) *correlations* are perfect concept correlations of the second degree: a macro-type is causally correlated with a union of micro-types. It amounts to the claim that the set of micro-tokens that causally realizes the macro-type coincides with the union of the sets of tokens that constitute the relevant micro-types by definition. Example: the causal correlation between a dominant phenotype and its two corresponding genotypes, a homozygote and a heterozygote.

Reductionistic, Holistic, and Mixed Strategies

An important question is the relation between the presented epistemological analysis of reduction of laws and concepts and all kinds of metaphysical positions and methodological research strategies. I restrict myself to *vertical reduction*, that is, reduction of concepts in one of the explicated senses, and reduction of laws based on aggregation and/or identification. This leaves reduction of laws based on approximation, that is, corrective reduction, apart as a kind of *horizontal reduction*. From here on in this section 'reduction' is to be read as 'vertical reduction', except when otherwise stated.

Following Looijen (2000), three methodological approaches or research strategies with respect to a certain macro-domain can be distinguished. In the *radical reductionistic strategy* all attention is directed to the reduction of macro-laws and concepts and the establishment of aggregate concepts and aggregated laws, that is, concepts and laws formulated in lower level terms. In the *radical holistic strategy* all attention is directed to the formation of macro-concepts, to discovering macro-laws, and finally to the explanation of macro-laws by encompassing macro-theories (hence, without using aggregation or identification). Finally, there is the *mixed strategy*, according to which all these activities take place alternately, depending on all kinds of considerations, such as the stage of the research, what seems possible, and, last but not least, what can stimulate what, etc. To be sure, the mixed strategy favors reduction when possible. Roughly speaking, in the mixed strategy one describes the macro-phenomena and their possible relations in macro-terms, and tries to explain them in micro-terms as far as possible, and hence in macro-terms as far as necessary.

This brings me to the formulation of three current philosophical (ontological-cum-epistemological) positions with respect to a certain "macro-domain." They are formulated as general statements about the possibility of reduction of the concepts and laws of that domain. *Radical reductionism* is the belief that every macro-concept and macro-law can be reduced. *Radical holism* is the belief that no (interesting) concepts and laws of the domain can be reduced. Finally, *restricted reductionism* (and holism!) is the belief that some concepts and laws may be reducible, but others may not be.

It is important to note that the terminologically corresponding strategies and positions are not strictly coupled, except perhaps that radical philosophical holism only leaves room for the radical holistic strategy. The converse is not self-evident. There are excellent examples of research according to the radical holistic strategy, e.g., phenomenological thermodynamics and macro-economics, where it is seen as a compatible but separate task to try to reduce the macro-concepts and -laws, i.e., to work according to the radical reductionistic strategy. To be sure, it has to be conceded that reductionistic strategies in general have been very successful in the history of science.

However, the radical reductionistic strategy often leads to impressive minute research, constituting perfect aggregate concepts and aggregated laws, for which, however, nobody is waiting. On the other hand, the radical holistic strategy frequently degenerates into hardly testable and transferable insights.

In line with these roughly formulated impressions, it is plausible to formulate the working hypothesis that in many cases the mixed strategy will be the best strategy. For, to reduce concepts and laws of a certain domain, they have first to be established. In its turn, it is frequently the case that the search for concepts and laws has been considerably stimulated by reductionistic questions.

We would like to conceive examples of the mixed strategy more specifically as important cases of interaction of research programs, as outlined in SiS Ch. 1. In the case of reductive interaction the guide program is of course a program on the macro-level, whereas at least one supply program is supposed to deal with the micro-level. When the supply program is, like the guide program, in the internal phase, the reductive interaction is symmetric, when it is in the external phase it is asymmetric.

Zandvoort's (1986) paradigm example of a supply program in the natural sciences is the NMR (nuclear magnetic resonance) program, engaged in asymmetric reductive co-operation with chemical and biological research programs. An important asymmetric example in the social sciences is the utility maximization or rational choice program. It has proved its strength in microeconomics and nowadays it co-operates with guide programs in macroeconomics (Janssen 1993) and macro-sociology (explanatory sociology). A historical example of the symmetric reductive type is the interaction between phenomenological thermodynamics and statistical mechanics.

The type of interdisciplinary research just indicated fits well with the following view on the different types of explanation. Beside the suggested "vertical" forms of reductive explanation, there are not only other vertical forms of non-reductive explanation (e.g., based on correlation), but also many types of "horizontal" explanation (e.g., based on approximation). As far as the nomological explanation of laws is concerned, all these types of explanation seem to fit well in my general decomposition model. However, as we have seen in the previous section, there have always been claims that there are types of explanation which are essentially different from nomological explanation. More specifically, I have argued that, e.g., intentional explanation and functional explanation (in biology) indeed constitute separate forms of explanation and that they fit into a general model of so-called *explanation by* (intentional, functional, etc.) *specification*, which even includes certain types of causal explanation. The relevance to the present context is the following. Explanation by specification may be of a horizontal or of a vertical nature. In particular, it may be the dominant way of horizontal explanation in a guide program. Consequently, instead of disqualifying explanation by specification, it may, for example, play a crucial role in reductive interdisciplinary mind-body research: it can open the way for interesting forms of reductive neurophysiological explanation, a way that would be indistinguishable in the overwhelming number of possibilities confronting the radical reductionistic strategy. Conversely, reductive neurophysiological explanation may very well suggest new ways for further development of psychological explanation, a possibility that is blocked by the radical holistic strategy.

6. *Levels, Styles, and Mind-Body Research*

SiS Ch. 1 indicates several goals and types of symmetric and asymmetric co-operation between research programs, where the programs may or may not deal with different levels of aggregation or even belong to different disciplines. Moreover, they may or may not use the same or different styles of description and explanation, that is, causal, functional or intentional. In SiS Ch. 6 these themes are further elaborated, with an emphasis on mind-body research and drawing heavily upon work by Bechtel, Burton, Darden, Mackor, Millikan and Panhuysen.

Section 6.1 sets out a matrix of four kinds of so-called interfield research, starting from a survey of possible interlevel research between epistemological levels in general and the two major divisions of them, viz., in ontological levels and epistemological styles, in particular. Section 6.2 deals with the mutual relations between the styles, with emphasis on interstyle monolevel research. Section 6.3 focuses on monostyle interlevel research, notably biophysical mind-body research, and indicates several kinds of reduction and correlation of concepts and laws. Section 6.4 broadens the scope by discussing some models and examples of interlevel and interstyle interfield mind-body research. The main example concerns two types of juvenile delinquency.

Finally, Section 6.5 gives some examples of interdisciplinary research, which is nevertheless of a monostyle monolevel nature.

IV. Confirmation and Empirical Progress

In the first six sections we have concentrated on (units of) description and explanation of various kinds, without bothering about details of testing and evaluation of the relevant hypotheses. This will be the subject of the next two sections.

In this part first a sketch will be given of the main ideas behind confirmation and falsification of a hypothesis by the so-called HD (hypothetico-deductive) method. Confirmation of a hypothesis, however, has the connotation that the hypothesis has not yet been falsified. Whatever the truth claim associated with a hypothesis, as soon as it has been falsified, the plausibility (or probability) that it is true becomes and remains zero. In this part I also elaborate how theories can nevertheless be evaluated after falsification.

HD testing attempts to give an answer to one of the questions in which one may be interested, *the truth question*, which may be qualified according to the relevant epistemological position. However, *the* (*theory*) *realist*, for instance, is not only interested in the truth question, but also in some other questions. To begin with, there is the more refined question of which (individual or general) facts the hypothesis explains (its explanatory successes) and which facts are in conflict with the hypothesis (its failures); *the success question* for short. I show in this part that the HD method can also be used in such a way that it is functional in (partially) answering this question. This method is called HD evaluation, and uses HD testing of test implications. Since the realist ultimately aims to approach the strongest true hypothesis, if any, i.e., *the* (*theoretical-cum-observational*) *truth* about the subject matter, the plausible third aim of the HD method is to help answer the question of how far a hypothesis is from the truth, *the truth approximation question*. Here the truth will be taken in a relatively modest sense, viz., relative to a given domain and conceptual frame. It can be argued (see in particular ICR, Ch. 7) that HD evaluation is also functional in answering the truth approximation question.

The other epistemological positions (see Section 3) are guided by two related, but more modest success and truth approximation questions, and it can be shown (see ICR, Ch. 9) that the HD method is also functional in answering these related questions. The *constructive empiricist* may not only be interested in the question of whether the theory is empirically adequate or observationally true; i.e., whether the observational theory implied by the full theory is true.

He may also be interested in the refined success question about what its true observational consequences and its observational failures are, and in the question of how far the implied observational theory is from the strongest true observational hypothesis, *the observational truth*. The *referential realist* may, in addition, be interested in the truth of the reference claims of the theory and how far it is from the strongest true reference claim, *the referential truth*. The *instrumentalist* phrases the first question of the empiricist more liberally: for what (sub-)domain is it observationally true? He retains the success question of the empiricist. Finally, he will reformulate the third question as follows: to what extent is it the best (and hence the most widely applicable) derivation instrument?

The method of HD evaluation will turn out, in this part, to be a direct way to answer the success question and in ICR it is shown to be an indirect way to answer the truth approximation question, in both cases for all four epistemological positions. The present part will primarily be presented in a relatively neutral terminology, with specific remarks relating to the various positions. The success question will be presented in terms of successes and counterexamples: what are the potential successes and counterexamples of the theory? Instead of speaking of counterexamples, which has a falsificationist flavor, one may prefer to speak of failures or problems.

In sum, two related ways of applying the HD method to theories can be distinguished. The first one is *HD testing*, which aims to answer the truth question. However, as soon as the theory is falsified, the realist with falsificationist leanings, i.e., advocating exclusively the method of HD testing, sees this as a disqualification of an explanatory success. The reason is that genuine explanation is supposed to presuppose the truth of the theory. Hence, from the realist-falsificationist point of view a falsified theory has to be abandoned and one has to look for a new one.

The second method to be distinguished, *HD evaluation*, keeps taking falsified theories seriously. It tries to answer the success question, the evaluation of a theory in terms of its successes and counterexamples (problems) (Laudan 1977). For the (non-falsificationist) realist, successes remain explanatory successes and, when evaluating a theory, they are counted as such, even if the theory is known to be false.

It is important to note that the term '(HD) evaluation' refers to the evaluation in terms of successes and counterexamples, and not in terms of truth approximation, despite the fact that the method of HD evaluation nevertheless turns out to be functional for truth approximation. Hence, the method of HD evaluation can be used meaningfully without any explicit interest in truth approximation and without even any substantial commitment to a particular epistemological position stronger than instrumentalism.

In my view, Chapters 7 and 8 of SiS present the core of the methodological practice in the various sciences, for which reason they are here relatively extensively summarized. They originate from ICR (in particular Ch. 5 and 6), for they provide the glue between confirmation and truth approximation. In addition to what I will present here, SiS Ch. 7 gives a survey of qualitative and quantitative notions of confirmation, based on Ch. 2-4 of ICR, and deals more in detail with "falsifying general hypotheses," which if accepted lead to general problems of theories. Moreover, it briefly deals with statistical test implications and it presents a systematic analysis of the complications that may arise in applying the HD method. Referring to Part III of ICR, SiS Ch. 8 indicates already why the evaluation methodology can be functional for truth approximation. Moreover it explains and justifies in detail the non-falsificationist practice of scientists, as opposed to the explicit falsificationist view of many of them, not only in terms of the fruitful dogmatism discovered by Kuhn and Lakatos, which can be distinguished from dogmatism in pseudoscience, but also in terms of truth approximation, for example, by the paradigmatic non-falsificationist method of idealization and concretization, as propagated by Nowak.

7. *Testing and Further Separate Evaluation of Theories*

This section starts with a brief exposition of HD testing, that is, the HD method of testing hypotheses, and continues with the separate evaluation of a theory by the HD method, even when it has been falsified.

HD Testing Leading to Confirmation or Falsification

HD testing attempts to give an answer to one of the questions that one may be interested in, *the truth question*, which may be qualified according to the relevant epistemological position. The HD method prescribes the derivation of test implications and testing them. In each particular case, this may either lead to confirmation or to falsification. Whereas the "language of falsification" is relatively clear, the "language of confirmation" is a matter of great dispute.

According to the leading expositions of the hypothetico-deductive (HD) method by Hempel (1966), Popper (1934/1959) and De Groot (1961/1969), the aim of the HD method is to determine whether a hypothesis is true or false, that is, it is a method of testing. On closer inspection, this formulation of the aim of the HD method is not only laden with the epistemological assumption of theory realism, according to which it generally makes sense to aim at true hypotheses, but it also mentions only one of the realist aims, i.e., answering the "truth question." Applying the HD method to this end will be called HD testing as distinct from HD evaluation, which has other primary aims.

For the moment, I will confine my attention to the HD method as a method of testing hypotheses. Though the realist has a clear aim in undertaking HD testing, as I have already indicated, this does not mean that HD testing is only useful from that epistemological point of view.

A test of a hypothesis may be experimental or natural. That is, a test may be an experiment, an active intervention in nature or culture, but it may also concern the passive registration of what is or was the case, or what happens or has happened. In the latter case of a so-called natural test, the registration may be a more or less complicated intervention, but is nevertheless supposed to have no serious effect on the course of events of interest.

According to the HD method a hypothesis H is tested by deriving test implications from it, and checking, if possible, whether they are true or false. Each test implication has to be formulated in terms that are considered to be observation terms. A test implication may or may not be general in nature. Usually there is background knowledge B, which is assumed to be true. Moreover, a test implication is frequently of a conditional nature, if C then F ($C \rightarrow F$). Here C denotes one or more "initial conditions" and F denotes a potential fact (event or state of affairs) predicted by H and C. If C and F are of an individual nature, F is called an individual test implication, and $C \rightarrow F$ a conditional test implication. When C is artificially realized, it is an experimental test, otherwise it is a natural test.

The basic logic of HD testing can be represented by some (valid) applications of Modus (Ponendo) Ponens (MP), where '$\models$' indicates logical entailment and where 'I' denotes a test implication:

$$\begin{array}{l} B, H \models I \\ \underline{B, H\quad} \\ I \end{array} \qquad\qquad \begin{array}{l} B, H \models C \rightarrow F \\ \underline{B, H, C} \\ F \end{array}$$

It should be stressed that $B, H \models I$ and $B, H \models C \rightarrow F$ are supposed to be deductive claims, i.e., claims of a logico-mathematical nature.

The remaining logic of hypothesis testing concerns the application of Modus (Tollendo) Tollens (MT). Neglecting complications that may arise, such as that B's or C's truth may be disputed, if the test implication is false, the hypothesis must be false, and therefore has been falsified, for the following arguments are deductively valid ('$\neg$' indicates negation):

$$\begin{array}{l} B, H \models I \\ \underline{B, \neg I} \\ \neg H \end{array} \qquad\qquad \begin{array}{l} B, H \models C \rightarrow F \\ \underline{B, C, \neg F} \\ \neg H \end{array}$$

When the test implication turns out to be true, the hypothesis has of course not been (conclusively) verified, for the following arguments are invalid, indicated by '-/-/-':

$B, H \vDash I$	$B, H \vDash C \rightarrow F$
B, I	B, C, F
-/-/-	-/-/-/-/
H	H

Since the evidence (*I* or *C&F*) is compatible with *H*, we may at least say that *H* may still be true. However, we can say more than that. Usually it is said that *H* has been confirmed. It is important to note that such confirmation by the HD method means more than mere compatibility; it is confirmation in the strong sense that *H* has obtained a *success* of a (conditional) deductive nature. By entailing the evidence, *H* makes the evidence as plausible as possible. This I call *the success perspective* on ((conditional) deductive) *confirmation*.

Falsification and confirmation have many complications, e.g., due to auxiliary hypotheses, see SiS Section 7.3.3. As already indicated, however, there is a great difference between falsification and confirmation. Whereas the "logical grammar" of falsification is not very problematic, the grammar of confirmation, i.e., the explication of the concept of confirmation, has been a subject of much dispute. For this subject the reader is referred to Section 7.1.2 of SiS for a survey and to Part I of ICR for a detailed treatment.

In the rest of this section it is shown that a decomposition of the HD method applied to theories is possible which naturally leads to an explication of the method of separate HD evaluation, using HD testing, even in terms of three models. Among other things, it will turn out that HD evaluation is effective and efficient in answering the success question. In the next section I use the separate HD evaluation of theories for their comparative HD evaluation.

HD Evaluation of a Theory Leading to an Evaluation Report

The core of the HD method for the evaluation of theories amounts to deriving from the theory in question, say *X*, General Test Implication (GTI's) and subsequently (HD) testing them. For every GTI *I* it holds that testing leads sooner or later either to a counterexample of *I*, and hence a counterexample of *X*, or to the (revocable) acceptance of *I*: a *success* of *X*. A counterexample, of course, implies the falsification of *I* and *X*. A success minimally means a "derivational success"; it depends on the circumstances whether it is a predictive success and it depends on one's epistemological beliefs whether or not one speaks of an explanatory success.

Now, it turns out to be very illuminating to write out in detail what is implicitly well-known from Hempel's and Popper's work, viz., that the HD method applied to theories is essentially a stratified, two-step method, based on a macro- and a micro-argument, with much room for complications. In the macro-step already indicated, one derives GTI's from the theory. In their turn, such GTI's are tested by deriving from them, in the micro-step, with the help of suitable initial conditions, testable individual statements, called Individual Test Implications (ITI's). The suggested decomposition amounts in some detail to the following.

For the macro-argument we get:

Theory: X
Logico-Mathematical Claim (LMC): if X then I
____ Modus Ponens (MP)
General Test Implication (GTI): I

A GTI is assumed formally to be of the form:

I: for all x in D [if $C(x)$ then $F(x)$]

that is, for all x in the domain D, satisfying the initial conditions $C(x)$, the fact $F(x)$ is "predicted." All specific claims about x are supposed to be formulated in observation terms.

Successive testing of a particular GTI I will lead to one of two mutually exclusive results. The one possibility is that sooner or later we get falsification of I by coming across a falsifying instance or counterexample of I. Although a counterexample of I is, strictly speaking, also a counterexample of X, I also call it, less dramatically, a *negative instance* of or an *individual problem* for X. The alternative possibility is that, despite variations in members of D and ways in which C can be satisfied, all our attempts to falsify I fail, i.e., lead to the predicted results. The conclusion attached to repeated success of I is of course that I is established as true, i.e., as a general (reproducible) fact. I will call such an I a (*general*) *success* of X. Finally, it may well be that certain GTI's of X have already been tested long before X was taken into consideration. The corresponding individual problems and general successes have to be included in the evaluation report of X (see below).

Recorded problems and successes are (partial) answers to the success question: what are the potential successes and problems of the theory? Hence, testing GTI's derived in accordance with the macro HD argument is effective in answering this question. Moreover, it is efficient, for it will never lead to irrelevant, neutral results, that is, results that are neither predicted by the theory nor in conflict with it. Neutral results for one theory only come into the picture when we take test results of other theories into consideration, that is, the comparative evaluation of two or more theories (see the next section).

I call the list of partial answers to the success question, which are available at a certain moment t, the *evaluation report* of X at t, consisting of the following two components:

the set of individual problems,
i.e., established counterexamples of GTI's of X,
the set of general successes,
i.e., the established GTI's of X, that is, general facts derivable from X.

Hence, the goal of separate theory evaluation can be explicated as aiming at such an evaluation report.

Models of HD Evaluation

Let us now have a closer look at the testing of a general test implication, the micro-step of the HD method, or, more generally, the testing of a *General Testable Conditional* (GTC). The micro HD argument amounts to:

General Test Conditional (GTC): G: for all x in D [if $C(x)$ then $F(x)$]
Relevance Condition: a in D
____Universal Instantiation (UI)
Individual Test Conditional: if $C(a)$ then $F(a)$
Initial Condition(s) (IC): $C(a)$
____Modus Ponens (MP)
Individual Test Implication (ITI): $F(a)$

If the specific prediction posed by the individual test implication turns out to be false, then the hypothesis G has been falsified. The relevant (description of the) object has been called a *counterexample* or a *negative instance* or an *individual problem* of G. If the specific prediction turns out to be true the relevant (description of) the object may be called a *positive instance* or an *individual success* of G. Besides positive and negative instances of G, we may want to speak of *neutral instances* or *neutral results*. They will not arise from testing G, but they may arise from testing other general test implications.

Consequently, the evaluation report of GTC's basically has two sides, like the evaluation reports of theories; one for problems and the other for successes. Again, they form partial answers to the success question now raised by the GTC. However, here the two sides list entities of the same kind: negative or positive instances, that is, individual problems and individual successes, respectively.

It is again clear that the micro HD argument for a GTC G is effective and efficient for making its evaluation report: each test of G either leads to a positive instance, and hence to an increase of G's individual successes, or it leads to a negative instance, and hence to an increase of G's individual problems. It does not result in neutral instances. Note that what I have described above is the micro HD argument for *evaluating* a GTC. When we confine our attention to establishing its truth-value, and hence stop with the first counterexample, it is the (micro) HD argument for *testing* the GTC.

Concatenation of the macro and micro HD argument gives the full argument for theory evaluation leading to individual problems and individual successes. Instead of the two-step concatenated account, theory evaluation can also be presented completely in terms of contracted HD evaluation, without the intermediate GTI's, leading directly to individual problems and individual successes.

Any application of the HD method (concatenated or contracted) leading to an evaluation report with individual problems and individual successes will be called an application of the *micro-model* of HD evaluation. It is clear that application of the micro-model is possible for all kinds of general hypotheses, from GTC's to theories with proper theoretical terms.

However, as far as theories which are not just GTC's are concerned, the macro-step also suggests the model of *asymmetric* HD evaluation of a theory, leading to an evaluation report with individual problems and general successes. In that case, GTI's are derived in the macro-step, and only tested, not evaluated, in the micro-step. In the micro-model of HD evaluation of theories, in particular when contraction is used, the intermediate general successes of theories may disappear from the picture. However, in scientific practice, these intermediate results frequently play an important role. The individual successes of theories are summarized, as far as possible, in general successes. These general successes relativize the dramatic role of falsification via other general test implications. As we shall see in the next section, they form a natural unit of merit for theory comparison, together with counterexamples, as the unit of (individual) problems. In the next section, the model of asymmetric HD evaluation plays a dominant role. The results it reports will then be called counterexamples and (general) successes.

However, individual problems can frequently be summarized in terms of "general problems." They amount to established "falsifying general hypotheses" in the sense of Popper. Hence, there is also room for a *macro-model* of HD evaluation, where, besides general successes, the evaluation report lists general problems as well. In this case, all individual successes and individual problems are left out of the picture as long as they do not fit into an established general success or problem. Note that there is also the possibility of a fourth model of HD evaluation of an asymmetric nature, with individual successes and general problems, but as far as I can see, it does not play a role in scientific practice.

The three interesting models of HD evaluation of theories can be ordered in terms of increasing refinement: the macro-model, the asymmetric model, and the micro-model.

It can be shown that the main lines of the analysis of testing and evaluation also apply when the test implications are of a statistical nature. However, for deterministic test implications there are already all kinds of complications of testing and evaluation, giving occasion to "dogmatic strategies" and suggesting a refined scheme of HD argumentation. Although such problems multiply when statistical test implications are concerned, I shall restrict myself to a brief indication of those in the deterministic case.

8. *Empirical Progress and Pseudoscience*

The analysis of separate HD evaluation has important consequences for theory comparison and theory selection. The momentary evaluation report of a theory immediately suggests a plausible way of comparing the success of different theories, of further testing the comparative hypothesis that a more successful theory will remain more successful and, finally, the rule of theory selection, prescribing its adoption, for the time being, if it has so far proven to be more successful. The suggested comparison and rule of selection will be based on the asymmetric model of evaluation in terms of general successes and individual problems. However, it will also be shown that the symmetric approach, in terms of either individual or general successes and problems, leads to an illuminating symmetric evaluation matrix, with corresponding rules of selection.

Asymmetric Theory Comparison

A central question for methodology is what makes a new theory better than an old one. The intuitive answer for the new theory being as good as the old is plausible enough. The new theory has at least to save the established strengths of the old one and not to add new weaknesses on the basis of the former tests. In principle, we can choose any combination of individual or general successes and problems to measure strengths and weaknesses. However, the combination of general successes and individual problems, i.e., the two results of the asymmetric model of (separate) HD evaluation, is the most attractive. First, this combination seems the closest to actual practice and, second, it turns out to be the most suitable one for a direct link with questions of truth approximation. For these reasons I will first deal with this alternative and come back to the two symmetric alternatives.

Given the present choice, the following definition is the obvious formal interpretation of the idea of (*prima facie*) progress, i.e., *increasing success*:

> Theory *Y* is (at time *t*) at least as successful as (more successful than or better than) theory *X* iff (at *t*)
> – all individual problems of *Y* are (individual) problems of *X*
> – all general successes of *X* are (general) successes of *Y*
> (– *Y* has extra general successes or *X* has extra individual problems)

The definition presupposes, of course, that for every recorded (individual) problem of one theory, it has been ascertained whether or not it is also a problem for the other, and similarly whether or not a (general) success of one is also a success of the other. The first clause may be called the "instantial clause" as appealing and relatively neutral. From the realist perspective it is plausible to call the second clause the "explanatory clause." From other epistemological perspectives one may choose another, perhaps more neutral name, such as, the general success clause. It is also obvious how one should define, in similar terms to those above, the general notion of "the most successful theory thus far among the available alternatives" or, simply, "the best (available) theory."

It should be stressed that the diagnosis that *Y* is more successful than *X* does not guarantee that this will remain the case. It is a *prima facie* diagnosis based only on facts established thus far, and new evidence may change the comparative judgment. But, assuming that established facts are not called into question, it is easy to check that the judgement cannot have to be reversed, i.e., that *X* becomes more successful than *Y* in the light of old and new evidence. For, whatever happens, *X* has extra individual problems or *Y* has extra general successes.

It should be conceded that it will frequently not be possible to establish the comparative claim, let alone that one theory is more successful than all its available alternatives. The reason is that these definitions do not guarantee a constant linear ordering, but only an evidence-dependent partial ordering of the relevant theories. In other words, in many cases there will be 'divided success': one theory has successes another theory does not have, and *vice versa*, and similarly for problems. Of course, one may interpret this as a challenge for refinements, e.g., by introducing different concepts of "relatively maximal" successful theories or by a quantitative approach. However, it will become clear that in case of "divided success" another heuristic-methodological approach, of a qualitative nature, is more plausible.

As a matter of fact, the core of HD evaluation amounts to several heuristic principles. The first principle says that, as long as there is no best theory, one may continue the separate HD evaluation of all available theories in order to explore the domain further in terms of general facts to be accounted for and individual problems to be overcome by an overall better theory. For the moment, I will concentrate on the second principle, applicable in the relatively rare case that one theory is more successful than another one, and hence in the case that one theory is the best.

Suppose theory *Y* is at *t* more successful than theory *X*. This condition is not yet a sufficient reason to prefer *Y* in some substantial sense. That would be a case of "instant rationality." However, when *Y* is at a certain moment more successful than *X*, this situation suggests the following comparative success hypothesis:

CSH: *Y* (is and) will remain more successful than *X*

CSH is an interesting hypothesis, even if *Y* is already falsified. Apart from the fact that *Y* is known to have some extra successes or *X* some extra individual problems at *t*, CSH amounts at *t* to two components, one about problems, and the other about successes:

CSH-P: all individual problems of *Y* are individual problems of *X*
CSH-S: all general successes of *X* are general successes of *Y*

where 'all' is to be read as 'all past and future'.

Although there may occasionally be restrictions of a fundamental or practical nature, these two components concern, in principle, testable generalizations. Hence, testing CSH requires application of the micro HD argument. Following CSH-P, we may derive a GTI from *Y* that does not follow from *X*, and test it. When we get a counterexample of this GTI, and hence an individual problem of *Y*, it may be ascertained if the problem is shared by *X*. If it is not, we have falsified CSH-P.

Alternatively, following CSH-S, we may derive a GTI from *X* which cannot be derived from *Y*, and test it. If it becomes accepted, its acceptance means falsification of CSH-S. Of course, in both cases, the opposite test result confirms the corresponding comparative subhypothesis, and hence CSH, and hence increases the registered success difference. In the following, for obvious reasons, I call (these two ways of) testing CSH *comparative HD evaluation*.

The plausible rule of theory selection is now the following:

Rule of Success (*RS*)
When *Y* has so far proven to be more successful than *X*, i.e., when CSH has been "sufficiently confirmed" to be accepted as true, eliminate *X* in favor of *Y*, at least for the time being.

RS does not speak of "remaining more successful," for that would imply the presupposition that the CSH could be completely verified (when true). Hence I use 'so far proven to be more successful' in the sense that CSH has been "sufficiently confirmed" to be accepted as true; that is, CSH is accepted as a (twofold) inductive generalization. The point at which CSH is "sufficiently confirmed" will be a matter of dispute. Be this as it may, the acceptance of CSH and consequent application of RS is the core idea of *empirical progress*, a new theory that is better than an old one. RS may even be considered as the (fallible) criterion and hallmark of scientific rationality, acceptable for the empiricist as well as for the realist.

As soon as CSH is (supposed to be) true, the relevance of further comparative HD evaluation is diminished. Applying RS, i.e., selecting the more successful theory, then means the following, whether or not that theory already has individual problems. One may concentrate on the further separate HD evaluation of the selected theory, or one may concentrate on the attempt to invent new interesting competitors, that is, competitors that are at least as successful as the selected one.

Given the tension between reducing the set of individual problems of a theory and increasing its (general observational) successes, it is not an easy task to find such interesting competitors. The search for such competitors cannot, of course, be guided by prescriptive rules, like RS, but there certainly are heuristic principles of which it is easy to see that they stimulate new applications of RS. Let me start by explicitly stating the two suggested principles leading to RS. First, there is the *principle of separate HD evaluation* (PSE): "Aim via general test implications to establish new laws which can be derived from your theory (general successes) or, equivalently, aim at new negative instances (individual problems) of your theory." Secondly, the *principle of comparative HD evaluation* (PCE): "Aim at HD testing of the comparative success hypothesis, when that hypothesis has not yet been convincingly falsified." In both cases, a typical Popperian aspect is that one should aim at deriving test implications, which are, in the light of the background knowledge, very unlikely or even impossible. The reason is, of course, that a (differential) success of this kind is more impressive than that of a more likely test implication. In view of the first comparative (confirmation) principle (P.1, see ICR, p. 24, or SiS, p. 207), such a success leads in case of PSE to more confirmation of a theory, assuming that that has not yet been falsified, and in case of PCE to more confirmation of the comparative success hypothesis in general.

As already suggested, RS presupposes previous application of PSE and PCE. But some additional heuristic principles, though not necessary, may also promote the application of RS. To begin with, the *principle of content* (PC) may do so: "Aim at success preserving, strengthening *or*, pace Popper, weakening of your theory." A stronger theory is likely to introduce new individual problems but gain new general successes. If the latter arise and the former do not materialize, RS can be applied. Something similar applies to a weaker theory. It may solve problems without sacrificing successes. I would also like to mention the *principle of dialectics* (PD) for two theories that escape RS because of divided success: "Aim at a success preserving synthesis of two RS-escaping theories." In ICR (Section 8.3), I explicate a number of dialectical notions in this direction. Of course, there may come a point at which further attempts to improve a theory and hence to discover new applications of RS are abandoned.

In sum, the asymmetric model of HD evaluation of theories naturally suggests the definition of 'more successful', the comparative success hypothesis, the testing of such a hypothesis, i.e., comparative HD evaluation, and the rule of success (RS) as the cornerstone of empirical progress. Separate and comparative HD evaluation provide the right ingredients for applying first the definition of 'more successful' and, after sufficient tests, that of RS, respectively. In short, separate and comparative HD evaluation are functional for RS, and HD testing evidently is functional for both types of HD evaluation. The method of HD evaluation of theories combined with RS and the principles stimulating the application of RS might well be called the instrumentalist methodology. In particular, it may be seen as a free interpretation or explication of Laudan's problem solving model (Laudan 1977), which is generally conceived as a paradigm specification of the idea of an instrumentalist methodology. However, it will also be called, more neutrally, the *evaluation methodology*. It will be said that RS governs this methodology. The claim is that this methodology governs the short-term dynamics of science, more specifically, the internal and competitive development of research programs.

Note that the evaluation methodology demonstrates continued interest in a falsified theory. The reasons behind it are easy to conceive. First, it is perfectly possible that the theory nevertheless passes other general test implications, leading to the establishment of new general successes. Second, even new tests leading to new individual problems are very useful, because they have to be overcome by a new theory. Hence, at least as long as no better theory has been invented, it remains useful to evaluate the old theory further in order to reach a better understanding of its strengths and weaknesses.

Symmetric Theory Comparison

The symmetric models of separate HD evaluation, i.e., the micro- and the macro-models, suggest a somewhat different approach to theory comparison. Although these approaches do not seem to be in use to the extent of the asymmetric one and can only indirectly be related to truth approximation, they lead to a very illuminating (comparative) evaluation matrix.

A better theory has to be at least as successful as the old one, and this fact suggests general conditions of adequacy for the definitions of a "success," of a "problem" and of a "neutral result." The asymmetric definition of 'at least as successful' presented above only deals explicitly with individual problems and general successes; neutral results remain hidden, but it is easy to check that they nevertheless play a role. The symmetric models take all three types of results explicitly into account. The macro-model focuses on such results of a general nature, the micro-model on such results of an individual nature.

The notions of general successes and general problems are not problematic. Moreover, general facts are neutral for a theory when they are neither a problem nor a success. A better theory retains general successes as (already tested) general test implications, and does not give rise to new general test implications of which testing leads to the establishment of new general problems. Moreover, general problems may be transformed into neutral facts or even successes, and neutral general facts may be transformed into successes. The notions of individual successes, individual problems and neutral results are not problematic either, as long as we list them in terms of positive, negative and neutral instances, respectively. A better theory keeps the positive instances as such; it does not lead to new negative instances, and neutral instances may remain neutral or become positive. However, if we want to list individual successes and/or individual problems in terms of statements, the situation becomes more complicated, but it is possible (see ICR, pp. 116-7).

Let us now look more specifically at the symmetric micro-model, counting in terms of individual problems, successes and neutral results, that is, negative, positive and neutral instances or (statements of) individual facts. Hence, in total, the two theories produce a matrix of nine combinations of possible instances or individual facts. In order that the matrix can also be made useful for the macro-model, I present it in terms of facts. For the moment, these facts are to be interpreted as individual facts. The entries represent the status of a fact with respect to the indicated theories *X* and *Y*.

		X		
		negative	neutral	positive
Y	negative	B4: 0	B2: –	B1: –
	neutral	B8: +	B5: 0	B3: –
	positive	B9: +	B7: +	B6: 0

The (comparative) evaluation matrix

From the perspective of *Y* the boxes B1/B2/B3 represent unfavorable facts (indicated by '–'), B4/B5/B6 (comparatively neutral or) indifferent facts (0), and B7/B8/B9 favorable facts (+). The numbering of the boxes, anticipating a possible quantitative use, was determined by three considerations: increasing number for increasingly favorable results for *Y*, a plausible form of symmetry with respect to the diagonal of indifferent facts, and increasing number for indifferent facts that are increasingly positive for both theories.

It is now highly plausible to define the idea that *Y* is *more successful* than *X* in the light of the available facts as follows: there are no unfavorable facts and there are some favorable facts, that is, B1/2/3 should be empty, and at least one of B7/8/9 non-empty. This state of affairs immediately suggests modified versions of the comparative success hypothesis and the rule of success.

It is also clear that, by replacing individual facts by general facts, we obtain macro-versions of the matrix, the notion of comparative success, the comparative success hypothesis and the rule of success. A general fact may be a general success, a general problem or a neutral general fact for a theory.

In all these variants, the situation of being more successful will again be rare, but it is certainly not excluded. In ICR (Chapter 11) I argue, for instance, that the theories of the atom developed by Rutherford, Bohr and Sommerfeld can be ordered in terms of general facts according to the symmetric definition.

		Light propagation experiments					Experiments from other fields							
Theories \ Experimental facts		Aberration	Fizeau convection coefficient	Michelson-Morley	Kennedy-Thorndike	Moving sources and mirrors	De Sitter spectroscopic binaries	Michelson-Morley, using sunlight	Variation of mass with velocity	General mass-energy equivalence	Radiation from moving charges	Meson decay at high velocity	Trouton-Noble	Unipolar induction
Ether theories	Stationary ether, no contraction	A	A	D	D	A	A	D	D	N	A	N	D	D
	Stationary ether, Lorentz contraction	A	A	A	D	A	A	A	A	N	A	N	A	D
	Ether attached to ponderable bodies	D	D	A	A	A	A	A	D	N	N	N	A	N
Emission theories	Original source	A	A	A	A	A	D	D	N	N	D	N	N	N
	Ballistic	A	N	A	A	D	D	D	N	N	D	N	N	N
	New source	A	N	A	A	D	D	A	N	N	D	N	N	N
Special theory of relativity		A	A	A	A	A	A	A	A	A	A	A	A	A

Comparison of experimental record of seven electrodynamic theories.
Legend: A: agreement, D: disagreement, N: not applicable

Another set of examples of this kind is provided by the table (adapted from: Panofsky and Phillips, 1962, p. 282), representing the records in the face of 13 general experimental facts of the special theory of relativity (STR) and six alternative electrodynamic theories, viz., three versions of the ether theory and three emission theories. According to this table, STR is more successful than any of the others; in fact it is maximally successful as far as the 13 experimental facts are concerned. Moreover, Lorentz's contraction version of the (stationary) ether theory is more successful than the contractionless version. Similarly, the ballistic version of the emission theory is more successful than the other two. However, it is also clear that many combinations lead to divided results. For instance, Lorentz's theory is more successful in certain respects (e.g., De Sitter's spectroscopic binaries) than the ballistic theory, but less successful in other respects (e.g., the Kennedy-Thorndike experiments).

In the present approach it is plausible to define, in general, one type of divided success as a liberal version of more successfulness. *Y* is *almost more successful* than *X* if, besides some favorable facts and (possibly) some indifferent facts, there are some unfavorable facts, but only of the B3-type, provided there are (favorable) B8- or B9-facts or the number of B3-facts is (much) smaller than that of their antipodes, that is, B7-facts. The provision clause guarantees that it remains an asymmetric relation. Crucial is the special treatment of B3-facts. They correspond to what is called *Kuhn-loss*: the new theory seems no longer to retain a success demonstrated by the old one. The idea behind their suggested relatively undramatic nature is the belief that further investigation may show that and how a B3-fact turns out to be a success after all, perhaps by adding an additional (non-problematic) hypothesis. In this case it becomes an (indifferent) B6-fact. Hence, the presence of B3-facts is first of all an invitation to further research. If this is unsuccessful, such a B3-fact becomes a case of recognized Kuhn-loss. Unfortunately, the table above does not contain an example of an almost more successful theory.

Cases of divided success may also be approached by some (quasi-) quantitative weighing of facts. Something like the following quantitative evaluation matrix is directly suggested by the same considerations that governed the number ordering of the boxes.

		X		
		negative	neutral	positive
Y	negative	B4: −1/−1	B2: −3/+3	B1: −4/+4
	neutral	B8: +3/−3	B5: 0/0	B3: −2/+2
	positive	B9: +4/−4	B7: +2/−2	B6: +1/+1

The quantitative (comparative) evaluation matrix

All qualitative success orderings of electrodynamic theories to which the table gives rise, remain intact on the basis of this quantitative matrix (which is not automatically the case). Moreover, we now of course get a linear ordering, with Lorentz's theory in the second position after STR and far ahead of the other alternatives. Of course, one may further refine such orderings by assigning different basic weights to the different facts, to be multiplied by the relative weights specified in the quantitative matrix.

Like a similar observation in the symmetric case, it is now possible to interpret the qualitative and the quantitative versions of the evaluation matrix as explications of some core aspects of Laudan's (1977) problem-solving model of scientific progress, at least as far as empirical problems and their solutions are concerned.

Our notion of comparative evaluation is governed by the notion of being "(almost) more successful." This is a rather strict strategy. In ICR I question the general usefulness of quantitative liberalizations of "successfulness," and for that matter, of "truthlikeness," mainly because they need real-valued distances between models, a requirement which is very unrealistic in most scientific contexts. Hence, the applicability of liberal notions may well be laden with arbitrariness. Be this as it may, it is important to stress that the strict strategy does not lead to void or almost void methodological principles. If there is divided success between theories we should try to apply the already mentioned Principle of Dialectics: "Aim at a success preserving synthesis of the two RS-escaping theories," which has, of course, a plausible program-bound version. Hence, the restricted applicability of the strict notion of comparative success does not exclude the possibility of clear challenges being formulated in cases where they do not apply, on the contrary.

V. Truth, Product, and Concept Approximation

HD testing and evaluation are primarily concerned with descriptive and explanatory research. In the previous part we saw how this leads to a plausible definition of empirical progress. Part V of SiS starts with a general chapter on how empirical progress is related to truth approximation, at least in descriptive and explanatory research that aims at "the nomic truth," and is therefore called nomological research. Moreover, it is shown that there is an interesting partial analogy with progress in the "product approximation" that occurs in design and explicative research. Design research programs are analyzed in detail in a separate chapter. In this synopsis the general chapter is merely summarized and the reader is referred to that chapter in SiS or, as far as truth approximation is concerned, to Chapters 7 and 9 of ICR. On the other hand, some of the main points of the chapter about design research are sketched.

9. *Progress in Nomological, Design, and Explicative Research*

Although nomological, design, and explicative research seem rather different at first sight, it is argued in Chapter 9 of SiS that they are partially analogous in that they can formally be presented in terms of either a target set of desired possibilities or a target set of desired features. This characteristic implies that their respective definitions of 'formal progress' essentially coincide. The differences between the three types of research are due to the fact that "determinable progress" requires specific definitions: empirical progress for nomological research and conceptual progress for explicative research. Only in the case of design research does determinable progress coincide with formal progress, as long as the target sets of desired and undesired features are determined beforehand. The analysis in this chapter is, as far as nomological research is concerned, essentially based on Chapters 7 and 9 of ICR, including the resulting explication of descriptive and nomological research programs. However, the presentation of the main results in terms of (un)desired possibilities and features is new and more transparent for purposes of application, extension, and comparison. This type of presentation is, for example, also used in Kuipers (2002) on the relation between beauty, empirical success and truth.

10. *Design Research Programs*

In this section we deal with design research programs in some detail. The previous sections have been dominated by descriptive and explanatory research programs on the object level, and by some explicative programs analyzing certain aspects of such programs, such as "explanation by specification." In this section I present the basic results of an explicative program with respect to research programs that aim at a certain product. The core idea is that design research programs attempt to bring together the properties of available materials and the demands derived from intended applications. The structure and development of such programs – in other words, their logic of problem states and state transitions, including assessment criteria and heuristic principles – is described in set-theoretic terms, starting with a naive model comprising an intended profile and the operational profile of a prototype. Drug research will provide the main example. In a first fundamental concretization, the useful distinction between structural and functional properties is built into the model.

Chapter 10 of SiS, moreover, deals with the diagnosis of a conceptual confusion in a first attempt to explicate the indicated core idea, the lattice model of Weeder et al. (Weeder and Kester 1982; Bodewitz, De Vries and Weeder 1988). Three further concretizations are also presented, dealing with potential applications, potential realizations, and potentially relevant properties. The partial analogy between "product" and "truth approximation" is elaborated next. It turns out that the differences are at least as important as the similarities. The chapter concludes with some indications of the usefulness of the models for the ways in which products reach the market, in comparison to the so-called social construction of technology approach.

The Naive Model of Problem States

Drawing upon a survey of the literature (Saren 1984), it is interesting to see that existing models of design research, in particular within firms, where it is usually called "innovation," tend to treat the actual design process as a kind of black box intermediate stage, within a more encompassing sequence of stages. In this section, I show how it is possible to analytically decompose this black box into a sequence of problem states and state transitions. Drug research (Vos 1991, Section 2.7) provided the main type of example for our joint analysis of design research in (Kuipers, Vos and Sie 1992), which forms the basis of the corresponding chapter in SiS.

Recall that design research programs are directed at the design or construction of certain products or processes, which have to satisfy previously determined demands and these demands are based on the intended applications. For brevity, I speak of products, also when processes are meant or the improvement of already existing products or processes. The hard core of such programs is frequently called the "lead": the basic idea about how the product is to be composed or how it has to work.

Let *RP* indicate the set of *relevant properties* for the product to be developed. For each element of *RP* it is assumed that its presence or absence is explicitly required in the specification of the intended product. Let the subset *W* of *RP* indicate the set of (*desired or*) *wished for properties* of the intended product; *W* will be called the *intended profile*. Of course, $RP - W$ is the set of undesired or unwanted properties.

For each concrete candidate product x, henceforth called prototype x, it is important to determine which properties in *RP* it actually has. Let the subset $O(x)$ indicate the set of these factual or *operational properties* of x; $O(x)$ is called the *operational profile* of x.

In drug research (Vos 1991, p. 62), the lead can be identified with the intended profile and some idea about how to realize it. More specifically, the "lead compound" comprises a chemical compound, with its operational profile, together with the intended profile. Assuming an interesting amount of overlap between the two profiles, but not yet a perfect correspondence, the challenge is to reduce the differences stepwise.

This type of problem situation in a certain state of development can clearly be depicted, viz., as the fact that the two profiles in the figure below do not coincide: the problems consist of the two starred sets, $W - O(x)$ and $O(x) - W$. $W - O(x)$ represents so to speak the unrealized "positive" desires, and $O(x) - W$ the realized "negative" desires.

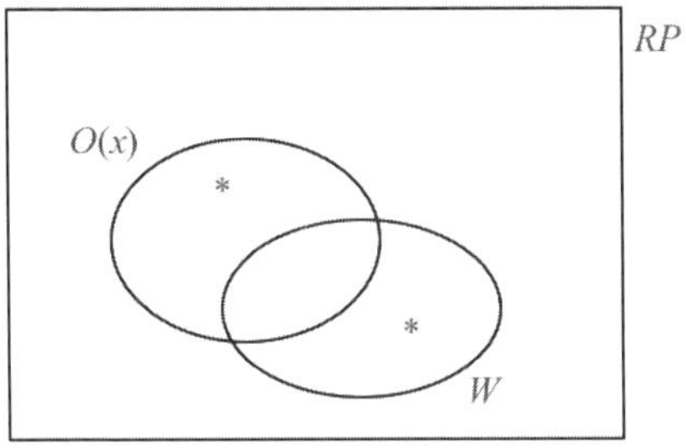

A problem state

Let us introduce the plausible basic formal notion for comparing any pair of profiles P and P^*, i.e., any pair of subsets of RP. It is the so-called *symmetric difference* $P \Delta P^*$, defined as the union $(P - P^*) \cup (P^* - P)$. $P\Delta P^*$ indicates an elementary type of qualitative distance or dissimilarity between P and P^*. Hence, the smaller this set is, the greater the similarity between P and P^*.

On certain occasions it will also be attractive to have quantitative formal notions, in which case we have to assume that RP is finite, or at least all profiles to be considered. $|P|$ indicates the number of elements in P, hence $|P\Delta P^*|$ is a quantitative measure of the dissimilarity of P and P^*.

By consequence, $\mathrm{O}(x)\Delta W$, the union of $W - \mathrm{O}(x)$ and $\mathrm{O}(x) - W$ (the two starred areas in the figure), represents the problem state in qualitative terms and $|\mathrm{O}(x)\Delta W|$ in quantitative terms. More precisely, $\mathrm{O}(x)\Delta W$ specifies the set of problems, i.e., the deviations from the claim that $\mathrm{O}(x) = W$, whereas $|\mathrm{O}(x)\Delta W|$ represents only the number of problems, i.e., the number of deviations.

Transitions of Problem States

The set of problems in the problem state, $\mathrm{O}(x)\Delta W$, forms the starting point for negotiation about what to do: trying to change x into some x' such that $\mathrm{O}(x')$ becomes more similar to W, or changing W into some W' such that W' becomes more similar to $\mathrm{O}(x)$, or both.

It is plausible to give a formal characterization of certain transitions of one problem state into another by the following definitions ('$\subset$' is the sign for proper subset).

Definition 1

(a) Prototype x_2 is a qualitative improvement of x_1 in view of W iff
$\mathrm{O}(x_2)\Delta W \subset \mathrm{O}(x_1)\Delta W$

(b) it is a quantitative improvement iff $|\mathrm{O}(x_2)\Delta W| < |\mathrm{O}(x_1)\Delta W|$

Definition 2

(a) Intended profile W_2 is a *qualitative concession* to prototype x compared with W_1 iff $O(x)\Delta W_2 \subset O(x)\Delta W_1$

(b) it is a quantitative concession iff $|O(x)\Delta W_2| < |O(x)\Delta W_1|$

These definitions provide the basic assessment criteria for state transitions. The first definition enables us to evaluate potential improvements of the prototype, i.e., transitions from one prototype to the other in the face of a fixed intended profile. The second definition specifies how to evaluate potential concessions, i.e., transitions from one intended profile to the other in the face of a fixed prototype.

The first type of transition is an answer to one particular specification of the problem state, viz., how to bring the prototype closer to the intended profile? The second type of transition is an answer to the remarkable conversion of this problem specification, viz., how to bring the intended profile closer to the prototype? The fact that this problem specification is realistic is documented by Vos (1991) and hinted at in the title of that book: *Drugs looking for diseases*.

It is evident that in both cases the qualitative judgement implies the quantitative one, but not the reverse. The quantitative definitions are so rough that they even lead to linear orderings for prototype and intended profile transitions. However, the qualitative definitions obviously lead only to partial orderings of both types of transition. But from the formal point of view, they represent the purest cases. They are depicted in the following figures, where in both cases the two #-areas are empty and at least one of the two *-areas is non-empty.

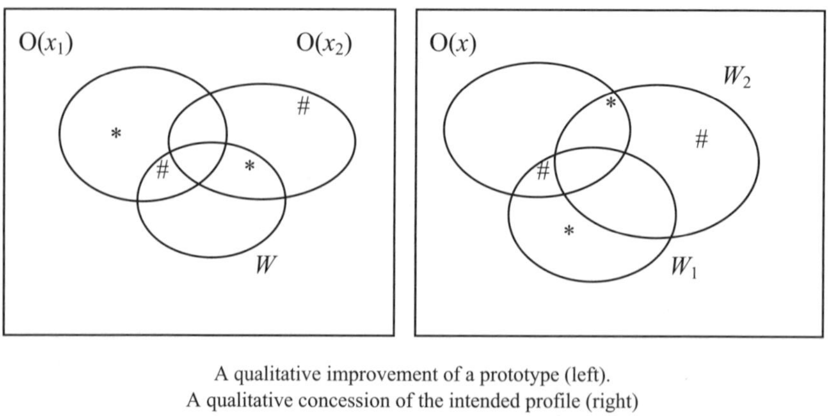

A qualitative improvement of a prototype (left).
A qualitative concession of the intended profile (right)

Some remarks about constraints for the domain *RP* of relevant properties have to be made. It is clear that there is no reason to have both a property in *RP* and its counterpart: if, for example, ‘flexible’ belongs to *RP*, and *W* or O(*x*) is supposed to contain ‘non-flexible’ one simply excludes ‘flexible’ from either *W* or O(*x*). This avoidance of duplication is even an advantage in that it restricts the formal constraints on profile pictures. An equally inconvenient formal constraint can be avoided by excluding the occurrence in *RP* of the combination of two properties in *RP* as a new property. If we did not exclude a combination of properties as a new property all profiles would have to be closed for combinations. There do not seem to be other plausible restrictions on *RP*. It is important to note that not every subset of *RP* needs to be nomically possible, i.e., not every conceptually conceivable profile needs to be realizable in reality. In other words, there will be all kinds of (unavoidable) causal connections between (subsets of) the properties in *RP*.

From the presented point of view the development of a design research program is a succession of problem states, where problem transitions will, as a rule, be quantitative, and ideally qualitative, improvements or concessions. There may be different types of specific research involved. Direct experimental test research is involved first. For any new prototype *x* and intended profile *W* the claim O(*x*) = *W* has to be evaluated by experiments. Besides this direct empirical research, descriptive, explanatory or explicative research that primarily belongs to other research programs may also be involved. In Zandvoort’s terms (1986), the design program operates as *guide program*, and the others as *supply programs*.

Possible Refinements

The presented set-theoretic model is a naive model in many respects and its value depends largely on the degree to which it can be concretized in order to adapt to all kinds of realistic complications. In practice, one or more of the following refinements will be required.

- R1) Instead of having a simple yes/no character, properties usually have to be construed as functions with a range of more than two values, possibly even infinitely many.
- R2) Some properties may be more important than others, without the latter being negligible.
- R3) It is usually not immediately clear whether a property is relevant or not. In the course of product development their relevance may become clear, and the different actors in the process may start to negotiate about them.
- R4) In many cases there is a plausible distinction between structural and functional properties, such that the intended profile is primarily

specified in terms of functional properties, whereas the available prototype is primarily known in terms of its structural properties.

R5) In some cases, it is very helpful to include a set of potential (intended) applications explicitly in the model, as suggested in the lattice model of Weeder et al.

R6) It may sometimes also be helpful to include a set of potential realizations, roughly corresponding to different materials in the lattice model.

The first two refinements are essentially technical. Vos (1995) (see also Vos 1991) describes in detail how the first refinement can be realized and suggests how to deal with the second. The refinements R3 and R4 are of fundamental conceptual importance. The refinement of the naive model with "potentially relevant properties" (R3) is described in SiS Section 10.5. The important distinction between structural and functional properties (R4) is introduced below. The importance of the further refinements R5 and R6 depends very much on the type of intended product. In SiS Section 10.4 it is shown that the set of potential applications (R5) is particularly relevant in the case of drug research, viz., diseases. SiS Section 10.4 also presents the introduction of a set of potential realizations (R6) for complex products, based on Sie (1989). The refinements R1/2/3 are essentially compatible with the refinements R4/5/6. Hence, there is a network of related models, with the naive model as the point of departure and the choice of a refined model to be determined by the context.

Structural versus Functional Properties

In many contexts it is possible and customary to divide the set of relevant properties *RP* into a subset of technical or structural properties *S* on the one side and a subset of service or functional properties *F* on the other. *S* and *F* do not overlap and they exhaust *RP*.

Saviotti (1996), for example, uses the distinction, in his terminology, between technical and service characteristics, as the basis for a description of technological development, i.e., the rise and fall of products or technologies on the market. In the description of aircraft, e.g., helicopter technology, there are on the one hand technical properties like length, rotor diameter, engine power, engine type, number of engines, geometry and, on the other, service properties like maximum takeoff power, maximum speed, range. Note that these properties, in their present formulation, are not of the simple yes/no type, but that is irrelevant for our present purposes. I use the same distinction in the description of the process of product development, but I prefer the more general terminology of structural versus functional properties.

It is plausible to call $O(x) \cap S$ the *operational structural profile* $OS(x)$ of prototype x and $O(x) \cap F$ the *operational functional profile* $OF(x)$. Of course, it follows that $O(x) = OS(x) \cup OF(x)$. It is also possible to make the same division in the intended profile W: the *intended structural profile* WS and the *intended functional profile* WF. But it is not evident that the latter distinction is also realistic. It requires us to specify W beforehand, with the consequence that WS is simply equal to $W \cap S$ and WF equal to $W \cap F$. In practice it is usually the other way around. As a rule, WF is provisionally determined first, for we start by asking what the product is supposed to do in the intended applications. As soon as we have fixed WF the next question is how WF can be realized. It is of course not guaranteed that there is just one subset of S causally implicating precisely these functional properties. In other words, there need not be a unique WS; we have to leave room for "functional equivalents." I shall call a set of structural properties *appropriate* or an *appropriate structural* profile for WF if it causally implies WF. Such a set will be indicated by $AS(WF)$, or simply AS when WF is clear from the context.

The foregoing discussion can be clarified by saying something more about the S/F-division. Of course, this distinction will have some arbitrariness. But it is plausible and helpful to assume that the S/F-division satisfies at least the following:

S/F-splitting principle of minimal causality
for all x and x': $OS(x) = OS(x')$ causally implies $OF(x) = OF(x')$,
(b) all elements of S are necessary to make (a) true in general.

Calling x and x' structurally equivalent if $OS(x) = OS(x')$ and functionally equivalent if $OF(x) = OF(x')$, (a) says that structural equivalence causally implies functional equivalence. In consequence, there is a function associating a unique set of functional properties with each set of structural properties. The reverse implication is not required; functional equivalence does not imply structural equivalence and hence there is no similar function. The reason for clause (b) is the following. Without (b), the splitting principle would allow S/F-divisions such that "S causally overdetermines F" and there is no reason for doing so. On the contrary, the larger F is, the more freedom exists in the choice of wanted and unwanted properties.

Note that we have not assumed that the splitting principle is sufficient for the determination of a unique division of RP in S and F. It leaves room for more than one possibility and the choice between them will be based on additional considerations. One further consideration may concern prices. It seems plausible to assume that the market price of a product is primarily determined by its functional properties, whereas the cost price is primarily determined by its structural properties. In other words, following the standard criticism of Marx's labor theory of value, it is assumed, and specified, that the cost price is determined by different aspects of the product than those which determine how much people want to pay for it. It may be useful to use such price considerations in the further determination of the *S/F*-division, for it is evident that cost and market price considerations play an important role in the R&D-process.

Let me now return to the intended and operational profiles. We have already noted that as a rule the desired functional properties of the intended profile are provisionally decided upon earlier than the desired structural properties. For the operational profile of a prototype it may be the other way around. To some extent its structural properties will be known beforehand, and the question is what additional structural properties it has, and what functional properties.

In this way I have differentiated two basic intuitions regarding design research: the distinction between factual and desired properties in the naive model on the one hand and the distinction between structural and functional properties in the present refined version on the other. The resulting asymmetric model will be called the *S/F* model and its characteristic problem state is depicted in the figure below, in which the arrow indicates the inherent asymmetry by the causal determination.

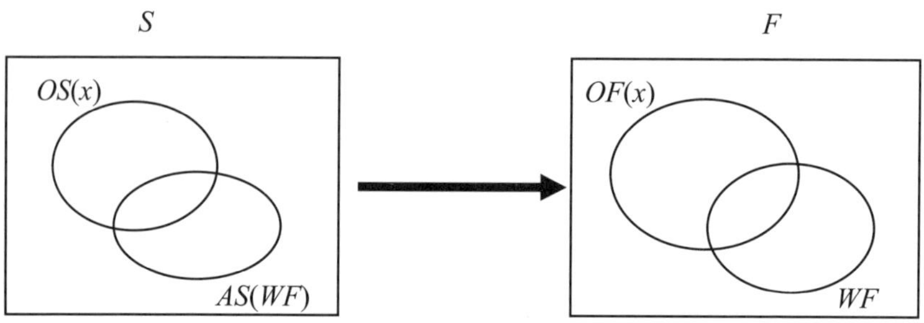

A problem state in the S/F model

It is plausible to assume that the comparison of prototypes in view of a fixed intended profile and of intended profiles in view of a fixed prototype is primarily a matter of comparing functional profiles. In consequence, it is also plausible to conceive the restrictions of the original definitions of the assessment notions of improvement and concession to functional properties as defining the basic assessment notions of the *S/F* model . Hence, from now on I assume for the *S/F* model that 'O' and '*W*' in *Definitions* 1 and 2 have been replaced by 'O*F*' and '*WF*', respectively.

Moreover, it is now plausible to assume that preliminary estimations of judgements of functional improvements are suggested by ideas about possible relations between structural and functional similarity. It is easy to formulate some principles of which it is not only evident that they are formally *in*valid, but also that they play an important heuristic role. The basic idea, of course, is that structural similarity implies functional similarity, and vice versa. Restricting our attention to the qualitative cases we get the following heuristic principles, in which *AS* indicates an arbitrary appropriate structural profile for *WF*.

Heuristic principles

*HP*1: if x_2 is structurally more similar to $AS(WF)$ than x_1 (i.e., $OS(x_2)\Delta AS \subset OS(x_1)\Delta AS$) then x_2 will (probably) be a functional improvement of x_1 in view of *WF* (i.e., $OF(x_2)\Delta WF \subset OF(x_1)\Delta WF$).

*HP*2: if x_2 is a functional improvement of x_1 in view of *WF* then x_2 will (probably) be structurally more similar to *AS* than x_1.

Intuitively speaking, *HP*1 states that increasing similarity with an appropriate structural profile is likely to lead to increasing similarity with the intended functional profile, and *HP*2 claims the converse. Because of the causal asymmetry *HP*2 will have more exceptions than *HP*1.

These heuristic principles concern functional improvements and not concessions. In view of the fact that the question of whether a change of intended functional profile is a functional concession to a fixed prototype is entirely a matter of comparing sets of functional properties, there are no comparable heuristic principles for functional concessions.

VI. Capita Selecta

In the final part of SiS I pay attention to some special subjects that merely illustrate the rich number of developments that are taking place in philosophy of science. The selection is very much determined by my personal research activities in the past. Subjects that are equally deserving of a chapter include: statistical hypotheses, causal analysis, default reasoning, measurement, systems theory, explicative research programs.

Two of three topics presented in SiS concern very important and productive research programs, viz., computational (design) programs in the philosophy of science, and the structuralist approach to scientific theories. Both will only be indicated by a summary of the relevant chapter in SiS. I close with a brief exploration of the famous norms of Merton as default norms for research ethics.

11. *Computational Philosophy of Science*

The first special topic concerns a particular type of design research program, known as computational philosophy of science. The products to be designed are computer programs that deal with the discovery, evaluation and revision of (empirical) hypotheses. This development may best be described as a recent research tradition, with several specific research programs. The binding ideas are, first, that discovery, contrary to traditional opinion in the philosophy of science, is accessible to methodological analysis, and, second, that discovery, evaluation and revision should be seen as special cases of problem solving, to be approached computationally in ways developed in cognitive psychology and artificial intelligence. Deviations may arise from differences in the main objectives one is after, where one has at least to choose between historical, psychological and philosophical adequacy, and practical relevance. Moreover, the specific techniques to be used may diverge considerably.

In SiS Ch. 11 I give impressions of some specific lines of research: the BACON family of programs, in search of quantitative laws in physics, a family of programs (GLAUBER, STAHL and DALTON) in search of qualitative laws in chemistry (both characterizations based on Langley et al. 1987), and Paul Thagard's program PI (Thagard 1988), of a rather different set-up, in which evaluation is at least as important as generation. Referring to (Shrager and Langley 1990), a variety of other programs, in which theory revision is dominant, is also indicated. Moreover, some straightforward connections with neo-classical philosophy of science are discussed.

In the second part of the chapter, I deal critically with the evaluative part of PI and Thagard's later program ECHO (Thagard 1992). I argue that implementation of the simple evaluation matrix, presented in Section 8, would lead to at least as good results with respect to the historical and philosophical adequacy of the resulting theory selections. All computer programs mentioned are partly competing and partly able to co-operate. Although it is too early for a general evaluation and integration of such programs, it is clear that this development is important as a stimulus for neo-classical philosophy of science and, possibly, for actual scientific research.

12. *The Structuralist Approach to Theories*

The second special chapter in SiS typically deals with an explicative research program in the philosophy of science, viz., the study of the structure of scientific theories. There are two main approaches to the structure of empirical theories. The statement approach conceives theories primarily as sets of statements. This approach has long been considered as the only and obvious approach, e.g., by Carnap and Popper. However, it is also possible to conceive theories primarily as sets of models. One version of this so-called semantic approach is the set-theoretic or structuralist approach. It was already introduced by Suppes and refined by Sneed, Stegmüller, Balzer, and Moulines. Its basic idea is that theories amount to the specification of classes of set-theoretic structures satisfying certain conditions. In SiS Ch.12, after briefly discussing the attractive features of the structuralist approach in general, the approach is introduced stepwise, first without the distinction between theoretical and non-theoretical terms, then with that distinction in order to avoid circularity or infinite regress in measurement. The basic outline of the resulting representation of three examples is given: classical particle mechanics, the periodic table of chemical elements, and psychoanalytic theory. Then some further refinements follow, viz., absolute versus relative empirical content, various ways of determining intended applications, relations between theories, theory-nets, and constraints. Finally, the usefulness of the structuralist approach for non-empirical theories is considered.

13. *"Default Norms" in Research Ethics*

In the final chapter of SiS I try to contribute to the discussion of incorruptible professional behavior in scientific research. For this purpose I take Merton's famous norms as a point of departure, not in order to note what so many noted, namely that scientists do not always conform to them (see e.g. Mulkay 1977, and Ziman 1994), but to conceive them as "default norms" by raising the question which deviations may be defensible and which ones are not. Here I pay roughly equal attention to the four norms. In SiS Ch. 13 the emphasis is on the norm of disinterestedness. Moreover, it concludes with a discussion of the possibility and usefulness of subsuming all this within a general professional code for scientific researchers.

Science and ethics have a complex relation. They cannot easily be separated. Although pleas for completely autonomous science as well as for extreme ethical dirigism still occur, an interaction model for science and ethics is increasingly widely propagated. This model is not only presented as characteristic of the factual relation, but also as desirable for future relations, that is, it is in the interests of science and society, of course, under rationally defensible conditions. It concerns in particular moral dilemmas with respect to:

- incorruptible professional behavior in scientific research
- the present and future wellbeing of humans and (laboratory, domestic, and wild) animals and the state of their environment
- the communication between science and society.

In the last twenty years all kinds of initiatives have been taken to shape the interaction between science and ethics on three levels. On the individual level, universities organize courses for students and postgraduates in an effort to activate their moral sensibility. On the institutional level codes of the professional behavior of scientists and codes governing the behavior of the scientific and non-scientific staff in private firms have been designed, suggesting or prescribing rules of behavior. On the societal level, public debates are organized between the laity and experts about possible and desirable scientific and social developments.

Merton's Norms Conceived as "Default Norms"

As is well known, Robert K. Merton (1942) formulated four norms which scientists, ideally speaking, should live up to and which have become known as the CUDOS norms, after the initials of their names: Communism, Universalism, Disinterestedness, and Organized Skepticism. I will briefly characterize them:

> "*Communism*": scientific knowledge is the product of common effort, for which reason newly acquired knowledge should be made public and should be considered as a collective good, with explicit recognition, of course, of the discoverer or inventor.
>
> *Universalism*: one should judge the work of others irrespective of persons ("blind refereeing") and sex, race, nationality and the like should not play a role in the selection among applicants for a position.
>
> *Disinterestedness*: personal and group interests should be subordinated to the interests of research, which excludes evident kinds of fraud, such as plagiarism, forgery, and fakery.
>
> *Organized skepticism*: new research results and methods should be presented to colleagues with an open mind for criticism.

It will be clear that these norms need not be exhaustive and that they will overlap now and then. Moreover, although Merton was of the opinion that the norms were in the interest of science, he was also well aware that they did not have rigorous *des*criptive validity. I will argue that they cannot even have rigorous *pre*scriptive validity, because, besides evidently non-defensible deviations, there are various kinds of conceivably well argued deviations. Moreover, and at least as important, there appears to be a large "gray area" where every researcher has to find his own way in every particular case. Precisely due to the existence of this gray area it is difficult to conceive of a detailed prescriptive professional code.

Merton's norms are clearly principles with a deontological flavor in the sense of deontological ethics: whatever the consequences in specific cases, they prescribe what to do or not to do. Defensible deviations are, of course, based on utilitarian considerations pertaining to the consequences of a dogmatic application of the norms in concrete cases. In other words, these norms are "default norms." Depending on whether or not the deviations are supposed to be catchable in subrules, default norms, besides their deontological reminiscence, also have the flavor of so-called rule-utilitarianism or act-utilitarianism. In the following I try, for each norm, to explore the possibilities for defensible and non-defensible deviations, though without presenting them in each case explicitly in terms of their consequences and without investigating whether or not defensible deviations can be caught in subrules.

"Communism"

With his "communism" norm Merton intended primarily that new scientific knowledge, as soon as it is sufficiently trustworthy, should be made public, not only for fellow researchers, but also for society.

This claim raises, first of all, the question of whether there can be good reasons for secrecy. Kaiser (1996) distinguishes three kinds of interest that might be considered as justifying secrecy: military, commercial, and public interest. Reservedness about new findings that might be used on a large scale for military purposes can at least count on good arguments of a utilitarian nature. The case for secrecy is less convincing when only commercial interests are involved, and commercial and contract research is a rapidly growing area of concern in this respect. Temporary secrecy, for example, waiting for the approval of a patent request, may be defensible. Permanent secrecy of a scientifically interesting finding, however, does not seem to be ever justifiable. We may hope that recipes, that are cautiously kept secret, such as the one for Coca-Cola, are not scientifically interesting.

That public interest does not easily justify secrecy may be illustrated by what happened in the early seventies in the Norwegian salmon fish farming industry, an example extensively discussed by Kaiser and here restricted to the crucial facts. A team of scientific researchers advising the salmon farming industry kept one of their discoveries secret within the team, at great risk to public health. They had discovered that a generally used food supplement for the salmon was carcinogenic to the human consumers of the salmon, despite initial good reasons to assume that this could not be the case. The team faced the dilemma between a general health interest and a general employment interest; the latter was especially important since so many people were directly or indirectly involved in the salmon industry. The team leader decided to let the employment interest overrule the health interest. More precisely, he decided that a functionally equivalent replacement should first be found or developed for the food supplement before use of the health endangering one would be discontinued. As Kaiser (1996, p. 220) reports, the leader's message to his team was: "I urge you to keep this strictly secret, until we have developed for substance *X* a replacement that is more desirable. Just imagine how this finding would stir up the public if it becomes known." This consideration was apparently convincing to the other members of the team; that is, at least no "whistle blowers" appeared. The new supplement was found and the old one replaced without public uproar. Despite our possible inclination to want to believe that such things could not happen anymore in the twenty-first century, the example suggests that more or less comparable cases could still occur.

It is clear that in most cases there are no good reasons for secrecy. Nevertheless, publication of new findings at the proper time and in the proper way is a matter of great concern. Regarding the proper manner of publication, Solomon (1996) stresses that, in case of experimental results, enough should be specified in order to enable the repetition of the experiment by others and that, in relevant cases, data, samples and, I add, computer programs should be made available upon request.

Regarding timely publication, it is important to distinguish between, first, the informal communication network of friendly colleagues, local and otherwise, and, second, the official scientific publication circuit and, finally, the media for the general public. The problematic cases of untimely publication concern primarily the general media. Notorious examples in the 20st century history of science are N-rays (around 1900), polywater (late sixties) and, more recently, cold nuclear fusion (Ponns and Fleischmann), memory molecules (Benveniste) and, quite a row, at least in The Netherlands, an alleged Aids-inhibiting substance (Buck and Goudsmit). It is important to note that untimely publication does not cause debates when it happens to become a success story. The first messages about superconductivity, for example, also came in the newspapers, with speculations about possible applications, before the phenomenon was considered to be well established by the community of physicists.

Untimely publication is encouraged, among other factors: by interest in being the first, and to be recognized as such, by the aim of obtaining new sources of money or other means for further research as soon as possible, by the sincere belief that the discovery renders society a good turn, by vanity – even to the extent of having Nobel prize expectations – and, last but not least, by dubious science journalism. The most important factor preventing untimely publication is likely to be the potential loss of face among colleagues when things go wrong. For the rest, standard checkpoints like the referee system, repetition of experiments and proof checks by oneself and others, prevent new results from being officially published in too early a stage. For the general media, adequate science journalism is a prerequisite. Good science news reporting prevents, first of all, the expectations of many vulnerable groups from being raised unjustifiably. Not only Buck and Goudsmit gave false hope to Aids patients, the Dutch public television news was also responsible for that. Moreover, in general, it would be a good thing if not only scientists but also science reporters and the media that publish their stories were to lose credibility whenever a lame-duck discovery is reported with a great fanfare.

Universalism

Strictly speaking, the use of authority arguments is a deviation from the universalism norm because somebody's prestige is brought into play. But in everyday and scientific practice we cannot live without such arguments. Consider, for example, the mutual trust one needs to have for doing efficient interdisciplinary research. However, I am inclined to make a difference between defensible and non-defensible authority arguments, for example, depending on whether or not the scientist in question makes a statement within the domain of his recognized competence. Well-known examples of problematic transgressions concern genuine and near Nobel prize winners making or supporting claims in a field in which they are incompetent, and, frequently, cannot be contradicted by real experts, as there are none in the area concerned. In the latter kinds of areas, e.g., regarding certain long-term future developments about which everyone is of course free to speculate, utterances by laureates are taken more seriously than those made by other, equally incompetent people.

Another type of problematic deviation from the universalism norm concerns all kinds of (supposed) discrimination, in particular, of women and ethnic minorities. The policies of "positive discrimination" or "positive action" are considered by its proponents as defensible in the face of the consequences of the normal "laissez faire" policy. One may compare it with a prohibition of political parties that are against democracy, in order to protect democracy against attacks from inside. In the US, however, positive action in favor of minorities is again being retracted in several states. The same development seems to be taking place in The Netherlands regarding extreme forms of positive discrimination in favor of women, such as appointing a sufficiently qualified female candidate to an academic position in the presence of (much) better qualified male candidates. Nevertheless, it has to be noted that, at least in The Netherlands, it turns out to be very difficult to get a reasonable number of women into higher academic positions without positive action. Besides the admission of hidden differences in treatment of female candidates and male-biased limitations to part-time work, flexible-time work and telework, the notion that differences in ambition also play a role seems to be gaining ground. For women with appropriate ambitions, however, besides fighting against all kinds of unequal treatment and male-biased limitations, new kinds of networks and mentorship may need to be found.

Organized Skepticism

Probably the best known relativization of the idea of (organized) rigorous skepticism comes from the historical findings of Kuhn and Lakatos, who concluded that some dogmatism has been rather productive in the history of science. However different their elaborations, they showed that tenacity in supporting a paradigm or research program should not always be considered as dubious dogmatic behavior, but may also be considered as evidence of fruitful perseverance, seemingly against better knowledge. "The function of dogma in science" is the telling title of an article by Kuhn. Of course, not all dogmatic research becomes respectable in this way. Within the boundaries, that is, the dogmas, of a research program, one may or may not systematically aim at empirical progress and even truth approximation. The presence of this striving is a characteristic difference between examples of fruitful "dogmatic science" and static pseudoscience. For example, the attempts and debates aiming at improving the specifications of the evolutionary theory have no serious analogue in "creation science" (see e.g. Sober 2000). For a general elaboration of the claimed contrast, see SiS Section 8.3.

This point is associated with a latent, very productive, division of labor among scientists. Besides the manifest division of labor between theoreticians and experimenters, without, of course, neglecting the double talented exceptions, there seems to be another division operative that is less well known. In particular, among theoreticians there are not only many constructive researchers, who build on their own research program or, more frequently, on the work of others, but also researchers whose strength it is to play a critical role. One may even question whether the current ideal type of a researcher, who is alternately constructive and destructive, is the proper ideal type for modern science. Precisely because science is a co-production of many people, it is more plausible that a functional division of labor is more fruitful than a homogeneous group of researchers, alternately playing the roles that have to be fulfilled. Instead of a deviation of Merton's fourth norm, we come here across a possibility for realizing it, which Merton himself did not seem to think of.

Disinterestedness and Its Challenges

Regarding offenses against the disinterestedness norm, one thinks in the first place of evident cases of fraud, such as plagiarism, forging of data, faking of experiments and their results. One frequently reads that, happily enough, such cases of fraud are rather exceptional. The classic *The Betrayers of Truth*

(Broad and Wade 1982) and the Dutch variant *Valse vooruitgang* (*False Progress*, Van Kolfschoten 1993) confirm this belief. Nevertheless, it is worthwhile paying attention to the possibility of such occurrences in academic education, because sooner or later one may be confronted, directly or indirectly, with such a case, and one will be forced to have an opinion about it. Moreover, there are all kinds of semi-fraud, in particular semi-plagiarism. Van Kolfschoten distinguishes three categories: conscious plagiarists, unconscious plagiarists, and synthesizers who give poor acknowledgements. Unconscious plagiarism, hence committed in good faith, not only occurs, but perhaps even more frequently than one is inclined to think. One way to minimize the risk is by answering the question to oneself, after finishing a manuscript, whether some of the ideas in it may have been taken from others, learned by reading or conference visiting, in particular in the period before actually working on the manuscript. Finally, when conceiving and writing a synthetic work, it is extremely difficult to indicate all the sources that may have played a role, let alone to specify that role. Apologizing in advance for this possibility, and thus relativizing ones own apparent originality, may give some comfort.

In assessing manuscripts, e.g., when writing referee reports, in editorial work, in advising or working for a publisher, all kinds of variants of fraud and semi-fraud occur (see Lafollette 1992). Referees may be tempted to lie in a referee report or, at least, to manipulate information, they may postpone the completion of a report unnecessarily long in view of some personal interest, and they can even steal ideas or even fragments from the manuscript. Members of editorial boards and publishers may fake or forge referee reports and they can lie about the referee process. All this occurs, probably not on a large scale, but there are variants, which are worthy of some further attention.

Besides evident cases of fraud, there is a large gray area of, in particular, strategic behavior aiming at the allocation of personnel and material means that is difficult to classify as objectionable or not. In these cases everybody will use his own standards, with the risk of being judged in a specific case by others, using other standards. SiS Ch. 13 deals in particular with strategic behavior concerning the number of publications and citations, writing research proposals and assessing them and the formation and extension of networks.

In popular empirical science research (Latour 1987, Woolgar 1988, and others) it is not only suggested, rightly, that strategic behavior plays a large role in circles of scientific research, but this observation is frequently presented in a way which unconditionally sanctions and encourages it, although at one's own risk. Such convictions even grant scientists a license to luxuriate in their own power. To be sure, scientists not only aim at cognitive goals like empirical success or even the truth of their theories, but they also have social aims like recognition and power, and hence the goal to obtain means to achieve such aims. And although these goals frequently reinforce each other, such convergences by no means imply that the conscious pursuit of these social goals is good for science.

However, it does not follow from this that a general code of conduct for scientific researchers is advisable, for if prescriptive it might do more harm than good, e.g. lead to a general avoidance of risky research. Hence, the question is whether a non-prescriptive code is possible as a useful point of reference in the interest of science and society.

Appendix 1: Table of Contents SiS

Structures in Science: Heuristic Patterns based on Cognitive Structures. An Advanced Textbook in Neo-Classical Philosophy of Science

Appendix 2: Outline Table of Contents ICR

From Instrumentalism to Constructive Realism: On Some Relations between Confirmation, Empirical Progress, and Truth Approximation

Appendix 3: Acronyms

CP	set of Conceptual Possibilities
CSH	Comparative Success Hypothesis
CUDOS norms	Communism, Universalism, Disinterestedness, Organized Skepticism
GTC	General Testable Conditional
GTI	General Test Implication
HD (evaluation, method, testing)	Hypothetico-Deductive (evaluation, method, testing)
IC	Initial Condition(s)
I&C	Idealization-and-Concretization
ICR	From Instrumentalism to Constructive *Realism*
IGL	Ideal Gas Law
ITI	Individual Test Implication
KTG	Kinetic Theory of Gases
LMC	Logico-Mathematical Claim
MP	Modus (Ponendo) Ponens
MT	Modus (Tollendo) Tollens
O(*x*)	the set of Operational properties of *x* (Operational profile of *x*)
OF(*x*)	Operational Functional profile of *x*
OS(*x*)	Operational Structural profile of *x*
PC	Principle of Content
PCE	Principle of Comparative HD evaluation
PD	Principle of Dialectics
PSE	Principle of Separate HD evaluation
RE	Rule of Elimination
RP	the set of Relevant Properties
RS	Rule of Success
S/F division	Structure/Function division
SiS	Structures in Science
STR	Special Theory of Relativity
UI	Universal Instantiation
W	the set of Wished for properties (intended profile)
WF	intended Functional profile
WS	intended Structural profile

University of Groningen
Department of Theoretical Philosophy
Oude Boteringestraat 52
9712 GL Groningen
The Netherlands
e-mail: t.a.f.kuipers@philos.rug.nl
http://www.rug.nl/filosofie/kuipers

REFERENCES

Bechtel, W. (1988). Philosophy of Science. An Overview for Cognitive Science. Hillsdale: Erlbaum.

Bickle, J. (1998). *Psychoneural Reduction. The New Wave*. Cambridge, MA: The MIT Press.

Bodewitz, H., G. de Vries and P. Weeder (1988). Toward a Cognitive Model for Technology-Oriented R&D Processes. *Research Policy* **17**, 213-224.

Broad, W. and N. Wade (1982). *Betrayers of the Truth*. New York: Simon en Schuster.

Causey, R. (1977). *Unity of Science*. Dordrecht: Reidel.

Dennett, D. (1987). *The Intentional Stance*. Cambridge, MA: The MIT Press.

Duschl, R. (1990). *Restructuring Science Teaching. The Importance of Theories and Their Development.* New York: Teachers College Press.

Groot, A. de (1961/1969). *Methodologie*. Den Haag: Mouton, Den Haag, 1961. Translated as: *Methodology* (New York: Mouton, 1969).

Hempel, C. (1966). *Philosophy of Natural Science*. Englewood Cliffs, NJ: Prentice-Hall.

Holland, J., K.J. Holyoak, R.E. Nisbett and P.R. Thagard. (1986). *Induction. Processes of Inference, Learning and Discovery*. Cambridge, MA: The MIT Press.

Janssen, M. (1993). *Micro-Foundations*. London: Routledge.

Kaiser, M. (1996). Towards more Secrecy in Science? – Comments on Some Structural Changes in Science and their Implications for an Ethics of Science. *Perspectives on Science* **4** (2), 207-230.

Kim, J. (1996). *Philosophy of Science*. Boulder, CO: Westview Press.

Kim, J. (2000). *Mind in a Physical World.* Cambridge, MA: The MIT Press.

Kolfschoten, F., van (1993). *Valse Vooruitgang*. Amsterdam: Veen.

Krajewski, W. (1977). *Correspondence Principle and Growth of Science*. Dordrecht: Reidel.

Kuipers, T.A.F. (1982). The Reduction of Phenomenological to Kinetic Thermostatics. *Philosophy of Science* **49** (1), 107-119.

Kuipers, T.A.F. (2000/ICR). *From Instrumentalism to Constructive Realism. On Some Relations between Confirmation, Empirical Progress, and Truth Approximation. Synthese Library*, vol. 287. Dordrecht: Kluwer Academic Publishers.

Kuipers, T.A.F. (2001/SiS). *Structures in Science. Heuristic Patterns Based on Cognitive Structures. An Advanced Textbook in Neo-Classical Philosophy of Science. Synthese Library*, vol. 301. Dordrecht: Kluwer Academic Publishers.

Kuipers, T.A.F. (2002). Beauty, a Road to The Truth. *Synthese* **131** (3), 291-328.

Kuipers, T.A.F. (forthcoming). Empirical and Conceptual Idealization and Concretization. The Case of Truth Approximation. Forthcoming in (English and Polish editions of) Liber Amicorum for Leszek Nowak.

Kuipers, T.A.F., R. Vos and H. Sie (1992). Design Research Programs and the Logic of Their Development. *Erkenntnis* **37** (1), 37-63.

Ladyman, J. (2002). *Understanding Philosophy of Science*. London: Routledge.

Lafollette, M. (1992) *Stealing into Print. Fraud, Plagiarism, and Misconduct in Scientific Publishing*. Berkeley: University of California Press.

Langley, P., H. Simon, G. Bradshaw and J. Zytkow (1987). *Scientific Discovery. Computational Explorations of the Creative Mind.* Cambridge, MA: The MIT Press.

Latour, B. (1987). *Science in Action*. Milton Keynes: Open University Press.

Laudan, L. (1977). *Progress and Its Problems*. Berkeley: University of California Press.

Looijen, R. (1998/2000). *Holism and Reductionism in Biology and Ecology. The Mutual Dependence of Higher and Lower Level Research Programmes*. Ph.D. dissertation. Revised version: *Episteme*, vol. 23, 2000. Dordrecht: Kluwer Academic Publishers.

Matthews, M. (1994). *Science Teaching. The Role of History and Philosophy of Science*. London: Routledge.

Merton, R. (1942). Science and Technology in a Democratic Order. *Journal of Legal and Political Sociology* **1**, 115-126. Reprinted under several titles, e.g. The Normative Structure of Science. In: R. Merton (ed.), *The Sociology of Science*, pp. 267-278. Chicago: The University of Chicago Press, 1973.

Mulkay, M. (1977). Some Connections between the Quantitative History of Science, the Social History of Science, and the Sociology of Science. In: P. Löppönen (ed.), *Proceedings of the International Seminar on Science Studies*, pp. 54-76. Helsinki: Academy of Finland.

Nagel, E. (1961). *The Structure of Science*. London: Routledge & Kegan Paul.

Newell, A. and Simon, H.A (1972). *Human Problem Solving*. Englewood Cliffs, NJ: Prentice-Hall.

Niiniluoto, I. (1999). *Critical Scientific Realism*. Oxford: Oxford University Press.

Nowak, L. (1974). Galileo of the Social Sciences. *Revolutionary World*, vol. **8**, 5-11.

Nowak, L. (1980). *The Structure of Idealization*. Dordrecht: Reidel.

Panofsky, W. and M. Phillips (1955/1962). *Classical Electricity and Magnetism*. Second edition. London: Addison-Wesley.

Pettit, Ph. (1996). Functional Explanation and Virtual Selection. *The British Journal for the Philosophy of Science* **47**, 291-302.

Popper, K.R. (1934/1959). *Logik der Forschung*. Vienna, 1934. Translated as *The Logic of Scientific Discovery*. London: Hutchinson, 1959.

Saren, M. (1984). A Classification and Review of Models of the Intra-Firm Innovation Process. *R&D Management* **14** (1), 11-24.

Saviotti, P. (1996). *Technological Evolution, Variety and Competition*. Cheltenham: Edward Elgar.

Selz, O. (1924). *Die Gesetze der Produktiven und Reproduktiven Geistestätigkeit. Kurzfgefasste Darstellung*. Bonn: Cohen.

Shrager, J. and P. Langley, eds. (1990). *Computational Models of Scientific Discovery and Theory Formation*. San Mateo.

Sie, H. (1989). *Industrieel Onderzoek en Haar Relatie tot Academisch Onderzoek*. Masters thesis philosophy of science, University of Groningen.

Sober, E. (2000). *Philosophy of Biology*. Second edition. Boulder/Oxford: Westview Press.

Solomon, M. (1996). Information and the Ethics of Information Control in Science. *Perspectives on Science* **4** (2), 195-206.

Thagard, P. (1988). *Computational Philosophy of Science*. Cambridge, MA: The MIT Press.

Thagard, P. (1992). *Conceptual Revolutions*. Princeton: Princeton University Press.

Vos, R. (1991), Drugs Looking for Diseases. Dordrecht: Kluwer.

Vos, R. (1995). The Logic and Epistemology of the Concept of Drug and Disease Profile. In: T.A.F. Kuipers and A. Mackor (eds.), *Cognitive Patterns in Science and Common Sense* (*Poznań Studies in the Philosophy of the Sciences and the Humanities*, vol. 45), pp. 69-86. Amsterdam: Rodopi.

Weeder, P. and D. Kester (1982). Variatie en Selectie: de Constructie van een Industrieel Produkt. Het geval Tenax. *Kennis en Methode* **VI** (3), 221-251.

Weinberg, S. (1993). *Dreams of a Final Theory*. London: Vintage.

Wolpert, L. (1992). *The Unnatural Nature of Science*. London: Faber & Faber.

Woolgar, S. (1988). *Science: The Very Idea*. London: Tavistock.

Zandvoort, H. (1986). *Models of Scientific Development and the Case of NMR*. Dordrecht: Reidel.

Ziman, J. (1994). *Prometheus Bound: Science in a Dynamic Steady State*. Cambridge: Cambridge University Press.

TYPES OF RESEARCH AND RESEARCH PROGRAMS

David Atkinson

A NEW METAPHYSICS
FINDING A NICHE FOR STRING THEORY

ABSTRACT. Theo Kuipers describes four kinds of research programs and the question is raised here as to whether string theory could be accommodated by one of them, or whether it should be classified in a new, fifth kind of research program.

In his *Structures in Science,* Theo Kuipers distinguishes four kinds of research programs: the descriptive, the explanatory, the design and the explicatory. Kuipers (2000) In this short communication, I propose to flesh out the first, the second and the fourth of these categories, illustrating them by means of our developing knowledge of the solar planetary system. This will lead us from the description of Kepler's laws via the explanation of Newtonian mechanics to the explication of Einstein's theory of relativity. I then broach the subject of string theory, which has the pretension of being an all-embracing Theory of Everything, subsuming Einstein's theory of gravitational force, and theories about all other forces as well. The question arises as to whether this theory, if it proves successful, could also be accommodated by one of Kuipers' programs, or whether it needs to be classified in a new, fifth kind of research program.

Working with the accurate observational data that the Danish nobleman Tycho Brahe had assiduously acquired, Johannes Kepler came finally to the conclusion that the orbits of the planets around the sun are not circles, nor circles inscribed upon circles (epicycles), but ellipses, with the sun at one of the two foci (Kepler 1619). The elliptical orbits of the different planets are nearly, but not quite, all in the same plane. An elliptical orbit in this plane may be specified by three numbers: the major axis (the largest diameter through the middle point), the minor axis (the smallest diameter through the middle point), and the angle between the major axis and some fixed cosmological direction (say the direction of some distant star).

The observational fact of the ellipticity of the planetary orbits constituted the first solid part of Kepler's descriptive program (his first law). The speed of each planet in its elliptical orbit around the sun is not constant, but is larger

In: R. Festa, A. Aliseda and J. Peijnenburg (eds.), *Cognitive Structures in Scientific Inquiry* (*Poznań Studies in the Philosophy of the Sciences and the Humanities,* vol. 84), pp. 95-102. Amsterdam/New York, NY: Rodopi, 2005.

when the planet is close to its perihelion (the point at which it is closest to the sun), and is smaller when it is close to its aphelion (when it is furthest from the sun). Kepler's second law can be explained as follows: consider an imaginary line drawn from the sun to the planet in question, and let *T* be a time interval that is short compared with the planet's period (its "year"). During such an interval, the planet moves, in our imagination trailing the line with it, and a roughly triangular area is swept out by this line. When the planet is near perihelion, the height of the triangle (the distance between the sun and the planet) is less than it is near aphelion; but on the other hand the base of the triangle (or more accurately the curved segment of the elliptical orbit that is traversed during time *T*) is longer when the planet is close to perihelion than when close to aphelion, for its speed is higher in the former case. Kepler's second law is the empirical finding that the areas of these roughly triangular shapes are the same, the shorter base being precisely compensated by the greater height. More generally, for *any fixed* time interval, not necessarily small compared with the planetary period, the area swept out by the line is the same, no matter where the planet is (close to perihelion, close to aphelion, or anywhere in between).

The first two of Kepler's laws have to do with any one planet; but his third law relates them all together. Kepler observed that the further a planet is from the sun, the longer is its period. By trial and error he found that there is a universal relation between the planet's period and the semi-major axis of its orbit (i.e. the average of the largest and the smallest distance of the planet to the sun). For all of the planetary orbits in our solar system, the period is proportional to the semi-major axis raised to the power 3/2.[1]

Such in essence are the results of the Brahe-Kepler program, applied to the planets that revolve around the sun. That the Keplerian laws are more than mere curiosae is intimated by the fact that they apply unchanged to the system of Jupiter's moons, and also to that of Saturn's moons (the largest of the Jovian and Saturnian satellites were known in Kepler's time). Nevertheless, these "laws" of Kepler should perhaps better be called "observed regularities"; they might be likened to a botanical classification. It is therefore apt that Kuipers calls the Brahe-Kepler program "descriptive," but of course this description cries out for an explanation.

[1] In some renditions of Kepler's third law, the *average distance* of the planet to the sun is stipulated instead of its semi-major axis. If this is understood as the average of the distances from the sun at perihelion and at aphelion, that is correct. However, it is incorrect if it is understood as the mean distance of the planet from the sun, computed over a complete period, for that is the integral over time of the distance, divided by the period. This mean distance turns out to be a function of *both* the minor and the major axes of the ellipse.

In his *Principia,* Newton (1683) proposed an explanation for all of Kepler's three laws. In Book I he calculated what the centripetal force on a planet would have to be, in order for it to move in an ellipse, and he found it to be proportional to the inverse square of the distance from the focus. It was only in Book III, *The System of the World,* that Newton formally applied this theorem, and others that he had proved in Book I, to the actual solar system. He showed that all three of Kepler's laws would be true if, and only if, the sun exerts an attraction on each planet that is inversely proportional to the square of the instantaneous distance between the sun and the planet in question.

Given that Kepler's third law states that the period of each planet is proportional to the 3/2 power of the semi-major axis, and that Saturn is about 10 times as far away from the sun as is the earth, its period is predicted to be $10\times\sqrt{10}$, or roughly 30 of our earthly years, which agrees well with observation.[2] Newton gives a table in *The System of the World, Phenomenon I*, showing how well the periods and the average distances of the four largest satellites of Jupiter, which had been measured by Giovanni Cassini, agree with the prediction of Kepler's third law.

We may truly call Newton's gravitational theory (coupled to his laws of motion) an explanatory program in the sense of Kuipers, encompassing in its purview the motions of the planets, of the moon as it endlessly falls toward the earth, and also of the proverbial apple. Nonetheless, two prominent contemporaries of Newton, namely Leibniz and Huygens, variously expressed their rejection of the "occult force of gravitation" that was supposed to act instantaneously through empty space. Newton's defence was twofold: first he agreed that no-one in his right mind would believe that such an occult force could exist, but secondly he recalled that he had calculated, in Book I of his *Principia,* what the force must be to induce the observed planetary motions. But is it true that the force of gravitation acts instantaneously? And is it true that it acts over empty space, with no medium to carry it? Is there a deeper level at which this mystery can be explained and perhaps explicated?

In his general theory of relativity, Einstein found such a deeper level. His coordinate-covariant equations for matter in arbitrary, accelerated motion, and his abandonment of Euclidean geometry for the description of space-time, may in a certain sense be called an explicatory program for gravitation.[3] There is no more talk of occult forces that act over empty space, indeed there is no talk of

---

[2] Kepler himself used this example as an illustration of his third law in Chapter IV of his *Epitome*, "De causis proportionis periodicorum temporum". (Kepler 1618)

[3] The *general* theory of relativity was completed ten years after the *special* theory of relativity. Special relativity had itself been seen as an explication of the concepts of time interval and spatial separation in terms of the space-time continuum. The general theory reduces to the special theory in the case that gravity is absent (or is weak enough to be considered negligible).

gravitational forces at all. The old dispute as to whether absolute space exists (Newton), or whether only matter is truly existent, space being merely separation between material bodies (Leibniz), is dissolved. Matter creates space-time and induces its very geometry, which is of a non-Eucidean, hyperbolic nature; but matter moves along the generalization of Euclidean straight lines, the shortest separation between the end-points in this curved, hyperbolic space-time. There is no difference between the motion of an astronaut with respect to a spaceship at rest on its launching pad, and his motion with respect to the same ship in outer space, on condition that it is accelerating at 1*g*. In either case the relative motion of astronaut and cabin can be described in terms of the same curvilinear coordinates; in either case, no internal measurements and no biological effects can distinguish one scenario from the other.[4]

Gravitation is posited to be simply the experienced effect of curved space-time; the geometry of space-time is induced by matter, and it is altered when the distribution of matter changes. The alterations of the gravitational field are not instantaneous, but are transmitted at the local speed of light (the speed itself being a function of the geometry). Einstein's theory of gravitation gives almost the same description of planetary motion as does Newton's; but there are tiny differences of detail. One of them is that the motion of a planet about the sun is predicted to be not exactly an ellipse (Kepler's first law), but rather a *precessing* ellipse. This means that the direction of the major axis of the quasi-ellipse does not after all remain fixed with respect to a distant star, but rather rotates very slowly in the plane of the orbit. This deviation from the Newtonian result is most pronounced for a highly eccentric orbit (i.e. one in which there is a large difference between the major and minor axes). Mercury, the planet nearest to the sun, has such a highly eccentric orbit, and Einstein's theory predicts that it should precess by 43 seconds of arc per century, or equivalently by one whole degree of arc in 8400 earthly years. In about 3 million years, the perihelion ought to have wandered all the way around the orbit back to its original place, a definite prediction of Einstein's theory that contradicts Newton's theory.[5]

---

[4] Strictly speaking, these statements are accurate only infinitesimally. In fact, since the earth, spaceship and astronaut have finite extent, and the gravitational field of the earth is necessarily inhomogeneous, measurements of Coriolis forces could distinguish the scenarios. The above account should therefore be read as a marginally vulgarized version of the statement that there is *locally* no distinction in general relativity between gravitation and acceleration.

[5] Actually, because of perturbing effects due to the attractions of the other planets on Mercury, as well as observational complications caused by the precession of the earth's axis of rotation, one finds, on the basis of Newtonian theory, that the apparent precession of Mercury's perihelion should be 5557 seconds of arc per century, whereas observation yields 5600 seconds, a difference of 43 seconds (Cushing 1998, p. 259).

In fact, Le Verrier had reported in 1859 that the perihelion of Mercury advances by 38 seconds per century "dû à quelque action inconnue" (the present estimate is 43.11 ± 0.45 seconds per century); and *ad hoc* attempts had been made by various people to explain this departure from the Newtonian prediction. When Einstein calculated from his theory that Mercury's perihelion should precess by 43 seconds per century "without the need of any special hypothesis," he was for a few days "beside myself with joyous excitement." He told De Haas he had had the feeling that something actually snapped in him. Nature had spoken (cited in Pais 1982, p. 253).

Einstein's theory of General Relativity is in part explicatory, in that it clarifies Newton's intuition of forces acting across empty space. In part, as mentioned above, it is also explanatory, for it predicts and explains new phenomena that go beyond Newton's theory. That Einstein's theory may be construed as being partly explicatory and partly explanatory is in accord with Kuipers' dictum that mixtures of research programs are the rule (Kuipers 2001, Section 1.1.1).

If the only fields in nature were those of gravity, Einstein's program could have been deemed the Theory of Everything, the end of physics. However, there are other fields, not only of the electromagnetic sort, but also of a nuclear nature (the so-called weak and strong forces). Thus a yet deeper layer of explanation and explication is called for. The modern candidate for a Theory of Everything is string theory, according to which the known fundamental particles, photons, electrons, quarks and so on, with their associated fields, are nothing more nor less than different frequencies of vibration of the postulated string. In other words, the building blocks of our universe are notes on a cosmic string, rather than autonomous elementary particles. The theory aims at a definitive unification of all known forces, including gravitation. According to string theory, the gravitational attractive force between two particles should increase more rapidly than do the other forces, as the particles approach one another, until the gravitational force is as strong as all other forces. However, to test string theory experimentally, one would have to penetrate to impractically tiny distances, or equivalently to accelerate particles to impossibly high energies before allowing them to collide with one another.

To give a rough idea how impractical it would be, let us consider briefly the history of particle accelerators at CERN, in Geneva. The first machine, a proton synchrotron (PS), came on-line in 1959 and accelerated protons to an energy of 28 GeV (the unit, the giga-electron-volt, corresponds roughly to the energy that would be produced if one proton were to be annihilated, turning all its rest-mass into energy). In 1980 the SPS ('S' for 'super') produced protons and antiprotons at 170 GeV per particle, and in 2005 the Large Hadron Collider (LHC), a machine with a diameter of 27 km, is expected to produce protons of energy 7000 GeV. An American project to build an even bigger machine in Texas was killed by the former president Clinton; and there are no plans anywhere to build bigger accelerators than LHC. The economic limit seems to have been reached. In forty-five years the maximum attainable energy per particle has risen by a factor of 250. To test string theory adequately one would have to produce energies that are ten to the power sixteen (ten thousand million million) *times higher* than those that LHC will produce in 2005. It seems safe to say that we will never be able to produce energies anywhere near this value, and that string theory can never be confronted with the crucial test of experiment.

Is string theory truly a scientific theory? String theory could be tested *in principle,* and in this it differs from unscientific world systems. But it will never be testable *in practice,* and in this it differs radically from Newton's or Einstein's theories of gravitation. Is string theory merely a chapter of mathematics, a science, but not a natural science? Is there perhaps another possibility? Suppose that string theory can be completed in a consistent manner; and suppose that it accommodates the Standard Theory of elementary particles, as well as Einstein's General Theory of Relativity, as a low-energy approximation. This future Theory of Everything would postulate certain new properties of gravity at very high energies, where these new properties *de facto* cannot be tested.

Ed Witten, the most prominent proponent of string theory, once said, on being asked about experimental support for the theory: "Things fall." The chain of implication may be reconstructed as follows: string theory contains Einstein's general relativistic theory of gravitation, Einstein's theory contains Newton's theory as an approximation, and Newton's theory describes quantitatively the falling of "things," like planets, moons and apples. The weakness of the answer is apparent if one reverses the order, and considers the nested inductions: Newton's explanatory theory is subsumed in Einstein's curved space-time explication-explanation, and this is further seen as a property of multidimensional strings. As for Einstein's inductive leap, we have seen that it predicts not only Mercury's precession, but does so with good numerical accuracy. This, and other successful predictions, support the realist's conviction that Einstein's leap was at least in the right direction. But what now of the string theorists' specific claims, as for example that space-time has ten dimensions (of which we can perceive only four)? Perhaps all that could be said, in the most favourable case, is that string theory is, or may come to be, one of the possible unifying logical systems relating General Relativity and the Standard Model. In this sense it would indeed be a scientific theory.

To which of the three programs that we have discussed does string theory belong? Clearly not to the descriptive program, however often Witten may claim that string theory describes how things fall. Nor does it belong to the explanatory program, for as we have seen, string theory does not make testable predictions. Edward Witten himself implied as much. For he said that, while General Relativity and presumably the Standard Model are included in string theory, they can hardly be claimed as predictions of that theory. At best one could call them *postdictions.* Witten (1999) But even that claim would be too generous. For although string theory predicts the particles of the Standard Model, it predicts also the existence of a host of other particles, of which there is no sign.

On the other hand, he did say that supersymmetry is a genuine prediction of string theory, and suggested that, if supersymmetric partners of ordinary particles were to be found in future high-energy experiments, that would constitute a confirmation of string theory. This claim is however overly enthusiastic. In the first place, supersymmetry antedates string theory, and its experimental observation, while being good news for string theorists, who need supersymmetry to avoid causal paradoxes involving *tachyons*, would not specifically favour string theory above the earlier supersymmetric descriptive programs. In the second place, since string theory does not place any lower bound on the masses of the postulated supersymmetric partners, a failure to detect any of them in experiments at LHC, for example, could always be shrugged off with the claim that the masses must then be higher, out of reach of the new machine. In short, string theory's prediction that supersymmetry exists is not falsifiable.

Would then string theory qualify as a purely explicatory program? As Kuipers explains, these programs are associated solely with the explication of concepts or judgements; their aim is merely to render certain informal concepts and intuitive judgements more precise. Although there do seem to be some reasons for classifying string theory as an explicatory program, it is certain that these reasons would be rejected out of hand by string theorists themselves. For such a classification would imply that string theory is merely a part of mathematics, and perhaps of philosophy. String theorists do not think of themselves as mathematicians, and certainly not as philosophers (it is still the case that, for many scientists, "philosophizing" is put on a par with daydreaming or sloppy reasoning). No, string theorists think of themselves as full-blooded physicists: they really want to say something about the furniture of the world, and I suspect that they would be inclined to regard their theories as constituting an explanatory program. They have scientific aspirations.

Could we honour such aspirations; and, if so, would this require the introduction of a fifth research program? Such a program might be delineated as one that encompasses an explanatory program, but in a manner different from the way in which an explicatory program in some cases does so. The latter can explicate a theory belonging squarely to an explanatory program, for example the Copenhagen Interpretation is an explicatory program that aims at making precise the distinction in quantum mechanics between observer and observation. A theory that belongs to the fifth program, however, is not concerned with explication, in the sense of making a certain intuition in an explanatory program *precise*. Rather, it aims at *unification*, that is the bringing together of apparently different explanations into one coherent logical or mathematical framework. If novel objects are involved, as is the case in string theory, then we cannot be talking of a purely explicatory program. If no new predictions can in practice be empirically tested, however, then we are not concerned with a purely explanatory program either. The fifth research program would imply a new sort of metaphysics: different empirically confirmed explanations could be underpinned by a mathematical theory whose essentially new ontological claims cannot be tested in the crucible of experiment.

c/o University of Groningen
Faculty of Philosophy
Oude Boteringestraat 52
9712 GL Groningen
The Netherlands

REFERENCES

Cushing, J.T. (1998). *Philosophical Concepts in Physics.* Cambridge: Cambridge University Press.

Kepler, J. (1618). *Epitome Astronomiæ Copernicanæ.* München: C.H. Beck'sche Verlagsbuchhandlung (1953, with notes in German).

Kepler, J. (1619). *Harmonice Mundi.* München: C.H. Beck'sche Verlagsbuchhandlung (1940, with notes in German).

Kuipers, T.A.F. (2001/SiS). *Structures in Science. Heuristic Patterns Based on Cognitive Structures. Synthese Library*, vol. 301. Dordrecht: Kluwer Academic Publishers.

Newton, I. (1687). *Philosophiæ Naturalis Principia Mathematica.* Berkeley: University of California Press (1966, Motte's translation, revised by Cajori).

Pais, A. (1982). *Subtle is the Lord.* Oxford and New York: Oxford University Press.

Witten, E. (1999). Duality, Spacetime and Quantum Mechanics. Seminar, http://doug-pc.itp.ucsb.edu/online/plecture/witten

Theo A. F. Kuipers

KEPLER, NEWTON, EINSTEIN AND THE STRING THEORY
REPLY TO DAVID ATKINSON

Apparently, the string theory raises the interesting question of the extent to which my typology of pure and hybrid research programs is exhaustive. Before I enter into this question, I take the opportunity to use David Atkinson's lucid survey of "pre-string" fundamental physics to indicate some further illustrations of cognitive structures explicated in SiS and ICR.

Brahe-Kepler, Newton, Einstein

Calling Brahe-Kepler's program descriptive is not only adequate, we may even split the contributions of Brahe and Kepler in terms of an "individual" (i.e. individual fact gathering) and an "inductive descriptive subprogram" (SiS, p. 6, ICR, pp. 171-2), for, as Atkinson indicates, Kepler obtained his three laws by inductively generalizing Brahe's observational data. Newton's theory of gravitation represents not only an evident explanatory program, it is also typically equipped with a well-known "evaluation report" (SiS, p. 216, ICR, p. 98) of general successes, notably the laws of Kepler, Galileo and the tides, and (generalized) individual problems, notably the precessing perihelion of Mercury. Moreover, its explanatory successes were not all of the postdictive kind; some included corrective predictive successes (SiS, p. 216, ICR, p. 98). For example, as elaborated in SiS (Sections 3.1.1, 3.3.2), Newton's explanation of Galileo's law of free fall is a pure case of "corrective reduction."

Finally, Einstein's general theory of relativity not only illustrates the notion of being (unequivocally, empirically) "more successful" (SiS, p. 230, ICR, p. 112), viz. relative to Newton's theory, but also that "concept explication" occurs, not only in philosophy and mathematics, but also in the empirical sciences (SiS, pp. 6-9), leading to hybrid explanatory-cum-explicatory programs. As a matter of fact, both Einstein's special and general theories of relativity are typically hybrid. The first is hybrid by explicating the notion of

In: R. Festa, A. Aliseda and J. Peijnenburg (eds.), *Cognitive Structures in Scientific Inquiry* (*Poznań Studies in the Philosophy of the Sciences and the Humanities,* vol. 84), pp. 103-105. Amsterdam/New York, NY: Rodopi, 2005.

simultaneity (see Note 3 of Atkinson's contribution for some details) and explaining light propagation and other experiments (unequivocally better than competing theories, SiS, p. 236-237, ICR, pp. 118-9). The general theory is hybrid in that it not only explicates away the "occult force of gravitation," but also outstrips Newton's theory in explanatory success, and not only postdictive (e.g. Mercury's behavior) but also predictive, e.g. the famous light bending which was (more or less) confirmed by Eddington's data. By stating later in his paper that "Einstein's leap was at least in the right direction," Atkinson even seems to support the idea that Einstein's theory is closer to the truth than Newton's, a claim which I see as an evident example that a theory of truth approximation should recognize as a (possible) case of truth approximation (ICR, p. 177). However, I know of Atkinson's reserves in this respect, in particular regarding of the existence of "the truth." As a matter of fact, making the characterizability postulate (ICR, p. 147) explicit and its subsequent relativisation, is something I owe to him.

There is also an interesting global point to make about the three examples of research programs, which should have been mentioned in SiS. Atkinson's short stories about (the core ideas underlying) Kepler's laws and Newton's and Einstein's theory of gravitation and the common practice to speak of their laws and theories instead of their programs, make clear that the development of these programs leading to the final theories was mainly a one-person affair. To be sure, if we focus on the (continued) "separate evaluation" of Newton's theory and on the application of his laws of motion to other forces than gravitation, it typically makes sense to speak of Newton's program that was elaborated by Newton and by others. That is, an evolving (real and virtual) coproduction of several researchers, which is typical of many programs, e.g. string theory, despite the fact that Edward Witten may rightly be called by Atkinson its "most prominent proponent."

The Nature of the String Program

When reading Atkinson's exposition, my attention was suddenly caught by the background music on the radio, when a live Proms violin concerto by Tchaikovsky (concerto in D, opus 25, August 7, 2001) was abruptly halted because of a broken string, which was subsequently repaired by the soloist Vadim Repin himself, accompanied by many humorous comments. If we interpret the breaking of the string as a metaphor for falsification of the string theory and its successful repair as a metaphor for an improved version, perhaps even one that comes closer to the truth, Atkinson's exposition claims that this sequence of events is merely possible *in principle*, not *in practice*. I must confess that I, as a non-specialist, always have doubts about such rigorous claims as these by specialists. However, for the sake of argument I should like to dwell upon the question of what type of research program it is if HD testing and evaluation will always remain practically impossible. To begin with, I have some reserves regarding Atkinson's hesitation to call it at least in some sense explanatory: "for … string theory does not make testable predictions". I would certainly qualify his chain of implication "string theory contains Einstein's general relativistic theory of gravitation, Einstein's theory contains Newton's theory as an approximation, and Newton's theory describes [and explains, I assume] quantitatively the falling of 'things', like planets, moons and apples" (p. 100) as a sequence of (partly deductive, partly corrective) reductive explanations of, in the end, experimental results. Hence, string theory has postdictive explanatory successes. To be sure, apparently none of them is an extra success, neither relative to General Relativity nor to the Standard Model. It is the specific claims (e.g. space-time has ten dimensions) and predictions of a host of particles not predicted by the Standard Model that seem impossible to test. Hence, I would call it an explanatory program, albeit of a very special kind in two senses, a negative and a positive one. In contrast to normal explanatory programs, it cannot be experimentally evaluated. On the other hand, as Atkinson strongly emphasizes, the string program has 'unification' of explanatory programs as its chief target, in particular of General Relativity and the Standard Model. In this respect, string theory is similar to other cases, e.g. "the modern synthesis" unifies and surpasses the theories of Darwin, Mendel, and Morgan, but in this case the unifying theory is testable, or at least no less testable than the theories it unifies. Hence, as long as there are no other examples of untestable explanatory programs of a unifying nature, I hesitate to speak of a fifth type. However, I am happy to agree that this is essentially a matter of words, not touching upon the really special character of the string program.

Let me conclude by briefly commenting upon Atkinson's bracketed remark "it is still the case that, for many scientists, 'philosophizing' is put on a par with daydreaming or sloppy reasoning." This reminds me of course of Weinberg's (1993) chapter "Against philosophy." It is regrettable in two respects. First, it illustrates quite convincingly that diehard positivist, that is, (epistemologically) instrumentalist, attitudes cannot only retard but may even become ridiculous. But, second, it also illustrates how one-sided it is to identify positivist philosophy (of science) with philosophy (of science) in general, including the many moderately realist representatives, such as Popper and many others, notably philosopher-scientists, from Einstein to Atkinson.

REFERENCE

Weinberg, S. (1993). *Dreams of a Final Theory*. London: Vintage.

Thomas Nickles

PROBLEM REDUCTION: SOME THOUGHTS[1]

ABSTRACT. Reduction was once a central topic in philosophy of science. I claim that it remains important, especially when applied to problems and problem-solutions rather than only to large theory-complexes. Without attempting a comprehensive classification, I discuss various kinds of problem reductions and similar relations, illustrating them, *inter alia*, in terms of the blackbody problem and early quantization problems. Kuhn's early work is suggestive here both for structuralist theory of science and for the line I prefer to take. My central claims in the paper are (1) that problem reduction is important in its own right and does not "reduce" to theory reduction and (2) that problem reduction is generally more important than theory reduction to methodology as the "control theory" of inquiry.

1. Introduction

The topic of reduction used to be a central topic of philosophy of science, whereas now an article appears only occasionally and is usually critical of the idea. Each of the first several Philosophy of Science Association meetings featured an obligatory symposium on "The Unity of Science," which typically called for a discussion of reduction. For reduction had become a central issue to professional, academic philosophy of science owing to the fact that the academic discipline of philosophy of science had crystallized while the logical positivists were dominant (and largely because of their influence) and the fact that defending the unity of science was central to the positivist program.

Apart from prior hints, such as William Whewell's talk of consilience of inductions, it was the positivists and their allies who made reduction an important topic for philosophy of science in the first place, just as they made explanation a central topic, alongside prediction and confirmation. Some of

[1] Theo Kuipers describes his work as neo-classical. As such, it is deliberately conservative, attempting to preserve whatever was valuable in the logical positivist tradition. Accordingly, I thought it appropriate to discuss the topic of reduction. Just as Kuipers' own views, in the end, move considerably beyond those of the positivists, so I wish to suggest an extension of the concept of reduction. Yet neither account (Kuipers' large-scale one and my small one here) flies utterly in the face of the positivist tradition. I take this opportunity to thank Theo and also Jeanne Peijnenburg for some comments on an earlier draft. Thanks also to the U.S. National Science Foundation for support of earlier research on which part of this paper is based.

In: R. Festa, A. Aliseda and J. Peijnenburg (eds.), *Cognitive Structures in Scientific Inquiry* (*Poznań Studies in the Philosophy of the Sciences and the Humanities,* vol. 84), pp. 107-133. Amsterdam/New York, NY: Rodopi, 2005.

these people are central figures in both enterprises. The publications that first come to mind are Carl G. Hempel (1942), Hempel and Paul Oppenheim (1948), Hilary Putnam and Oppenheim (1958), and Ernest Nagel (1949, 1961). It is no accident that reduction came to be linked with explanation, for reduction in the primary sense was conceived as reduction of *theories*, and this in turn was taken to be the *explanation* of one theory by another.

The positivists already recognized other kinds of reduction, of course. They aimed for unity at three distinct levels: linguistic-conceptual, doctrinal, and methodological. Accordingly, they raised reduction questions at all three levels. First, there were questions about the reducibility of one *language*, sublanguage, or linguistic expression to another, e.g., physical language to phenomenal language, theoretical language to observational language, intensional discourse to extensional discourse, chemical language to physical, and the reduction of all significant language to physical thing language. Rudolf Carnap (1936-7) even used the label "reduction sentences" at one stage of this development. Linguistic reduction was a way to eliminate metaphysics, unclarity, and logically weird discourse, namely discourse that is nonextensional or that otherwise resists formulation in the first-order predicate calculus.

At the *doctrinal* level, there were problems about the reduction of one theory to or by another (where theories are assertions in some appropriate language), e.g., geometric "ray" optics by physical optics and physical optics in turn by electromagnetic theory, thermodynamics by statistical mechanics, classical mechanics by relativity and quantum mechanics, and classical genetics by molecular genetics. Theory reduction was important to the positivists' concern to establish science as a progressive, more-or-less linear-cumulative enterprise, based upon a physicalist conception of the universe.

Finally, there was *methodological* reduction, although the term 'reduction' was not as likely to be employed here. For intellectual and social reasons that we need not pursue, some of the positivists possessed an almost essentialist view that all genuine sciences *must* employ the same scientific method, which it was their business to articulate. Thus, many of the positivists held that all mature sciences should produce theories formalizable as partially interpreted logical calculi that can be tested against experience in a hypothetico-deductive manner and that provide deductive-nomological explanations. In his well-known 1942 article, and later, Hempel contended that all legitimate explanation in all sciences at all times must have a deductive- (or inductive-) nomological form. Historical explanations were admissible only insofar as they could be viewed as legitimate "explanation sketches," that is, insofar as they were *reducible* to the pattern of explanation supposedly employed in physics. This deductive framework was later applied to intertheoretic reduction, notably by Nagel (1949, 1961) in his classic account.

Today, reduction at none of these three levels remains a central, disciplinary aim of general philosophy of science, although the structuralists and a few others still devote considerable attention to it (see below). While various reduction issues are locally important in the special fields of philosophy of science, today the emphasis is on the *dis*unity of science (Dupré 1993, Galison and Stump 1996). In some circles, "reductivist" has become a term of abuse.

There are various reasons for this decline, some weighty, some questionable. Among them are these. First, although they made important advances on several particular points, all the larger linguistic reduction programs eventually foundered. Second, at the level of doctrinal reduction, the positivists attempted to find relatively simple logical relations between large theory complexes. As subsequent work in history of science and its successor, "science studies," has shown, the dynamics of scientific change is vastly more complex than this. Hull (1972) and Sklar (1976) had an early, major impact on the philosophical discussion. Today philosophers appreciate that there can be many kinds of interesting relations between and among theories, with deductive reduction being only one of these – and a virtually unrealizable ideal at that. Science just does not progress by a simple logical unfolding of this sort. Furthermore, many scientific disciplines, even many specialties in physics, such as solid state physics, do not feature big, overarching theories of the sort that have dominated philosophical discussion. Accordingly, today's literature displays a serious reaction against "theory-centered" accounts of science and, more generally, against physics as *the* model science.

Third, the positivists' attempt to capture and bottle the methodological "essence of science," so to speak, failed to appreciate the great diversity of serious scientific pursuits. An additional point is that the shift toward the study of scientific practice, especially experimental practice, has tended to demote the centrality of logical-structural issues in science studies, even in philosophy of science. Historically, philosophers, including philosophers of science, have been far more curious about finding the correct theory or world picture than in the activities that produce such pictures. Specifically, the positivists were primarily interested in the finished products of science rather than in the processes that produce them. Empirical study of the details of ongoing scientific activity was not attractive to philosophers, who saw their business as "higher" and purely conceptual rather than empirical. Someone recently characterized the positivist position as the idea that everybody *else* should be an empiricist!

As we know, the idea of philosophy of science as purely logical or conceptual analysis was challenged within the positivist-pragmatist movement itself, long before the arrival of contemporary science studies. Quine, Hempel, and others contended that philosophy cannot be demarcated by a sharp conceptual/empirical boundary.[2] The new history and sociology of science pounded this point home. And insofar as scientific practice becomes the central focus, it is no longer clear where reduction fits into the picture. This realization was expressed already in the title of Kenneth Schaffner's (1974): "The Peripherality of Reductionism in the Development of Molecular Biology." This paper registers Schaffner's recognition that issues centrally important to philosophers are not necessarily of central importance to working scientists.

So is the general topic of reduction dead in philosophy of science? In *Structures in Science*, Theo Kuipers writes:

> Scientific practice is almost by definition a matter of reduction, either in the transformational sense of inferring something (deductively or otherwise) from something else, or in the idealizational sense of concentrating on some aspects and provisionally neglecting others, to be accounted for later. (p. 89)

Clearly reduction is not dead for Kuipers, and I agree. I find his broad conception(s) of reduction congenial. Science clearly has to do with abstracting from the bloomin', buzzin' confusion of the world as commonly perceived and with compacting information in various illuminating and useful ways. In this paper I shall try to broaden the usual treatment of reduction even more by focusing not on theory structures but on problems. Hence my discussion has to do with methodology of science and the heuristics of problem solving, but not unity of method in the positivist sense. Moreover, problem reduction need have nothing to do with explanation: it is of independent interest. Not all scientific problems are explanation problems, and rather few are problems of ontological economy. Working scientists, with some important exceptions, are not as concerned with ontological economy as philosophers are. However, all scientists in all fields are intensely interested in economy *of research*, certainly including problem-solving economy.

The structure and dynamics of problems never became a central issue for the positivists, nor has it for most philosophers since.[3] Today most of the

---

[2] As Hempel (reprinted as 1965a, p. 113) put it, "Concept formation and theory formation go hand in hand; neither can be carried on successfully in isolation from the other." Quine's (1951) conclusions were more radical. His strongest claims in "Two Dogmas" remain open to debate, but relatively few of the new generation of philosophers defend the existence of a sharp empirical-conceptual boundary and conceive philosophy as a purely conceptual, "analytic" discipline.

[3] In this respect they did not continue the interesting work of Ernst Mach, e.g. Mach (1976, chap. XV).

explicit methodological work on problems and problem solving is found in theory of computability, artificial intelligence (AI), and computer-information science generally. Here reduction, broadly construed, remains a crucial concern. Within philosophy, I agree with Matti Sintonen (1996) that structuralist and semantic or model-theoretic approaches to science are in a better position to address problems and their relations than was the axiomatic-calculus approach.

As Sintonen notes, by focusing on problem reduction in the sciences, we also do more justice to scientific practice. One can argue that methodology of science itself is more properly pitched at this pragmatic level, on control-theoretic grounds. (Philosophers have often focused on issues, such as the truth of theories, that seem to have little controlling influence in actual scientific research, although they may be of philosophical interest.) But then it is especially incumbent upon those who pursue a problem-solving account of scientific inquiry to inquire into problems themselves and their relations to one another.

2. What Is Problem Reduction? Some Historical Illustrations

Scientists and mathematicians speak of problem reduction in a variety of ways, sometimes completely informal ways and sometimes mathematically formalized to some degree. I believe that paying attention to this talk can yield philosophical insights, especially when it is a fairly common way of speaking. I do not advocate "ordinary language" philosophy of science in the sense that common modes of locution are taken uncritically. However, noticing how scientists talk is a valuable heuristic device, and it does help to keep philosophy of science in touch with science as practiced. It was paying attention to a fairly common parlance of this sort that led me to the concept of domain-preserving, limiting-case reductions-*to* (in which a *later* or *successor* theory reduces to a *predecessor*, under some operation such as taking a limit), as an alternative to Nagelian, domain-combining reduction-*of* T_1 by T_2 (in which the later theory typically belongs to an initially different domain than the reduced theory and reduces the earlier if time-order is relevant).[4] Wimsatt

[4] See Nickles (1973). Honest disagreement is possible about how common a locution this is. Sahotra Sarkar (1992, note 24) reports Abner Shimony's view that my use of reduction-*to* is not standard in physical science. But surely it is fairly common for physicists to say that one theory or model reduces to another one (sometimes an earlier one) under a limit or other approximating operation, e.g., as velocity or Planck's constant goes to zero or as mass becomes large. Bohr's correspondence principle employs this idea as do the methodological correspondence principles of Popper and Post (Post 1971).

(1976) unpacked this distinction in terms of "intra-level" versus "inter-level" reductions.

The basic idea of analyzing complex problems into subproblems and subsequently synthesizing their solutions into a complete solution is very old. It has been an explicit part of problem-solving methodology at least since the ancient Greeks. Recall Plato's method of division and how Aristotle divided the sphere of knowledge into distinct subject matters, not to mention the famous "method of analysis" of the Greek mathematicians. Centuries later, analysis and synthesis or *resolutio* and *compositio* formed the core of Galileo's method as well as that of Descartes, as expounded in his *Discourse on Method* (1637) and other methodological works. Or consider this still later formulation by John Herschel, which also amounts to a method of discovery:

> Complicated phenomena, in which several causes concurring, opposing, or quite independent of each other, operate at once, so as to produce a compound effect, may be simplified by subducting the effect of all the known causes, as well as the nature of the case permits, either by deductive reasoning or by appeal to experience, and thus leaving, as it were, a *residual phenomenon* to be explained. It is by this process, in fact, that science, in its present advanced state, is chiefly promoted. Most of the phenomena which nature presents are very complicated; and when the effects of all known causes are estimated with exactness, and subducted, the residual facts are constantly appearing in the form of phenomena altogether new, and leading to the most important conclusions. (1830, §158, Herschel's emphasis)

Herschel's form of the method of analysis and synthesis applies especially well to analytical chemistry and other fields in which scientists can physically manipulate the components.

An ambiguity tends to run through this tradition, one that is often harmless. Writers sometimes speak of problem reduction as the decomposition of complex *problems* into simples,[5] but sometimes they speak of the decomposition of complex *entities* or phenomena or situations into simples. Thus in his early *Rules for the Direction of Mind*, Descartes wrote of decomposing entities into their "simple natures" in the same breath that he spoke of the reduction of complex problems. This eventually led him into

---

I agree that one can just as well say that the limit relations show classical mechanics to be a special case (in some sense) of relativity theory and that (in this sense) relativity theory reduces classical mechanics, by absorption. However, there is still a difference of emphasis at stake here. Those who stress Nagel-type reductions are usually interested primarily in theory of justification, and in ontological reduction, whereas I am primarily interested in the heuristic use of reductive relationships for purposes of solving new problems in short, for "scientific discovery." Historically, Bohr's correspondence principle became a powerful heuristic.

[5] For a contemporary example, see George (1980), p. 20. As one solution method he lists "Problem-reduction solutions (e.g., reduce to subproblems: if you want to go from Beaconsfield to Cambridge (the problem), to go from Beaconsfield to Amersham is a sub-problem)." Computer science is replete with this sort of talk.

trouble, since there was no reason to assume that what is simplest or most basic in nature, that is, in order of being, is also simplest in epistemology, or order of knowing. He eventually dropped talk of simple natures while retaining the idea of basic problems. In my view it is an improvement to treat problem reduction and ontological reduction as distinct but sometimes overlapping subject-matters.

One of the first exemplars of modern, mathematical mechanics was Galileo's showing that projectile motion is compounded of two distinguishable components, each of which can be described separately (the law of falling bodies and the law of "inertial" motion). When combined, the two equations immediately yield the equation of a parabola. Before long mechanicians could break down more complex motions such as a that of a chair thrown across a room – by reducing the problem to a center-of-gravity problem and adding to the Galilean treatment the rotational degrees of freedom around the center of gravity. Thus a seemingly complex problem can be reduced to a few, standard, simple ones.

The Galileo example illustrates the double efficiency of solving a new problem by analysis and synthesis. Not only is a complex problem made tractable but also the individual solution components can often be methodized into standard solution procedures or at least "recycled" as models of what solutions to similar problems will look like. Once a student has learned to solve a few basic kinds of motion problem, she can solve any of a infinite variety of similar problems through case-based reasoning. I return to this Kuhnian point below. My claim here is that problem reduction of some kind or other is the main route to routinized problem solving in a scientific domain, and routine problem solving techniques bring us about as close to a uniform, workable *method* as we can expect to get.

In quite a different field, Darwin was finally able to make real headway on a theory of evolution when he succeeded in separating out, as distinct mechanisms, a mechanism of variation, a mechanism of selection (namely, natural selection), and a mechanism of retention or transmission. Fortunately nature cooperated with his scheme sufficiently to get his program off the ground. The reduction of the overall problem to three distinct problems was crucial since Darwin had little idea what the variation and transmission mechanisms were. Yet given that such mechanisms did exist, his theory of natural selection was enough to provide a powerful theory (a "phenomenological" theory, so to speak) or theory schema as the basis for a promising research program.

Notice that reductions of this sort presuppose that the various simple problems are independent of one another (in a sense the very opposite of reduction) or at least only very weakly related to one another. This is the problem counterpart of what Herbert Simon (1981) termed a weakly decomposable or nearly decomposable system. The interaction terms among the components are weak.

But why, initially, should anyone suppose that analysis or problem reduction in the form of decomposition is a good methodological strategy? Why should we suppose that the world is amenable to such treatment? Have scientists simply been lucky? Well, in a sense they have been. Descartes and other early investigators could only assume a priori that this method would work. The world could have turned out to be too complex for us to fathom. Rescher somewhere asks us to imagine that we were aquatic creatures immersed in the ocean instead of human beings that can observe simple astronomical patterns and the flight of projectiles in the thin fluid medium of the atmosphere. Were we aquatic creatures, hydrodynamics rather than ordinary mechanics would be desirable as our basic science, but what is the likelihood that we could have launched such a science? The Cartesian method breaks down for nonlinear dynamical systems.

Simon (1981) provides an argument that the reductive-analytic, divide-and-conquer strategy is *empirically* justified in a wide range of cases. His central idea is that evolving systems that are hierarchically organized into nearly decomposable subsystems will tend to evolve orders of magnitude more rapidly than entities that are not so organized. Thus the nearly decomposable systems will come to dominate the universe. Hence, we should *expect* such an analyzable structure for any entities that have evolved. In response, Wimsatt (1980) shows that things can be more complicated than this, given such phenomena as co-evolution. In fact, natural systems often seem to be far more economical, far less modular, than the analytical model would suggest, since their components may be part of a network utilized in multiple processing tasks. Think of pleiotropy and polygenicity.

There are many and diverse kinds of problem reduction, as further examples will help to bring out. Hence, we should not look for an "essence" of problem reduction. Here is one sort of example. Sometimes scientists speak of problem reduction when a new theory (or modification of an old one) removes part of a difficulty. For example, the so-called hierarchy problem in the search for a unified theory of the four physical forces is the fact that gravity is extremely weak compared to the other forces. The electro-weak scale differs by a factor of 10^{16} from the so-called Planck energy and Planck length scale that characterizes gravity. As Arkani-Hamed, Dimopoulos, and Dvali (2000) report, recent theoretical work modifying the inverse-*r*-squared law of gravitation over short distances by altering the dimensional structure of space-time "reduces" the hierarchy problem by many orders of magnitude. In fact, if we postulate enough extra dimensions, the hierarchy problem disappears completely.

One could perhaps generalize this idea in terms of the problem space representation of Newell and Simon (1972), in which problem solving activity involves determining the "distance" between an initial state (or one's current state) and the goal state. Any move that reduces this distance reduces the problem.

Sometimes problem reductions are really equivalences. The two problems can be construed as alternative formulations of the same problem. But since reductions are usually considered unidirectional, asymmetric, why call such a case *reduction*? One answer is based simply on time order of solution. If A is solved and B is later discovered to be identical with A, then we tend to say that B reduces to A, and that their solutions so reduce.

Another answer is that an asymmetry can often be reintroduced at an epistemological level. This may happen because one form of the problem is more tractable than another, perhaps because certain problem solving techniques (e.g., mathematical or computational capabilities) are more developed than others; or it may be that one representation of the problem is easier for human investigators to grasp or more heuristic in pointing toward its solution. After all, when it comes to cognition, there are *no* necessary equivalences. Even logically equivalent expressions are not cognitively equivalent, since a person may fail to recognize their equivalence and may even assert one while denying the other. The laws of logic are not the same as the "laws" of thought! Just as one theory formulation may be more useful, perhaps more "intuitive," than a logically equivalent formulation, so may alternative problem formulations differ. In some cases a transformed representation can render a seemingly difficult problem trivial.

There are many "insight" puzzles that illustrate this last point. Here is one. A standard chess board consists of 64 squares (half dark, half light) and so can be covered by 32 domino-like rectangles, each of which covers two squares. Now cut out two squares on opposite corners of the board. The question is whether 31 dominoes can be arranged to cover the remaining 62 squares. The problem is easily solved, without calculation or fitting exercises, once we realize that each domino covers one light *and* one dark square, yet removing opposite corners always removes two squares of the same shade.

3. Partial Reduction and a Historical Example

Summarizing their earlier work, Newell and Simon (1972) characterized problem solving as search. Searching for the solution to a problem is, in effect, to explore a search space, which often turns out to have a tree structure arising out of the possible "moves" or operations that may be applied to elements such as symbol strings. In the sciences the search operations in such a space need not be limited to transformations governed by strict, logical rules but will often include fallible heuristics that cut down the search space to a more manageable size. This move expands problem-solving methodology to include heuristics in order to meet the demand for economy of research. Later developments in computer science provide resources for comparing the difficulty of various formally characterizable problems and also for studying the relative efficiency of different problem solving strategies (see below).

Efficient methods for solving novel problems are highly desirable. Equally important is the efficient use of results already obtained. We met this idea in the Galileo example. Building up a classification of different types of problems and a stock of standard solutions to them in effect *methodizes* what was once a series of individual, historical discoveries. Many writers have noted that the mark of a mature science is the existence of routine problems, and this means developing the ability to see "new" problems as variations on old ones. That is, the new problems and sought solutions *reduce* to old ones, *mutatis mutandis*. I return to this sort of problem reduction in the next section.

Whether or not the researchers searching for a problem solution are aware of the fact that they are exploring a search space or are cognizant of the boundaries or general structure of the space depends on how much domain knowledge they possess. When little is known, they can only search blindly and unsystematically (Campbell 1974; Nickles 2003c). When much is known, the search can be more focused, meaning that a smaller space needs to be searched. In other words, acquiring new knowledge of the structure of the space can reduce the size of the problem in an obvious sense. When it does so, we can speak of the new knowledge as constituting a *constraint* on the problem. Finding or constructing additional constraints reduces the problem by constraining the search space. To be more precise, we should distinguish the search space as it really is and the search space as the investigators conceive it to be. (Newell and Simon make such a distinction.) In some cases, strong enough constraints will be available to determine the solution completely, although the investigators may not recognize this. Not all problems are this neat, of course. Sometimes there is not one unique solution, and sometimes we need not bother, for the purposes at hand, to seek the optimal solution among the many possible solutions. Researchers in some fields distinguish hard from soft constraints, the idea being to find or construct a solution that satisfies the hard constraints and, in addition, as many soft constraints as possible, within the means available.

Observe that on this AI view of problem solving, it makes sense to speak of a *partial* reduction of problems. Indeed, each paring down of the search space, each pruning of the tree, amounts to a partial reduction. We can view heuristics that trim search trees in various ways as more-or-less risky, hypothetical constraints, and therefore hypothetical problem reductions. If problem reduction has been little noticed by philosophers, partial reduction has received hardly any attention at all. Yet many scientific advances can be construed as partial reductions of problems. Gustav Kirchhoff's work on blackbody radiation is a famous example, which I briefly summarize from Nickles (1978).

It was Kirchhoff who first sharply formulated the so-called blackbody problem and established its importance. In a series of papers published around 1860, he showed, by means of equilibrium arguments, that for each given frequency ν and temperature T, the ratio of emissive to absorptive power is the same for all bodies, a result later known as "Kirchhoff's law." He then considered a perfectly absorbing, "black" body and deduced that (1) the emissivity of a blackbody depends on ν and T alone and is independent of the material constitution of the bodies and of their environment; that is, the energy density ρ of blackbody radiation is a function of ν and T alone; and that (2) blackbody radiation is equivalent to "cavity radiation," the radiation inside a black-walled cavity, at the same temperature (later even a cavity with reflecting walls, as long as a bit of carbon, e.g., lamp black, is present); and that (3) cavity radiation not only depends on ν and T alone, therefore, but also is independent of both the material nature and the shape of the cavity.

In establishing (1), Kirchhoff in effect "reduced" the blackbody distribution problem to that of determining the explicit form of a function of just two variables, $\rho(\nu, T)$, for cavity radiation. Two decades later, results (2) and (3) provided the theoretical basis for transforming both the experimental and the theoretical blackbody problems. Experimentally, a radiating cavity with a small opening is a much better source of blackbody radiation than a hot surface. Theoretically, (2) and (3) transformed the problem into one of cavity radiation and permitted simplifying assumptions about the structure of the walls, as in Planck's later work, or rendered matter-theoretic assumptions altogether unnecessary, as in the still-later approaches of Rayleigh, Ehrenfest, and Debye.

Meanwhile, in 1884 Boltzmann theoretically proved, via the second law of thermodynamics, an earlier conjecture of Stefan that the integral of ρ over all frequencies is proportional to T^4. This "Stefan-Boltzmann law" therefore imposed an important additional constraint on the function ρ. Ten years later Wien proved that $\rho(\nu, T) = \nu^3 f(\nu/T)$, a result called "Wien's displacement law" (*Verschiebungsgesetz*) and that yields the Stefan-Boltzmann law by integration. As Einstein and others remarked, Wien's displacement law reduces the blackbody problem to one of determining the form of a function of a single variable, v/T (cf. Jammer 1966, p. 9).

We find in this series of examples a rich variety of results and "moves" worthy of being called problem reduction and partial reduction.

4. Is Problem Reduction More Fundamental Than Theory Reduction?

Here are five arguments why problem reduction is more fundamental than theory reduction. I shall state the arguments and then appraise them.

(1) Science is primarily a problem-solving activity. Solving problems is the direct or proximate goal of inquiry, whatever other, long-range goals there might be (truth, predictive adequacy, explanation, ontological reduction, systematic theory of the domain, practical applications, etc.). Thus any methodology of science that pretends to remain in touch with science as practiced should focus (although not exclusively) on problem solving. That is, any such methodology of science should aim to provide or describe a "control theory," insofar as that is possible, and the primary locus of control in science is to be found at the level of problems and problem solving rather than at the level of finished theories or at the level of formulating the general aims of science. Hence it is more important for methodology of science to study problems, problem representations and their various relations, and problem-solving algorithms and heuristics than to study the logical structure of finished theories.

(2) Specifically, problem reductions, and anticipated reductions, are often of great heuristic value in directing scientific research. Showing that problems reduce to one another or are equivalent under easily accomplished transformations can alter research priorities. A problem is the more important to solve the better it represents a large and/or significant class of problems, the more paradigmatic or exemplary it looks. Solve one and you've solved them all! Scientists and engineers tend to work on these problems first and then to methodize and streamline successful solution procedures. In short, problem reduction, broadly understood, is a key component in *heuristic appraisal*, a kind of appraisal just as important to science as epistemic appraisal (confirmation, disconfirmation), since heuristic appraisal directs or controls research.

According to (1) and (2), problem reductions alter scientific practice within an established domain more than do theory reductions. Most large "theory reductions" are "in principle" reductions that alter scientific practice only on the periphery. (Recall Schaffner's peripherality thesis concerning reduction in molecular biology.) Except at very high velocities, one continues to use classical mechanics. Only when dealing with extremely large masses do scientists employ general relativity. The reduction of classical mechanics to quantum mechanics (supposing that it *is* a reduction) rarely affects problem solutions involving objects above the micro scale. And so on. The "reverse" concept of reduction-*to* explains why this is so, for (to take a familiar example) in the limit of low velocities, special relativity theory is negligibly different from classical mechanics. As people often say, relativity theory "reduces to" classical mechanics in this domain.

(3) The traditional focus on theory reduction is too theory centered. Not all scientific specialty areas aim to construct systematic theories of the sort familiar in mechanics and electromagnetic theory. But problem solving is always a central concern. Since problem reductions achieve complete or partial problem solutions, problem reduction is of wider interest than theory reduction.

(4) Even in theory-rich disciplines, we should understand theories erotetically, as answers to questions or solutions to problems. In the history of science a theory typically grows up slowly, around one or more key problem solutions. Hence problem reductions are logically prior to theory reductions. That is, theory reduction can always be understood in terms of one or more problem reductions. Typically, a large theory reduction amounts to achieving multiple problem reductions in one swoop.

(5) The final argument makes the stronger claim that any problem reduction is automatically a theory reduction.

How good are these arguments? Since I am partial to a pragmatic, problem-solving approach to understanding science, (1) and (2) carry considerable weight with me. On the pragmatic view, theories are important to working scientists mainly insofar as they are useful to their ongoing work of problem solving, as opposed to giving broad philosophical perspectives on the universe. Scientists often find it fruitful to work with theories and models they know to be false as long as these commitments generate and help solve interesting new problems. Conversely, scientists sometimes decide not to pursue a theory in their specialty area that they believe may be true, if it does not provide sufficient calculative resources (Pickering 1984).

Furthermore, I am convinced that heuristic appraisal – appraisal of the future feritility or promise of a line of research – is just as important as the epistemic appraisal (more commonly called "confirmation theory"), that has dominated philosophical accounts. It is heuristic appraisal, more than anything, that guides large and small research decisions.

(3) makes an important point, namely that we should not forget the erotetic basis of scientific inquiry, as theory-centered approaches sometimes do.[6] However, (3) is too strong if intended to be descriptive of scientific practice. One point, as I argue below, is that there can be problem reductions without theory reductions and hence reducing one problem to another does not necessarily solve the former problem. Another point is that sometimes it is easier to see logico-mathematical relations among theories or their equations than to recognize relations directly among problems. In the case of domain combining reductions such as physical optics to electromagnetic theory, the two theory programs originally had different problem-solving aims (to explain different domains of phenomena by different techniques). To oversimplify a complex story, it was eventually noted that electromagnetic theory could be extended to cover radiant energy of "other" kinds. Whewell would have termed this reduction a striking "consilience of inductions." The point is, the reductive possibilities became most apparent after the theories had been considerably developed.

In the case of domain-preserving reductions (e.g., relativity theory to classical mechanics, or the reduction of classical mechanics by relativity theory), the new theory is typically developed to handle problems that the old theory could not. When a new theoretical framework is available, it is then easily shown that it "reduces to" the old theory under the relevant limiting conditions (Post 1971).

In both cases it is plausible to say that a successful theory reduction accomplishes a problem reduction, or rather, multiple problem reductions. However, logically speaking, this amounts to saying that theory reductions entail problem reductions, not vice versa. Although entailment amounts to a kind of logical presupposition, it is incorrect to say that, in the order of knowing, problem reduction necessarily comes first.

(4) is basically correct, subject to the qualifications just mentioned. The lack of attention by philosophers to problem reduction is surprising given the centrality of problems to inquiry. Most philosophers of science writing during the heyday of reduction were more interested in big world pictures than in providing a descriptively adequate account of scientific activity itself. This is fine as long as the philosophers did not confuse their own aims with those of scientists in their specialty areas. Today a number of people working in science

---

[6] Popper and Kuhn (see below) adopted an erotetic approach, as did Lakatos (1976) and Laudan (1977). For the loss of an erotetic approach in the history of philosophy, see the articles in Meyer (1988), including Nickles (1988). Hintikka has long pursued a formal erotetic approach to inquiry, including scientific discovery. See also Kleiner (1993) and Sintonen (1996). Wiśniewski (1995) is an important recent work that investigates inferential relations between questions and between statements and questions. There is also a good deal of work in computer science.

studies do aim for descriptive adequacy, but their accounts of scientific work are usually so detailed as to make the big, simple reductive questions look naïve, especially if unity of science is a motivation. (Again, Hull 1972 and Sklar 1976 were among the first to do this.) In the information sciences and especially in AI, on the other hand, usable problem-solving methods, including reductive methods, are the central concern.

(5) is false as it stands. In fact, it contradicts the point of (3). For not all problem solutions are theories in anything like the usual sense. (5) is especially off the mark, given my broad interpretation of problem reduction, where the "mere" change of problem representation can make a difference.

My overall conclusion is that problem reduction does not "reduce" to theory reduction but is of independent interest. In some respects, although not in all, problem reduction is of more general interest than theory reduction. When it comes to detailed study of scientific activity, I claim that problem reduction remains a centrally important topic – more important than theory reduction. The most obvious respect in which theory reduction holds its separate interest is that it tends to provide wide world views, including support of ontological reduction.

The counterargument that theory reduction is more important than problem reduction to *confirmation* in science cannot be sustained, for problem reductions can also make significant contributions to the warrant of a theory or research program. In fact, Laudan (1977) attempted to replace traditional confirmation theory by an accounting system based on problem-solving success, a view that can already be found, in different forms, in Kuhn and in Lakatos. Traditional confirmation theory focuses on the true and the probable and neglects the useful and the fruitful. Here I repeat my above claim that, when push comes to shove, scientists in their ongoing practice value utility and heuristic fertility over truth.

5. Problem Reduction Without Theory Reduction: More Examples

I now provide a historical example that illustrates the importance of problem reduction in the absence of theory reduction, namely Ehrenfest's work on the early quantum theory, particularly his attempt to quantize the nonlinear oscillator.[7]

Ehrenfest's search for the really essential elements of the early quantum hypotheses led him to the *adiabatic principle*, which soon became one of the central dogmas of quantum mechanics. The basic idea is that adiabatic changes

[7] The detailed case study and references can be found in Nickles (1976).

in a quantized system do not alter its quantum state, nor, therefore, the set of quantized energy levels available to the system. In other words, Ehrenfest proved that the quantum conditions are adiabatic invariants. This principle was of great heuristic value because it helped solve *the* central family of problems of the new quantum theoretic research program, namely, how to quantize a larger domain of physical systems. Planck, Einstein, and others had shown how to quantize the simple harmonic oscillator, and Bohr had done the same for the atom (especially hydrogen and the lighter atoms); but how quantize other systems, even simple variations such as the anharmonic oscillator? The reason the adiabatic principle was heuristically powerful now becomes clear, for any system that could be adiabatically transformed into a known system must obey the same quantum rules.

An interesting point here is that a quantization problem reduction can be accomplished in the absence of an extant theory reduction. Consider two systems S_1 and S_2. If one can show that they are adiabatically related, then one has shown that they obey the same quantum conditions. In other words, the problem of determining the quantum conditions for one reduces (symmetrically) to the problem of determining the quantum conditions for the other. *This holds even when the community has not yet solved the problem of how to quantize either one, that is, when a quantum "theory" or model of neither system is available.*[8]

The problem situation of the anharmonic oscillator was slightly different. Peter Debye had suggested quantum rules by analogy with the harmonic oscillator. The small physics community working on such problems at the time thought this extension was plausible but not well founded. Debye himself considered it arbitrary, a "trick" or "artifice" (*Kunstgriff*). Subsequently, Ehrenfest provided that foundation by showing that the anharmonic oscillator can be adiabatically transformed into a harmonic oscillator. In this case he did accomplish a (small) theory reduction *by means of* his more general and routinized technique of problem reduction. This application of the adiabatic principle confirmed Debye's conjecture (rather than serving the above-mentioned heuristic function) but it did more than that. For Debye's conjecture no longer had the status of a distinct, confirmed hypothesis, as the hypothetico-deductive model of science would suggest. Rather, it was now securely founded on the theory of the harmonic oscillator.

It would be easy to produce other examples from the history of science and still easier from the history of mathematics. Leibniz, for example, focused on basic problems and families of problems such as differentiation and integration

---

[8] Of course, problem reductions themselves cannot be accomplished in a mature science apart from a body of background "theory." However, this sort of problem reduction remains different from a direct theory reduction in the sense of both Nagel (1961) and Nickles (1973).

because he realized that a great many rate-of-change problems reduced to the former whereas many others reduced to determining an area under a curve by integration. (He also noted that differentiation and integration were inverse operations.) It often happens in mathematics and related disciplines that one problem reduces to another.

Whether or not we wish to speak of *reduction* to the *same* problem will, of course, depend upon how we identify and individuate problems, and that will often depend on the level of abstraction. As already noted, the relation between such problems is sometimes symmetrical. An important contemporary example is the set of **NP**-complete problems, including traveling salesman-type problems but also a large variety of others (Lawler *et al.* 1985). It has been shown that an algorithm that solves any one of the hard problems in this set in polynomial time can be transformed into algorithms that solve each of the others in polynomial time. The problems are said to be *equivalent* under a polynomial-time transformation. One problem (or "language") is said to be *polynomial-time reducible* to another if there exists a polynomial-time function mapping symbol strings in one to strings in the other (and satisfying basic conditions). The fact that none of the problems *has* been solved so far, in polynomial time, is evidence to most mathematicians that the entire set is unsolvable in that manner. In any case, this extended concept of problem reduction is routinely employed to classify problems in terms of their degree of difficulty and, correspondingly, to classify algorithms.

The Ehrenfest and **NP** examples suggest a generalization of our usual concept of problem identification, one that relativizes problem identification to a transformation of a suitable kind. We can then say that two problems are *identical relative to transformation T* if they remain invariant under T in some specified respect. As the **NP**-complete example indicates, such a conception of problems is already employed in scientific and mathematical practice. A problem is said to be **NP**-*complete* if it is in **NP** and all the problems in **NP** are "polynomial-time reducible" to it.

6. "Kuhnian" Rhetorical Problem Reductions

Kuhn's *Structure of Scientific Revolutions* is one of the works that supposedly killed the traditional idea of theory reduction. The fact that historical theories and their successors often are mutually incompatible, the meaning change objection, and the problem of so-called Kuhn loss all helped to discredit the strict Nagelian model of reduction as the deductive explanation of one theory by another. According to Kuhn and company, scientific change is too revolutionary, too incommensurable, to be captured in terms of deductive logical relations. However, Kuhn's own account of normal science in effect introduced a new kind of problem reduction or problem equivalence that bears developing, because if Kuhn is on the right track, this sort of reductive achievement is fundamental to scientific research as well as to the way students learn.

I am referring to Kuhn's account of exemplars. Science is basically problem solving, Kuhn says, or, more precisely, puzzle solving, given the heavy constraints that normal science places upon acceptable solutions. On his account, puzzle solving proceeds by directly modeling new problems-plus-solutions upon the exemplary ones already available, rather than by the application of sets of methodological rules. Moreover, normal science "guarantees" that its research puzzles are solvable in terms of the resources it provides. This means, to employ a metaphor that Kuhn himself does not use, that the set of available exemplars of a mature normal science "spans" the problem space or domain of puzzles of that scientific specialty.[9] Research puzzles are typically challenging, nonetheless, so this level of problem solving should be distinguished from the routine problem solving concerning problems of an already mastered type. While the paradigm guarantees that problems of both degrees of difficulty can be solved by recycling the resources already available within the paradigm, in the case of challenging new puzzles, no one in the relevant historical scientific community has yet shown how to do so.

Normal scientific research amounts to a kind of analysis and synthesis. Sometimes (especially in the mature, mathematical sciences that Kuhn was talking about), this will be analysis and synthesis of a traditional sort, neatly dividing complex problems into simpler subproblems. But sometimes it will not be. Especially at the frontier of research, where no firm theory or model is available, direct modeling will tend to be less analytical in the Cartesian sense because more holistic in directly modeling complex problems in context. In these cases normal research nonetheless can provide analysis, but of a non-Cartesian variety. The first pass will imperfectly match the new problem to a problem already solved, at least empirically or by approximation. The imperfections will highlight those respects in which the current problem differs from the exemplars, and these respects become subproblems, which may be tackled by a second pass of direct modeling; and so on. Many writers have noted this pattern in casuistic reasoning and in legal reasoning in particular.[10] A similar phenomenon is often encountered in case-based reasoning in AI. This process sometimes involves a mutual fitting of the new case to the old, so that the standard interpretation of the old case evolves over time (Nickles 2000).

[9] This statement leaves untouched the problem of how the "base" exemplars themselves are found in the first place.

[10] Recall, also, Kuipers' remark quoted near the end of my §1. Cf. the Herschel quote above.

How is this mutual fitting accomplished? More dramatically, how is it that a scientific expert (or a Kuhnian paradigm) can solve a potential infinity of domain problems, on the basis of presumably finite resources? This is a centrally important problem[11] that we can consider only briefly, one that calls for an evolutionary-adaptive solution – the production of variants and the selection of those with higher fitness (Nickles 2003c). The problem is basically a generalized version of the problem of Plato's *Meno*, construed as a problem of recognition or identification.

Kuhn attempts to solve the problem by introducing an "acquired similarity relation" that enables investigators to recognize any of a potential infinity of problem variants as just that, variants on a given exemplar or combination of exemplars. Similarity comes in degrees, so problem recognition and quasi-reduction (as we might call it) are not all-or-nothing affairs. This suggests that exemplars serve as "basins of attraction," to employ a non-Kuhnian metaphor. In any case, Kuhn contends that his rhetorical account, in place of traditional rules accounts, gains in flexibility what it loses in logical precision.[12] The acquired similarity relation supposedly handles both routine problem solving and the more challenging research puzzles of normal science, as distinguished above.

In fact, this flexibility extends well beyond the usual concept of problem identification and problem reduction. For it allows for the kinds of *genealogies* of problems to which Kuhn calls attention in the Postscript to *Structure*. His most detailed example links Galileo's problem of describing balls rolling down and up inclined planes to the point pendulum problem to Huygens' problem of the physical pendulum to Daniel Bernoulli's problem of water flowing from an orifice.

Now are such genealogies cases of successive problem reductions? Well, no and yes. There are crucial formal identities or at least similarities, at bottom, but the acquired physical intuitions are just as important, on Kuhn's account. In effect, they supply the relevant transformation under which the problems are seen to be identical, or nearly so. Yet, as physical problems, the

---

[11] This problem is analogous to the problem of how the vertebrate immune system, with finite means, can respond appropriately to a potential infinity of invading antigens. There are parallels to many other problems, too, e.g., problems of categorization.

[12] In depending so heavily upon the trope of similarity rather than upon strictly logical rules, Kuhn's epistemology or cognitive theory is rhetorical rather than logical. It is noteworthy that Schaffner (1976, p. 618), in order to handle the above-mentioned objections to the Nagel model of reduction, makes the reduction relation hold between the reduced theory and a corrected version of the so-called reduced theory. So what is the relation between the original reduced theory and the corrected version? His answer is that it is one of "strong analogy." Sarkar (1992, p. 174) credits my reduction-*to* with helping to articulate Schaffner's notion of strong analogy, but I suspect that the rhetorical tropes here are ineliminable.

various problems in the genealogy are not identical, nor is one always a special case of its successor. On the old issue of whether analogy, similarity, and other modeling relations in science can be analyzed in purely mathematical-structural terms, Kuhn clearly sides with the opposition. Physical analogy is more than mathematical structural identity. Although both are important, acquired physical intuition does not reduce to mathematical intuition. If Kuhn is right, then even the most mature forms of scientific thought often retain a degree of physical concreteness unexpected from reading Piaget. On the other hand, we could be dealing with a two-stage process in which physical intuition helps to fix on a relevant analogy but then structural identities or analogies take over.

Although the issues are too large to pursue here, I am inclined to think that Kuhn's rather vague account is on the right track (or at least a more suggestive track than other approaches) in trying to understand how scientists solve problems in actual practice, and how they acquire their scientific intuitions in the first place. Accordingly, I think it important to understand modeling relationships among problems. If we think of problem *identity* (or at least problem equivalence) as *mutual reduction*, then modeling in Kuhn's narrowest sense is literal problem reduction. That is, completely routine problem solving amounts to recognizing or identifying the given problem as just another instance of an exemplar already in hand, say a damped harmonic oscillator. But more innovative work stretches the similarity relation so that water from an orifice can be modeled on Huygens' physical pendulum. Once one sees this, the new problem can be solved if the exemplar can. We might term this procedure *analogical* problem reduction. Having made the distinction, we should immediately blur it, for on Kuhn's account even completely routine problem solving frequently involves strong similarity rather than complete identity.

Since Kuhn's acquired similarity relation is paradigm relative, we are on safer ground to speak of "perceptual reduction or equivalence" or "cognitive equivalence" of problems rather than of reduction or equivalence in some timeless, absolute sense. Problem solving becomes a kind of high-level perceptual categorization task. Some will find the term 'reduction' especially inappropriate for the genealogy sort of case. However, it seems to me that there remains a sense in which one problem "cognitively" reduces to the other, that this sense often involves scientific intuition beyond the intuition related to mathematical structure, and that it is important for understanding scientific practice.

A more general reason for thinking that Kuhn's sort of cognitive approach is on the right track is that forward-looking theory of inquiry, as opposed to backward-looking epistemic justification, must take seriously the distinction between perceived equivalence and actual equivalence. As noted before, even logically equivalent problems may not be perceived as equivalent. It is precisely the task of researchers to recognize and exploit these and other types of equivalences.

7. The Semantic Interpretation of Theories

As is well known, Kuhn's work is one among several sources of inspiration for the so-called semantic interpretation of theories, certainly including Wolfgang Stegmüller's structuralist formulation.[13] For Kuhn, it is the collection or library of exemplars and the acquired similarity relation that carry the cognitive burden. In his account, theory generalizations such as $F = ma$ are merely schematic frameworks that help to link the exemplars in various ways but do not stand to the exemplars in the relation of premises to conclusions. Paradoxically stated, it is the basic *applications* of the theory rather than the theory itself that carry the weight. Or rather, the primary applications, taken together, constitute the theory itself. The semantic conception of theories, in its various forms, exploits this insight; and this surely goes a long way toward explaining why Kuhn himself, to the surprise of critics, was receptive of Stegmüller's highly formal development of the semantic approach.

There are several versions of the semantic approach. As witnessed by his *Structure in Science* (chap. 12), Theo Kuipers is associated with the formal, European approach, based originally on the work of Sneed and Stegmüller and more recently on Balzer, Moulines, and Sneed's *Architectonic for Science* (1987). At the opposite extreme, Ronald Giere (1994, 1996) attempts to capture Kuhnian insights in a manner that is so informal as to be only loosely identifiable with the formal semantics origins of the idea.[14] Meanwhile, Nancy Cartwright (1983, p. 159) prefers not to call her account of theories a version of the semantic interpretation. Indeed, it is doubtful whether Kuhn's own view of theories squarely fits the semantic interpretation, since he denies that a theory is really a deductive system at all, or can be put in correspondence with one. Insofar as this is true, we cannot apply to theories the standard tools of proof theory and perhaps not even model theory. To pose the problem in a Wittgensteinean manner: Insofar as we regard a theory as a Kuhnian family of

---

[13] I happened to be present, as a graduate student, when Joseph Sneed lectured at Princeton in the late 1960s and, if memory serves me, first attracted Thomas Kuhn's attention.

[14] In recent work Giere (1996) links his liberal version of the semantic theory to the idea of mental models in psychology. While this does probably come closer to capturing Kuhnian insights in cognitive terms, I can imagine a formalist critic complaining that Giere is guilty of "psychologism," that the semantic theory belongs to logic not to psychology.

models, aren't we faced with the old problem of family resemblance, and hence rhetorical rather than (or in addition to) logical relations?

Be that as it may, the structuralists can make precise claims about reductive and other relations. A necessary condition is that each model of the reduced theory is also a model of the reducing theory. Additional requirements are needed, since not all formal models are even relevant to the empirical subject-matter in question. The structuralists address this problem partly in terms of what they call "constraints."

In my view it is a major advantage of the semantic approach, with its emphasis on standard applications, that it can deal with problems, problem reduction, and other problem relations better than can the positivist construal of a theory as a partially interpreted logical system. It would be philosophically interesting to see these advantages spelled out more fully. So far this has not been done, as far as I know. Accordingly, I "second the motion" of Matti Sintonen (1996), who points out that most work on the semantic approach still focuses on theory structure rather than on the more erotetic aspects of inquiry. Now that the various versions of the model-theoretic interpretation are highly developed, as concerns theory structure, I join Sintonen in urging more attention to problems and problem solving.

More controversially, for reasons suggested above, it also seems to me that the more informal versions of the semantic approach come closer to saving Kuhnian insights than do the highly formal versions. My guess is that the Meno problem of recognition, applied to problems, cannot be solved in strictly formal logical terms, including formal semantical terms, but will require a nonformal, "rhetorical" concept of similarity or the like. That is ultimately what Kuhn's position comes down to, as I interpret his account of normal science. As Kuhn views it, direct modeling comports with the rhetorical tropes of similarity, analogy, and perhaps metaphor[15] and not with logical rules couched in terms of necessary and sufficient conditions or even in terms of mathematical approximations. To be sure, the structuralists abandon the positivist view of theories as a logical calculus, but some of them do continue to speak of a correspondence between the model theoretic and positivist proof theoretic conception of theories, which implies more deductive integration

---

[15] I am thinking of such moves as Newton's treating the moon as a projectile. In Nickles (1998, and 2003a) I argue that case-based reasoning and model-based reasoning (both informal and formal, as in new-generation AI) better capture Kuhn's conception of normal science and its flexibility than does rule-based reasoning. Case-based reasoning can be rule-based at one level but interestingly example-based at another. Model-based reasoning (in the cognitive psychological sense) may do a still better job of capturing Kuhn's insights (Nersessian 2003).

than Kuhn endorsed.[16] So it is not clear to me how the more formal structuralists can handle this version of the Meno problem. But of course this problem remains a severe challenge for everyone, not only structuralists!

8. Concluding Remarks

My two principal claims are (1) that problem reduction (and related problem relations), is not reducible to theory reduction (and related theory relations) and (2) that problem reduction is more important to methodology of science, provided that we understand methodology as the "control theory" of ongoing scientific research rather than as a philosophical interpretation of the current theoretical results of science. Clearly, my emphasis lies on the process rather than the product. Not everyone agrees with this emphasis. Perhaps few philosophers do.

In his thoroughgoing and insightful studies of reduction, Clifford Hooker has maintained that all reduction is imbedded in theories. According to Hooker (1979, pp. 83, 85) there are three main motives for reduction "on any account," namely "explanatory-theoretical unification and ontological unification (hence ontological economy)... [and] achievement of an intelligible or plausible world view." A few pages later he adds, "Reduction involves explanatory and ontological unification and this is not guaranteed by derivation" (p. 85).

Hooker (1981, p. 44) does emphasize the diversity of reductions:

> 'Reduction', like so many other concepts, fragments into a logical diversity under analysis. This diversity can only be expected to further diversify as more historical examples are studied in detail.

Accordingly, it is rather surprising that in the next paragraph (and in the series of articles as a whole), he defends a *general* theory of reduction.

> By way of specifying the theory domain, then, for reduction theory I shall restrict it to inter-theoretical reductions. I assume all other cases embeddable, and requiring embedding, in these. Specifically, I shall ignore all formal cases (if there are any), subsume commonsense cases under scientific theory cases and construe all so-called philosophical reductions as species of inter-theoretical reduction. (p. 44)

---

[16] The difference is one of degree, I suppose, since Kuhn left intact deductive pieces of a theory and its applications. At the most liberal end of the model-theoretic spectrum, Giere totally gives up the need for elaborate theory structure and even law-like claims. His is a semantic interpretation of theories that does not require either theories or laws! Insofar as such a view is defensible, it overcomes the objection that the semantic view of theories is, by its very nature, *theory* centered. This would be another major advantage of the semantic approach over the old, positivist proof-theory approach. The semantic view becomes a more general theory of science rather than specifically a view about theories.

To be sure, the very term 'reduction' implies a process of eliminating or compacting so that one ends up with less, in some sense, than that with which one started (or rather, as Theo Kuipers and I agree, one can do just as much or more with less). If the more and the less must always be explicit theoretical or ontological claims, then Hooker is surely correct. But why must they? Why, a priori, is theory reduction the only kind of genuine reduction? Why not include problem reduction, whether or not theoretical-ontological issues or meaning-change issues, are immediately at stake?

A critic may reply that my dispute is merely semantic, that Hooker and others simply want to provide a separate treatment for strictly reductive than for merely "derivational" relations. My response is that their account of reduction (and of science in general) is too theory-centered, even too philosophical – and also too historical in sometimes tying itself in knots over the slightest shift in meaning. Streamlining of research processes is surely at least as important as streamlining the current theoretical conclusions of that research. On a nuts-and-bolts, pragmatic problem-solving account of science, problem-solving is where the primary action usually is for scientists working at the frontier of research. Theoretical interpretation (or reinterpretation) often comes later, to an interesting degree. To overstate my point, for emphasis: Generally speaking, methodology of science should concern itself with heuristic, problem-solving issues, with ongoing scientific practice, as much as with anything else, and for this reason it should not be divorced from problem solving in mathematics and computer science and wedded only to grand philosophical worldviews. As I remarked above, today most of the methodological work on problem solving, that is, most of the work on methodology that is potentially useful to working scientists, has been taken over by computer scientists working in such fields as formal learning theory and evolutionary computation. But there is no reason why philosophers should not be deeply engaged in such work, as, in fact, Hooker and others are in some of their other projects.

Another way to bring out the difference is this. The older discussions of reduction of the Nagel type were, quite naturally, concerned primarily with theory of justification and ontological unification – justifying the claim that light is just electromagnetic radiation, that gases are just swarms of molecules, and so on. That is all fine. However, I emphasize the heuristic use of reductive relationships for purposes of solving new problems, that is, for scientific discovery or innovation. Again, I have overstated this difference, since much of what I have said about derivational problem reduction can also be relevant to justification.

In one respect my informal-historical approach to reduction issues, from a pragmatic problem-solving perspective, and the formal semantic approach of

Kuipers *et al.,* are in much the same boat in that we have both been subjected to the same criticism – that formal or derivational relations are not enough. In this discussion I have given reasons why this objection does not particularly bother me. Perhaps it should bother Kuipers and company somewhat more, insofar as they remain centrally focused on the reduction of empirical theories, but there is no reason why they cannot extend their approach more aggressively into an erotetic account of science as a question-answering or problem-solving activity.

University of Nevada
Department of Philosophy (102)
Reno, NV 89557-0056
U.S.A.

REFERENCES

Arkani-Hamed, N., S. Dimopoulos and G. Dvali (2000). The Universe's Unseen Dimensions. *Scientific American*, August, pp. 62-69.

Balzer, W., C. Ulises Moulines and J. Sneed (1987). *An Architectonic for Science*. Dordrecht: Kluwer.

Campbell, D. (1974). Evolutionary Epistemology. In: P. A. Schilpp (ed.), *The Philosophy of Karl R. Popper*, pp. 413-463. LaSalle, Ill.: Open Court.

Carnap, R. (1936-7). Testability and Meaning. *Philosophy of Science* **3**, 420-468, and **4**, 1-40.

Cartwright, N. (1983). *How the Laws of Physics Lie*. New York: Oxford University Press.

Cohen, R. S., C. A. Hooker, A. C. Michalos and J. W. Van Evra, eds. (1976). *PSA 1974*. Dordrecht: Reidel.

Dupré, J. (1993). *The Disunity of Science*. Cambridge, MA: Harvard University Press.

Galison, P. and D. Stump, eds. (1996). *The Disunity of Science*. Palo Alto, CA: Stanford University Press.

George, F. H. (1980). *Problem Solving*. London: Duckworth.

Giere, R. (1994). The Cognitive Structure of Scientific Theories. *Philosophy of Science* **61**, 276-96.

Giere, R. (1996). The Scientist as Adult (comment on Alison Gopnik). *Philosophy of Science* **63**, 538-541.

Hempel, C. G. (1942). The Function of General Laws in History. *The Journal of Philosophy* **39**. Reprinted, with slight modifications, in: Hempel (1965b), pp. 231-243.

Hempel, C. G. (1965a). Empiricist Criteria of Cognitive Significance: Problems and Changes (a conflation of two articles published in 1950 and 1951). In: Hempel (1965b), pp. 101-119.

Hempel, C. G. (1965b). *Aspects of Scientific Explanation and Other Essays*. New York: Free Press.

Hempel, C. G. and P. Oppenheim (1948). Studies in the Logic of Explanation. *Philosophy of Science* **15**, 135-175. Reprinted with "Postscript 1964" in: Hempel (1965b), pp. 245-295.

Herschel, J. (1830). *A Preliminary Discourse on the Study of Natural Philosophy*. Reprinted by the University of Chicago Press, 1987.

Hooker, C. (1979). Critical Notice of R. M. Yoshida: Reduction in the Physical Sciences. *Dialogue* **18**, 81-99.

Hooker, C. (1981). Towards a General Theory of Reduction. *Dialogue* **20**, 38-59, 201-236, and 496-529.

Hull, D. (1972). Reduction in Genetics – Biology or Philosophy? *Philosophy of Science* **39**, 491-499.

Jammer, M. (1966). *The Conceptual Development of Quantum Mechanics*. New York: McGraw-Hill.

Kleiner, S. (1993). *The Logic of Discovery*. Dordrecht: Kluwer.

Kuipers, T. A. F. (SiS/2001). *Structures in Science.* Dordrecht: Kluwer Academic Publishers.

Lakatos, I. (1976). *Proofs and Refutations*. Cambridge: Cambridge University Press. Originally published as a series of articles in *The British Journal for the Philosophy of Science*, **14** (1963-64).

Laudan, L. (1977). *Progress and Its Problems*. Berkeley: University of California Press.

Lawler, E. L., J.K. Lenstra, A.H.G. Rinnooy Kan and D.B. Shmoys, eds. (1985). *The Traveling Salesman Problem*. New York: Wiley.

Mach, E. (1976). *Knowledge and Error*. Dordrecht: Reidel. English translation of *Erkenntnis und Irrtum*, 1905.

Meyer, M, ed. (1988). *Questions and Questioning*. Berlin: de Gruyter.

Nagel, E. (1949). The Meaning of Reduction in the Natural Sciences. In: R. Stauffer (ed.), *Science and Civilization*, pp. 99-135. Madison: University of Wisconsin Press.

Nagel, E. (1961). *The Structure of Science*. New York: Harcourt, Brace.

Nersessian, N. (2003). Kuhn, Conceptual Change, and Cognitive Science. In: Nickles (2003b), pp. 178-211.

Newell, A. and H. Simon. (1972). *Human Problem Solving*. Englewood Cliffs, N.J.: Prentice-Hall.

Nickles, T. (1973). Two Concepts of Inter-Theoretic Reduction. *Journal of Philosophy* **70**, 181-201.

Nickles, T. (1976). Theory Generalization, Problem Reduction and the Unity of Science. In: R. S. Cohen *et al.* (1976), pp. 33-75.

Nickles, T. (1978). Scientific Problems and Constraints. *PSA 1978*, vol. 1, pp. 134-148.

Nickles, T. (1988). Questioning and Problems in Philosophy of Science: Problem-Solving Versus Directly Truth-Seeking Epistemologies. In: Meyer (1988), pp. 43-67.

Nickles, T. (1998). Kuhn, Historical Philosophy of Science, and Case-Based Reasoning. *Configurations* **6** (special issue on Thomas Kuhn), 51-85.

Nickles, T. (2000). Kuhnian Puzzle Solving and Schema Theory. *Philosophy of Science* **67** (special PSA 1998 conference issue), S242-S255.

Nickles, T. (2003a). Normal Science: From Logic to Case-Based and Model-Based Reasoning. In: Nickles (2003b), pp. 142-177.

Nickles, T. (2003b). *Thomas Kuhn. Contemporary Philosophers in Focus series*. Cambridge: Cambridge University Press.

Nickles, T. (2003c). Evolutionary Models of Innovation and the Meno Problem. In: L. Shavanina (ed.), *The International Handbook of Innovation*, pp. 54-78. Amsterdam: Elsevier Science.

Pickering, A. (1984). *Constructing Quarks*. Chicago: University of Chicago Press.

Post, H. (1971). Correspondence, Invariance, and Heuristics. *Studies in History and Philosophy of Science* **2**, 213-255.

Putnam, H. and P. Oppenheim (1958). The Unity of Science as a Working Hypothesis. In: H. Feigl, M. Scriven and G. Maxwell (eds.), *Minnesota Studies in the Philosophy of Science*, vol. 2, pp. 3-36. Minneapolis: University of Minnesota Press.

Quine, W. V. O. (1951). Two Dogmas of Empiricism. *Philosophical Review* **60**, 20-43. Reprinted with changes in: *From a Logical Point of View* (Cambridge, Mass.: Harvard University Press, 1953), pp. 20-46.

Sarkar, S. (1992). Models of Reduction and Categories of Reductionism. *Synthese* **91**, 167-194.

Schaffner, K. (1974). The Peripherality of Reductionism in the Development of Molecular Biology. *Journal for the History of Biology* **7**, 111-29.

Schaffner, K. (1976). Reductionism in Biology: Prospects and Problems. In: R. S. Cohen, *et al.* (1976), pp. 613-632.

Simon, H. (1981). The Architecture of Complexity. In: *The Sciences of the Artificial*, 3rd ed., pp.183-216. Cambridge, MA: The MIT Press.

Sintonen, M. (1996). Structuralism and the Interrogative Model of Inquiry. In: W. Balzer and C. U. Moulines (eds.), *Structuralist Theory of Science*, pp. 45-74. Berlin: de Gruyter.

Sklar, L. (1976). Thermodynamics, Statistical Mechanics and the Complexity of Reductions. In: R. S. Cohen *et al.* (1976), pp. 15-32.

Wimsatt, W. (1976). Reductive Explanation: A Functional Account. In: R. S. Cohen *et al.* (1976), pp. 671-710.

Wimsatt, W. (1980). Reductionistic Research Strategies and their Biases in the Units of Selection Controversy. In T. Nickles (ed.), *Scientific Discovery: Case Studies*, pp. 213-259. Dordrecht: Reidel.

Wiśniewski, A. (1995). *The Posing of Questions: Logical Foundations of Erotetic Inferences. Synthese Library*, vol. 252. Dordrecht: Kluwer.

Theo A. F. Kuipers

PROBLEM REDUCTION AND ITS RELEVANCE
REPLY TO THOMAS NICKLES

The perspective of problem reduction, as presented by Thomas Nickles, is in many respects a stimulating way to look at my own work. I will mention a number of points, and elaborate two of them somewhat.

(1) I like Nickles' talk, in Section 1, about reduction on three levels, that of language, theory and method, and I agree about the foundering of far-reaching reduction programs regarding them. However, in SiS I try to show that modest reduction claims regarding concepts, laws and explanatory methods are very defensible (SiS, Ch. 5, 3, 4, respectively).

(2) With Kim (2000, p. 89, also quoted in SiS, p. 134), I regret very much that it is almost politically incorrect to use the term 'reduction' for evident successful cases. Perhaps Nickles' and Sintonen's convincing extension – if not focussing – the idea of reduction to the category of problems may help to re-establish the use of the term 'reduction' when relevant.

(3) Perhaps to Nickles' surprise, there is a direct link between problem reduction and truth approximation, *see below*.

(4) As Nickles points out at the end of Section 2, problem reductions may be based on logical equivalencies, which are epistemologically asymmetric. A special case is the transformation of problem representations. Below I indicate that the general phenomenon also occurs in philosophy and that the special case relates to aesthetic considerations.

(5) Nickles distinguishes in Section 3 the search space "as it really is (in something like Popper's Third World)" and "as the investigators conceive it to be." Although I feel affinity with Popper's Third World, I doubt whether it leaves room for a search space as it really is. With Newell and Simon, I think that the search space needs to be defined before we can sharply formulate a problem, let alone search for a solution. As ICR illustrates, even to explicate the idea of truth approximation we need to assume a domain and a vocabulary, for otherwise "the truth" is not defined.

In: R. Festa, A. Aliseda and J. Peijnenburg (eds.), *Cognitive Structures in Scientific Inquiry* (*Poznań Studies in the Philosophy of the Sciences and the Humanities,* vol. 84), pp. 134-137. Amsterdam/New York, NY: Rodopi, 2005.

(6) As will be clear from Ch. 12 of SiS, I am of course very happy with Nickles appreciation of the semantic (or structuralist) interpretation of theories, as appears from his Section 6. However, in view of Nickles' critical discussion and my liberal usage of structuralist terminology in the rest of SiS and in ICR, it might be useful to formulate different degrees of structuralism. For it is one thing to view theorizing as primarily an attempt to represent (natural or artificial) systems in terms of (mathematical) structures, called models of the theory; it is another to suppose that the relevant classes of models can be set-theoretically axiomatized in the sense explicated by Suppes. It would be still more specific to think of a logical axiomatization in some formalized language, with a first order language, without a serious distinction between observational and theoretical terms, as the logical-empiricist ideal.

Problem Reduction and Truth Approximation

I cannot resist the temptation to quote the last paragraph of Section 1 of Nickles' paper in full:

> As Sintonen notes, by focusing on problem reduction in the sciences, we also do more justice to scientific practice. One can argue that methodology of science itself is more properly pitched at this pragmatic level, on control-theoretic grounds. (Philosophers have often focused on issues such as the truth of theories that seem to have little controlling influence in actual scientific research, although they may be of philosophical interest.) But then it is especially incumbent upon those who pursue a problem-solving account of scientific inquiry to inquire into problems themselves and their relations to one another.

This paragraph can be seen as a perfect characterization of the two main messages of ICR. First, in line with Laudan (1977), both realists and instrumentalists try to improve the problem-solving merits of theories; that is, they use the instrumentalist rather than the falsificationist method. Second, however, *pace* Laudan, whether they like it or not, this method is functional for truth approximation. Of course, there are various ways of improving the problem-solving merits of theories, including various kinds of problem reduction. Hence, I read Nickles' contribution as a systematic and historical underpinning of (partial or complete) problem solving by problem reduction. In particular, his Section 4 documents this. However, Nickles is inclined to classify theory reduction, including law reduction, as something quite distinct from problem reduction, whereas in my view the successful reduction of a law by a theory (as analyzed in SiS, Section 3.3) is just *one* typical kind of problem reduction, for the theory, together with the relevant auxiliary hypotheses, explains the (domain of validity of the) law, that is, the law is a general success of the theory of a very special kind, even if the theory is false. Hence, in contrast to Nickles' suggestion at the end of Section 4, stressing the importance of law reduction is not against a Laudan-like problem solving view and in favor of a confirmationist view or, I would like to add, a falsificationist view.

Problem Reduction Guided by Epistemological and Aesthetic Considerations

Problem reductions may be based on logical equivalencies, which are epistemologically asymmetric. This also occurs in philosophical matters. For example, in dealing with non-zero probabilities for universal generalizations, the epistemologically attractive but methodologically impracticable systems developed by Hintikka and Niiniluoto turned out to be equivalent to *prima facie* epistemologically dubious, but very practical systems (Kuipers 1978). Another example is the logical equivalence of three different formulations of or foundations for the same definition of truthlikeness: solely in terms of consequences of theories, solely in terms of models of theories, or partially in terms of consequences and partially in terms of models. As I argue in ICR (Ch. 8), the last one, the dual foundation, nicely fits the refined HD method for the evaluation of theories.

A special case of problem reduction by equivalencies is indeed that "a transformed representation can render a seemingly difficult problem trivial," as Nickles illustrates with a well known chess board problem at the end of Section 2. There is even a straightforward link with beauty in science here, for such problem reductions are frequently mentioned as typically beautiful. As a matter of fact, in science we come across two kinds of beauty. We speak of the beauty of methods of proof and of problem solving on the one hand, and of results such as propositions, laws, theories, and truths on the other. The so-called diagonal proof of the non-denumerability of real numbers is an example of a method that strikes almost everyone for its simplicity and inventiveness. It typically reduces the problem by representing it in a very special way. In Kuipers (1991) I have collected ten examples of beautiful problem-solving methods for quite mundane problems such as the quest for the resulting concentration after mixing red and white wine twice in a certain way (see my reply to Van Bendegem). However, as suggested, solutions themselves may also be considered as beautiful, like new results in general. Moreover, regarding results themselves, we might distinguish between new results that are found beautiful because they are surprising, perhaps because they open new perspectives, and results that are found beautiful because they fit into the current "aesthetic canon," a term introduced by McAllister (1996). In Kuipers (2002) I address the last type of aesthetic considerations (Thagard and Miller comment on that paper, in this volume and the companion volume, respectively). It would be interesting to investigate the extent to which the "aesthetic canon" functions as a means of problem reduction, by stimulating the search for certain kinds of solutions of problems.

REFERENCES

Kim, J. (2000). *Mind in a Physical World.* Cambridge, MA: The MIT Press.

Kuipers, T. (1978). *Studies in Inductive Probability and Rational Expectation.* Dordrecht: Reidel.

Kuipers, T. (1991). Dat Vind Ik Nou Mooi. In: R. Segers (ed.), *Visies op Cultuur en Literatuur.* Opstellen naar Aanleiding van het Werk van J.J.A. Mooij, pp. 69-75. Amsterdam/Atlanta: Rodopi.

Kuipers, T. (2002). Beauty, a Road to The Truth. *Synthese* **131** (3), 291-328.

Laudan, L. (1977). *Progress and Its Problems.* Berkeley: University of California Press.

McAllister, J. (1996). *Beauty and Revolution in Science.* Ithaca, NY: Cornell University Press.

Maarten Franssen

DESIGN RESEARCH PROGRAMS

ABSTRACT. In this paper Kuipers' set-theoretic approach to scientific research programs as applied to design research programs is reviewed. The main criticism is that this approach, through its conception of properties as "atomic," cannot do justice to the fact that most properties that matter in design problems come in degrees. Thus the approach offers no help with a main difficulty in design problems: that of evaluating different design concepts or prototypes when multiple features or properties, each of which giving rise to a comparative ordering of the concepts or prototypes, have to be taken into account. This problem is argued to be isomorphic to the well-known problem of social choice and therefore, in view of Arrow's theorem, a "deep" problem.

1. Introduction

In his book *Structures in Science* (2001), Kuipers develops a set-theoretic approach to scientific research programs, which has as an interesting feature the incorporation of design in the framework of philosophy of science. The design of a new product is looked upon as the matching of a set of actual properties possessed by the product and a set of desired properties. The characteristic aspect of design is that there is such a set of desired properties. In the account propagated by Kuipers this set plays the a role comparable to the role of the "true theory" in descriptive and explanatory research.

Kuipers himself notices some obvious differences between, on the one hand, design research programs and, on the other, descriptive and explanatory research programs. In the latter, the true theory is not known. To establish the extent to which a given theory matches the truth, the complicated apparatus of inductive logic and a theory of confirmation are necessary, and it cannot be said that this apparatus is at present sufficiently developed to guarantee that its use will lead to a perfect match. The set of desired properties, on the other hand, *is* known: the delineation of that set is our own work. However, and that is a second major difference, that set is not fixed once and for all. Desired properties articulated in one phase of the design process may be traded for different properties in a later phase, for instance when it has become clear that some properties are impossible to realize or mutually incompatible, or when

In: R. Festa, A. Aliseda and J. Peijnenburg (eds.), *Cognitive Structures in Scientific Inquiry* (*Poznań Studies in the Philosophy of the Sciences and the Humanities,* vol. 84), pp. 139-153. Amsterdam/New York, NY: Rodopi, 2005.

our assessment of the situation in which the end-product is going to function has changed. The truth, on the other hand, is generally assumed to be stable.

These differences, however, are not in themselves objections against the view that in both design research programs and descriptive and explanatory research programs the prime objective is to increase the match between a working set and a target set. In this contribution I will take the view that design research programs can be so described as a starting point and restrict myself to a discussion of some of the difficulties – several smaller ones and one major difficulty – that lie on the way of this approach. Whether these are also difficulties for descriptive and explanatory programs, and if so, in which form, and what this means for the extent of the analogy between the two types of research programs, will not be of my concern, apart from some side remarks.

2. Design Research Programs Aimed at the Resolution of Problem States

Kuipers views a design program as directed at the creation of a product or process with a particular profile of properties. The creation of this profile is the goal of the design program. Such a desired profile can be represented by a set W – the set of wished-for properties – in the space RP of all relevant properties. At each stage the outcome of the design program so far will be a prototype[1] x with a certain profile of properties, that can be represented by the set $O(x)$ in RP. The mismatch of $O(x)$ and W indicates the extent to which the goal is not yet reached: x still has a number of undesired properties, corresponding to the set $O(x) - W$, and there are a number of desired properties that x still lacks, corresponding to the set $W - O(x)$. The union of these two sets represents the total mismatch between the current prototype x and the goal of the design program, and is termed by Kuipers the problem state of the design program. Such a problem state can be (partially) resolved in two ways: either by replacing the prototype x by a new prototype x' or by changing the goal of the design program from W into W′. In both cases one clearly wants the change to be in the direction of a better match between prototype and design goal. One case in which this can unproblematically be taken to be the case is when the new prototype has less undesired properties and more desired properties than the old prototype, or, in the case of a change of goal, when the undesired properties of the prototype are fewer with respect to the new goal, and its desired properties more, than with respect to the old goal. In set-theoretic language, we can say that a change from prototype x to prototype x'

[1] The term 'prototype' is used here to indicate all of the different forms an outcome at an intermediate state can take: a design proposal – or a concept, as such a proposal is called by design methodologist –, a blueprint or an actually realized prototype.

constitutes an improvement if $(O(x')-W)\cup(W-O(x')) \subset (O(x)-W)\cup(W-O(x))$. Similarly, we can say that a change of the goal from W to W′ is an improvement if $(O(x)-W')\cup(W'-O(x)) \subset (O(x)-W)\cup(W-O(x))$. An example of such a sequence where prototype x' is an improvement to prototype x is drawn in the right figure below.

RP

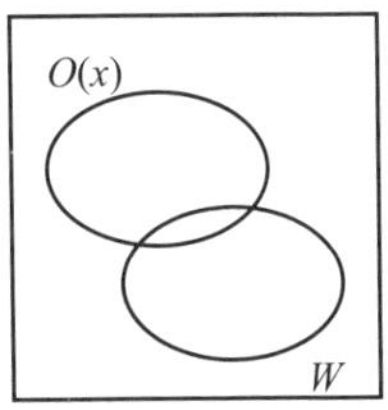

RP

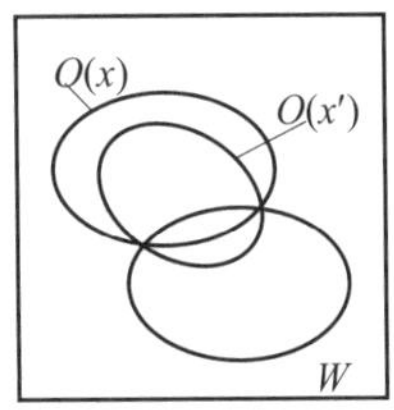

Note that the 'if' in the above definitions of improvement or progress cannot be replaced by 'if and only if'. It may very well be that a change of x to x' is considered an improvement even if the above relation does not hold. At least, there is no reason to exclude this *a priori*. In the case where we would be able to "count" properties, such that it would be meaningful to speak in a quantitative sense of the "volume" or the "content" $|(O(x)-W)\cup(W-O(x))|$ of the set $(O(x)-W)\cup(W-O(x))$, we could define all transitions of x to x' are improvements for which $|(O(x')-W)\cup(W-O(x'))| < |(O(x)-W)\cup(W-O(x))|$.[2] But again there seems to be no justification for restricting progressive changes to those changes for which the above inequality holds.

The view of design research programs just sketched is presented by Kuipers more or less as an extension or an application of the analysis of the epistemic core of the enterprise of science, nomological research programs. These are directed at characterizing, among the totality of conceptual possibilities, those that are nomically possible, i.e. possible given the laws of nature. The set of all nomic possibilities T – the "truth" – is the goal of a nomological research program, whereas a particular characterization – a theory X – generally characterizes the truth only partially. The connection with design research programs becomes clearer if the business of nomological research programs is also formulated in terms of properties – rather to be called features, to avoid confusion – of conceptual possibilities and, correspondingly, of the theories characterizing them. Set-theoretically, such a feature is represented by the set of all conceptual possibilities possessing it. The truth

[2] Note that this criterion is weaker than the qualitative one $(O(x')-W)\cup(W-O(x'))\subset (O(x)-W)\cup(W-O(x))$.

and the theories reaching for it are then represented by sets of features, or sets of sets of conceptual possibilities. The criterion for progress, comparable to the one discussed above for design research programs, is then the following. A theory X' constitutes progress with respect to theory X if (1) more features shared by all conceptual possibilities in X' are also true of all conceptual possibilities in T than features shared by all conceptual possibilities in X, and/or (2) less features shared by all conceptual possibilities in X' are also true of all conceptual possibilities not in T than features shared by all conceptual possibilities in X. This way of stating the comparison between nomological research programs and design research programs suggests that the set-theoretical approach fits the latter perhaps better than the former. Whether it makes sense to speak of a conceptual space comprising all the properties of specific entities depends both on the way these entities and their properties are conceived and individuated. The individuation of prototypes and (ideal) products-to-be-designed does not seem problematic. This cannot be said of the conceptual and nomic possibilities of nomological research programs, however. It is not clear whether these are to be read as possible worlds, as situations, as events, or as again some things different. On one occasion Kuipers states that the set of features of a theory X can also be viewed as the set of all statements entailed by X. This suggests a model-theoretic interpretation of 'conceptual possibilities'. On this interpretation, however, one would expect the truth to be a singleton, which seems incompatible with the definition of progress employed.

This leaves the conception and individuation of properties, the discussion of which I will restrict to design research programs. Kuipers realizes that some restrictions are in order. In particular it seems sensible not to allow logical operations on properties. If p and q are properties, then *not-p*, *p-and-q* and *p-or-q* are not – at least not *ipso facto* – properties.[3] If they were, grave difficulties would arise in the case where p is a desired property and q an undesired one. This requires a careful consideration of the individuation of properties, something which I will not undertake here, however.

A more important restriction is that not all properties relevant to design are of the same quality. In design it is customary to distinguish between functional and structural properties, and also to distinguish between a product's material make-up and the more abstract properties related to its use. To give a simple example illustrating the differences: the ability to use it to travel through the air would be a functional property of a particular product, to achieve this by making use of the lift produced by moving horizontal surfaces attached to the travel compartment with high speed through the air, and to produce this speed

[3] This emphasizes the non-model-theoretic orientation of the current set-theoretic approach to design research programs.

with a combustion engine operating a propeller would be its structural properties, and the precise physical make-up of the whole contraption, the actual materials, construction techniques and dimensions used, would be its material properties. Apart from this subdivision in kinds of properties, the introduction into the design program of the potential applications op the product-to-be-designed is important. These potential applications motivate the design goal and possible adaptations of the design goal. Without them, a reformulation of the design goal *W*(*F*) in answer to a problem state would be completely *ad hoc* and the "progress" achieved in this way would be empty.

These considerations lead to the following, more sophisticated overview of the problem state for a particular design research program. Here *M*(*R*) is the set of material properties of potential realizations of products or processes; *S* the set of structural properties of such products; *F* the set of their functional properties, and *C*(*A*) the set of characteristics of all potential applications of a designed product or process. Within *C*(*A*), *C*(*y*) is the set of characteristics, or the profile, of an application of product-to-be-designed *y*. Within *F*, *OF*(*x*) is the functional profile of prototype *x* and *WF*(*y*) the profile of desired or wished-for properties of the design goal *y*. Similarly within *S*, *OS*(*x*) is the structural profile of *x* and *AS*(*y*) a structural profile of *y* that is appropriate for *WF*(*y*). Within *M*(*R*), M(*x*) is the material profile of *x* and *AM*(*y*) a material profile that is adequate for *AS*(*y*). The black arrows are causal arrows: a material profile in *M*(*R*) causally determines a corresponding structural profile in *S*, which again causally determines a functional profile in *F*. The white arrow is not a causal arrow: a particular application one has in mind for a product or process does determine the functional profile that will be the goal of the design process, but responsible for this process of determination are the mental states of the designer, not the causal efficacy of the application.

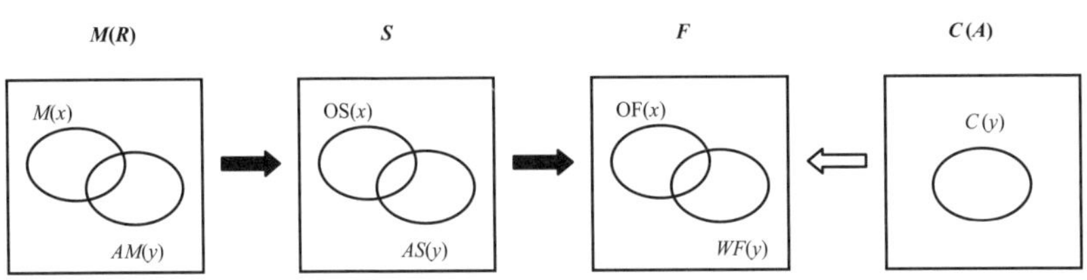

3. Structural and Functional Properties in Design

It is Kuipers' claim that the criterion of progressive design steps, i.e., decreasing the mismatch between prototype and goal, operates only at the level of functional properties. If the profile WF is interpreted in such a way that all properties that are not in WF are automatically undesired properties, then the mismatch between $OF(x)$ and WF can be defined as the set $(O(x)-WF) \cup (WF-O(x))$, or equivalently, the set $(O(x)\cup WF)-(O(x)\cap WF)$. It is also possible, however, to distinguish, apart from the profile of desired functional properties WF, both a profile of non-desired functional properties NF and a profile of neutral or indifferent functional properties IF, where $F=WF \cup NF \cup IF$. In that case a problem state is defined as the set $(O(x)\cap NF)\cup(WF-O(x))$. This allows for more flexibility when articulating the design specifications, in the light of changed expectations about the product's application.

If the assessment of design results takes place only at the level of functional properties, this makes one wonder what the role of the sets $AM(y)$ and $AS(y)$ is in the above figure. The aim of the design research program is to create something – realize a material object or process – that has as its functional properties all those properties and only those properties that are elements of $WF(y)$. Although the functional properties in WF are "known" beforehand, the structural and material properties that realize these functional properties are not. The problem of designing is, *prima facie*, exactly to find at least one set of material properties that realizes the "known" functional properties. The sets $AM(y)$ and $AS(y)$ are thus precisely what is *not* known. If they were known, the design problem would have been solved already.

Actually, more often than not a material realization that, on the functional level, comes a large way in the direction of $WF(y)$, or that may even coincide with it, is already in existence, marketed by a competing firm. Then the design task is rather to find a different material realization that answers to the functional requirements, or answers more fully to them, because the existing realization will generally be protected by patents. This existing product may serve as the input for a heuristic rule that recommends "vary on an already existing and functioning product or prototype." Such a heuristic rule is suggested by the set-theoretical representation, as given above, of a typical problem state for a design research program, according to which structurally similar prototypes promise to be functionally similar as well. Such trial-and-error methods are common in the design of materials, such as the design of drugs, an area from which Kuipers has taken many of his examples. As the patents often concern the method of synthesization rather than the product itself, it is enough to discover a new route to the known realization. However, in product design, innovation is often directed at one particular component of the product, for which an improved design or an alternative design is sought. For these cases heuristic rules of the form "vary on an existing product or prototype" are of little help. The required step away from the existing product is too large. Heuristic rules actually used by design teams may either try to exploit the expertise of the firm by testing solutions that have already been developed for their usefulness in solving quite different design problems, or may be directed at other desired properties than the performance of the function the component was designed for, for instance "try to reduce the number of components."

As a general point, it is important to recognize that the distinction between functional and structural properties, or between service properties and technical properties, as Saviotti seems originally to have construed the distinction, is not univocal. The activities within a design research program may be far removed from the level of applications, although the (potential) existence of such applications is never questioned. The intricacies involved may be illustrated by two research programs directed at fundamental innovation, namely the construction of the turbojet engine (see Constant 1980) and of the first masers and lasers (see Bromberg 1991).

To start with the former case: a potential market for turbojet engines for airplanes was recognized by the few pioneers who developed the engine. Frank Whittle, for instance, realized in the 1930s that the only obstacle to a large increase of the speed of airplanes were the propeller engines, as all kinds of aerodynamic difficulties would forbid speeds not all that much larger than the speeds of the best planes of his day through propeller propagation. The functional goals of the subsequent design research program aimed at the development of a turbojet engine were rather trivially dictated by potential applications: obtaining a large enough thrust, given the weight of the motor and the test plane. Design research proved to be mainly occupied with controlling structural aspects such as turbine-blade fracture and steady combustion of the fuel used. These aspects were of course in an obvious way linked to the desired functional property of trustworthiness of operation. However, to learn about the behavior of the materials used, the necessary mutual adjustment of the components, the pros and cons of various design choices, is just as much an objective of such a research program as is the creation of a prototype. Nonetheless, a few properties that are uncontroversially functional or "service" may be seen as having driven the design research.

On the other hand, structural properties seemed to dominate in the design research programs aimed at the construction of the first masers and lasers. Part of the background of these programs was the idea that there was bound to be a market for sources of very narrow-bandwidth microwave and optical radiation and for very low-noise amplifiers of such radiation, especially for military application. Mixed with this, however, was a purely scientific motive of establishing the existence of the mechanism of stimulated emission. The design programs themselves, during the 1950s, were aimed at getting working prototypes that used the physical principle of stimulated emission to generate radiation. The functional properties did not state more than just that. One was still too far removed from applications to be able to include, for instance, as a specification whether the radiation should be pulsed or continuous, such that both types emerged from the laboratories, depending on the material used. Actually, with masers it turned out that they were indeed greatly superior to existing sources of microwave radiation and existing amplifiers regarding bandwidth and noise level, but they were never used since it proved impossible to have them operate at temperatures other than a few Kelvin. This was not seen, however, as a failure of the maser design research program. The experience gained proved vital for the development of the laser, which proved to be feasible at room temperature. But the first laser design research programs were exclusively directed at getting a radiation signal, which basically meant finding a material with the right pattern of energy levels and solving the problem of generating standing waves smaller than a micron. What, therefore, was seen as the end property, or the goal set of properties, of the initial research program was only in the meagerest of senses a functional property and would be considered a set of structural properties in the construction of lasers for specific applications, such as either amplifiers or sources of small-bandwidth radiation.

Thus the distinction between structural and functional properties and the form of the assessment or evaluation process thus differ greatly, depending on the type of design. In what may be termed as fundamental design or innovation, very often only a few properties matter. The majority of functional properties – concerning range, efficiency, service life, operating conditions, dimensions and the like – are considered indifferent. The design activity is directed at properties that in later research programs will be looked upon as structural. In product design, on the other hand, for a market where already many products exist, functional properties that are linked in a transparent way to the application of the product are what matters and very few properties can probably be considered as indifferent.

4. Properties with a Range of Possible Values

The way various sets can be compared to each other regarding their amount of match with a target set is also involved with a second major difficulty, namely that of quantification. Design is hardly ever a question of counting the number of properties a design proposal or a prototype possesses or lacks, and of subsequently comparing different proposals or prototypes with respect to these numbers. Different prototypes and different products are rather seen as relevantly different with respect to the score on these properties. For instance, both prototypes *A* and *B* may have property *X*, and what is relevant in choosing between them is whether *A* scores better than or worse than *B* on the *amount* of *X* possessed. *A* may be cheaper or less cheap than *B*, heavier or lighter than *B*, more or less robust than *B*, *et cetera*. It does not make much sense, therefore, to say that "cost," "weight" or "solidity" are desired properties. Nor does it make more sense to say that "cheapness," "lightness" (or "heaviness"), and "robustness" are desired properties. We are dealing with ordinal scales here: a product is never expensive or heavy as such, only relatively expensive or relatively heavy.

This comparative nature of relevant properties generally overrules the simple counting procedure. This can be nicely demonstrated by an example from the area of medical-drug design. The following is derived from Vos (1991). In the early 1960s, the British pharmaceutical corporation ICI synthesized initially pronethalol and later propranolol as the first beta-blocking cardiac drugs on the market. Both had the beta-blocking property (a desired property) and both had the property of being lipophylic and of therefore easily crossing the blood-brain barrier and causing side effects such as nausea and dizziness (an undesired property). What was relevant in the comparison of the two, however, was that their tendency to cause neurological side effects was more or less equally strong, whereas the beta-blocking property of propranolol was ten times as strong as that of pronethalol. This caused therapeutic doses of propranolol to lead to unwanted side effects to a much lower degree than therapeutic doses of pronethalol. Both the relative strength of the beta-blocking property possessed and of the lipophylic property possessed are relevant in considering that propranolol is favored over pronethalol. It are thus degrees of properties possessed that matter, not properties as such.

It might be argued that the yes/no property "without undesired side effects in therapeutically effective doses" discriminates between propranolol and pronethalol. However, this remark would miss the point. Both therapeutic effectiveness and seriousness of side effects come in degrees, so there remains a general problem. To properly evaluate design proposals or prototypes, it will therefore be necessary to establish as a minimum requirement ordinal scales for each property (desired as well as undesired) and to develop a general method for comparing ordinal and quantitative scores for properties.[4]

---

[4] There is here a further analogy with nomological research programs, where the idea of a measure of the distance of a particular theory from the truth has been discussed since Popper. The recognition that there are two conflicting intuitions involved in proposals for such a measure (see Zwart 2001) suggests that the discussion on multi-criteria choice that follows below is potentially relevant to nomological research.

Nor can the issue be circumvented by raising each value a particular property can assume to the status of a property in itself. This is not so much due to the fact that most properties assume values on a continuum – after all there is a limit to (desired) accuracy of measurement procedures – but rather to the fact that the desired functional profile of a product cannot be stated as a wish for point values of the relevant properties. At most the design requirements contain point values as end points, like "a weight less than x kg" or "a cruising velocity of at least y miles/hour." At the time of the formulation of the design requirements it is often unknown what is feasible. In fact, the design research itself often alters the situation regarding feasibility, especially where the design goal is to improve upon an existing product. This approach of conceptualizing properties at a lower level would therefore make it very hard to hold on to the view of design research programs in terms of desired and achieved profiles of properties.

The establishment of a measurement scale for a property cannot be considered such a straightforward matter as it may initially seem. Admittedly, the knowledge made use of in design often involves concepts for which an operationalization on a quantitative scale is already available. Even if that is the case, however, it is not obvious that the measurable behavior of the prototype on this scale is related in the right manner to what is desired. The desired properties of a product are often formulated on a different functional level compared to the directly measurable properties of a prototype. The relation between these levels is not always the same; some desired properties are formulated as functional-in-use properties, others as functional-in-theory properties. To stick to the illustrative case of cardiac drugs to elucidate this, desired properties of drugs were partly formulated as therapeutic – e.g. relieving the pain of angina pectoris, stabilizing the rhythm of an arhythmic heart – partly as theoretical – e.g. occupying the receptor sites in the heart tissue of the adrenergic system. It depended on the amount of insight into the working of the heart and the circulatory system to establish to what extent these functional-in-theory and functional-in-use properties are related. It is hardly a straightforward matter to establish whether a particular quantitatively-measurable property – e.g. the extent to which a certain amount of a drug slows down a heart rate that is artificially increased by the admonition of a specific amount of isoprenaline – is a good indicator of its therapeutic success in relieving a particular symptom of cardiac malfunction.

The problem how to operationalize a particular functional requirement is itself an important aspect of design programs. A nice example of this can be found in the area of early airplane design (see Vincenti 1990, ch. 3). An important design requirement about which there was confusion during a considerable period was the longitudinal stability of an airplane. This stability concerned the reaction of a plane, when flying with a certain speed at a certain angle, to sudden changes in the angle, for instance caused by gusts. If, without any interference by the pilot, a plane returned to the original position it was stable, if the deviation tended to increase it was unstable. Almost all early planes were unstable, but this in itself posed no problem, because the pilot was there to interfere. That it was required of the pilot that he continuously corrected the flight of his airplane by adjusting the controls was not yet considered as an undesirable property in those pioneering days. When both the complexity of airplanes and the duration of flights started to increase, however, stable airplanes came to be preferred. On the other hand, a too stable airplane would be very hard to maneuver. The desired amount of stability therefore also differed between types of planes (e.g. between bomber and fighter planes). It took a lot of experimenting with different airplanes and checking with the experiences of pilots while flying these planes before in 1941 the first reliable design criteria for airplane stability could be proposed. For longitudinal stability this amounted to an acceptable range of values for the gradient of the functional relation between the stick force necessary to steady a plane and the plane's speed. For a plane's stability in maneuvering it took the form of an acceptable range of values for the stick force required to realize a transversal acceleration of one *g*.

5. Design as a Multiple-Criteria Decision-Making Problem

But let us suppose that all problems of comparative and quantitative operationalization have been solved. The remaining design problem is then to establish a global ordering of design concepts or of prototypes. This is obviously the problem remaining when there are several "ultimate" functional requirements for each of which an ordinal or quantitative scale is available. But even in the (hypothetical) case where there is only a single or one overriding functional requirement, or the case where the functional requirements are of a present/absent kind, the problem may pose itself on a lower, structural level, where the issue is to decide which of several proposed solutions or possible designs is the most promising.

What makes this problem both interesting and difficult is that it is isomorphic to a problem situation studied extensively in so-called social-choice theory. Suppose a group of human individuals is to decide which of several incompatible options is to be chosen for execution. Each individual supposedly is able to order the various options with respect to their desirability from his or her own point of view. That would be minimally required (allowing some simplification) to enable this individual to choose an option if he or she could decide the issue on his or her own, since the existence of a preference order ensures that there is a most preferred option, or a number of most preferred options among which the individual is indifferent. However, it is the group's predicament that the choice must be a collective choice. In line with the point of view of rational-choice theory from which the whole situation is defined, the problem would be solved if a procedure would be available that would translate the individual preference orderings into a collective preference ordering. In 1950 it was shown by the economist Arrow that this route toward a solution is closed (see Arrow 1951). As is well known, he proved that there is no function that transforms a profile of two or more individual preference orderings on a set of three or more options into a collective preference ordering on these options if this function is to satisfy the following five conditions:

1. *Collective rationality*: The function must lead to a (complete and transitive) ordering for any profile of individual orderings serving as input.
2. *Unrestricted domain*: There are no restrictions on the individual preference orderings, as long as they are orderings.
3. *Pareto optimality*: If all individuals prefer option x to option y, x must be preferred to y in the collective ordering.
4. *Independence of irrelevant alternatives*: If a particular option is removed from the set, the collective ordering that remains when this option is crossed out must be identical to the collective ordering that is the image of the profile of individual orderings in each of which this particular option is crossed out.
5. *Nonexistence of a dictator*: There must not be one specific individual to whose individual preference ordering the collective ordering is always identical.

Each of these five conditions was considered by Arrow to be highly desirable for a process of collective decision-making. (Note that there is no suggestion that they are jointly exhaustive!) Whether that is indeed the case is a point of much discussion in social-choice theory that need not detain us here. The point at issue is that they neatly transpose to the multi-property choice situation at hand in design research programs.

First it has to be recognized that the problem situation is isomorphic. The role of the options under choice is played by the various design proposals, prototypes or end-products to be globally compared. The role of the individual actors is played – less obviously perhaps – by the different properties or, more generally, the criteria that go into determining the global performance of a design proposal, prototype or product. The input situation is that for each of the properties an ordering of the various proposals etc. is given. (Note that the quantitative measurability of a property ensures the existence of such an ordering.) The task is to translate profiles of property scores into overall rankings of proposals or prototypes. Adherence to the five conditions stated by Arrow – *mutatis mutandis* – would block the existence of a general solution to this problem as soon as the number of relevant properties is two or more and the number of proposals to be compared three or more. That it is reasonable to adhere to these five conditions can be easily seen:

1. *Global rationality*: The analogue of Arrow's first condition is obvious and, just as in social-choice theory, merely reflects the fact that our initial aim is to find a procedure that never fails to deliver an overall ordering of proposals or prototypes, taking all properties into account.
2. *Unrestricted domain*: This condition is arguably even less controversial than it might be in the case of collective choice. We cannot *a priori* rule out any distribution of properties occurring in a particular design prototype or product.
3. *Pareto optimality*: This is as unproblematic as in the social-choice case. If product x scores better than product y on every property considered relevant, x must be ranked overall better than y.
4. *Independence or irrelevant alternatives*: The analogue here is straightforward. The absence or presence of a particular proposal or prototype z should not influence our opinion whether proposal x is overall better or worse than proposal y.
5. *Nonexistence of a dictator*: This translates into the condition that no single property determines the overall ranking. Satisfaction of this condition is necessary for any overall weighing scheme to deserve that name.

Although textbooks on design methodology emphasize the central place of the evaluation of various design proposals or design concepts in order to arrive at the overall best choice, they do not testify to any awareness of the difficulties involved in developing a general solution procedure for this problem. Let me give two examples of this lack of awareness. In his *Total design* (1991), Stuart Pugh presents the following method of evaluating various design concepts. One particular design proposal is chosen as a reference and for each of the other proposals three numbers are determined: the number of criteria for which this particular proposal is considered to be better as the reference proposal, the number of criteria for which it is considered equal in quality and the number of criteria for which it is worse. Although it is left in the open exactly how a ranking of the various proposals is to be established on the basis of these scoring triplets, it follows from Arrow's theorem that, given the ways the triplets can be used to establish a rank order, violations of either condition 1 or condition 4 from the above conditions can never be ruled out. The most obvious danger is that of intransitivities in the global ranking (a violation of condition 1). The reader receives no warning of this danger, however.

Nigel Cross, on the other hand, in his *Engineering design methods* (1989), proposes to solve the evaluation problem by calculating the overall "utility" of each design proposal as a weighted average of the number of "utility" points scored by the proposal on each criterion. It is assumed that the same "utility" range (say, from 1 to 7 points) is used for all criteria. The weighing factors express the relative importance of the criteria. Establishing these factors is a major task in itself, if one wants to avoid arbitrariness. However, this method is liable to violations of condition 4: the rank ordering depends on the set of alternatives considered, such that adding or removing a proposal can change the order. Removal of the worst proposal from the set could mean a reversal of what is considered to be the best overall proposal. Again the reader is not warned to take care.

Even when occasionally an awareness of the relevance of Arrow's theorem to the evaluation of design proposals or concepts is shown, the discussion betrays a lack of understanding of the way the isomorphism between the cases of multiple-criteria analysis and social choice is to be construed (e.g. Scott and Antonsson 1999).

6. Conclusion

Summarizing, it is seen that the set-theoretic account of research programs is able to capture a number of aspects typical of design research. At the same time it must be kept in mind that the division of the total set of properties in functional, structural and material, which serves to define the level at which the match between the design proposal or prototype and the design goal is evaluated, is generally context-dependent. However, it turns out that the taking into account of the ordinal aspects of both desired and undesired properties of design proposals and prototypes, over and above their mere presence or absence, is both necessary and adds a serious complication to the design problem situation, which seems to lack a general solution. Of course this is not the end of it. Rather, a whole new line of study is opened for the topic of design research programs. Each of the conditions posed by Arrow can be looked upon more closely, to see whether they can be weakened. The condition of global rationality is perhaps too strong; we might be satisfied if a procedure just pointed out an overall best proposal. The condition of independence of irrelevant alternatives implicitly contains the supposition that the input information is purely ordinal; perhaps the situation changes if use can be made of quantitative input. An enormous literature on social choice is available to be of help – literature that includes, *inter alia*, proofs that seem to show that the above suggestions do not serve to solve the problem in a general way.

But this is not the place to investigate this problem situation in greater detail. My aim is merely to point out this amazingly little recognized predicament of design research programs. With respect to the way design research programs fit into the set-theoretical scheme, it seems to be a problem no design research program can completely avoid. It would equally emerge in descriptive or explanatory programs, however, if there would be two or more criteria that would go into evaluating the amount of fit between a descriptive or explanatory theory and the "truth." Zwart (2001) has argued that this is indeed the case. This interesting issue merits a separate discussion.

Delft University of Technology
Dept. of Philosophy
Jaffalaan 5
2628 BX Delft
The Netherlands

REFERENCES

Arrow, K.J. (1951). *Social Choice and Individual Values.* New York: John Wiley.

Bromberg, J.L. (1991). *The Laser in America, 1950-1970.* Cambridge, Mass./London: The MIT Press.

Constant, E.W. (1980). *The Origins of the Turbojet Revolution.* Baltimore/London: Johns Hopkins University Press.

Cross, N. (1989). *Engineering Design Methods.* Chichester: John Wiley.

Kuipers, T.A.F. (2001/SiS). *Structures in Science: Heuristic Patterns Based on Cognitive Structures*. Dordrecht: Kluwer Academic Publishers.

Kuipers, T.A.F., R. Vos and H. Sie (1992). Design Research Programs and the Logic of their Development. *Erkenntnis* **37**, 37-63.

Pugh, S. (1991). *Total Design: Integrated Methods for Successful Product Engineering.* Workingham: Addison-Wesley.

Scott, M.J. and E.K. Antonsson (1999). Arrow's Theorem and Engineering Design Decision Making. *Research in Engineering Design* **11**, 218-228.

Vincenti, W.G. (1990). *What Engineers Know and How They Know It: Analytical Studies from Aeronautical History.* Baltimore/London: Johns Hopkins University Press.

Vos, R. (1991). *Drugs Looking for Diseases: Innovative Drug Research and the Development of the Beta Blockers and Calcium Antagonists*. Dordrecht: Kluwer Academic Publishers.

Zwart, S.D. (2001). *Refined Verisimilitude.* Dordrecht: Kluwer Academic Publishers.

Theo A. F. Kuipers

COMPARING PROPERTIES AND PROFILES

REPLY TO MAARTEN FRANSSEN

I was happy to learn from the contribution of Maarten Franssen that at least one participant in the mega-research program *The dual nature of technical artifacts* of the Delft University of Technology has taken notice of the study of design research undertaken by Rein Vos, Hauke Sie and myself, that uses (medical) drug research as a paradigm example. In so far as his paper is an exposition of our analysis, it is very adequate. More importantly, it raises two very interesting issues, among them a very surprising analogy with Arrow's paradox. But let me start with two minor points. First, although SiS may not be entirely clear on this point, in ICR no doubt is left that "the truth" searched for in nomological research is not a "singleton," as long as the relevant set of nomic possibilities happens to contain at least two (non-isomorphic) possibilities. Second, it was very illuminating to read about the dependence of our distinction between structural and functional properties on the type of design, notably "fundamental design" (or innovation) versus "product design." Although one of our refinements already dealt with "indifferent properties," Vos, Sie and I certainly did not realize that they were more important for fundamental design than for our drug examples of product design.

The Double Comparative Nature of Properties

Franssen rightly notes that the model of intended and operational profiles characterized by subsets of relevant properties is rather naïve. However, I am happy to draw attention to the fact that Rein Vos (Vos 1991, Sections 6.2.3/4 and Appendix I; 1995, Sections 3.2/3) had already elaborated the following two refinements. First, instead of having a simple yes/no character, properties are here construed as functions with a range of more than two values, possibly even infinitely many. Second, instead of counting all relevant properties as equally important, some properties may here be more important than others,

In: R. Festa, A. Aliseda and J. Peijnenburg (eds.), *Cognitive Structures in Scientific Inquiry* (*Poznań Studies in the Philosophy of the Sciences and the Humanities,* vol. 84), pp. 154-156. Amsterdam/New York, NY: Rodopi, 2005.

without the latter being negligible. The first refinement introduces the comparison of values of properties, the second the comparison of properties.

The first refinement starts with construing value spaces, that is, the Cartesian product of the ranges of values of the distinct properties. For further details the reader should consult the references, but one non-technical point is worth stressing here. Vos documents that, at least in medical drug research, it frequently occurs that one starts by grouping specific disease profiles into global disease profiles: for example, besides the normal condition, a mild, a moderate and a serious type of heart failure are distinguished, with similar groupings of intended and operational drug profiles. The second refinement – introducing relative degrees of importance of properties – may also be seen as a sophisticated way of dealing with the distinction between relevant and irrelevant properties. A general set-theoretical approach to this problem seems to become technically very complicated. However, Vos managed the technical elaboration (in Appendix I of his 1991) of the special case of one "dominant" characteristic, which, however, does not need to be present, among a number of relevant characteristics. E.g. a disease may have a "pathognomonic sign," that is, "a feature which is so typical for a certain disease that the physician will diagnose a patient with that feature, immediately and without doubt, as suffering from that disease" (Vos 1991, p. 353).

Analogy with Arrow's Paradox

Maarten Franssen is certainly right in pointing out that there is a strong formal analogy, even isomorphy, between ordering profiles and the construction of a group preference out of the preference orderings of its members, leading to Arrow's famous paradox, according to which it is, in general, impossible to realize a set of five very plausible conditions of adequacy. The analogy starts by comparing the preference ordering of one individual with the ordering of profiles relative to their scores on the set of values of one property. As soon as all the one-property orderings of profiles are not uniform, Arrow's problem may arise. Although I have nothing like a solution to offer, I would like to relativize the problem in two respects.

First, as Vos points out in the above publications, at least in the context of medical drug research this type of problem does not seem to occur frequently for two reasons. In particular profiles of actually occurring diseases and, as a consequence, the corresponding intended drug profiles, by aiming to counteract the problematic disease characteristics, can frequently be ordered in a uniform way. Moreover, this frequency may even be enlarged by the globalization of profiles indicated above. To be sure, Vos' proposal of ordering is limited to such uniform cases and does not (implicitly) claim to handle Arrow-like problematic cases. It should be noted, however, that this relativization might not hold so easily for the operational profiles of various prototypes. But it is precisely for their comparison that a second relativization is important.

As a matter of fact, this second relativization is analogous to a relativization of the need for forcing an ordering of two theories when the one is not straightforwardly more successful (and hence, probably, more truthlike) than the other, a need that is presupposed by Zwart (cf. Note 4 in Franssen's contribution). However, in my view, the main research task in the case of divided success of theories is not to force an ordering, but to aim at a "dialectical synthesis," that is, to improve upon both. Similarly, in the case that operational (drug) profiles score dividedly relative to an accepted unique intended (drug) profile, the ultimate task is to improve upon both prototypes. Of course, for practical purposes, forcing some ordering of theories and prototypes (conceived as products) may have to be undertaken. Be this as it may, in all these four cases (nomological and design research and their application) there is a strong disanalogy with group preferences. In the latter case there is – at least we may hope so – no target ordering that is independent of the existing preferences, whereas the former cases are guided by such a target, whether it is known, as in the case of design research, or hidden, as in the case of nomological research.

REFERENCES

Vos, R. (1991). *Drugs Looking for Diseases*. Dordrecht: Kluwer Academic Publishers.

Vos, R. (1995). The Logic and Epistemology of the Concept of Drug and Disease Profile. In: T. Kuipers and A. R. Mackor (eds.), *Cognitive Patterns in Science and Common Sense*, pp. 69-86. Amsterdam/Atlanta: Rodopi.

Jean Paul Van Bendegem

PROOFS AND ARGUMENTS
THE SPECIAL CASE OF MATHEMATICS

ABSTRACT. Most philosophers still tend to believe that mathematics is basically about producing formal proofs. A consequence of this view is that some aspects of mathematical practice are entirely lost from view. My contention is that it is precisely in those aspects that similarities can be found between practices in the exact sciences and in mathematics. Hence, if we are looking for a (more) unified treatment of science and mathematics it is necessary to incorporate these elements into our view of what mathematics is about. As a helpful tool I introduce the notion of a mathematical argument as a more liberalized version of the notion of mathematical proof.

1. Introduction

In *Structures in Science*, chapter 13, entitled "'Default-Norms' in Research Ethics," Theo Kuipers defends the idea that the Merton norms for the ideal scientific community – summarized as the CUDOS norms, viz. *C* for Communism, *U* for Universalism, *D* for Disinterestedness, and *OS* for Organized Skepticism – are best seen as standards against which one can measure deviations in actual scientific practice, so-called defaults, rather than norms that are actually implemented. I feel quite sympathetic to this kind of approach as I have on another occasion[1] developed a similar view, the difference being that I made no reference to Merton's vocabulary – I rather wrote about an ideal community – and, perhaps more importantly, my subject was mathematics.

The reason I consider the latter element more important is that it opens up the possibility for a (more) unified treatment of the (exact) sciences as well as mathematics. Needless to say, the subject matters may be quite different –

[1] See Van Bendegem (1993). There is a rather nice form of continuity present here. That paper was written on the occasion of the retirement of Else Barth, also from Groningen in the same department where Theo Kuipers is at present at work. Needless to say, the author of this paper has intentionally forced the continuity.

In: R. Festa, A. Aliseda and J. Peijnenburg (eds.), *Cognitive Structures in Scientific Inquiry* (*Poznań Studies in the Philosophy of the Sciences and the Humanities,* vol. 84), pp.157-169. Amsterdam/New York, NY: Rodopi, 2005.

electrons, molecules, genes, species, etc. on the one hand, and algebraic, topological, geometrical, etc. structures on the other – but that does not preclude that in other respects, e.g., as a (professional) activity, they are sufficiently similar to be treated in a uniform way. I would assume that Theo Kuipers would welcome such an approach since in chapter 6, "Empirical Progress and Pseudoscience" of *From Instrumentalism to Constructive Realism* he considers the extension of his approach to philosophy and theology.

First I want to say a few things about the ideal picture of mathematics and what is absent from it, then I present my evidence that what is missing is essential to mathematical practice and I close with some observations of a more formal nature on the new "real" picture and how it relates to scientific practice.

2. Proof is a Proof is a Proof is a Proof

Ask the question "What it is that mathematicians do all day long?" and the answer will be: "Looking for proofs." (Depending on your philosophical view "looking for" can be made more precise by expressions such as "discovering" or "constructing".) What a proof is, is clear to all: a connected series of statements, the last one being the statement to be proved and every step in the proof to be justified either because it is an axiom or the result of the application of one of the logical rules. Surely there is no room for experiments here – what could a mathematical experiment look like? Surely there is no room for induction (no matter of what kind), unless the author of this paper is suggesting that there is a similarity between scientific induction and mathematical induction, which he definitely is not. Surely there is no room for "competing" proofs, whatever that may be supposed to be.

In addition, the idea that we have a clear idea of what proofs are and what they are about is supported by the fact that it is possible to have formal versions of the notion of proof. Let Proof(α, A) stand for the relation that expresses that α is a mathematical proof of A. Then all kinds of formal statements can be written down that reflect properties that "good" proofs have and should have, such as:

If Proof(α, A) then A

or If Proof(α, A) and Proof(β, $A \supset B$) then Proof(γ, B)
(where γ is the result of joining together α and β)

and so forth. This is precisely what is being studied in the field of provability logics.[2] End of story.

As must be clear, this is not, as far as I am concerned, the end of the story. Two considerations will, I hope, make clear why one should have doubts. The first consideration is that most, if not nearly all *interesting* mathematical proofs do not satisfy the formal standards. It is sufficient to take any textbook on any mathematical topic whatever and turn to the middle of the book; it is then immediately obvious that the "proofs" presented there are not proofs in the formal sense[3]. Of course, one might argue that these "proofs" can always be rewritten in the formal style. Apart from the fact that the practical feasibility may be seriously doubted – how long would the formal counterpart of Andrew Wiles's proof of Fermat's Last Theorem be? I guess very, very long – for the purpose of this paper it is sufficient to note that mathematicians themselves do not do it. They do not invest part of their time in rewriting existing "proofs" as formal, correct ones. Why don't they?

This brings me to my second consideration. A closer look at what mathematicians do reveals that they spend quite some time on activities that seem strange from the formal proof perspective. Why, to give but one example, waste time on proving special cases of a universal statement? Why, e.g., prove that $x^3 + y^3 \neq z^3$, for x, y and z integers, when the statement to be proved is $(\forall n > 2)(x^n + y^n \neq z^n)$, for x, y and z integers?

The thesis of this paper can now be reformulated thus: it is precisely in those aspects of the activities of mathematicians that "disappear from view" when seen from the formal proof perspective, that mathematics is quite similar to the sciences. Moreover, formal proofs and not "proofs" are, all things considered, rather atypical, thus it seems quite appropriate to focus on the non-formal aspects.

Yet another way of presenting what I have in mind is this: instead of talking about "proofs," let me introduce the notion of a *mathematical argument*. Just as in the case of formal proof, we could imagine a relation Arg(α, A) to indicate that α is a mathematical argument for or in support of the statement A. The Arg relation can be seen as a weaker (and hence more general) notion than the Proof relation. If α happens to be a formal proof for A, then it is obvious that we want that:

---

[2] See, e.g., Boolos and Jeffrey (1989), chapter 27.

[3] The reason that I mention the middle of the textbook is that in the introduction of a textbook, where the first few elements and concepts are introduced, "proofs" tend to be rather simple and so it is possible to spell them out in full detail, hence they can be considered to be formal proofs. Often it is believed that the same holds throughout the book, but that is precisely the point I am denying.

If Proof(α, A) then Arg(α, A),

But if α is not a formal proof, then surely we do not want the inverse:

If Arg(α, A) then Proof(α, A).

This is all rather trivial – at the end of the paper I return to the question of a formal analysis of the Arg relation – but as a tool of thought I do believe that the Arg relation is rather useful. First, it retains the connection with the (formal) proof relation, which now appears as an extreme case, one end of a continuum but, secondly, there is no longer any reason to expect certainty of mathematical statements, since an argument *supports* a statement but does not (necessarily) prove it. In short, it presents mathematical activity as a fallible activity (though I am not claiming it is a form of fallibilism in the sense of Lakatos[4]), thereby reducing the philosophical importance attached to such questions as what the source of mathematical certainty can be. Along more modest lines, this approach at least helps to bridge the unfortunately still existing gap between formal-mathematical reasoning on the one hand and informal-argumentative reasoning on the other.

All this having been said and done, I assume that the reader is anxious to know what mathematical arguments could be. The next section presents a summary of such possible candidates for α, some of which I merely mention either because they are quite evident or because I have written about them in other places.[5]

3. Presenting the Evidence for Mathematical Arguments

(a) Obviously the first candidates for α are real proofs as they appear in the journals or as they are presented at conferences. It is a nice challenge for any student to rewrite a real proof in formally precise terms. Just one example: take the following problem.[6] Given $2n$ consecutive natural numbers, when a random selection of $n+1$ numbers is made then two of these are necessarily relatively prime. Argument: when $n+1$ numbers are selected from $2n$ consecutive numbers, two of them are necessarily neighbors, hence relatively prime. QED. Given the formal language of Peano Arithmetic (and perhaps a bit of set theory), it is not a trivial task to spell out this argument in detail. Do note, however, that as a mathematical problem, it is rather trivial.

---

[4] See, e.g., Koetsier (1991), who has further developed the Lakatosian fallibilist approach to mathematics.

[5] See, e.g., Van Bendegem (2000) and (2001).

[6] See Aigner and Ziegler (1998), p. 123. This little problem was suggested by Paul Erdös.

(b) A second candidate for α are so-called "informal" proofs. These proofs are to be distinguished from real proofs where one believes that it is possible to rewrite the proof in all formal detail. In short, a real proof is an instance of correct reasoning. Informal proofs have the property that they are basically not (formally) correct, yet lead to a correct result. As the reader might wonder why mathematicians would waste their time doing such a thing, the answer is this: it gives the mathematician at least some idea of what the result could be.

The most often quoted example is Euler's famous argument for the sum of the inverses of the squares, namely the argument that $\Sigma\ 1/n^2 = \pi^2/6$ (the summation taken over the natural numbers). I shall not present the full argument but its general structure. Euler first reasons about polynomials of finite even degree $2n$, of the following form:

$$b_0 - b_1x^2 + b_2x^4 - \ldots + (-1)^n b_n x^{2n} = 0$$
with roots: $r_1, -r_1, r_2, -r_2, \ldots, r_n, -r_n$.

He shows that the following holds:

$$b_1 = b_0(1/r_1^2 + 1/r_2^2 + \ldots + 1/r_n^2). \qquad (*)$$

All of this is quite regular mathematics. He then assumes that the same line of reasoning applies to polynomials of infinite degree. It is at this point that the reasoning goes astray, for there is no reason to suppose that the same result will hold for the infinite case. Thus the polynomial:

$$1 - x^2/3! + x^4/5! - x^6/7! + \ldots = 0,$$
with roots: $\pi, -\pi, 2\pi, -2\pi, 3\pi, -3\pi, \ldots$,

(as it is the series expansion of $\sin(x)/x$), will satisfy (*), thus

$$1/3! = 1/\pi^2 + 1/4\pi^2 + 1/9\pi^2 + \ldots,$$

or: $1 + 1/4 + 1/9 + \ldots = \pi^2/6$. QED (?)

This is not some kind of outlandish curiosity, for, as Dunham (1990, pp.207-222) shows, the same line of reasoning can be used for other summations as well (as Euler actually did), such as :

$\Sigma\ 1/(2n)^2 = \pi^2/24$, i.e., the sum of the reciprocals of all even squares,

$\Sigma\ 1/(2n+1)^2 = \pi^2/8$, i.e., the sum of the reciprocals of all odd squares,

and $\Sigma\ 1/n^4 = \pi^4/90$, i.e., the sum of the reciprocals of all fourth powers.

It is worth emphasizing that all these results turned out to be correct, hence these arguments can be rightfully called arguments.

(c) A third candidate is so-called "career induction." I have already mentioned Fermat's Last Theorem. Another famous example is Goldbach's Conjecture, i.e., the statement that every even natural number is the sum of two prime numbers. Career induction is the idea that, if you have to prove a universal statement of the form $(\forall n)A(n)$, then it is worthwhile investigating $A(1)$, $A(2)$, up to some finite number k. Formally speaking there could only be one case where such an approach is interesting, namely, if it turns out that one of the special cases does not hold, i.e., one proves $\sim A(m)$, for some particular number m, thus refuting $(\forall n)A(n)$. But in cases such as Fermat and Goldbach, this has not happened. One could certainly suggest that, although one was looking for a counterexample, one ended up (almost by accident or unintentionally) proving the cases.

What does seem to be the case, however, is that by searching for proofs for special cases the mathematician gains some insight into the kind of proof elements and proof concepts that will be needed if a proof of the universal statement is ever to be found. In the case of Fermat, this is clear: the method of infinite descent was used in the special cases and it turned out to be a powerful method for dealing with the general case.[7] As to Goldbach, here the problem is open as we do not have a proof at the present moment. However, as the excellent paper of Echeverria (1996) makes clear, even without a real proof of Goldbach, it is clear that the numerical evidence that has been gathered, together with other considerations, has convinced most if not all mathematicians that Goldbach's conjecture is true.[8]

(d) A fourth candidate is so-called mathematical "experiments" including visualizations and computer graphics (see Hege and Polthier 1997, 1998). Such "experiments" cannot be considered formal proofs because, as we all know, the translation of a mathematical problem involving infinite domains (such as the real or complex numbers) to the computer screen consisting of a

---

[7] Although I must add straight away that in the final proof by Andrew Wiles it is hard to see that this is a paper about number theory. Elliptic curves, group representations, Galois fields, …, those are the ingredients needed to prove the statement, hence there is no direct use for infinite descent here, as it is a typical number-theoretic idea: if a solution in natural numbers exists, then a solution exists that is strictly smaller; this is impossible, because one would then have an infinite number of solutions, hence there is no solution. Infinite descent is very closely related (in some cases equivalent) to mathematical induction.

[8] Although not essential to the thesis of this paper, it is worth mentioning that Georg Cantor, the mathematician responsible for transfinite set theory, also spent some time on Goldbach's conjecture. The standard story is that a nervous breakdown made it impossible to work on serious matters, so Cantor "wasted" his time calculating decompositions in two primes for all even numbers up to 1000. However, as Echeverria shows, the contribution was very important. What Cantor studied was the function $G(2n)$, i.e. the number of ways that an even number $2n$ can be written as the sum of two primes. The cases studied by Cantor showed that $G(2n)$ is an increasing function, that is, if $n > m$, then $G(2n) > G(2m)$. If this could be proved in general, it proves Goldbach, because $G(2n) \geq 1$ is sufficient.

finite set of pixels must involve approximations. To be specific, suppose that a three-dimensional object, an algebraic description of which is given, is visualized on the computer screen and the visual object has certain properties, then it would not be correct to conclude that the object actually does have that property. In fact, as the literature shows, it is always necessary to establish estimations of the errors involved, but that needs to be proved mathematically, so, therefore, the image cannot add anything new. Or can it?

It is undoubtedly the case that an image can "reveal" certain aspects of a mathematical object. Seeing a mathematical object (or an approximation of one) does provide information in a different format. It is rather tempting to give a semiotic analysis at this point,[9] but the fact that a formal text and a picture are not the same can hardly be a point of discussion. Even if it turns out that a property of the visualization is a computer artefact, this might still provide some insight.

Another type of mathematical "experiment" consists of number crunching, but that is nothing other than a technological version of the previous type of argument, namely career induction. It simply consists of checking a finite number of cases of a universal statement $(\forall n)A(n)$ by direct computation, the only difference being that more cases can be checked than can be done by hand.

There is, however, one other interesting case, different from the previous ones. Ivars Peterson (1988) discusses the Plateau problem: given a boundary curve B, what is the minimum surface S having B as its boundary? Mathematically this is a profound and difficult problem. Analytical methods are often insufficient. There is, however, a simple way to find solutions, though not necessarily the set of all solutions. Construct the boundary B in metal wire. Dip it in soapy water and a film will form having B as its boundary. Physics tells us that this film is a minimum surface. Hence, Peterson says: "They can explore shapes that are often too complicated to describe mathematically in a precise way. They can solve by experiment numerous mathematical problems associated with surfaces and contours." (p. 48) The relations between such experiments and mathematics is actually quite a profound philosophical issue.[10]

(e) A recent candidate is mathematical arguments that involve probabilistic considerations. It is important to be rather precise about what such a type of argument looks like. On the one hand, what one presents is a real proof in the

---

[9] I am thinking here of authors such as Michael Otte, see his (1997) or Brian Rotman, see his (2000).

[10] I refer the reader to Van Bendegem (1998) for more details.

sense of type (a), discussed above. On the other hand, however, what the proof says involves probabilities. Examples of such proofs are typically to be found in number theory, concerning theorems such as: Given a number n, if a test T, involving a random choice of k numbers all less than or equal to n, is performed on n, and the answer is yes to T, then the number n is prime with a probability of $1 - 1/4^k$.

If the problem we want to solve is to know whether or not the number n is a prime with certainty, then it is obvious that such an argument supports that idea. In that sense it is an argument for the statement that the number n is indeed prime. It is quite interesting to note that this means that some proof of some statement can be an argument for a closely related statement. There is an additional element, which I shall not elaborate further in this paper, why mathematicians are interested in such probabilistic statements. If one wants certainty, then the computational costs of actually checking whether the number is prime or not is exponential (or it is not known whether a polynomial procedure exists) in time or space needed, whereas the probabilistic approach runs in polynomial time or space. (See Ribenboim 1989, pp. 107-120 for further details.)

(f) Another recent candidate is proofs involving the use of computers. Without any doubt, the famous example of this case is the four-color theorem. The theorem states that four colors are sufficient to color any planar map in such a way that neighboring areas are colored differently. The first published proof consisted of two parts. The first part was a "classical" mathematical proof, in which it is shown that the set of all possible maps can be reduced to a finite set such that if all maps in that finite set can be colored, so can the full set. It is actually a beautiful piece of mathematics. But the second part consists of a computer listing, presenting the details of a computer program that has actually colored all the maps and said, "yes, I have colored them all" at the end of the day. Obviously, according to the definition of proof given at the beginning of this paper, this is not a proof, definitely not a formal proof. But it is obviously a mathematical argument because it shows that the theorem is very likely to be true. In fact, as a mathematical argument it is clear that this result is far better than if a human mathematician had actually colored all the maps as humans are more likely to commit errors than computer programs. (For a discussion of this type of argument, see Tymoczko 1986.)

Computer programs also play a part in checking existing mathematical (real) proofs. Something quite curious is happening here. Suppose we have sufficiently complex computer programs – there are some good candidates around at the present moment – that can rewrite real proofs as formal proofs. One can imagine that a real proof of fifty lines in a rewritten form turns into a formal proof of a couple of thousand lines (probably presented in some type of clausal form[11]). The program tells us that the proof is formally correct. Since we have used a computer program, what we have here is a mathematical argument. Hence, a mathematical argument can help convince us that a real proof is correct because the formal counterpart has been declared all right by the program. It shows that real proofs and arguments can support one another in quite complex ways.

(g) Finally one should take into consideration arguments that are not "purely" mathematical, but involve non-mathematical elements. A very fine example in this connection is the use of foundational-philosophical arguments to arrive at the probable truth or falsity of a particular mathematical statement. These are most certainly not mathematical proofs, whether formal or real, but they do help to decide certain questions or, at least, to give an orientation to the proof search.

An example would be a situation whereby a statement is believed to be false because it has implications that, although strictly mathematically speaking are without fault, nevertheless are considered to be paradoxical on philosophical and/or non-mathematical grounds. Two specific examples are:

(i) The Banach-Tarski paradox that throws doubt (for some) on the axiom of choice as it is used in set theory. The paradox states that it is always possible to decompose in three-dimensional space a ball of volume V into two balls of volume V, using only rigid motions (translations and rotations). There is no mathematical problem here, but the paradoxical character of the result is clear and serves as an argument against the axiom of choice.[12]

(ii) The Continuum Hypothesis (CH) as it is discussed in the writings of Kurt Gödel.[13] On the basis of philosophical arguments, Gödel had the profound conviction (at least during a specific period of his career) that CH cannot be the case, i.e., $2^{\aleph_0} \neq \aleph_1$. This conviction was, in his own words, based on the fact that CH had implausible consequences. This is particularly interesting because he himself had already produced a result that shows that there exists a model of the set-theoretical axioms wherein CH is actually true and, hence, is

---

[11] This is the favorite way of representing statements in automated reasoning. It is based on the fact that every statement in a classical logical system, such as first-order predicate logic, can be rewritten in a standard format in which all quantifiers occur at the beginning of the statement and the quantifier-free part can be rewritten in terms of conjunctions, the members of which are disjunctions of atomic formulas with or without a negation in front of them. Thus $(\forall x)(Px \supset (Qx \& Rx))$ is rewritten as $(\forall x)((\sim Px \square Qx) \& (\sim Px \square Rx))$. These disjunctions are called clauses.

[12] See Moore (1982) for details.

[13] CH is the statement that between the countable infinite, $\aleph_0$, and the infinity corresponding to the continuum or the set of reals, $2^{\aleph_0}$, there are no other infinities, hence, the "next" infinity $\aleph_1$ must be equal to $2^{\aleph_0}$. See Feferman et al. (1990) for the full details, in particular the introductory note by Gregory Moore, pp. 154-175.

not refutable! So this left only two possibilities: either CH is provable or it is undecidable (which turned out to be the case). As the former case was excluded for Gödel, one would expect that he concluded that CH is undecidable. Which indeed he did, but the undecidability for him meant that we had to look for additional axioms that would decide CH, in his case in the negative. To avoid any confusion, Gödel was not being incoherent here. The additional axioms would exclude the model that Gödel considered to be quite artificial and that was constructed with the sole purpose of showing CH to be true in it.

Although I do not claim any completeness for the above list of types of mathematical arguments, I do believe that it shows, firstly, that mathematical arguments are different from formal-mathematical proofs; secondly, that such arguments are abundant; and thirdly, that such arguments do allow mathematicians to convince themselves of the truth, falsity, provability or refutability of particular mathematical statements.

4. Mathematical Arguments and Empirical Evidence

Let me return to the relation Arg(α, A) in a more general setting. At first sight, it seems that not much can be said about this relation in general. It is definitely not the case that

If Arg(α, A) then A

for obvious reasons. Nor that

If Arg(α, A) and Arg(β, $A \supset B$) then there is a γ, such that Arg(γ, B)

for the simple reason that α and β can be of an entirely different nature. How can one compare a philosophically motivated argument for A with a career induction argument?

There is, however, one important idea that is worth exploring. It seems quite reasonable to claim that if there is a mathematical argument α that supports A, then we are willing to express a commitment about A. It also seems reasonable that we could imagine a scale between 0 and 1 (these values of course being arbitrary), and a function P that assigns to a statement A a value $P(A)$ between 0 and 1. $P(A)$ expresses our degree of confidence in or our commitment to the fact that A is indeed correct or true. Three conditions seem extremely plausible:

If Proof(α, A) then $P(A) = 1$
If Proof(α, $\sim A$) then $P(A) = 0$
If Arg(α, A) and not Proof(α, A) then $0 < P(A) < 1$.

Furthermore, what a mathematical argument does is to increase the degree of our commitment, hence the following principle is defensible:

If $P_{\text{before}}(A)$ is given and $\text{Arg}(\alpha, A)$ then $P_{\text{after}}(A) \geq P_{\text{before}}(A)$

This strongly suggests that mathematical arguments behave, generally speaking, in the same way as empirical evidence for a scientific hypothesis. I am not claiming that the function *P* should be a probability function,[14] but there are some properties of *P* that are common both to mathematical arguments and empirical evidence:

(a) A statement is supported more strongly if the number of mathematical arguments for it increases. Note that this also holds for real proofs. It is a common practice among mathematicians to find more than one real proof for a mathematical statement.

(b) Arguments that are independent of one another are more interesting than mutually dependent arguments. Note again that for real proofs if another proof is found it should be different (usually meaning either using a different proof method or using a different mathematical domain) to increase the support.

(c) Unexpected arguments have a greater impact than expected arguments. This feature too is typical of real proofs. When a proof is unexpected – an example would be the use of a proof technique from another mathematical domain that one did not expect – this counts as more important and/or convincing than a "regular" real proof.

No doubt this list could be further extended, but my aim here was only to show that the similarities are not superficial, justifying the conclusion that I do believe – and I assume that Theo Kuipers would join me in this – that at least in some aspects of mathematical practice the way we build up our support of mathematical statements is quite similar to the corresponding way it is done in the sciences. It therefore also justifies the hope that this paper might be a small contribution to a unified treatment of science, philosophy, theology *and* mathematics.

[14] A good argument against a probability interpretation is that conditional probability is lacking in this presentation. In the best of cases we could talk about such expressions as $P(A, \alpha)$ – i.e., the degree of commitment to *A* given argument α – but this runs counter to a definition of $P(A, \alpha)$ in terms of $P(A\&\alpha)$, as $A\&\alpha$ is a "mixed" expression.

Vrije Universiteit Brussel
Centrum voor Logica en Wetenschapsfilosofie
Pleinlaan 2, B-1050 Brussel
Belgium
e-mail: jpvbende@vub.ac.be
http://www.vub.ac.be/CLWF/

REFERENCES

Aigner, M. and G. Ziegler (1998). *Proofs from THE BOOK.* New York: Springer.

Bendegem, J.P., van (1993). Real-Life Mathematics versus Ideal Mathematics: The Ugly Truth. In: E.C.W. Krabbe, R.J. Dalitz and P.A. Smit (eds.), *Empirical Logic and Public Debate. Essays in Honour of Else M. Barth*, pp. 263-272. Amsterdam: Rodopi.

Bendegem, J.P., van (1998). What, If Anything, is an Experiment in Mathematics? In: D. Anapolitanos, A. Baltas and S. Tsinorema (eds.), *Philosophy and the Many Faces of Science*, pp. 172-182. London: Rowman & Littlefield.

Bendegem, J.P., van (2000). Analogy and Metaphor as Essentials Tools for the Working Mathematician. In: F. Hallyn (ed.). *Metaphor and Analogy in the Sciences* (*Origins: Studies in the Sources of Scientific Creativity*), pp. 105-123. Dordrecht: Kluwer Academic.

Bendegem, J.P., van (2001). The Creative Growth of Mathematics. *Philosophica*, vol. 63, **1**, 1999 (date of publication: 2001), 119-152.

Boolos, G.S. and R.C. Jeffrey (1989). *Computability and Logic.* Third edition. Cambridge: Cambridge University Press.

Dunham, W. (1990). *Journey Through Genius. The Great Theorems of Mathematics.* New York: Wiley.

Echeverria, J. (1996). Empirical Methods in Mathematics. A Case-Study: Goldbach's Conjecture. In: G. Munévar (ed.), *Spanish Studies in the Philosophy of Science*, pp.19-55 Dordrecht: Kluwer.

Feferman, S., J.W. Dawson, Jr., W. Goldfarb, C. Parsons and R. N. Solovay, eds. (1990). *Kurt Gödel. Collected Works.* Vol. II: *Publications 1938-1974.* Oxford: Oxford University Press.

Hege, H.C. and K. Polthier, eds. (1997). *Visualization and Mathematics. Experiments, Simulations and Environments.* New York: Springer.

Hege, H.C. and K. Polthier, eds. (1998). *Mathematical Visualization. Algorithms, Applications and Numerics.* New York: Springer.

Koetsier, T. (1991). *Lakatos' Philosophy of Mathematics. A Historical Approach. Studies in the History and Philosophy of Mathematics*, vol. 3. New York/Amsterdam: North-Holland.

Moore, G.H. (1982). *Zermelo's Axiom of Choice. Its Origins, Development, and Influence.* New York: Springer.

Otte, M. (1997). Mathematik und Verallgemeinerung. Peirce' semiotisch-pragmatische Sicht. *Philosophia Naturalis* **34** (2), 175-222.

Peterson, I. (1988). *The Mathematical Tourist. Snapshots of Modern Mathematics*. New York: Freeman.

Ribenboim, P. (1989). *The Book of Prime Number Records*. New York: Springer.

Rotman, B. (2000). *Mathematics as Sign. Writing, Imagining, Counting*. Stanford: Stanford University Press.

Tymoczko, T. (1986). *New Directions in the Philosophy of Mathematics*. Stuttgart/Boston: Birkhauser.

Theo A.F. Kuipers

MATHEMATICS AND EXPLICATION
REPLY TO JEAN PAUL VAN BENDEGEM

Both specific claims of Jean Paul Van Bendegem are very plausible. First, there are many convincing mathematical arguments that are no genuine mathematical proofs and, second, the way in which these arguments build up our support of mathematical statements is quite similar to the way it is done in the empirical sciences. Since Van Bendegem is, like me, in general very much interested in the similarities between "science, philosophy, theology, and mathematics," I will start this reply by summarizing my view on the similarities and differences. Next I deal with his specific claims.

Mathematical Research as Concept Explication

Leaving theology here aside, I would like to claim that the basic similarity between philosophy and mathematics is the focus on the explication of informal concepts. In SiS I wrote (p. 8):

> For philosophy and mathematics the fourth type of program, the *explicative* research program, is the most important type. Such programs are directed at concept explication, i.e., the construction of a simple, precise and useful concept, which is, in addition, similar to a given informal concept (cf. Carnap 1963, pp. 1-18). For example, the concepts of 'logical consequence' and 'probability' have given rise to very successful explicative programs in the borderland between philosophy and mathematics. One of the main explicative programs dealt with in ICR is intended to explicate the intuitive idea of 'truthlikeness'. Although several analyses in the present book [SiS] could have been explicitly presented as examples of concept explication, we have made this identification in only a few chapters, and not even very rigorously at that, … .
>
> The strategy of concept explication is the following. From the intuitive concept to be explicated one tries to derive conditions of adequacy that the explicated concept will have to satisfy, and evident examples and counter-examples that the explicated concept has to include or exclude.

Let me also mention that explication may go further than the explication of informal concepts, it may also aim at the explication of intuitive judgments,

In: R. Festa, A. Aliseda and J. Peijnenburg (eds.), *Cognitive Structures in Scientific Inquiry* (*Poznań Studies in the Philosophy of the Sciences and the Humanities,* vol. 84), pp. 170-173. Amsterdam/New York, NY: Rodopi, 2005.

i.e., intuitions, including their justification, demystification or even undermining. A main example in ICR concerns the intuition about the functionality of choosing empirically more successful theories in order to enhance truth approximation. Another, certainly demystifying, example is the intuition that beauty may be an indication of the truth (Kuipers 2002). The strategy of "intuition explication" is a plausible extension of that involving concept explication.

So far the special nature of mathematical and philosophical research has been highlighted somewhat, but the similarity between concept explication and empirical research may not yet be very convincing. However, in several branches of mathematics the domain of research is quasi-empirical. For example, as Lakatos (1976) has demonstrated so beautifully, the history of the explication of the idea of a regular polyhedron and Euler's conjecture for all polyhedra that the number of vertices plus the number of faces equals the number of edges plus two, is such a quasi-empirical story. However, not all mathematical concepts and theorems have this feature. For example, the informal logico-mathematical notion of a group is primarily an abstract notion with, to be sure, many evident examples and non-examples. There is nevertheless quite some similarity between progress in explicative research on the one hand and (descriptive and explanatory) nomological and design research on the other. In SiS, pp. 263-4, I wrote:

> Like nomological research, this [explicative] task may be represented in terms of conceptual possibilities. Let us further assume first that there is a unique solution and hence a unique set of desired possibilities. Let a provisional explication also be conceived as (determining) a unique extension of conceptual possibilities. Then it is plausible to define formal progress in explicative research formally in the same way as in the case of nomological research.
>
> In real-life explicative research, however, the resemblance of a provisional explication has to be evaluated in other terms. In particular, such evaluation takes place in terms of evident examples, that is, evidently desired possibilities, evident 'non-examples', that is, evidently undesired possibilities, and, finally, so-called conditions of adequacy, that is, conditions to be fulfilled and which correspond to desired features. Hence, the definition of 'conceptual progress' in explicative research is straightforward. Provisional explication Y is better than provisional explication X, roughly speaking, if and only if Y treats more evident examples and non-examples properly and/or fulfills more conditions of adequacy.

The partial analogy between nomological and design research on the one hand and explicative research on the other is obvious, including the possibility of functionally equivalent explications. However, at least two differences with nomological research are very interesting. Whereas evident non-examples play an important role in explicative research, there is no nomological analogue for them, as this would require the realization of nomic impossibilities. Moreover, nomological research is more or less bound to a unique solution, whereas explicative research may well lead to the conclusion that two or more interesting explications can be given, which are functionally equivalent relative to the desired features but nevertheless mutually exclude each other.

Design research shares with explicative research this possibility of more than one useful solution. However, as in nomological research, there does not seem to be an analogue for evident non-examples in the case of design research. Moreover, it is argued in some detail in SiS, Chapter 9, that formal progress in design research is relatively easy to determine, but not so in concept explication. Hence, although there is also a strong analogy between design and explicative research, i.e., both aim at a certain product, the analogy is not perfect.

Let me use the opportunity to stress something that I forgot to do in SiS. Although I mention in SiS that explicative research can also pertain to crucial terms in the empirical sciences, I forgot to emphasize that in this case conditions of adequacy and evident examples and non-examples should agree as much as possible with up to date empirical, in particular nomological, research. As Hempel (1952, p. 12) already put it: "An explication of a given set of terms, then, combines essential aspects of meaning analysis and empirical analysis." Ideally, concept explication in the empirical sciences leads to an improved conceptual framework for further empirical research.

Mathematical Arguments

Van Bendegem's paper nicely illustrates that the transition from mathematics to philosophy of mathematics is methodologically not a big step. Its main aim is to start the explication of the informal notion of a mathematical argument, as a much weaker notion than that of a mathematical proof. As a matter of fact, the examples (b)-(g) are at least in part intended as evident cases of mathematical arguments not qualifying as proofs. Moreover, some of the general statements evidently function as a condition of adequacy or, very interestingly, as a "non-condition" of adequacy. In particular, the second claim in Section 4, according to which the combination of arguments for A and $A \supset B$ need not be an argument for B, is an intriguing and perhaps disputable but nevertheless clear illustration of an intended non-condition of adequacy. In that section, Van Bendegem also sets the stage for an explication of the way arguments change our degree of confidence in a mathematical statement.

Let me close with a question that intrigues me on the basis of reading Van Bendegem's paper: what are the differences and similarities between arguments that can be transformed into, or replaced by, genuine proofs and arguments which cannot? More specifically, I mean the following. As is well-known, in many cases it is possible to prove a claim without really calculating and deducing the conclusion, but by giving a very elegant argument, of a mathematical and/or empirical nature, that immediately convinces everybody who starts to understand it. In Kuipers (1991), I collected 10 such examples. One of them deals with mixing white and red wine. Two identical bottles contain five glasses of wine, the one white, the other red. The bottles can contain six glasses. Now one pours a glassful from the bottle with red wine into the one with white wine, shakes the latter very well, and then pours a glass of the mixture back into the red wine bottle, and again shakes very well. Which bottle has the highest concentration of wine that originally comes from the other bottle? Of course, one can make a calculation, leading to the conclusion that the concentrations are the same. However, one may also immediately see this solution by realizing that otherwise the total amount of white and red wine would have changed. Assuming that this in itself cannot count as a proof but can be transformed into an indirect proof, it seems to be an argument of a third kind: it is neither a proof in the strict sense nor an argument that merely increases our degree of conviction.

REFERENCES

Carnap, R. (1963). *Logical Foundations of Probability*. Chicago: The University of Chicago Press.

Hempel, C. (1952). *Fundamentals of Concept Formation in Empirical Science*. Chicago: The University of Chicago Press.

Kuipers, T. (1991). Dat Vind Ik Nou Mooi. In: R. Segers (ed.), *Visies op Cultuur en Literatuur. Opstellen naar Aanleiding van het Werk van J.J.A. Mooij*, pp. 69-75. Amsterdam/Atlanta: Rodopi.

Kuipers, T. (2002). Beauty, a Road to The Truth. *Synthese* **131** (3), 291-328.

Lakatos, I. (1976). *Proofs and Refutations: The Logic of Mathematical Discovery*. Cambridge: Cambridge University Press.

TYPES OF EXPLANATION

Erik Weber and Helena De Preester

MICRO-EXPLANATIONS OF LAWS

ABSTRACT. After a brief introduction to Kuipers' views on explanations of laws we argue that micro-explanations of laws can have two formats: they work either by aggregation and transformation (as Kuipers suggests) or by means of function ascriptions (Kuipers neglects this possibility). We compare both types from an epistemic point of view (which information is needed to construct the explanation?) and from a means-end perspective (do both types serve the same purposes? are they equally good?).

1. Introduction

1.1. *Kuipers on Micro-Explanations*

Theo Kuipers (SiS/2001, pp. 82-104) claims that explanations of laws have different forms, but contain only steps of five types: application steps, aggregation steps, identification steps, correlation steps, and approximation steps. As a rule, these steps occur in this order, but exceptions are possible. Not all explanations contain steps of all types, and some types may occur more than once.

Kuipers defines reductions as a subclass of explanations of laws: reductions contain at least one aggregation, identification or approximation step. So explanations of laws that contain only an application and a correlation step (Kuipers gives explanations of interbreeding laws in classical Mendelian genetics as examples) are not reductions.

Kuipers defines micro-reductions as explanations that contain at least one aggregation step. Let us look at one of Kuipers' examples, the ideal gas law (IGL). The IGL tells us that for one mole of gas in a container, the product of volume V and (macroscopic) pressure P is proportional to the empirical absolute temperature T, such that the proportionality constant is the same for all gases (viz. R):

$$PV = RT$$

According to the kinetic theory of gases an isolated quantity of gas consists of molecules which move and collide with each other and with the container wall

In: R. Festa, A. Aliseda and J. Peijnenburg (eds.), *Cognitive Structures in Scientific Inquiry* (*Poznań Studies in the Philosophy of the Sciences and the Humanities*, vol. 84), pp. 177-186. Amsterdam/New York, NY: Rodopi, 2005.

in accordance with Newton's laws of motion. In the application step these laws are applied to one molecule colliding with the container wall. In this step we use the auxiliary hypothesis that the collision is elastic. The result is the "individual law" which states that the momentum exchange q equals $2mv_w$ (where m is the mass and v_w the velocity in the wall direction). The second step, the aggregation step, leads, by means of some auxiliary statistical hypotheses, to the "aggregated law" that the product of the kinetic pressure on the container wall p and the volume V is equal to $(2/3)N\bar{u}$ (where N is Avogadro's standard number of molecules in a mole of gas, and $\bar{u}$ the mean kinetic energy of the molecules). In the third and last step, two identity hypotheses ($p = P$ and $\bar{u} = (3/2)(R/N)T$) are used to derive the IGL.

It is important to notice that explaining the macroscopic law requires both the aggregation and the identification step: the aggregation step alone allows us only to explain the intermediate law about the collective behavior of the molecules ($pV = (2/3)N\bar{u}$). In Kuipers' view, a micro-reduction of a macroscopic law (as opposed to a law about the collective behavior of the micro-level objects) requires an aggregation step and a *transformation* step (a general term introduced to denote both identification and correlation steps). In the explanation of the IGL, the transformation step is an identification step. In Kuipers' other example of micro-reduction (the explanation of Olson's law about collective goods) the transformation step is a correlation step.

1.2. *Aims and Structure of this Article*

The first aim of this article is to show that, contrary to what Kuipers claims, not all micro-explanations of laws contain an aggregation and transformation step. In Section 2 we give an elaborate example of what we call an AT explanation (micro-explanation of a law by means of aggregation and transformation). In Section 3 we show that laws can be micro-explained in a completely different way, viz. by means of function ascriptions. Explanations that use function ascriptions will be called FN explanations (the meaning of the N will become clear in that section).

Our second aim is to show that, from an epistemic point of view, FN explanations are superior: they require less information and are therefore easier to construct than AT explanations. The argument for this epistemic advantage will be given in Section 4.

The last aim of this contribution is to investigate the uses of AT and FN explanations. Micro-explanations are sometimes sought for practical reasons (prediction, manipulation) and sometimes for theoretical reasons (the desire to understand how the law fits into our worldview and how it relates to other laws). In Section 5 we argue that, if the motivation is practical, FN explanations are complementary to AT explanations. If the motivation is theoretical, FN explanations are useless.

2. AT Explanations of Laws

2.1. *Example*

Consider the following electrical circuit, which we call *C*:

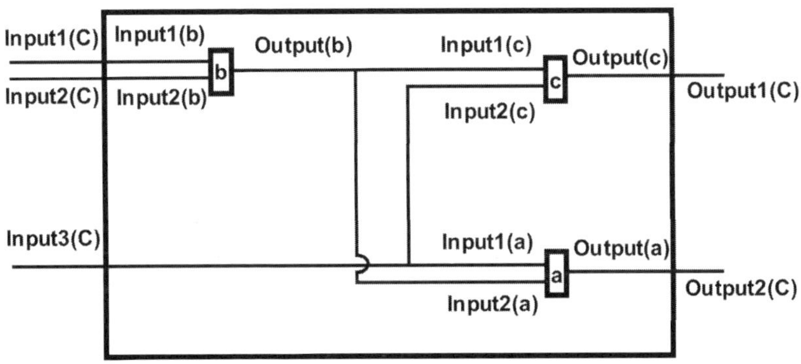

Assume that everything inside the large rectangle is contained in an opaque box, so that only the three input wires and two output wires are visible. Assume also that we can somehow measure whether these wires are charged or not. Then an experiment can be performed to see whether there is a law connecting the states of the input wires with the states of the output wires. Suppose that such an experiment yields the following law:

> *L*: If $\text{input}_1(C)=1$, $\text{input}_2(C)=0$ and $\text{input}_3(C)=1$, then $\text{output}_1(C)=0$ and $\text{output}_2(C)=1$.

'$\text{Input}_1(C)=1$' is shorthand for 'The first input wire of *C* is charged', '$\text{input}_2(C)=0$' for 'The second input wire of *C* is not charged', etc.

One way to explain this law is to derive it from an explanatory model by means of an aggregation and a transformation step. Performing those steps presupposes that the explanatory model from which we start consists of ontological claims, fundamental laws, interaction principles and bridge principles. In 2.2 we take a closer look at the content of the explanatory model, while in 2.3 we discuss the structure of the derivation. In section 3 we will present an FN explanation of the same law *L*.

2.2. *The Content of the Explanatory Model*

The *ontological claims* specify which micro-elements are contained in the macro-system (in this case: the circuit). In order to explain L in our example, we have to open the box. If the box is open, we can observe that the following ontological claim holds:

O_1: Circuit C contains three binary gates (a, b and c).

Each of the gates can be taken out of the circuit, so we can investigate their individual behavior. Assume that such test gives the following results:

F_1: a is an AND gate.
F_2: b is an XOR gate.
F_3: c is an XOR gate.

An AND gate has output 1 if and only if both inputs are 1. And XOR gate (exclusive OR) has output 1 if and only if the values of the inputs are different. We call F_1-F_3 *fundamental laws* because they describe the individual capacities of the elements of which the macro-system is composed: they are laws at a lower, more fundamental level than the law we want to explain.

Interaction principles provide information on the *relations* between the components of the system (unlike the fundamental laws, which give information about isolated components). In our example the interaction principles are:

I_1: The circuit is wired such that output(b) = input$_2$(a).
I_2: The circuit is wired such that output(b) = input$_1$(c).

Finally, there are the *bridge principles*. When the box is opened, we can observe the relations between properties of the circuit as a whole (the states of its input and output wires) and properties of components of the system (the states of the input and output wires of the three gates). More specifically, we can observe that:

B_1: Input$_1$(C) = input$_1$(b).
B_2: Input$_2$(C) = input$_2$(b).
B_3: Input$_3$(C) = input$_1$(a) = input$_2$(c).
B_4: Output$_1$(C) = output(c).
B_5: Output$_2$(C) = output(a).

We call B_1-B_5 bridge principles because they connect properties of the system as a whole with properties of its components.

2.3. *The Structure of the Derivation*

An AT explanation for *L* is obtained by deriving it from the explanatory model in two steps: first we derive an intermediate result *R* from the fundamental laws and the interaction principles (the aggregation step); then we use the bridge principles to derive *L* from *R* (the transformation step). Note that the ontological claims are not explicitly used in the derivation: they are conditions that must be fulfilled to make the claims of the other types meaningful. For instance, it does not make sense to claim that *a* is an AND gate if *a* does not exist.

In our example, the intermediate result obtained in the aggregation step is:

R: If $\text{input}_1(b)=1$, $\text{input}_2(b)=0$, $\text{input}_1(a)=1$ and $\text{input}_2(c)=1$, then $\text{output}(c)=0$ and $\text{output}(a)=1$.

This intermediate result differs from *L* in that it is *not* a regularity about *C*, and from the fundamental laws and interaction principles in that it is a *collective* law about *all* the components of the system. In this respect, *R* is analogous to the intermediate law obtained in the explanation of the IGL.

The aggregation step goes like this:

(1)	If $\text{input}_1(b)=1$ and $\text{input}_2(b)=0$, then $\text{output}(b)=1$	[F_2]
(2)	$\text{Output}(b) = \text{input}_2(a)$	[I_1]
(3)	If $\text{input}_1(b)=1$ and $\text{input}_2(b)=0$, then $\text{input}_2(a)=1$	[1,2]
(4)	If $\text{input}_1(a)= 1$ and $\text{input}_2\ (a)=1$, then $\text{output}(a)=1$	[F_1]
(5)	If $\text{input}_1(a)=1$, $\text{input}_1(b)=1$ and $\text{input}_2(b)=0$, then $\text{output}(a)=1$	[3,4]
(6)	$\text{Output}(b) = \text{input}_1(c)$	[I_2]
(7)	If $\text{input}_1(b)=1$ and $\text{input}_2(b)=0$, then $\text{input}_1(c)=1$	[1,6]
(8)	If $\text{input}_1(c)=1$ and $\text{input}_2(c)=1$, then $\text{output}(c)=0$	[F_3]
(9)	If $\text{input}_2(c)=1$, $\text{input}_1(b)=1$ and $\text{input}_2(b)=0$, then $\text{output}(c)=0$	[7,8]
(10)	If $\text{input}_1(b)=1$, $\text{input}_2(b)=0$, $\text{input}_1(a)=1$ and $\text{input}_2(c)=1$, then $\text{output}(c)=0$ and $\text{output}(a)=1$	[5,9]

Since (10) is identical to *R*, the aggregation step is completed. In the transformation step we transform the antecedent and consequent conditions of *R* by means of bridge principles:

(1)	If $\text{input}_1(b)=1$, $\text{input}_2(b)=0$, $\text{input}_1(a)=1$ and $\text{input}_2(c)=1$, then $\text{output}(c)=0$ and $\text{output}(a)=1$.	[R]
(2)	$\text{Input}_1(C) = \text{input}_1(b)$.	[B_1]
(3)	If $\text{input}_1(C)=1$, $\text{input}_2(b)=0$, $\text{input}_1(a)=1$ and $\text{input}_2(c)=1$, then $\text{output}(c)=0$ and $\text{output}(a)=1$.	[1,2]
(4)	$\text{Input}_2(C) = \text{input}_2(b)$.	[B_2]

(5) If $input_1(C)=1$, $input_2(C)=0$ and $input_1(a)=1$ and $input_2(c)=1$, then $output(c)=0$ and $output(a)=1$. [3,4]
(6) $Input_3(C) = input_1(a) = input_2(c)$ [B_3]
(7) If $input_1(C)=1$, $input_2(C)=0$ and $input_3(C)=1$, then $output(c)=0$ and $output(a)=1$. [5,6]
(8) $Output_1(C) = output(c)$ [B_4]
(9) If $input_1(C)=1$, $input_2(C)=0$ and $input_3(C)=1$, then $output_1(C)=0$ and $output(a)=1$. [7,8]
(10) $Output_2(C) = output(a)$. [B_5]
(11) If $input_1(C)=1$, $input_2(C)=0$ and $input_3(C)=1$, then $output_1(C)=0$ and $output_2(C)=1$. [9,10]

Step (11) is the law L, so the derivation is complete.

3. FN Explanations of Laws

The derivation given in 2.3 is not the only way to explain L. There is an alternative, in which the explanatory model contains the ontological claim that the circuit contains three binary gates (O_1), three function ascriptions, and three claims about well-functioning. The function ascriptions are:

F_a: The function of gate a is to ensure that $output_2(C)=1$ if and only if $input_3(C)=1$ and $output(b)=1$.

F_b: The function of gate b is to ensure that its output is 1 if and only if $input_1(C) \neq input_2(C)$.

F_c: The function of gate c is to ensure that $output_1(C)=1$ if and only if $output(b) \neq input_3(C)$.

L does not follow from these three function ascriptions alone. We need three claims which state that the gates function well:

N_a: Gate a functions normally.
N_b: Gate b functions normally.
N_c: Gate c functions normally.

In general, the explanatory model of an FN explanation contains ontological claims (which, as in AT explanations, serve as background knowledge in the derivation), function ascriptions (hence the F) and claims about normal functioning (hence the N).

It is important to note that FN explanations require an interested explainer, who has an ideal about how the system and its components must work. Without such an ideal, function ascriptions are impossible. The ideal of the explainer may differ from the ideal of the designer of the system (if there is one). AT explanations do not require such an ideal: the explainer can remain neutral.

4. FN Explanations Require Less Information

The functional explanation in section 3 is compatible with the AT explanation in 2.3. As already mentioned, the ontological presuppositions are identical. Furthermore, N_a is logically entailed by F_a together with F_1, I_1, B_3 and B_5. From these four last statements we can derive:

R Output$_2$ (C)=1 if and only if input$_3$(C)=1 and output(b)=1.

F_a says that the function of gate a is to ensure that R holds. This means that, if F_a is part of the explainer's ideal of how the system must work, he/she must conclude that a functions normally if he/she is convinced that F_1, I_1, B_3 and B_5 are true. Analogously, N_b is entailed by F_b together with F_2, B_1 and B_2; and N_c by F_c together with F_3, I_2, B_3 and B_4.

However, each of the gates would also function normally if the structure of the circuit were to differ slightly from its actual structure. So all the functions have multiple material realizations. For instance, N_a would also be true if F_1 and B_5 hold, input$_3$(C)=input$_2$(a) (instead of B_3) and output(b)=input$_1$(a) (instead of I_1). Similarly, b will still perform the function ascribed to it in F_b if it is an XOR gate with input$_1$(C) = input$_2$(b) and input$_2$(C) = input$_1$(b). For c the alternative material realisation is input$_3$(C)=input$_1$(c) and output(b)=input$_2$(c) (F_3 and B_4 remaining identical).

In our circuit example the FN explanation is easier to arrive at than the AT explanation: the functions can have multiple realizations, so the information we need about the system is less detailed. For instance, the AT explanation requires that we know that input$_3$(C) = input$_1$(a). The FN explanation only requires that we know that input$_3$(C) = input$_1$(a) *or* input$_3$(C) = input$_2$(a): this is sufficient to claim that a functions normally. This is the epistemic advantage of FN explanations: if the functions have multiple realizations, they require less knowledge than the AT explanations that are compatible with them. This epistemic advantage increases with the number of possible realisations of the function.

5. The Uses of AT and FN Explanations

5.1. *The Pragmatic Perspective*

Micro-explanations can be used to *change* the observed macro-level relation and to *predict* whether this relation (assuming that we do not intervene) will still hold at a later time. We discuss the cases in this order.

Suppose we want our circuit *C* to behave as follows:

L : If $input_1(C)=1$, $input_2(C)=0$ and $input_3(C)=1$, then $output_1(C)=1$ and $output_2(C)=0$.

The AT explanation in section 2 suggests a number of possible changes: we can change the wires so that one of the bridge principles or interaction principles changes, or we can replace one of the gates by a gate of a different type. By means of aggregation and transformation steps similar to the ones in the explanation we can calculate which (set of) change(s) is sufficient to obtain the desired result. The FN explanation, if available, shortens the calculations: by showing that some changes do not affect the normal functioning of any of the elements, we can show that no change at the macro-level will occur (and thus that the desired result will not be obtained). For instance, we can calculate that changing the circuit such that $input_1(C) = input_2(b)$ and $input_2(C) = input_1(b)$ (other things remaining the same) will not affect the normal functioning of *b*, nor of any of the other gates. So we can eliminate this set of changes without going through the complete aggregation and transformation procedure.

This example shows that, from the point of view of manipulation, the FN explanations are useful complements to AT explanations. However, FN explanations are not "autonomous": they become useful only if an AT explanation is available too.

In the case of prediction, FN explanations provide a similar shortcut. The AT explanation tells us where to look for changes at the micro-level. The significance of these changes (do they imply a different law at the macro-level?) can be evaluated by means of a calculation involving an aggregation and transformation procedure. If an FN explanation is available, a simpler calculation is possible by showing that the change(s) do or do not affect the normal functioning of one of the components.

5.2. *The Theoretical Perspective*

If the motivation for explaining a law is purely theoretical (i.e. if the aim is to understand how the law fits into our worldview and how it relates to other laws), FN explanations are useless. The IGL example shows that in some cases the fundamental laws and interaction principles that are used in an AT explanation are obtained be specifying a general theory. This specification procedure, which Kuipers calls the application step, consists in adding appropriate auxiliary hypotheses to the theory. The theory and the auxiliary hypotheses together entail the fundamental laws and interaction principles. An AT explanation which is constructed in this way has unificatory power: it shows how the explained law fits into our general world view, and (together with explanations of other laws that start from the same theory) how the law relates to other laws. FN explanations cannot have a similar unificatory power, because function ascriptions do not follow from any general theory.

6. Conclusion

We have shown that Kuipers' analysis of micro-explanations wrongly neglects functional explanations. Our contribution is not only a reaction to Kuipers, but also to authors who make the opposite mistake. An example of the latter category is Robert Cummins. In Cummins 1975 he argues that capacities (e.g. the circuit's capacity to produce certain outputs given certain inputs) can be explained by following the subsumption strategy or the analytical strategy. The *subsumption strategy* consists in showing that the capacity is a manifestation of one or more general laws, i.e. laws governing the behaviour of things generally, not just things having the specific capacity to be explained. For instance, we can explain why an object *a* has the capacity to rise in water of its own accord by invoking Archimedes' principle. This principle is applicable to all objects and all fluids, not just the object *a* and water. So the specific case (object *a*, capacity to rise in water) is explained by showing that it can be expected on the basis of a general principle. The *analytical strategy* proceeds by analyzing a capacity of *a* into a number of other capacities of *a* or components of *a*. As examples, Cummins mentions assembly-line production, schematic diagrams in electronics, and explanations of biological capacities:

> The biologically significant capacities of an entire organism are explained by analyzing the organism into a number of "systems" – the circulatory system, the digestive system, the nervous system, etc., – each of which has its characteristic capacities. These capacities are in turn analyzed into capacities of component organs and structures. (1975, pp. 760-761).

Obviously, micro-explanations are a subclass of what Cummins calls analytical explanations. Since Cummins regards function ascriptions as indispensable ingredients of *all* analytical explanations, this entails that for Cummins all micro-explanations must use function ascriptions. In other words: for Cummins all micro-explanations are FN explanations. We have shown that this position is wrong: FN explanations do not have unificatory power, and their pragmatic function presupposes AT explanations.

ACKNOWLEDGMENTS

Helena De Preester is Research Assistant of the Fund for Scientific Research – Flanders (F.W.O. Vlaanderen). The research for this paper was supported by F.W.O. Vlaanderen through research project G.0015.99 ("Intentional and Functional Explanations: A Philosophical Analysis"). We thank the members of the Centre for Logic and Philosophy of Science of Ghent University for their comments on previous versions of this paper.

Ghent University
Department of Philosophy
Blandijnberg 2
B-9000 Gent, Belgium
e-mail: Erik.Weber@Ugent.be
Helena.DePreester@Ugent.be

REFERENCES

Cummins, R. (1975). Functional Analysis. *Journal of Philosophy* **72**, 741-765.

Kuipers, T. (2001/SiS). *Structures in Science. Heuristic Patterns Based on Cognitive Structures*. Dordrecht: Kluwer Academic Publishers.

Theo A. F. Kuipers

KINDS OF MICRO-EXPLANATION
REPLY TO ERIK WEBER AND HELENA DE PREESTER

The paper by Erik Weber and Helena De Preester is in at least two respects very stimulating. First, it nicely illustrates how my five-steps model of explanation can be adapted to what I would like to call structural explanations of system laws. Second, it provides a clear sight of a (compatible) kind of functional explanation of such laws that is not touched upon in SiS.

Structural Explanation of System Laws

The explanation elaborated by Weber and De Preester in Section 2 for a circuit law on the basis of my five-steps model of explanation (SiS, Ch. 3) is an excellent and totally unexpected kind of use of that model. More specifically, they convincingly show that an input-output law (L) characterizing the observable behavior of the circuit can be deduced from three "fundamental (individual) laws," two "interaction principles" and five "bridge principles." The deduction comprises an aggregation (A) step, followed by a transformation (T) step, hence their speaking of an AT explanation. I have nothing to add to this lucid analysis, just some additional remarks that may further exploit the example.

(1) The aggregation step is a nice example of what was intended with the second (italicized below) half of my elucidation (SiS, p. 87): "the total effect of the individual law for many objects is calculated by a suitable addition, *or composition (or synthesis) if more than one type of individual law is involved*," since the three individual laws, characterizing the gates, concern two types: AND and XOR gates. Moreover, only in the case of uniformly sequential or parallel grouping of a number of gates of the same type would it be adequate to speak of (straightforward) aggregation.

(2) As a matter of fact, the authors show that the five-steps model, which was primarily intended for the explanation of a law by a theory (indicated in Weber and De Preester's Section 5.2), can easily be adapted to a model for the

In: R. Festa, A. Aliseda and J. Peijnenburg (eds.), *Cognitive Structures in Scientific Inquiry* (*Poznań Studies in the Philosophy of the Sciences and the Humanities,* vol. 84), pp. 187-190. Amsterdam/New York, NY: Rodopi, 2005.

explanation of a "system law" starting with, instead of a "(theory-) application step," an "observation step," as I would like to call it, for it amounts to the establishment of observational laws regarding components of the system. Their fundamental laws amount essentially to thus established laws for the gates in the circuit.

(3) The interaction and bridge principles are essentially identity claims. The first ones claim that certain gate outputs are identical to certain gate inputs. The bridge (or transformation) principles claim that certain gate inputs and outputs are identical to certain circuit inputs and outputs, respectively.

(4) The fundamental laws (or individual gate laws) and the bridge principles (gate identities) constitute together the internal or micro-principles used by the micro-explanation.

(5) The resulting explanation may not only be called a case of micro-reduction due to the (complex) aggregation step, but also a case of identificatory reduction, due to the identity nature of all transformation principles. Hence, apart from the different nature of the (hidden) first step, the example is essentially similar to the reduction of the ideal gas law.

(6) It seems plausible to call the present type of explanation of circuit laws and, more generally, system laws, structural (reductive) explanations, in particular when they are opposed to what Weber and De Preester call functional explanations of such laws, to be discussed now.

Functional Explanation of System Laws

Inspired by Cummins (1975), Weber and De Preester also give a kind of functional explanation of the same system law, viz. an explanation in terms of function ascriptions to the three gates in the circuit and the assumption that these gates function normally. Again I would just like to make a couple of remarks.

(1) Talking about functions should not hide the fact that the normal functioning of a gate can be described by hybrid behavioral laws, e.g. F_a and N_a together imply the law (NF_a): "*output*$_2$(*C*) = 1 iff *input*$_3$(*C*) = 1 and *output*(*b*) = 1," and the resulting three laws together imply the system law to be explained.

(2) Such hybrid laws can easily be redescribed as bridge principles, from gates to the system or vice versa. For example, NF_a is equivalent to "if *output*(*b*) = 1, then *output*$_2$(*C*) = 1 iff *input*$_3$(*C*) = 1, and if *output*(*b*) = 0, then *output*$_2$(*C*) = 0." Hence, the resulting explanation fits into the five-steps model in the sense that these laws are in fact transformation principles of the causal correlation type such that the explanation amounts to a number of correlation

steps, with the peculiar fact that they do not start from individual laws of a substantial nature but of a (context-relative) tautological nature, such as "*output*(*b*) = 1 or *output* (*b*) = 0." In view of the fact that neither Weber and De Preester's version nor the indicated nonfunctional version use individual laws, let alone micro-principles giving rise to such laws, it seems less appropriate to talk in this case about micro-explanations, as Weber and De Preester in fact do.

(3) The foregoing remarks are not intended to play down the practical usefulness of normal function talk. As in the case of biological functions, following Ruth Millikan (see SiS, Sections 4.2 and 6.2), if adapted, function talk makes also perfect sense in the case of artificial functions. As a matter of fact, whereas ascriptions of biological functions are essentially based on at least two causal components, one of a proximate and one of an ultimate nature, artificial function ascriptions may be based on merely one type of causal laws, as NF_a illustrates. Weber and De Preester's contribution strongly suggests that a further general analysis of functional explanations related to artificial systems along the lines of "explanation by specification" as developed in Chapter 4 of SiS would be very interesting.

(4) Very illuminating is the "multiple realizability" that Weber and De Preester discuss in Section 4. The same functions that are played by the gates can be realized in other ways than the particular "material realizations" in the sample circuit. I would like to add that the reductive explanation of the circuit law in Section 2 provides a perfect "artificial" illustration of the compatibility of multiple realizability and reductive explanations. As suggested in SiS (e.g. pp. 154-5) with some examples from natural science, the popular claim in functionalist philosophy of mind by Fodor and Putnam that multiple realizability is a blockade for reduction, is due to a lack of understanding of successful reductive arguments in the natural sciences.

(5) Finally, Weber and De Preester go as far as to claim that a functional explanation of the circuit law is not theoretically interesting and only practically useful, i.e. useful for manipulation of the circuit and, I would like to add, diagnostic reasoning about it, if the relevant structural explanation is available as well. This agrees with my claim (SiS, p. 126): "Explanation by a certain type of specification [intentional, functional, causal] automatically leads to a corresponding type of description, in particular classification." In other words, (isolated) functional explanations are in a sense merely a kind of description. However, as indicated in the contribution of Grobler and Wiśniewski and my reply to them, in special cases, they may play an important role in the evaluation of theories.

REFERENCE

Cummins, R. (1975). Functional Analysis. *Journal of Philosophy* **72**, 741-765.

Eric R. Scerri

ON THE FORMALIZATION OF THE PERIODIC TABLE

ABSTRACT. A critique is given of the attempt by Hettema and Kuipers to formalize the periodic table. In particular I dispute their notions of identifying a naïve periodic table with tables having a constant periodicity of eight elements and their views on the different conceptions of the atom by chemists and physicists. The views of Hettema and Kuipers on the reduction of the periodic system to atomic physics are also considered critically.

1. Introduction

In 1988 Theo Kuipers and a young colleague, Hinne Hettema published an article in which they claimed to have formalized the periodic system of the chemical elements and to have arrived at some conclusions regarding the reduction of the periodic system (Hettema and Kuipers 1988). In 1997 I published an extensive critique of this article in which I claimed that the formalization had been carried out inappropriately and that any subsequent conclusions concerning the reduction of the periodic system by these authors were unfounded (Scerri 1997a).

A few of my criticisms have been addressed by Kuipers and Hettema in their more recent article (Hettema and Kuipers 2000). In addition I have been kindly invited to contribute to the Kuipers volume in view of my earlier critique. What I hope to do in the present article is to put my objections in a clearer manner than I had before. In addition I will attempt to respond to what the authors have said in their initial responses to me in their more recent publication. I believe I now understand the intentions of Hettema and Kuipers more clearly than I did originally and can therefore make my critique altogether more pertinent.

2. Critique of the New Article of 2000

In order to consider the new article I will proceed systematically through its text and will pause to make comments whenever I consider that they are

In: R. Festa, A. Aliseda and J. Peijnenburg (eds.), *Cognitive Structures in Scientific Inquiry* (*Poznań Studies in the Philosophy of the Sciences and the Humanities,* vol. 84), pp. 191-210. Amsterdam/New York, NY: Rodopi, 2005.

warranted. Just as they did in their 1988 article the authors are claiming to have carried out a formalization of the periodic system of the elements. More specifically they claim to have obtained a structuralist reconstruction of the development of the periodic system which allows them to discuss the question of the theoretical status of the periodic system. Among the most important issues addressed is "whether the Periodic Table is a proper theory or merely an empirical law." Mendeleev's periodic table and the modern version of the table are referred to as the naïve and sophisticated versions of the table respectively. The difference between them is explored in detail since the authors believe these differences to be "crucial" to their further discussion.

First of all whereas in their original article the authors cited some rather obscure sources on the history of the periodic system, I was gratified to see some attempt to redress the balance which I pointed out was lacking. The more recent article cites the book by Brock. Although this represents an improvement, Brock's excellent overview of the history of the whole of chemistry is not sufficiently comprehensive to be considered as an authoritative treatment of the periodic system. In spite of my earlier suggestion the authors have failed to consult or cite the only existing authoritative book-length treatment of the history of the periodic system by their fellow countryman Johan van Spronsen.

In their introductory section the authors state that one of their main aims is to discuss whether the Periodic table is a proper theory or an empirical law. As I indicated in my earlier comments I fail to see why this question should even arise since as far as I am aware, nobody has ever considered the periodic table as any kind of theory. The periodic table is a mere representation of what is frequently referred to as the periodic law. Whereas the principal discoverer of the periodic system consistently referred to the "periodic law" it is fair to say that modern treatments show less inclination to accord it a law-like status due to the presumed reduction of this law by quantum mechanics. Be that as it may there has never been any question, either at the time when the periodic system originated or in modern times, of regarding the periodic system as any form of theory.

Another issue that is announced in the introduction is that the authors intend to maintain their previously drawn distinction between what they term the naïve periodic law and the sophisticated periodic law. While agreeing that there may be grounds for making such a distinction I disagreed with the way in which the authors sought to make this distinction in my first critique. Since this is one of the points on which the authors seem to have addressed my remarks I will enter into some details and try to take the debate a little further.

According to Hettema and Kuipers the important feature which distinguishes the original periodic tables of Mendeleev and others from the modern periodic law is that the older versions called for a periodicity of eight elements while the modern form recognizes the fact that the lengths of successive periods can vary. In my first critique I pointed out that Mendeleev had devised many periodic tables and that few of these required that the periodicity of eight be maintained throughout the periodic system. I pointed out that Mendeleev's famous table of 1871 showed periodicities of 7,7,17,17. Hettema and Kuipers now respond by saying that they have consulted the book by Posin and that they have found that there are many different periodic tables. But they still seem to want to focus on one particular periodic table authored by Mendeleev in 1871 which does indeed show a regular periodicity of eight. They also deny my statement regarding the variation of periodicity that I attributed to Mendeleev and for the same year. This confusion is partly my fault since I should perhaps have stated that there were more than one table published by Mendeleev in the year 1871. However by referring the authors to the book by Van Spronsen I had hoped that they would discover this fact for themselves. Instead they have concentrated their attention on yet another marginal source, as far as serious scholarship on the periodic system is concerned, and they seem to have formed the opinion that there was only one Mendeleev table of 1871.[1]

As I also pointed out, Mendeleev's very famous first published table of 1869 does not show the elements distributed into periods of eight.[2] Indeed even in the cases where Mendeleev appears to be implying a periodicity of eight, if one looks at his table closely one notices that some of the columns consist of two columns of elements offset from each other. To maintain the notion of a periodicity of eight one would need to consider all such elements within a column as chemically analogous. Mendeleev was far too sophisticated a chemist to make such a mistake, a fact that is quite evident from the passage I quote below from his textbook, The Principles of Chemistry.

> Notwithstanding the resemblance in the atomic composition of the cuprous compounds, CuX, and the silver compounds, AgX, with the compound of the alkali metals, KX, NaX, there is a considerable degree of difference between these two series of elements. The difference is clearly seen in the fact that the alkali metals belong to those elements which combine with extreme facility with oxygen, decompose water, and form the most alkaline bases; whilst silver and copper are oxidised with difficulty, form less energetic oxides, and do not decompose water, even at a rather high temperature; they even displace

[1] I might just add that any scholar who has worked on the periodic system is well aware of the rather limited value of Posin's highly over-imaginative account of the life of Mendeleev. For example Greenaway's entry on the history of the periodic system in the Encyclopedia Brittanica warns the reader that Posin's book is a "fanciful and romanticized version."

[2] All I can do is to refer Hettema and Kuipers once again to the standard reference on the history of the periodic system, namely Van Spronsen's excellent book which contains diagrams of all the tables I have mentioned.

> hydrogen from very few acids. The difference between them is seen in the dissimilarity of the properties of many of the corresponding compounds. Thus cuprous oxide, Cu_2O, and silver oxide Ag_2O are insoluble in water; the cuprous and silver carbonates, chlorides and sulphates are also sparingly soluble in water. The oxides of silver and copper are also readily reduced to the metal. This difference in properties is in intimate relation with that difference in the density of the metals which exist in this case. The alkali metals belong to the lightest, and copper and silver to the heaviest therefore the distance between the molecules in these metals is very dissimilar- it is greater for the former than the latter. From the point of view of the periodic law this difference between copper and silver and such elements of the I group as potassium and rubidium is clearly seen from the fact that copper and silver stand in the middle of *those large periods* (for example, K, Ca, Sc, Ti, V, Cr, Mn, Fe, Co, Ni, Cu, Zn, Ga, Ge, As, Se, Br) which start with the true metals of the alkalis - that is to say, the analogy and differences between potassium and copper is of the same nature as that between chromium and selenium, or vanadium and arsenic. (Mendeleev 1891, vol. II, pp. 372-3) [my italics].

The reader will also note in passing that Mendeleev explicitly refers to "those large periods," a feature which flatly contradicts Hettema and Kuipers contention that Mendeleev regarded periodicity as unchanging in length. Furthermore, if one considers the periodic tables of other authors during the pioneering days of the periodic table, one arrives at the same conclusion. For example, most of the tables published by Lothar Meyer, Mendeleev's main contemporary, also contain periods of unequal lengths. The reason why I have labored this point is because the main part of the subsequent program of Hettema and Kuipers is predicated on their incorrect distinction between a naïve periodic system which requires unchanging periodicity and the sophisticated version in which a periodicity of eight is just one of many possible values.

3. The Periodic Table in Chemistry

The second section of the recent article by Hettema and Kuipers is titled "The Periodic Table in Chemistry." Here the authors touch very briefly on the history of the periodic system and claim that three important steps had occurred in chemistry which proved fundamental to the construction of the periodic table. These are,

> First of all there was a working concept of the chemical elements.
>
> Secondly it was known that all the chemical elements had different masses, even though it was only possible, initially, to ascribe crude relative weights to each of the elements.
>
> Thirdly it was known that some elements had very similar chemical behaviour. (Hettema and Kuipers 2000, p. 287-8)

The authors claim that these criteria proved sufficient for the construction of the Periodic Table. I would like to comment in passing that each of these three points, as stated, contains some truth but also serves to mask the historical situation due to the manner in which it is expressed in the above list. First of all it is by no means clear that there was a common working concept of the chemical elements. It has been claimed for example, that Mendeleev was able to make more progress than others in developing a successful periodic system precisely because he had a different conception of the nature of the elements (Scerri 2000a). In any case I believe the authors owe it to the reader to explain what they might mean by this working concept of the elements which they say was in place. Are they referring to the concept devised by Boyle or Lavoisier or Mendeleev himself?

The second point is only partly correct in that it was indeed recognized, largely as a result of the work of Dalton, that different elements possessed different masses. But this recognition hardly sufficed to allow the periodic system to emerge. The real problem lay in the conflicting schemes which were used to obtain the relative weights of the different elements. This crisis reached such proportions that in 1860 the first ever international chemical conference was convened in Karlsruhe, following which some semblance of order began to emerge. Only then did it become possible for a coherent periodic system to be assembled.

The third point is essentially correct except for the inclusion of the word 'very' in describing the similarities between some of the elements. There are many examples of groups in the periodic table, even among those recognized in the early days of its history, which are in fact rather different. The elements carbon, silicon, germanium, tin and lead serve as good examples. Carbon is a hard, black, non-metal or a gem stone in one of its other allotropes called diamond. Silicon and germanium are both semi-metallic and also semi-conductors of electricity and heat. Tin and lead are examples of metals which have been known since antiquity. These elements cannot really be said to be very similar.

But the objection to "very similar chemical behaviour" being a necessary condition for the establishment has a more profound aspect. Mendeleev is known to have had a rather complex view of the nature of the elements whereby he did not base his periodic system on the similarity of the elements as simple substances but on what he referred to as "basic substances." For example if one were to consider fluorine, chlorine, bromine and iodine one would probably not see any great similarities among them qua "simple substances" since they are two gases, a red liquid and a violet-black solid respectively. The similarity only emerges if one considers the compounds formed by these elements with sodium, for example, and the fact that they all form crystalline white solids.[3] According to Mendeleev it is the unobservable elements, in the sense of basic substances, rather than the simple substances which can be isolated, that form the basis of the periodic system.

4. Formalization

In the third section of their recent article Hettema and Kuipers develop the machinery which they hope will permit them to discuss the status of the periodic law and whether the periodic law is reduced to modern atomic theory. One step taken involves the assumption of a finite set E, representing the set of chemical elements. I cannot help wondering why they need to assume that the set of all elements will be finite, since there is no experimental indication that we are close to reaching the last of the elements or that this limit is even on the foreseeable horizon. If anything, the news from the laboratories which specialize in the synthesis of superheavy elements has been rather encouraging for those who believe that there are many more elements left to be synthesized (Armbruster, Hessberger 1998). Unless Hettema and Kuipers can adduce some convincing arguments from nuclear physics as to why they believe that the list of elements must necessarily be finite I do not see why they are entitled to make this assumption within their formal scheme.

After outlining five points, which form their definition 1, the authors state that z or atomic number is the only theoretical term. They add to this the claim that "no experimental problems arose in either the measurement of the atomic mass function m or the chemical similarity function ~ which also feature in their definition. Unless the authors are attributing a highly specific and unexplained meaning to these claims it is difficult to agree with them. There were in fact many severe problems associated with the definition and measurement of atomic weights. As I mentioned above, in passing, this was the main reason why the development of the periodic table was delayed until the late 1860s even though the other "sufficient conditions," as they are termed by the authors, were already in place by the beginning of the nineteenth century. In addition, to claim that chemical similarity could be established with "no experimental problems" is simply incorrect.[4] There are many cases of elements whose properties were very well known and which had been isolated

[3] I have carried out this fairly detailed analysis of the phrase "very similar chemical behavior" because it is one which recurs frequently in the recent article by Hettema and Kuipers.

[4] The further claim by Hettema and Kuipers is that the measurement of *m* was possible by making use of the "ideal gas law". Although this was true of some elements, not all elements can be easily vaporized with the result that their atomic weight remained undetermined, or mistaken, until another method was devised by Dulong and Petit and quite a different technique by Mitscherlich.

in adequate amounts but whose chemical similarities remained ambiguous for long periods of time.

5. Beryllium

The case of the placement of beryllium is a historically significant one because it involved a controversy that lasted a considerable period of time. The question was whether the element should be assigned a valency of 2 or 3 that would affect its atomic weight and it would, in turn, govern the position it took up in the periodic table.

In the case of metallic elements, Cannizzaro's method for determining atomic weights was not easy to apply, as it required volatile compounds. Instead, other methods such as one based on Dulong and Petit's law of atomic heats continued to be used. Furthermore, the chemical characteristics of the oxides generally provided an indication of the valency of the metal concerned. These rules are summarized below.

low valency oxides	MO & MO_2	strongly basic
intermediate valency oxides	M_2O_3	weakly basic
high valency oxides	MO_2, M_2O_5, MO_3	acidic

The metal beryllium provided one of the most severe tests for Mendeleev's system. The question was whether to place beryllium in group II above magnesium or in group III above aluminum. Its measured specific heat of 0.4079 indicated an atomic weight of approximately 14, which would place beryllium in the same group as the tri-valent aluminum. Furthermore, beryllium oxide is weakly basic, the lattice structure of the metal is unlike that of magnesium and beryllium chloride is volatile just like aluminum chloride. Taking these facts together, the association of beryllium with aluminum appears to be compelling.

In spite of all this evidence Mendeleev supported the view that beryllium is di-valent using arguments which were purely chemical, as well as arguments based on the periodic system. He pointed out that beryllium sulfate presents a greater similarity to magnesium sulfate than to aluminum sulfate and that whereas the analogues of aluminum form alums, beryllium fails to do so. He argued that if the atomic weight of beryllium were 14, it would not find a place in the periodic system. Mendeleev noted that such an atomic weight would place beryllium near to nitrogen where it should show distinctly acidic properties as well as having higher oxides of the type Be_2O_5 and BeO_3 which is not the case. Instead Mendeleev argued that the atomic weight of beryllium might be approximately 9, which would place it between lithium (7) and boron (11) in the periodic table.

In 1885 the issue was conclusively settled in favor of Mendeleev by measurements of the specific heat of beryllium at elevated temperatures. These experiments pointed to an atomic weight of 9.0 in reasonable agreement with Dulong and Petit's law and supported the di-valency of the element. The difficulties involved in this case, and others like it, demonstrate that chemical similarity is far from a trivial matter to establish.

6. The Placement of Lutetium and Lawrencium in the Periodic Table

The case that will be examined in this section involves a change to the periodic classification that has only been carried out in the last twenty years and, which to judge from the vast majority of chemistry and physics textbooks, has yet to be widely assimilated. The debate is over which two elements, both lanthanum and actinium or both lutetium and lawrencium should be placed under scandium and yttrium in group 3 of the periodic table.[5] A considerable amount of physical and chemical evidence has now been established to show quite convincingly that the more correct placement implies that group 3 consists of scandium, yttrium, lutetium and lawrencium.

Until relatively recently the use of electronic configurations dictated that the elements lanthanum and actinium should appear in these positions instead of lutetium and lawrencium. In order to appreciate this situation the electronic configurations which were formerly supposed to occur in the atoms of ytterbium (atomic number 70) as well as lutetium (atomic number 71) must be considered.

Ytterbium	Yb	[Xe] $4f^{13}$ $5d^{1}$ $6s^{2}$
Lutetium	Lu	[Xe] $4f^{14}$ $5d^{1}$ $6s^{2}$

According to this assignment the differentiating electron, that is the final electron to enter the atom of lutetium, is regarded as an *f* electron. This suggests that lutetium should be the final element in the first row of the rare earth elements, in which *f* electrons are progressively filled, and not a transition element as had previously been believed by chemists.

As a result of more recent spectroscopic experiments, however, the configuration of ytterbium has been altered to (Jensen, 1982).

Ytterbium	Yb	[Xe] $4f^{14}$ $5d^{0}$ $6s^{2}$

---

[5] The IUPAC numbering scheme for the groups of the periodic table, which run from 1 to 18, has been used. The older systems denote this as group IIIB in the US and group IIIA in Europe.

while that of lutetium remains unchanged. Ytterbium therefore now appears to mark the end of the rare earths. The subsequent element lutetium shows a differentiating electron labeled d, spectroscopically, which makes it an equally good candidate as lanthanum, of configuration [Xe] $5d^1$ $6s^2$, for the role of the first element in the third transition series.

Renewed chemical and physical measurements have shown conclusively that lutetium rather than lanthanum bears a close similarity with scandium and yttrium (Jensen 1982). Here then is another example that clearly shows some rather difficult experimental problems concerning the placement of the elements in the periodic system. The statement by Hettema and Kuipers that chemical similarity can be established without experimental problems is patently false, especially as such problems have persisted up to the present time. Moreover, the above mentioned reassignment of lutetium and lawrencium is by no means universally accepted even at the time of writing (Nelson 1996).[6]

7. The Periodic Law

I return to the article by Hettema and Kuipers in order to consider the next section which is entitled "The Periodic Law." Here it is stated that the naïve version of the periodic law (NPL) is due to Mendeleev and that the present-day (sophisticated) one (SPL) was developed in the early 1900s. However neither at this point nor anywhere else do the authors explain precisely what they take to be the developments which underlie what they term the sophisticated periodic law although they claim that "SPL has been developed in close contact with Atomic Theory." As I suggested in my previous critique it is rather important for them to address this question since many developments might be held to be responsible for the change.[7] If we accept the authors'

---

[6] The reason why most chemistry and physics textbooks have not adopted the new assignments is not because their authors dispute them but simply that they are not aware of them. The popular Internet periodic table pages which are maintained by Mark Winter of Sheffield University does feature the new arrangement of elements in group 3. Web page http://www.webelements.com/index.html.

[7] Hettema and Kuipers have made some response to my question. At one point in their more recent article they attribute the sophisticated periodic law to the work of Niels Bohr. First they state that "the expression $2n^2$ corresponds to the principal quantum number" but then immediately add "It is however not identical to it." In fact none of the possible developments in Atomic Theory which the authors might be alluding to had any influence on the realization that the lengths of periods vary according to the formula $2n^2$. The latter follows entirely from chemical similarities. All that atomic theory provided was successive explanations of the periodicity. The periodicity itself and the points at which it occurs are chemical, and empirical, phenomena.

contention that the sophisticated periodic law is one that embodies the varying lengths of periods, according to the formula $2n^2$ then we can try to identify when such tables came into existence and try to see whether they did indeed follow any theoretical developments. However, this immediately raises a problem since many rather early periodic tables already displayed the characteristic of varying period lengths.

As I mentioned earlier many of the tables of Mendeleev, Lothar Meyer, Newlands, and other discoverers of the periodic system displayed varying period lengths. One of the earliest periodic tables which not only shows varying period lengths but which consists of the now familiar medium-long form of the periodic table, with varying lengths given by the formula $2n^2$, was published by the inorganic chemist Werner in 1905. But Werner's table was developed quite independently of any theoretical developments such as quantum theory or "Atomic Theory" of any kind.[8] The sophisticated form of the periodic table, in the sense of varying periodicity, did not require any input from atomic theory whatsoever but developed from empirical chemical observations regarding chemical similarities.

The authors then discuss what they term "three requirements which have to be included in the definition of the Periodic Table." These requirements are designated as monotomicity, surjection and injection and are directly connected with ordering according to atomic weights, the lack of empty spaces in the table and a one-one relationship between the order of the elements respectively.[9] While accepting that some of these requirements show exceptions in the sense of pair reversals and the existence of isotopes the authors consider that these requirements are nonetheless useful for establishing the relationship between Mendeleev's naïve periodic law and the sophisticated periodic law. The naïve periodic law is formalized as,

$$\text{(NPL)} \qquad e \sim e' \text{ iff } \quad |\, z(e) - z(e') \,| \text{ is a multiple of 8.}$$

The formalization given to the sophisticated periodic law is more complicated in that it consists of two parts, the second of which itself is comprised of two parts. The conclusion which Hettema and Kuipers reach is that "It is easy to check that SPL reduces to NPL if n is fixed at the value of $n = 2$ (and hence $2n^2 = 8$)." I would like to suggest that this reduction which the authors claim to have established is trivial and that its establishment does not require any formalization of the periodic system. Surely one could simply state that whereas the older periodic tables had envisaged periods of eight elements, the

[8] A diagram and account of Werner's table can be found in Van Spronsen's book (Van Spronsen 1969, p. 152-154).

[9] This implies that if two elements have the same order number, they are necessarily the same element.

new version now generalized the lengths of periods to any value conforming to the formula $2n^2$ of which 8 is one special case.[10] Why do we require an elaborate formalization in order to establish this trivial connection?

8. The Atomic Theory and the Periodic Table

In the same section the authors draw another distinction which I began to dispute in my earlier critique. Hettema and Kuipers try to distinguish between what they call the chemical and the physical conception of the atom. Although such an argument could be made in principle, I believe that the version proposed by the authors is erroneous. The authors' first attempt at making this distinction takes the form of,

> …for the chemist, an 'atom' is viewed as an inherent part of a molecule, for the physicist, an atom represents first and foremost a nucleus surrounded by a cloud of electrons. The behaviour of the latter is described by quantum mechanics. (Hettema and Kuipers 2000, p. 296)

If such views might ever have distinguished chemist from physicist then I strongly suggest that such differences have ceased to exist since the advent of quantum mechanics into chemistry. Any casual examination of chemistry textbooks will show that modern chemistry is entirely based on the model of the atom as a nucleus surrounded by a cloud of electrons. Moreover it is not just physicists seeking to understand the atom who draw on quantum mechanics. The theory has become an essential part of any elementary high school or undergraduate course in general chemistry, to say nothing of further study and research in chemistry.

Hettema and Kuipers continue by expanding on the claimed distinction between the chemical and physical conceptions of the atom. They claim that chemists are accustomed to using a qualitative version of the physical picture of the atom whose full implications involves the use of quantum mechanics and computation. After citing some of my previous criticisms approvingly,[11] the authors claim that these qualitative versions used by chemists that involve separation of electrons into core and valence shells,

> can be called 'chemical' again since they deal with the functional definition of the atom as part of a molecule. (Hettema and Kuipers 2000, p. 297)

[10] However, I am disputing this way of characterizing the naïve periodic system.

[11] Hettema and Kuipers cite me as saying, "such explanations are indeed frowned upon by physicists as being of a typically picturesque and naïve kind, typical of chemists." What Hettema and Kuipers may not have realized was that the main culprit I had in mind was precisely the view of certain numbers of electron in particular shells, the model which they devote so much attention to in their articles.

I regret to say that I also propose to dispute this identification. The separation between core and valence electrons is made as much by physicists as it is by chemists. Indeed the highly successful physics sub-discipline of spectroscopy is dominated by the assumption that the outer electrons are responsible for observed spectroscopic transitions (Condon and Shortley 1935). In addition the use of this approximation throughout atomic spectroscopy confirms that it is in no way linked to the definition of an atom as part of a molecule as Hettema and Kuipers claim.

Also, as I tried to emphasize in my earlier comment, chemists are as much concerned with atoms and the properties of pure elements as they are with those of molecules and compounds. A good case in point is the study of the periodic classification of the elements itself. In establishing or studying the periodic system chemists are concerned with atoms and elements. This is regardless of whether they were early pioneers who depended on observable properties or modern chemists who make reference to electron shells and quantum mechanics to make sense of the periodic classification.

Before leaving this section I would like to cite again a paragraph from Hettema and Kuipers for convenience.

> For most chemists, the correct functional definition of an atom is 'part of a molecule' and the most important property of the atom is its chemical valency. For most physicists, an 'atom' is primarily a nucleus surrounded by a cloud of electrons, the latter being described by quantum mechanics. (Hettema and Kuipers 2000, pp. 297)

The implication that chemists do not use quantum mechanics is mistaken. In fact it is well known to science educators that quantum mechanics is made greater use of in courses in chemistry than it is in physics. In physics courses quantum mechanics represents just one of many topics such as electromagnetism, classical mechanics and relativity, whereas in chemistry quantum mechanics is the dominant theory which is used to explain the properties of all forms of matter. While it may be correct to say that chemists use watered-down versions of quantum mechanics it is an exaggeration to imply that they do not really use quantum mechanics at all as the authors seem to be doing here.[12]

As in the claimed distinction between the naïve and the sophisticated periodic laws Hettema and Kuipers proceed to try to establish the relationship between their conception of the chemist's and the physicist's atom. This part of the project begins with the statement that,

> In the chemical picture of the atom for instance, 'chemical similarity' includes 'having the same valency' whereas in the physical picture, 'chemical similarity' can be related to

[12] This is true even if one accepts the authors' claim that they are merely representing the extreme positions that differentiate chemists from physicists.

> similarities in the electronic configurations (in some cases the valence electrons). This means automatically that the concept of valence itself can be related to 'outer electron configuration'. (Hettema and Kuipers 2000, p. 297-8)

But this sense of chemical valency is one that is more characteristic of modern chemists than physicists. It has become one of the main paradigms of modern chemistry that valency is governed by outer electrons. What the authors are describing as the physical conception of valency and the atom is in fact the modern chemical conception. Meanwhile, to attribute such a view to physicists is mistaken, precisely because physicists go well beyond the independent-electron approximation that assumes that we can speak of a particular number of electrons in any particular shell.

This is a central point that I would like to impress upon Hettema and Kuipers since my previous attempt to do so seems to have failed. The view of particular numbers of electrons in shells around the nucleus dates from the Bohr model of 1913 and further developments in 1922. With the advent of the Pauli Exclusion Principle in 1925 it was realized that individual electrons are not in stationary states although the atom as a whole does still possess stationary states. Calculations on the energies of atoms and molecules must necessarily consider the mixing of electronic configurations if they are to recover anywhere near to the experimental energies of such systems. But even such an interpretation of the calculations involves a partial return to the notion of particular numbers of electrons in shells. In fact all talk of electrons in shells is banished in accurate calculations. The physicist goes beyond the orbital approximation of particular electrons in shells around the nucleus. Rather than being characteristic of the physicists conception of the atom the latter interpretation has been bequeathed precisely to the chemist!

Because of these limitations of the independent-electron model the explanation of the periodic system that Hettema and Kuipers claim can be obtained in terms of the number of outer-shell electrons is somewhat approximate. Nevertheless it does give a post facto explanation of the lengths of successive periods and thus of the $2n^2$ rule which was featured earlier.[13] The authors now acknowledge that the explanation is not complete since the electron shells do not fill sequentially as I argued in my earlier critique. But rather than facing the theoretical problems which this feature raises they merely refer the reader to a standard textbook on quantum mechanics. The

[13] Hettema and Kuipers state that the old quantum theory of Bohr is sufficient to explain the $2n^2$ rule and that the advent of quantum mechanics as developed by Shrödinger and Heisenberg "does not alter this interpretation." In saying this they fail to mention that the crucial step in the understanding of the periodic table in terms of numbers of electrons in shells and quantum numbers was provided by the Pauli Exclusion Principle and his postulation of a fourth quantum number. Without this development the old quantum theory failed to explain the form of the periodic system.

problem is that many textbooks do not acknowledge that the explanation for the periodic system given in terms of quantum numbers is only successful within the limitations of the homely model of electrons in shells.

When faced with the question of the reduction of the periodic table the modern theoretical physicist requires a deeper level of explanation. He or she is more likely to seek a quantitative prediction of some atomic property or other which shows periodicity and whose experimental values may be compared with calculated values. Such a property consists in first ionization energy, for example, and there are indeed good theoretical predictions of ionization energies that can be obtained from quantum mechanics. But this kind of approach requires going beyond the independent-electron approximation, and the associated notion of specific numbers of electrons in shells. Even then some physicists claim that such a reduction is not sufficiently deductive since the Schrödinger equation for each atom must be solved individually for each atom (Ostrovsky 2001). Such *ab initio* quantum chemistry carried out using linear expansions of terms made up of electronic configurations does not provide a general solution for any atom or molecule. The Schrödinger equation for each atom must be solved from first principles using a basis set (linear combination of electronic configurations) which incidentally is still chosen by reference to the Aufbau principle and not deduced from first principles.[14]

A more satisfactory reduction, but still approximate, can be achieved by using density functional theory. In 1926 Thomas proposed treating the electrons in an atom by analogy to a statistical gas of particles. No electron shells are envisaged in this model although electrons may still possess values for angular momentum as they do in the electron shell model. The method was independently rediscovered by Fermi two years later, and is now called the Thomas-Fermi method. For many years it was regarded as a mathematical curiosity without much hope of application since the results it yielded were inferior to those obtained by the method based on electron orbitals or methods based on orbital expansions.

Gradually the Thomas-Fermi method, or its descendants that have become known as density functional theories, has become as powerful as methods based on orbitals and in many cases can outstrip the orbital approaches in terms of computational accuracy (Gill 1998). The reason why these approaches may be considered more genuinely ab initio, or a deeper form of reduction, if I may speak loosely, is that one obtains a global solution for all the atoms in the

[14] The point I am making here is that although the calculation of the ground state energy of an atom, or its ionization energy, appears to be carried out rigorously from first principles, the choice of the basis set cannot be deduced from first principles. There is a strong sense in which so-called ab initio calculations are not strictly ab initio because of this feature (Scerri 1998, 1999, 2000d).

periodic table and even elements not yet discovered. The solution is expressed in terms of the variable Z which represents atomic number, and is the crucial feature which distinguishes one kind of atom from that of any other element. One does not need to repeat the calculation separately for each atom since the equation is solved once and for all for all possible atoms.

Incidentally, there is an important conceptual or even philosophical difference between the orbital methods and these density functional methods. It is that in the former case the theoretical entities are as a matter of principle completely unobservable whereas electron density invoked by density functional theories is a genuine observable. Experiments to observe electron densities have been routinely conducted since the development of X-ray and other diffraction techniques (Coppens 1997). This is why I and some others have been agitating about the recent reports, starting in Nature magazine in September 1999, that atomic orbitals had been directly observed (Scerri 2000b). This is simply impossible. Orbitals cannot be observed either directly, indirectly or in any other way since they have no physical reality. This state of affairs is dictated by quantum mechanics. Electron density is altogether different, as I have indicated, since it is a genuine quantum mechanical observable. I have tried to stress the educational implications of the claims for the observation of orbitals in other articles and will not dwell on the issue here (Scerri 2000c).[15]

9. A Case of Reduction

Hettema and Kuipers further claim that the case of the periodic table, concerning the relationship between the naïve and sophisticated versions, can be considered as an interesting case of a reductive explanation or reduction for short. They draw upon an account of explanation that requires what they term aggregation, identification and approximation. On the question of identification the authors state that,

> The necessary link between chemical similarity and 'equal outer electron configuration' states that the latter causes the former (Hettema and Kuipers 2000, p. 300).

This is a point which I touched on in my earlier critique but which I need to emphasize further. As I stated before the possession of any particular electronic configuration by an element is neither necessary nor sufficient for chemical similarity with another element. It is rather easy to generate counter

[15] Indeed as time has passed the best of both approaches have been blended together. Many computations are now performed by a careful mixture of the orbital and density functional approaches that are used within the same calculation scheme.

examples that show quite convincingly the lack of necessity or sufficiency. The element helium has two outer-shell electrons that might lead one to think that it would necessarily be similar to alkaline earth elements such as calcium or magnesium. In fact nothing could be further from the truth since helium is the single least reactive element in the entire periodic system while magnesium and calcium are reactive metals.[16] One need only consider the vigorous reaction that occurs when a few calcium pieces are placed into a beaker containing water. So much for sufficiency.

Similarly the hope of any necessary connection which the authors believe exists also suffers from serious counter examples. The elements nickel, palladium and platinum are placed in the same column of the periodic table, namely group 10, because of their close chemical similarities. However no two elements within this group of three share the same electronic configurations in its two outermost orbitals. They are respectively, Ni $3d^8 4s^2$; Pd $4d^{10} 5s^0$; Pt $5d^9 6s^1$. Of course this is quite apart from the problems alluded to earlier concerning exactly what is meant by chemical similarity. There is no clear-cut notion of this concept in chemistry as shown by the difficulties in the placement of certain elements such as beryllium, lawrencium and lutetium into the periodic system.

10. Is the Periodic Table a True Theory?

Hettema and Kuipers begin the section with the above sub-title by conceding that,[17]

> Most books on the subject of practical chemistry treat the Periodic Table as a *table*, and do not mention the word theory. (Hettema and Kuipers 2000, p. 300)

But they go on to propose that in fact the naïve version of the periodic table associated with Mendeleev's table should be regarded as a theory and that the sophisticated periodic law must be regarded as an empirical law because of the explanation which is provided by atomic theory. The basis of this claim seems to be the particular analysis that the authors have utilized concerning the "proper theories" and "empirical laws" and the relationship between them.

> A theory is a proper theory if it has at least one T-theoretical term. It is an empirical law (in the strict sense) if it has none. Hence an empirical law is an improper theory, i.e. a theory without theoretical terms of its own (Hettema and Kuipers 2000, p. 301).

[16] The only similarity might be the observed pattern of splitting of spectral lines in the presence of a magnetic field. However this can by no means be referred to as a chemical similarity.

[17] The authors would have been more correct in saying that there is not a single example of any book or article, either in chemistry or philosophy of science, apart from their own work, which has ever suggested that the periodic table should be regarded as a theory.

The crucial term that the authors take to distinguish Mendeleev's periodic table from the sophisticated version is atomic number. The authors claim that Mendeleev used atomic number implicitly while it still had no experimental underpinning and could therefore be regarded as a theoretical notion. The presence of this alleged theoretical term is thus taken to render Mendeleev's table, or the naïve periodic table, into a theory. By contrast the sophisticated periodic table also draws on atomic number as the ordering principle. But because of the theoretical account that is provided by atomic theory, Hettema and Kuipers conclude that atomic number is no longer a theoretical term. It follows in their view that the sophisticated periodic table cannot be regarded as a theory but must be regarded as an empirical law.

I see at least one major flaw in this way of looking at the periodic table. I believe that the assumption that Mendeleev used atomic number implicitly cannot be sustained. What he used, as is well known, was atomic weight. The fact that atomic weight and atomic number are well correlated throughout the periodic system does not allow one to make the identification which Hettema and Kuipers wish to make. The use of atomic number has the virtue of solving the problem of remaining gaps in the periodic system in a definitive manner. Once Moseley had carried out his famous X-ray experiments it became possible to determine precisely which elements remained to be discovered or where any remaining gaps existed in the periodic table. However, Mendeleev and other pioneers of the early periodic system did not share this luxury. They had the difficulty of trying to estimate where any gaps might lie and where to place the known elements within columns of the table. What renders this task particularly difficult is that the increase in atomic weights of the elements is far from regular.

This can be illustrated by considering the values of the atomic weights of the first row of the rare earth elements for example.[18]

La	Ce	Pr	Nd	Pm	Sm	Eu	
138.9	140.1	140,9	144.2	(145)	150.4	152.0	
Gd	Tb	Dy	Ho	Er	Tm	Yb	Lu
157.3	158.9	162.5	164.9	167.3	168.9	173.0	175.0

Rather than a smooth progression in atomic weights one notices a virtual twinning of elements by increasing atomic weights. It is mainly because of the irregularity in the gaps between their atomic weights that the rare earths proved to be so difficult to place in the periodic system. But even when some

[18] Modern values of atomic weights are used and rounded to one decimal place. The value in parentheses refers to the weight of the most stable isotope of the element.

of these atomic weights became available it was not possible to infer which elements were still missing, a feat that only became possible following the discovery of atomic number. It is therefore rather far-fetched to claim that Mendeleev implicitly used atomic number as an ordering scheme.

As a matter of historical fact, the only pioneer of the periodic system who might be said to have anticipated atomic number was John Newlands, who did not receive much credit for his published periodic systems. Nevertheless, his use of ordinal number, rather than values of atomic weight, did not allow him to produce better periodic systems than his contemporaries. Just like Mendeleev, and others who worked with atomic weight, Newlands did not know what gaps to leave. Of course the ordinal numbers that he associated with successive elements, which were known at the time, do not correspond with the modern atomic numbers that are given by the number of protons in the nuclei of the various atoms in question.[19]

In the same section of their paper Hettema and Kuipers claim that

> ...Mendeleev was willing to admit that global satisfaction of the naïve empirical claim at least required acceptance of some local exceptions (Hettema and Kuipers 2000, p. 302).

This is unfortunately not quite the case although a view that is propagated by many textbooks on chemistry. Although Mendeleev reversed the elements iodine or that of tellurium on chemical grounds he did not consider these to be exceptions to the ordering principle of increasing atomic weights. Mendeleev maintained throughout his life that either the atomic weight of iodine or that of tellurium had been incorrectly determined and encouraged experimenters to re-determine the weights of these two elements. But despite strenuous efforts, on the part of many chemists, the order in the atomic weights of these two elements remained unchanged. Tellurium does indeed have a higher atomic weight and yet must be placed before iodine on chemical grounds. Whereas Mendeleev repeatedly stressed that there would be no exceptions to the ordering of elements according to strictly increasing atomic weights the subsequent discovery of ordering based on atomic numbers has shown that he was incorrect.[20]

11. Conclusion

After devoting so much space to criticizing the views of Hettema and Kuipers I would like to conclude by saying that they are to be applauded for undertaking

[19] The view that Newlands, in some sense, anticipated atomic numbers is not universally accepted and has been recently disputed by Giunta (1999).

[20] It is now known that tellurium is correctly placed before iodine because it has one fewer protons in the nuclei of its atoms. The lower atomic weight of iodine atoms is due to the fact that most common isotopes possesses fewer neutrons than the most common isotopes of tellurium.

the very difficult problem of the reduction of chemistry. Whereas one frequently hears complaints that chemistry has been sadly neglected in philosophy of science I believe that this situation is as much due to the difficulty of the problems it presents rather than mere avoidance on the part of philosophers. I hope that my comments will spur Hettema and Kuipers and others to renewed attempts towards the reduction of the periodic system. As Popper once wrote, reduction is not always successful but attempts to carry it through invariably deepen our knowledge of the phenomena concerned in unexpected ways (Popper 1974).

UCLA
Department of Chemistry and Biochemistry
Los Angeles, CA 90095, USA.
e-mail: scerri@chem.ucla.edu

REFERENCES

Armbruster, P. and F.P. Hessberger (1998). Making New Elements. *Scientific American* **279**, 72-77.

Brakel, J. van (1999). On the Neglect of Philosophy of Chemistry. *Foundations of Chemistry* **1**, 111-174.

Brock, W. (1992). *Fontana History of Chemistry*. London: Fontana.

Condon, E.U. and G.H. Shortley (1935). *The Theory of Atomic Spectra.* Cambridge: Cambridge University Press.

Coppens, P. (1997). *X-ray Charge Densities and Chemical Bonding*. Oxford: Oxford University Press.

Gill, M.W. (1998). Density Functional Theory. In: von Ragué Schleyer (ed.), *Encyclopedia of Computational Chemistry*, vol.1, pp.678-689. Chichester: Wiley.

Giunta, C. J. (1999). J.A.R. Newlands' Classification of the Elements: Periodicity, But No System, *Bulletin for the History of Chemistry* **24**, 24-31.

Hettema, H. and T.A.F. Kuipers (1988). The Periodic Table – Its Formalisation, Status, and Relation to Atomic Theory. *Erkenntnis* **28**, 387-408.

Hettema, H. and T.A.F. Kuipers (2000). The Formalisation of the Periodic Table, In: W. Balzer, J. Sneed, C. Ulises Moulines (eds.), *Structuralist Knowledge Representation. Paradigmatic examples,* pp. 285-305. *Poznań Studies in the Philosophy of the Sciences and the Humanities*, vol. 75. Amsterdam/Atlanta: Rodopi.

Jensen, W. B. (1982). The Positions of Lanthanum (Actinium) and Lutetium (Lawrencium) in the Periodic Table. *Journal of Chemical Education* **59**, 634-636.

Mendeleev, D. I. (1891). *The Principles of Chemistry.* Translated by G. Kemansky. London and New York: Longmans, Green and Co.

Nelson, P.G. (1996). Periodic Relations. *Education in Chemistry*, November, 149-149.

Ostrovsky, V.N. (2001). What and How Physics Contributes to Understanding the Periodic Law. *Foundations of Chemistry* **3**, 183-195.

Popper, K.R. (1974). Scientific Reduction and the Essential Incompleteness in All Science. In: F.L. Ayala and T. Dobhzansky (eds.), *Studies in the Philosophy of Biology*, pp. 259-284. Berkeley, CA: Berkeley University Press.

Scerri, E.R. (1991). Electronic Configurations, Quantum Mechanics and Reduction. *British Journal for the Philosophy of Science* **42**, 309-325.

Scerri, E.R. and L. McIntyre (1997). The Case for the Philosophy of Chemistry. *Synthese* **111**, 213-232.

Scerri, E.R. (1997). Has the Periodic Table Been Successfully Axiomatized? *Erkenntnis* **47**, 229-243.

Scerri, E.R. (1998). Popper's Naturalized Approach to the Reduction of Chemistry. *International Studies in Philosophy of Science* **12**, 33-44.

Scerri, E.R. (1999). Response to Needham. *International Studies in Philosophy of Science* **13**, 185-192.

Scerri, E.R. (2000a). Naive Realism, Reduction and the 'Intermediate Position'. In: N. Bhushan, S. Rosenfeld (eds.), *Of Minds and Molecules*. New York: Oxford University Press.

Scerri, E.R. (2000b). Have Orbitals Really Been Observed? *Journal of Chemical Education* **77**, 1492-1494.

Scerri, E.R.. (2000c). The Failure of Reduction and How to Resist the Disunity of Science in Chemical Education. *Science and Education* **9**, 405-425.

Scerri, E.R. (2000d). Second Response to Needham. *International Studies in Philosophy of Science* **14**, 307-315.

Spronsen, J.W. van. (1969). *The Periodic System of the Chemical Elements*. Amsterdam: Elsevier.

Theo A. F. Kuipers

ON DESIGNING HISTORICALLY ADEQUATE FORMAL RECONSTRUCTIONS

REPLY TO ERIC SCERRI

Scerri's review of the first (1988) paper by Hinne Hettema and me[21] on the periodic table ended with the statement: "To conclude, I believe that the periodic table of the elements has yet to be axiomatized successfully, although the bold attempt by Hettema and Kuipers has raised a number of key issues in the philosophy of chemistry" (p. 239). The concluding section of his present review of our revised version, of 2000, begins with: "After devoting so much space to criticizing the views of Hettema and Kuipers I would like to conclude by saying that they are to be applauded for undertaking the very difficult problem of the reduction of chemistry." In view of Scerri's even more severe criticisms in the second review, its concluding statement is an even more generous statement than that of the first review. Hence, we have at least to concede that even the second version of our reconstruction and further discussion of the periodic table leaves much to be desired.

As Scerri has noticed, we did not respond in all relevant respects to his first criticisms. Apart from the fact that there was not much time between the appearance of Scerri's paper and the deadline for the revised version, it was also clear that a lot of new research would be necessary for a thorough revision. We should like to thank Scerri for his effort to clarify and elaborate a number of points in his present review. Together with the other critical points of his first review, they will certainly be of great help to somebody who might want to undertake a third attempt, in particular regarding the reduction of the periodic table.

Certainly, the main lesson to be drawn is that designing historically adequate formal reconstructions is an enormous job, not in the least "due to the difficulty of the problems it [chemistry] presents" as Scerri remarks. But it is worthwhile, for it will "deepen our knowledge of the phenomena" as he also

[21] This reply continues in the 'we' form for it is also on behalf of Hettema, although he maintains that this reply is too apologetic.

In: R. Festa, A. Aliseda and J. Peijnenburg (eds.), *Cognitive Structures in Scientific Inquiry* (*Poznań Studies in the Philosophy of the Sciences and the Humanities*, vol. 84), pp. 211-216. Amsterdam/New York, NY: Rodopi, 2005.

suggests. Moreover, we would like to add, although our reconstructions of a naïve and a refined version of the periodic table were in many historical respects problematic, conceived in purely formal terms they can still perfectly illustrate how such structuralist reconstructions may appear like and how they can be mutually related and related to other theories. That is, they may at least be conceived as toy structuralist examples with heuristic use-value. More generally, it is fortunate that Scerri's paper and some others (e.g. those of Van den Bosch and Causey in this volume and of Hamminga in the companion volume) draw attention to the structuralist approach. Unfortunately, Wolfgang Balzer, one of the pioneers of this approach, was not able to write his intended contribution on the relation between the HD-method and the structuralist approach.

For now, we should like briefly to discuss some specific points raised by Scerri, and conclude somewhat more extensively on the topic of the epistemological status of the periodic table.

Some Main Points of Full or Qualified Agreement, Sectionwise

On Section 2. It is indeed overly simplified to claim that Mendeleev designed his table with a periodicity of 8, for he was well aware of the possibility of, or even the need for different periodicities. In view of Scerri's Note 1, we are afraid that, stimulated by D. Posin's 1948 book on Mendeleev, we committed "formal romanticization," the main temptation of formal reconstruction. Moreover, despite some practical obstacles, we should have consulted Van Spronsen (1969).

On Section 3. Our presentation of the three conditions that were sufficient for the construction of the periodic table should have been amended in two respects: they were conditions as Mendeleev perceived them, with the important qualification on the notion of "similar chemical behavior" as indicated by Scerri.

On Section 4. Indeed, we did not justify our assumption that the set of chemical elements is finite. In view of our Note 2, which leaves room for elements to be discovered or artificially created, this assumption seems to become even stranger. However, as we should have mentioned, there are justifications. To understand these, one only has to investigate whether there is a certain value for the atomic number Z at which the electronic theory breaks down in the sense that a stable electronic behavior is no longer possible.[22] There are no hard and fast rules to determine how heavy a nucleus can get before a quantum theory of its electron cloud becomes problematic. The point is most easily seen for the one electron atom, where relativistic quantum mechanics, for example, indicates that the heaviest atom theoretically possible has an atomic number of the order of 137.[23] Beyond this point, depending on the theory chosen to describe the electronic behavior, the ground state becomes unstable. For instance, according to Bohr's semi-classical method, the inner electrons of an atom with a higher number would have to travel at a speed that exceeds that of light, which is physically impossible. Similar boundaries arise in modern versions of relativistic quantum mechanics.

On Sections 4, 5 and 6. Yes, our claim that no experimental problems arose in measuring atomic mass and establishing chemical similarity is clearly overstated. From the context, however, it was also clear that we only wanted to hint at the absence of circularity problems in interpreting the relevant experiments related to the periodic table. The beryllium case shows that even this claim is false as far as Mendeleev himself is concerned for he apparently used arguments from his table. However, as Scerri also reports, in the end experiments that were independent of the table settled the valency of beryllium. Moreover, as far as we can judge from Scerri's description, the experimental problems around lutetium and lawrencium that have arisen recently are also not intrinsically related to the periodic table. To be sure, in all cases, the experiments and criticisms are certainly guided by the table, but as explained in Ch. 2 of SiS, the periodic table provides perfect illustrations of the important distinction between (intrinsically) theory-laden and (merely) theory-guided observation.

---

[22] This leaves aside the discussion of the stability of the nucleus, which is a different theory and a different discussion altogether. In the context of this discussion it is conceivable, for instance, that a "stable" nucleus with a high Z value, such as 150 or more, would be stable for some time (which could be anywhere between a microsecond to a couple of seconds). While such a discovery would be highly exciting, the question of whether a stable electronic "cloud" could form around this nucleus, and whether therefore a meaningful chemistry would be possible with these heavy atoms, is yet another issue.

[23] The number of 137 is not gospel *per se*. It is, to be precise, derived from either Bohr's semiclassical theory of the atom, or the one-electron Dirac equation with a point nucleus. For instance, the Klein-Gordon equation (which does not take spin into account) has a catastrophe even when $Z > 137/2$ (see Itzykson and Zuber 1980). On the other hand, for the one electron atom, taking the physical extension of the nucleus into account pushes this point of instability to about $Z = 175$ (Itzykson and Zuber 1980, p. 83).

On Section 7. Regarding the sophisticated periodic law, it is important to make a distinction between its discovery and its status. Scerri is right in claiming that it was discovered, by Werner, independently of atomic theory. However, in view of the fact that the latter explained the former, with the consequence that table-independent measurement of the atomic number became possible, it lost its status as a proper theory in the sense of no longer having a proper theoretical term, viz. atomic number. Incidentally, we did not claim that the elaborate formalization was set up in order "to establish this trivial connection" between the naïve and the sophisticated law.

On Section 8. Scerri relativizes our distinction between a chemical and a physical conception of the atom, to some extent convincingly. However, he might have stressed our remark (reported in his Note 12) that we were sketching "the extremes of a gradual transition." Moreover, regarding *ab initio* quantum chemistry we would claim that that does not solve the Schrödinger equation atom by atom and that the relevant basis set has a very tenuous relationship with the Aufbau principle.

Let us very briefly sketch the practice of ab initio quantum chemistry to elucidate this point. In ab initio quantum chemistry the aim is to solve the electronic structure problem for either an atom or a molecule. In the most commonly practiced method, one chooses a basis set for each atom, and then proceeds to compute the overlap, potential, kinetic (1-particle) and coulomb and exchange (2-particle) integrals over the functions of the basis set. Using these integrals, a Fock matrix is constructed, which is used to iteratively solve the Fock equation until the solution is self-consistent. The point is that the wavefunction is computationally expressed as a linear combination of orbitals (a Slater determinant), which in turn are expressed as a combination of basis set functions. The electron correlation problem is generally solved on top of this Self-Consistent Field (SCF) wavefunction by either Many Body Perturbation Theory (MBPT), Configuration Interaction (CI) or more sophisticated methods such as Coupled Cluster (CC). The choice of basis set is thus pivotal to the overall quality of the calculation. If a wave function exhibits certain properties, these will only be found in the calculation if the original basis set was "rich" enough to express these properties.

The practical problem is that even a simple SCF calculation grows in complexity with the fourth power of the number of basis functions, while correlated calculations typically grow in complexity with the fifth or sixth order of the number of basis functions. The choice of a large basis set, while theoretically desirable, will therefore always present practical problems.

To sum up, we find it hard to see how Scerri's point that the Schrödinger equation is solved atom by atom can be sustained – the integrals that form the basis of the calculation by their very definition extend over the whole molecule. If Scerri means to say that basis sets are found atom by atom then this is true in the main (though it is neither necessary nor always done). The relationship between the basis set and the Aufbau principle has also become clear: the basis set needs to furnish, at a minimum, functions of sufficient complexity for orbitals with the required properties to be created as determined by the expected results of the computational model (correlated calculations can require more extensive basis sets to deliver accurate answers than non-correlated ones; and the calculation of certain electronic properties such as dipole moments or polarizabilities requires a different creation of the basis sets yet again).

Finally, the reader who is interested in a more documented defense of the distinction between the two conceptions of the atom is referred to Hettema (2000).

On Section 9. We should have mentioned that our claimed causal correlation between "equal outer electron configuration" is an idealization; it is neither necessary nor sufficient, as Scerri documents with relevant counter examples. Fortunately, such a similar remark is not made regarding the claimed identification of the charge of the nucleus and the atomic number. As is clear from our presentation, this identity is the core of our reduction claim.

Status as Observational Law or Proper Theory

Scerri's Section 10, "Is the periodic table a true theory?", and some remarks in Section 2, give rise to a couple of remarks. Already in the first version we claimed that there had been an important transformation of the status of the periodic table: from a true theory (in the sense of a proper theory) to an observational law. Although Scerri does not agree with us that the table ever had the status of a proper theory, we are pleased to note that by stating this with such emphasis, he underwrites the existence of this distinction. The recognition of the epistemological and methodological importance of this distinction almost got lost in philosophy of science, probably due to its apparent dependence on an absolute distinction between observational and theoretical terms. In Ch. 2 of SiS it is argued extensively, in line with Nagel's original exposition, and using ideas of Hempel and Sneed, that these distinctions are independent.

In Section 2 Scerri even goes so far as to question the law-like status of the periodic table in our time in view of "the presumed reduction of this law by quantum mechanics." However, in our view, this claim merely reflects problematic terminology. The idea is, of course, that a general observational fact loses its status as an *independent* law when it can be reduced to a theory. However, as Scerri rightly suggests, physical scientists are sometimes inclined to withdraw the law-like status altogether as soon as a law can be derived from a theory in a certain way. This is unfortunate terminology, because it suggests that reduction is a kind of elimination, whereas speaking of a "derived law," after a successful reduction, is the plausible thing to do.

In Section 10 Scerri elaborates his criticism of our claim that Mendeleev implicitly used the notion of an atomic number. Here we are inclined to disagree. Of course, by writing 'implicitly' we wanted to suggest that, although he was not using numbers, as we noted by this remark, he was using something that can be represented by numbers. To be precise, Mendeleev used a relation, chemical similarity, which was independent of atomic mass. This generated his very idea of gaps in the table based on atomic mass and the chemical similarity of known elements. The notion of a gap is a theoretical term in the sense that the existence of a gap cannot be established without using the very ideas underlying the table. And as soon as gaps are postulated, the known and unknown chemical elements can be successively numbered. To be sure, we should have made this point more explicit.

REFERENCES

Hettema, H. (2000). Philosophical and Historical Introduction. In: H. Hettema, *Quantum Chemistry. Classical Scientific Papers*, pp. xvii-xxxix. Singapore/River Edge/ London: World Scientific Publishing.

Hettema, H. and T. Kuipers (1988). The Periodic Table – Its Formalisation, Status, and Relation to Atomic Theory. *Erkenntnis* 28, 387-408.

Hettema, H. and T. Kuipers (2000). The Formalisation of the Periodic Table, In: W. Balzer, J. Sneed, C. Ulises Moulines (eds.), *Structuralist Knowledge Representation. Paradigmatic examples. Poznań Studies in the Philosophy of the Sciences and the Humanities*, vol. 75, pp. 285-305. Amsterdam/Atlanta: Rodopi.

Itzykson, C. and J.-B. Zuber (1980). *Quantum Field Theory*. New York: McGraw-Hill.

Posin, D. (1948). *Mendeleyev: The Story of a Great Scientist*. New York: McGraw-Hill.

Spronsen, J., van (1969). *The Periodic System of the Chemical Elements*. Amsterdam: Elsevier.

Jeanne Peijnenburg

CLASSICAL, NONCLASSICAL AND NEOCLASSICAL INTENTIONS

ABSTRACT. Kuipers' model of action explanation is compared, first with that of Anscombe, and then with models in the post-Anscombian tradition. Whereas Kuipers and Anscombe differ on the question of the first-person view, the difference with post-Anscombian writers concerns the so-called intentional statement. Kuipers criticizes the models of both Hempel and von Wright for their lack of an intentional statement. Kuipers' own model seems immune to this criticism, since it contains no less than *two* intentional statements, a "specific" and an "unspecific" one. I argue that, contrary to appearances, it is not so immune. The call for intentional statements is in fact a call for intentions that are irreducible to beliefs and desires. Kuipers' intentional statements, however, are about intentions that can be so reduced.

0. Introduction

It is hard to imagine a contemporary discussion about action explanation that makes no reference at all to Anscombe's *Intention* (1957). In discussing Theo Kuipers' views on action explanation, I too will take Anscombe's work as a starting point. I describe Anscombe's views in Section 1 and post-Anscombian views in Section 2. Then I start making comparisons: between Anscombe and Kuipers in Section 3 and between Kuipers and post-Anscombian philosophers in Section 4. Finally, in Section 5, I locate Kuipers' model of action explanation amongst other models in the field.

1. Anscombe

Anscombe (1957) introduced a famous distinction between three major contexts in which the concept 'intention' occurs:

(a) intentional action
(b) intention with which an action is performed
(c) expression of an intention for the future

In: R. Festa, A. Aliseda and J. Peijnenburg (eds.), *Cognitive Structures in Scientific Inquiry* (*Poznań Studies in the Philosophy of the Sciences and the Humanities*, vol. 84), pp. 217-233. Amsterdam/New York, NY: Rodopi, 2005.

(Anscombe 1957, p. 1, cf. pp. 24-25). The chief problem in *Intention* involves the relation between (a), (b), and (c). Before going into their mutual relations, let us first recall what (a), (b) and (c) are.

A recurrent theme in Anscombe's work is that the term 'intentional' refers to a *form* of descriptions of events (pp. 84-85). This form is such that the description can trigger a particular 'Why'-question, namely one to which the answer is something like 'In order to establish such-and-so'. Thus *intentional actions* are "actions to which a certain sense of the question 'Why' is given application" (p. 9). This 'Why'-question, Anscombe says, is *not* applicable if the agent is unaware of what he did, or can only reconstruct what he did on the basis of observing his own behavior, or is aware of his action but unable to give an account of it (pp. 11*ff*). In other words, an action is intentional if and only if the agent has immediate knowledge of his reasons for the action, i.e. he can explain his action directly in terms of his beliefs and desires and does not need a third person view from which to make reconstructions on the basis of his own observable behavior. For example, my raising my hand is an intentional action, for I can explain it directly by stating that I want to greet my neighbor and believe that by raising my hand my neighbor will indeed be greeted. If, unknown to me, my neighbor is Jack the Ripper, then by the same token I am greeting Jack the Ripper, albeit unintentionally. As Anscombe sees it, the greeting of my neighbor and the greeting of Jack the Ripper are two different descriptions of one and the same action: under the former description my action is intentional, whereas under the latter it is not.

Some descriptions under which an action is intentional express *intentions with which the action is performed*. Thus I raise my hand with the intention of greeting my neighbor, but not with the intention of greeting the Ripper. Each intention with which the action is performed can in turn give rise to a further 'Why'-question. Why are you greeting your neigbor? Because I wish to be polite. Why do you wish to be polite? Because I want to make daily life agreeable, etc. In this way an entire chain of answers to 'Why'-questions results: I raise my hand with the intention of greeting my neighbor, and I greet my neighbor with the intention of being polite, and I am polite with the intention of making daily life agreeable etc. All these answers express different intentions with which the action is performed, and they are all different descriptions under which one and the same action is intentional. It would be a mistake, however, to think that among these different descriptions there is one that is *the* true description of the action. Anscombe stresses that there are many descriptions under which the action is truly intentional, and that typically these descriptions display a chain or an *order*, in the sense that "each description is introduced as dependent on the previous one, though independent of the following one" (p. 45).

In some cases, the description of an intentional action employs the first-person and the future tense. In those cases, Anscombe states, the description is an *expression of an intention for the future*. Examples: 'I book a flight with the intention of going to France next month', 'I greet my neighbor with the intention of asking him a favor tomorrow', or simply: 'I intend to go to France next month', 'I intend to ask my neighbour a favor tomorrow'.

What is the relation between (a), (b), and (c)? More particularly, can the three contexts be reduced to each other or do they represent totally different meanings of 'intention'? Framed as such a dilemma, the question appears to be ill-phrased, for Anscombe seems to reject each of the two horns. On the one hand, she stresses that (a), (b) and (c) should *not* be seen as reflecting three totally different meanings of 'intention'. In particular, she intimates that (a) gives the primary meaning of which not only (b), but also (c) is derived. On the other hand, however, she argues that the three contexts cannot be reduced to each other. Especially the reduction of (c) to (a) or (b) seems to be forbidden. Anscombe discusses the example of St. Peter, who, while Jesus was led away to Annas and Caiaphas, "did not change his mind about denying Christ, and was not prevented from carrying out his resolution not to, and yet did deny him" (Anscombe 1957, p. 93). Thus "St. Peter could do what he intended not to, without changing his mind, and yet do it intentionally" (Anscombe 1957, p. 94). Apparently, this example is supposed to show that it is possible to intend not-*A* at time *t* and yet perform *A* intentionally at *t*. But this is difficult to understand, given Anscombe's earlier claim that intentions as expressions for the future are fully derivable from intentional actions.

Anscombe's subtle two-track policy is not easy to sustain. Perhaps that is the reason why post-Anscombian philosophy of action developed further in two different directions, a reductive and a nonreductive one, each more or less corresponding to a horn of the dilemma that Anscombe so deftly tried to avoid. In Section 2 I will discuss each of the directions in detail, but not before having stressed a feature that is common to both.

2. Post-Anscombian Philosophy of Action

Post-Anscombian philosophy of action (whether of the reductive or the nonreductive kind – see below) completely changed the meaning of Anscombe's context (c). Whereas for Anscombe the intentions in the context (c), just as the intentions in (a) and (b), refer to certain *descriptions* of *actions*, contemporary philosophers of action started to see intentions in (c) as referring *directly* to *mental states*. The very idea would probably have been abhorrent for Anscombe, who consistently emphasised that the definition of intention lies in the particular description of actions, and who always opposed the idea of intention as a distinct state of mind. In fact, one of Anscombe's motives for writing *Intention* was precisely to bring an end to the mistaken thought, regularly surfacing from Plato onwards, that intentions are mental states that exist relatively autonomously from language or behavior.

The contemporary deviations from Anscombe's framework are unmistakable, yet they often occur tacitly and unconsciously: one seems not even to notice that one is changing the meaning of Anscombe's concepts and is thereby veering away from what she had in mind. Michael Bratman may serve as an example here. His important book *Intention, Plans, and Practical Reason* (1987) opens as follows:

> Much of our understanding of ourselves and others is rooted in a commonsense psychological framework, one that sees intention as central. Within this framework we use the notion of intention to characterize both people's actions and their minds. Thus, I might *intentionally* pump the water into the house, and pump it *with the intention* of poisoning the inhabitants. HERE INTENTION CHARACTERIZES MY ACTION. But I might also *intend* this morning to pump the water (and poison the inhabitants) when I get to the pump this afternoon. AND HERE INTENTION CHARACTERIZES MY MIND. (p. 1 – italics by the author, small caps by me).

Here not only the triplet *intentionally*, *with the intention*, and *intend* comes from Anscombe, but also the example about pumping water and poisoning the household. Moreover, in a footnote to this excerpt, Bratman suggests that the distinction between intentions as characterizations of *actions* and as characterizations of *minds* stems from Anscombe too. Although this suggestion is doubtful, Bratman apparently did not realise the difference.[1]

This departure from Anscombe, implicit as they may be, did not remain without consequence. In fact, it initiated a sea change in the programme of philosophy of action. The important question for Anscombe, as we have seen, was 'What is the relation between (a), (b), and (c)?', where (a), (b), and (c) are different *descriptions*, namely of *actions* that are done with an intention. We saw that Anscombe, in answering that question, tried to manoeuvre between the devil of total reduction and the deep sea of complete autonomy: although she thinks that (a) gives the core meaning of 'intention', she also denies that (b) and (c) can be plainly reduced to (a). But the question that occupies post-Anscombian philosophers of action is different. It can be stated as: "What is the relation

[1] Bratman's suggestion that Anscombe distinguishes between intentions as characterizations of actions and of minds becomes a definite claim in other texts by Bratman. For instance, in (Bratman 1995, p. 243) we read (emphasis by Bratman): "There are two relevant aspects of intention: (1) a characteristic of action, as when one acts *intentionally* or *with a certain intention*; (2) a feature of one's mind, as when one *intends* (*has an intention*) to act in a certain way now or in the future. An important question is: how are (1) and (2) related? (See Anscombe 1963.)" Here, 'Anscombe 1963' refers the second edition of *Intention* (Anscombe 1957).

between (c) on the one hand and (b) and (a) on the other?", where (a) and (b), unlike Anscombe's view, are *direct characterizations* of actions, and where (c), even more un-Anscombian, sometimes is a direct characterization of a certain *state of mind.*

To the latter question, two answers have been given, each corresponding to a particular direction in post-Anscombian philosophy. In the reductive answer, it is denied that intention has an independent existence apart from reasons and actions. The nonreductive answer, on the other hand, tries to give intention a place of its own, distinct from reasons. Let us now take a closer look at each of the two answers.

The first direction in post-Anscombian philosophy of action is paved with classical theories like those of Hempel, von Wright, and Davidson in his early articles. I call it the reductive way because it does not distinguish between intentions and reasons. In this approach, actions are explained by reasons, where, roughly, actions are comparable to Anscombe's (a) and reasons to Anscombe's (b). Furthermore, reasons are pairs consisting of a belief and a desire, and intentions are synonymous with reasons. Thus the action of Jane opening the window is explained by giving Jane's reason or intention, namely that she had the desire of cooling the room and believed that by opening the window the room would be cooled. Internal variations are allowed: the explanation in question can be seen as a dispositional explanation (as Hempel did), a logical explanation (von Wright's view), or a causal explanation (Davidson 1963). These differences are, however, all of minor importance. The major point is that in all these cases actions (Anscombe's (a)) are explained by reasons (Anscombe's (b)), that reasons are synonymous with intentions (Anscombe's (c)) and that reasons or intentions are complexes of beliefs and desires. These cases are therefore instances of what is sometimes called *the belief-desire model* of intention.

Within this model, future intentions are not what Anscombe thinks they are, namely descriptions of actions using the first-person and the future tense. But neither are they distinctive psychological attitudes. It is exactly with respect to the latter point that the first direction in post-Anscombian philosophy differs from the second.

The second direction in post-Anscombian philosophy does regard a future intention as an attitude that differs fundamentally from believing, desiring, let alone from acting. Moreover, it states that a future intention gives the primary meaning of 'intention', on which intentional actions (Anscombe's (a)) as well as reasons (Anscombe's (b)) somehow depend. As a consequence, it rejects the belief-desire model of action explanation and offers alternative models.

A prominent proponent of the second direction is Donald Davidson in his later articles. After having effectively defended the belief-desire model in Davidson (1963), Davidson distanced himself from it in Davidson (1978). In the

Introduction to *Essays on Action and Events* (1980) he gives the following comment on that change:

> When I wrote [Davidson 1963] I believed that of the three main uses of the concept of intention distinguished by Anscombe (acting with an intention, acting intentionally, and intending to act), the first was the most basic. Acting intentionally, I argued in [Davidson 1963], was just acting with some intention. That left intending, which I somehow thought would be simple to understand in terms of the others. I was wrong. When I finally came to work on it, I found it the hardest of the three; contrary to my original view, it came to seem the basic notion on which the others depend (p. xiii).

What then is this basic notion? How to reconstruct intending to act, i.e. the notion that supplants Anscombe's intention in sense (c)? Davidson's final answer (so far) is well known: an intention to act is an unconditional or "all-out" judgement that a certain action is the best to perform.[2]

This answer shows that there are three things which, according to Davidson, an intention is *not*. First, it is not a conditional or "prima facie" judgement of the form "Action *A* is the best for me to perform, given that I want *X* and believe that *A* is necessary (useful, the best etc.) for achieving *X*." Such a judgement is typically the conclusion of a practical syllogism. It does not say that *A* is best *tout court* and hence it is noncommittal: no strings are attached to it. Since an intention *is* committal, it cannot be a conditional judgement. Second, an intention is not a reason (a pair of a belief and desire). For clearly, a reason is even more noncommittal than a conditional judgement, since it constitutes the condition mentioned *in* that very judgement. Third, an intention is not an action. The argument for that is simple: intentions as all-out judgements can exist in the absence of intended actions. We might for example intend to trap a tiger, but never live up to that intention; and even if we were to live up to it, the intention must guide our ensuing actions over time, and thus must have an existence that is somehow independent of the distinct individual actions. Davidson stresses, however, that this independence is logical and not ontological:

> We are stuck, it now seems to me, with states of intending which are independent of our reasons for intending and of our actions. I say this without, I hope, committing myself one way or another to an ontology of states. The independence of intentions is logical: it does not follow from the existence of reasons for an action and a corresponding action that there was an intention based on those reasons that explains the action (though if the action was performed for those reasons, the intention must have existed). (Davidson 1985, pp. 196-197)

Davidson is by no means the only philosopher who finally turned his back on the belief-desire model because it neglects the relatively autonomous role of

[2] Cf. "... intentions are distinguished by their all-out or unconditional form" (Davidson 1978, p. 102); "... intentions are 'all-out' positive evaluations of a way of acting ..." (Davidson 1985, p. 214); "An all-out judgement that some action is more desirable than any available alternative, is not distinct from the intention: it is identical with it." (Davidson 1985, p. 197).

intention as a characterization of the mind. Michael Bratman is another one. Like Davidson, Bratman denies that intentions are actions, reasons, or conclusions of practical syllogisms. In the end, however, his idea of intentions is quite different from Davidson's. At least two differences leap to the eye. First, whereas Davidson eschews committance to intentions as ontologically distinct states of mind, Bratman seems to have no such fear. Second and perhaps more importantly, Davidson construes intentions as unconditional all-out judgements, whereas Bratman associates them with something in the future. For Bratman, having an intention is essentially part of having a *plan*:

> ... our commonsense conception of intention is inextricably tied to the phenomena of plans and planning (Bratman 1987, p. 2).

> Our understanding of intention is in large part a matter of our understanding of future-directed intention. ... Why do we bother forming intentions concerning the future? Why don't we just cross bridges when we come to them? An adequate answer to this question must return to the central fact that we are planning creatures. We form future-directed intentions as parts of larger plans, plans which play characteristic roles in coordination and ongoing practical reasoning; plans which allow us to extend the influence of present deliberation to the future. Intentions are, so to speak, the building blocks of such plans; and plans are intentions writ large. (Bratman 1987, pp. 7-8).

The belief-desire model, says Bratman, cannot shed much light on this planning dimension of intention, for it wrongly focuses on intentional actions. Those actions may well involve the *execution* of prior plans, but that is not enough, for "plans are not merely executed. They are formed, retained, combined, constrained by other plans, filled in, modified, reconsidered, and so on. Such processes and activities are central to our understanding of plans, and to our understanding of intention." (p. 8). According to Bratman, the belief-desire model fails to do justice to these processes and activities.

Interesting as the differences between Bratman and Davidson may be, I will not dwell upon them any further. After all, my concern is the resemblance between their construals of intention, not the contrast. The contrast that I do take interest in is between Bratman and Davidson on the one hand and Theo Kuipers on the other.

It is high time to look at Kuipers' model for action explanation. I will first compare Kuipers' model with Anscombe's ideas, only to conclude that the difference is considerable (Section 3). My next step is to locate Kuipers' model in the post-Anscombian philosophy of action (Section 4).

3. Kuipers and Anscombe

Kuipers' model of action explanation is part of his explicative program, i.e. the last of the four research programs that he introduces in *Structures in Science*

(Part I, Chapter 1). According to this program, explaining actions is an instance, not of explanation by subsumption, but of explanation by specification (Part II, Chapter 4, Section 1). Hence it is primarily an interpretative or detailing affair, and this idea sits well with Anscombe's suggestion that to explain an action is a way of redescribing it (in terms of the intention with which the action is performed), rather than of deducing it from premises or subsuming it under a law.

There are more resemblances between Kuipers and Anscombe. An important part of *Intention* is devoted to Aristotle's theory of the practical syllogism, which Anscombe claims to have been widely misunderstood. In her view, essentially all interpretors regard Aristotle's practical syllogism as a deductive argument culminating in the conclusion that a certain thing must be done. This is wrong, as can be illustrated by the following example of a practical syllogism[3]:

	Green clothing suits any red-haired person
	Dutch army clothing is green
PS	I am a red-haired person
	This is an article of Dutch army clothing
	Ergo, this clothing suits me.

PS is deductive and valid of course, but it has the disadvantage that nothing follows about doing anything. A statement about performing a particular action *would* follow if the practical syllogism had the imperative form, for example:

	Do everything that suits your red hair
PS′	Doing such-and-such will suit my red hair
	Ergo, do such-and-such.

However, PS′ does not bring us any further, since its first premise is absurd. As Anscombe indicates, it is not only ridiculous, but even logically impossible to obey the first premise: there are a hundred different and incompatible ways of doing things that would suit my red hair, such as wearing a green dress, a black dress, army clothes, non-army clothes, walk in daylight, walk under artificial light, etc (cf. Anscombe 1957, p. 59). If I am correct, Anscombe's objection here is rooted in the very same intuition that brought Kuipers to his criticism of the Logical Connection Argument, that calls the following general statement, G, a meaning postulate:

G: for all x, y and z and all occasions, if $D(x,z)$ and $BN(x,y,z)$ then $P(x,y)$,

where $D(x,z)$= 'x desired goal z', $BN(x,y,z)$= 'x believed action y to be necessary to approach goal z', and $P(x,y)$= 'x performed y' (Kuipers 2001, p. 99). According to Kuipers, G is not a meaning postulate, for this would mean that G "connects a

[3] The example is a blend of Aristotle's famous "Dry food suits any human" example, Anscombe's example about a certain dress in a shop window, and some trivial facts about the Dutch army.

specific action, as a consequence of meaning relations, to specific mental states" (Kuipers 2001, p. 100). Kuipers objects to such a "magical" connection, as he calls it, not because "some primarily mental concepts have behavioral connotations, e.g., 'hot-tempered' certainly has such connotations, but [because] statements using only such concepts, even if they are quite specific, *might imply a specific action*" (p. 100, my italics). The resemblance is, I think, clear: like Anscombe, Kuipers realizes that a piece of practical reasoning can never necessitate the conclusion that a particular action must be performed.

Still another similarity between Kuipers' approach and that of Anscombe should be mentioned here: Kuipers' idea of an internal goal has much in common with Anscombe's concept of an intentional action. According to Kuipers, the statement "*x* performed action *y* with the intention of approaching goal *z*" mentions in fact two goals. The one is *z*, called the external goal of *y*; the other is the internal goal of *y* ("that is, the goal of action *y* according to the description used for it" – p. 102). For example, if *y* is 'opening the door', then the internal goal of *y* is 'having the door open' whereas its external goal might be 'cooling the room' or 'letting the tame canary out'. As Kuipers observed, it is wise to assume that in "*x* performed action *y* with the intention of approaching goal *z*" goal *z* is always external:

> The reason for excluding the internal goal from being *z* is that we consider it as a trivial meaning component of ['*x* performed *y*'] that *x* performed *y* with the intention of approaching or even realizing the internal goal of *y*. Hence, we concentrate further by definition on explanatory goals that are not as a matter of meaning related to the relevant action. On the other hand, ... in order to explain an action intentionally, it is sufficient to explain the internal goal of that action intentionally. (Kuipers 2001, p. 102)

The parallel between an internal goal and Anscombe's idea of an intentional action (a description that makes an action intentional) is, I hope, obvious.

However, the similarities between Kuipers' and Anscombe's approaches are outweighed by the differences. To begin with, there is a striking divergence in style. Anscombe's book is slender and abstruse, Kuipers's is considerably in weight yet very clear. Anscombe's texts are perfused with Wittgenstein II, so that, in the words of Richard Jeffrey, some might say they look like gibberish at first sight (Jeffrey 1989, p. 252). Kuipers' articles, on the other hand, are conceived in the best didactic tradition of classical philosophy of science: clear, straightforward, and unambiguous. Style, however, is not really at issue here. There is a more substantial difference.

Both Anscome and Kuipers profess an ability to deal with simple, everyday life cases of action explanation. As a consequence, both claim to preserve the *order* of the actual thought process by which an action is explained (cf. Kuipers 2001, p. 99, Anscombe, Sections 23, 26, 42). Yet there is a vital contrast here. Anscombe seems to be interested in *two* thought processes and hence *two* orders.

Kuipers, on the other hand, is thinking of only one order of one thought process. This can perhaps be best explained by recalling Anscombe's famous example of the shopping list.

Imagine a man walking around in a supermarket with a shopping list in his hand. He is followed by a detective who makes a record of what the man puts in his trolley. Here there are two relations: one between the man's list and the articles in the trolley, and one between the articles in the trolley and the detective's record. The difference between the two relations becomes clear as soon as a discrepancy occurs. If the list and the articles do not agree, then the mistake is not in the list but in the man's performance.[4] However, if the detective's record and the articles do not agree, then the mistake is in the record.

An important and also rather complex theme of *Intention* is the connection between both relations. In what sense other than that of order do they differ? Is the one relation more important than the other? Can they perhaps be reduced to one another?[5] As I see it, Anscombe believes that any adequate analysis of actions being explained by intentions must consider *both* relations. It must consider the thought process that starts with the list (corresponding to the first-person view) as well as the thought process that has the contents of the trolley as bench-mark (third-person view). One of the most difficult parts of Anscombe's book concerns the question how exactly the two processes hang together. Anscombe seems to be criticizing modern philosophy of action on the ground that it focuses on the latter relation (starting with the articles in the trolley), thereby wrongly trying to understand the other relation in terms of it. However that may be, the point that I wish to make is that the interaction between the two relations, and hence between first- and third-person views, is certainly not Kuipers' main concern. Kuipers is interested in only one thought process, namely the one that results in a written record on the basis of the products observed in the trolley. This follows immediately from the two steps that make up Kuipers' model of action explanation (pp. 102-104).

The first step is the introduction of two meaning postulates, MP-1_i and MP-2_i, both fixing the meaning of an intentional statement. MP-1_i fixes the meaning of a *specific* intentional statement $IP(x,y,z)$, MP-2_i defines an *unspecific* intentional statement $IP(x,y)$:

MP-1_i : $IP(x,y,z) \approx P(x,y)$ & $D(x,z)$ & $BU(x,y,z)$
MP-2_i : $IP(x,y) \approx$ there is a goal τ such that $IP(x,y,\tau)$,

---

[4] Anscombe illustrates this situation wittily: "if his wife were to say: 'Look, it says butter and you have bought margarine', he would hardly reply: 'What a mistake! we must put that right' and alter the word on the list to 'margarine'" (Anscombe 1957, p. 56).

[5] Of course, these questions are not at all independent of the main theme of Anscombe's book: what is the relation between intention in the contexts (a), (b), and (c)? More particularly, how is (c), the description using the first-person and the future tense, related to (a) and (b)?

where:

> *IP*(*x*,*y*,*z*) = *x* performed *y* with the intention to approach goal *z*
> *P*(*x*,*y*) = *x* performed action *y*
> *D*(*x*,*z*) = *x* desired goal *z*
> *BU*(*x*,*y*,*z*) = *x* believed *y* to be useful to approach *z*
> *IP*(*x*,*y*) = *x* performed *y* intentionally.

The second step is the reconstruction, on the basis of these meaning postulates, of the thought process that constitutes the model in the strict sense (for the convenience of the reader, I have added the meanings of the abbreviations in the column on the right):

THOUGHT PROCESS

(1)	verified action statement	*P*(*x*,*y*)!	*x* performed *y*!
(2)	*question*	*why P*(*x*,*y*)?	why did *x* perform *y*?
(3)	unspecific intentional statement	'by PoI'	'by principle of intentionality':
	as hypothesis	*IP*(*x*,*y*)?	*x* performed *y* intentionally?
(4)	specific intentional statement	'by idea'	
	as hypothesis	*IP*(*x*,*y*,*z*)?	*x* performed *y* with the intention to approach goal *z*?
(5)	non-trivial implications to be		
	tested	hence, by MP-1_i	
	desire hypothesis	*D*(*x*,*z*)?	*x* desired goal *z*?
	belief hypothesis	*BU*(*x*,*y*,*z*)?	*x* believed *y* to be useful for *z*?
(6a)	falsification of one or both,	not-*IP*(*x*,*y*,*z*)!	*x* did not perform *y* with the intention to approach goal *z*!
	hence, go back to step (3)		
(6b)	or, no clear results,	*IP*(*x*,*y*,z)?	*x* performed *y* with the intention to approach goal *z*?
	go to step (3) or (8)		
(6c)	or, verification of both,		
	hence *answer*: verified specific intentional statement	*IP*(*x*,*y*,*z*)!	*x* performed *y* with the intention to approach goal *z*!
(7)	now, conclude first as a side step:	by MP-2_i and existential generalization	
	verified unspecific intentional statement	*IP*(*x*,*y*)!	*x* performed *y* intentionally!
(8)	then go to new, related why- and how-questions	why?/how?	

The main product of this thought process is the verified specific intentional statement, $IP(x,y,z)!$, in (6c). It is the finishing point of a route that started with the observation of an action in (1), a corresponding Why-question in (2), and a tentative answer, via (3), in (4). This answer is tested in (5), and (6a)-(6c) reflect the possible outcomes of this test. The direction of the thought process is obvious: it goes from observed performance to the origin of the performance. Hence it is closer to the detective who records what the man put in the trolley than to the man who puts in the trolley what he intended to put in.

To conclude, Kuipers' model of action explanation, notwithstanding some important similarities, does not well agree with Anscombe's analysis. A vital difference keeps the two apart: Kuipers focuses on the third person view and seems not to be interested in what preoccupies Anscombe, viz., the interplay between the first and the third person view.

4. Kuipers and Post-Anscombian Philosophy of Action

Obviously, Kuipers' model belongs to the post-Anscombian era, not only in time (the first version of the model appeared in Kuipers 1985), but in content too. Given that post-Anscombian philosophy of action is in fact a coalition of two different factions (see Section 2), the question rises: to which of the factions does Kuipers belong? Is he thinking along the lines of what we have called the first direction, reducing intentions to reasons (i.e. belief-desire pairs), thus propagating a belief-desire model in the manner of Hempel, von Wright or the early Davidson? Or does he try to give intention a place of its own, distinct from reasons, thus promoting an alternative model as part of the second direction?

At first sight, the answer seems clear enough. Kuipers explicitly presents his model as an alternative to the standard explications of Hempel and von Wright. Indeed, the very reason for Kuipers to invent a new model for action explanation is precisely that the models of Hempel and von Wright cannot stand up to the many objections. Kuipers mentions no less than six objections to the standard explications of Hempel and von Wright, and he then notes:

> Both standard explications have been modified in order to cope with objections like those above. However, these attempts have not been convincing, for none of the two has received general acceptance. Hence, there is room for a third alternative. (Kuipers 2001, p. 100-101).

Kuipers believes that his own model as summarized above can be this third alternative, since it is able to cope with each of the six objections listed by him. Thus it seems that indeed Kuipers sides with the philosophers of the second direction, revolting as they do against standard explications and trying to give intentions and intentional statements the role they deserve.

This conclusion seems to be supported by a closer study of the objections on Kuipers' list. Whereas the first objection is that the models of Hempel and von

Wright lack "clear correspondence with research practice," the second states that they fail to give an "explicit role for specific intentional statements" (Kuipers 2001, p. 99). To be sure, such objections were precisely the reason for Bratman, Brand, Mele, Harman, Searle, Davidson and many others to turn their backs on the traditional belief-desire model and try to devise less deficient models. Bratman, for example, stresses time and again that the belief-desire model cannot account for the notion of intention as it is used in commonsense psychology, because it denies the distinctive role of intentions and intentional statements.

However, you cannot tell a book by its cover: any resemblance of Kuipers' model with the ideas of philosophers in the second direction is mere appearance; actually, the model is a far cry from what for instance Bratman or the later Davidson had in mind. When Bratman, Davidson, and all the other philosophers of the alternative trail declare that an explicit role should be given to intendings or intentions, they mean, of course, that intendings or intentions *are somehow independent of reasons and of actions*. Consequently, *statements* about intentions – intentional statements – cannot be plainly reduced to statements about actions or about reasons (belief-desire pairs). Bratman goes so far as to speak of intentions as separate mental states ("Intentions are distinctive states of mind, not to be reduced to clusters of desires and beliefs" – Bratman 1984, p. 376), but as we have seen that is not necessary for taking part in the second direction. One could also side with Davidson in acknowledging only the logical autonomy of intentions.

The difference between such views and Kuipers' approach becomes clear as soon as we consider the way in which Kuipers tries to cope with the objections against Hempel and von Wright, in particular the objection that Hempel and von Wright ignore the "explicit role for intentional statements." For what are the intentional statements in Kuipers' model? To be sure, there are two sorts: the unspecific intentional statement, $IP(x,y)$, and the specific intentional statement, $IP(x,y,z)$. The meaning of $IP(x,y)$ and $IP(x,y,z)$ is set down neatly in the two meaning postulates, MP-1_i and MP-2_i, and they leave no room for doubt: Kuipers intentional statements, the specific as well as the unspecific, are *completely reducible to statements about reasons and actions*. But what is the use of distinct intentional statements, next to statements about reasons and actions, if the former can be fully reduced to the latter? What is the need for a distinct concept of intention, next to reasons and actions, if reasons and actions make up the content of intentions?

5. Concluding Remarks

I think we must conclude that Kuipers' intentional statements are not intentional statements as they feature in the common objection against Hempel and von Wright. When people like Bratman and Davidson accuse the standard models of failing to account for intentional statements, they are talking about statements that are irreducible to statements about the agent's beliefs and desires. Kuipers' intentional statements, however, have no such character.

Don't they? One might object and state that Kuipers' intentional statements *are ir*reducible to beliefs, desires, and actions. After all, Kuipers says about the definition of *IP*(*x*,*y*,*z*) that it is only "a first approximation in the sense that the three components *do not exhaust [its] meaning*" (p. 102, my italics). On the same page and on pages 107*ff* he gives two examples of statements that might be added to the three meaning components of *IP*(*x*,*y*,*z*): the "time statement" (as it may be called), stating that "*D*(*x*,*z*) and *BU*(*x*,*y*,*z*) may not 'start' later than *P*(*x*,*y*)" (p. 102), and the causal statement, requiring "that the belief and desire component were causally effective" (p. 102). However, my point is that adding these statements to the meaning of *IP*(*x*,*y*,*z*) does not make *IP*(*x*,*y*,*z*) irreducible to statements about reasons and actions. For the time statement and the causal statement merely claim something about the reasons and actions themselves: they solely make reasons, actions and their relation more precise. But rendering reasons and actions more precise does not show that *IP*(*x*,*y*,*z*) is irreducible to statements about reasons and actions. It only shows that *IP*(*x*,*y*,*z*) might be reduced to *more precise* statements about reasons and actions.

But perhaps I am splitting hairs. Perhaps we should focus on the entire framework that surrounds Kuipers' intentional statements rather than criticizing these statements themselves. For specific as well as unspecific intentional statements feature in what Kuipers calls "an intentional context," i.e. the whole process of searching an intentional explanation for an action. The principle guiding the intentional context is the *principle of intentionality*:

PoI: if *P*(*x*,*y*) then *IP*(*x*,*y*),

that is, "if someone performs (or has performed) an action, he will do (have done) that intentionally" (Kuipers 2001, p. 103). PoI is a heuristic-methodological principle, that serves as a searchlight for anybody who tries to explain an action by invoking reasons. Being only a heuristic instrument, PoI does not state an analytic truth: it might well turn out, in a particular case, that *P*(*x*,*y*) is true and *IP*(*x*,*y*) is false. In such a case, of course, a person performed an action only for the sake of the internal goal of that action, not aiming to achieve a further, external, goal.

The fact that we sometimes perform actions for their own sake is of course familiar: we often execute an action without aiming at, let alone realizing, an external goal. But is it also possible to perform an action without realizing, or even aiming at, the *internal* goal of that action? According to Kuipers, this is clearly *not* possible. As we have seen, Kuipers considers it "a trivial meaning component of $P(x,y)$ that x performed y with the intention of approaching or even realizing the internal goal of y" (Kuipers 2001, p. 110). Yet, several philosophers in the second direction think this is mistaken. They take great pains to demonstrate that intending is a distinct phenomenon, not to be confused with intentional actions or with reasons for actions. They believe it *is* possible to perform y intentionally without intending to achieve the internal goal of y, just as Anscombe believes that Peter intentionally denied Christ without intending to do so.

The position defended by Kuipers they call the Simple View; basically, it states that intentionally y-ing entails intending to y (Bratman 1984, p. 376; Bratman 1987, p. 112). Despite its initial plausibility, the Simple View has been under vigorous attack (although it has been valiantly defended too, for instance by McCann 1991). The main argument against it is presented by Bratman in a much-discussed example that roughly goes as follows (Bratman 1987, pp. 113-114; the example is inspired by an example sketched in Audi 1973, p. 401).

Imagine a video-game in which a virtual target must be hit by virtual missiles; when the target is hit, the game is over. Success in this game depends partly on skill and partly on chance: even excellent shooting does not guarantee a hit. Vincent is a very skilled player of these games; indeed, his command of the medium is so great that he can play two games at the same time using two different machines. As it happens, the two machines are linked in such a way that it is impossible to hit both targets at the same time: if target 1 on machine 1 and target 2 on machine 2 are about to be hit simultaneously, both machines shut down before any target could be hit. Vincent knows that he can hit either target 1 or target 2, but not both of them. Skilled as he is, he increases his chances by simultaneously trying to hit target 1 on machine 1 with his left hand and target 2 on machine 2 with his right hand. If, under these circumstances, target 1 is hit, then Vincent hit target 1 *intentionally*. Hence, on the Simple View, he must have *intended* to hit target 1. But "given the symmetry of the case," as Bratman phrases it, Vincent must also have intended to hit target 2; after all, his attempts at hitting target 2 are not essentially different from his attempts at hitting target 1. Thus, on the Simple View, Vincent had both intentions, viz. to hit target 1 and to hit target 2. But that is not true, for our video-virtuoso knew perfectly well that he could not hit both targets. Bratman concludes that Vincent had neither intention, and this shows that one can hit a target intentionally without intending to hit it. Hence the Simple View is wrong.

Whatever one may think of examples like these (for instance, could not one say that Vincent's intention was "hit either target 1 or target 2 but not both"?), they seem to challenge a presupposition of PoI, namely that performing an action successfully implies realizing the internal goal of that action. Hence they do form a problem for PoI itself and even for the entire intentional context of which PoI is the guiding principle.

It seems that, on the basis of the criterion that gives rise to the watershed between two directions within post-Anscombian philosophy, Kuipers' explication of action explanation belongs to the first rather than to the second. At the end of the day, Kuipers' model stands in the time-honored tradition of Hempel and von Wright rather than in the current school of Bratman or Davidson. And perhaps that should not surprise us. For what else could we have expected from a model in "An Advanced Textbook in *Neo-classical* Philosophy of Science"?

University of Groningen
Faculty of Philosophy
Oude Boteringestraat 52
9712 GL Groningen
The Netherlands

REFERENCES

Anscombe, G.E.M. (1957). *Intention*. Oxford: Basil Blackwell. Second edition 1963. Reprinted 1968.

Audi, R. (1973). Intending. *The Journal of Philosophy* **70**, 387-403.

Bratman, M.E. (1984). Two Faces of Intention. *The Philosophical Review* **93**, 375-405

Bratman, M.E. (1985). Davidson's Theory of Intention. In: Vermazen and Hintikka (1985), pp. 13-26. Reprinted in 1988 with an added appendix in: E. LePore, B.P. McLaughlin (eds.), *Actions and Events. Perspectives on the Philosophy of Donald Davidson* (Oxford: Basil Blackwell, 1985), pp. 14-28.

Bratman, M.E. (1987). *Intention, Plans, and Practical Reason*. Cambridge, Mass.: Harvard University Press. Reprinted in 1999 by the Center for the Study of Language and Information (CSLI) at Stanford as CSLI-Publication in *The David Hume Series of Philosophy and Cognitive Science Reissues*.

Bratman, M.E. (1995). Intention. In: J. Kim, E. Sosa (eds.), *A Companion to Metaphysics*, p. 243. Oxford/Malden, MA: Blackwell.

Davidson, D. (1963). Actions, Reasons, and Causes. *Journal of Philosophy* **60**, 685-700. Reprinted in: Davidson (1980), pp. 3-19.

Davidson, D. (1978). Intending. In: Y. Yovel (ed.), *Philosophy of History and Action*. Dordrecht: D. Reidel Publishing Company. Reprinted in: Davidson (1980), pp. 83-102.

Davidson, D. (1980). *Essays on Actions and Events*. Oxford: Oxford University Press. Reprinted with corrections: 1982, 1985, 1986.

Davidson, D. (1985). Replies. In: Vermazen and Hintikka (1985), pp. 195-229 and 242-254. Jeffrey, R.C. (1989). Coming True. In: C. Diamond, J. Teichman (eds.), *Intention and Intentionality. Essays in Honour of G.E.M. Anscombe*, p. 251-260. Brighton: The Harvester Press.

Kuipers, T.A.F. (1985). The Logic of Intentional Explanation. *Communication and Cognition* **18** (1-2), 177-198.

Kuipers, T.A.F. (2001/SiS). *Structures in Science. Heuristic Patterns Based on Cognitive Structures. An Advanced Textbook in Neo-Classical Philosophy of Science*. Dordrecht: Kluwer.

McCann, H.J. (1991). Settled Objectives and Rational Constraints. *American Philosophical Quarterly* **28**, 25-36.

Vermazen, B. and M.B. Hintikka, (eds.) (1985). *Essays on Davidson. Actions and Events*. Oxford: Clarendon Press.

Theo A. F. Kuipers

INTENDING IN TERMS OF REASONS FOR ACTIONS

REPLY TO JEANNE PEIJNENBURG

Some texts are more representative of the analytic tradition than others. As usual, Jeanne Peijnenburg contributes an essay which could teach many so-called analytic philosophers, despite their popularity, what a genuine analytic style is. In her own paper on analytic philosophy (Peijnenburg 2000) she is just too mild about the flourishing of anti-analytic styles in circles pretending to stand on the shoulders of analytic giants. In the present paper she argues, quite convincingly, that my analysis of intentional explanations has affinities with and deviations from the three dominant approaches, that is, the behavioral one of Anscombe and two alternative successors who take the mind into account, viz. the reductive, belief-desire model (notably, Hempel, von Wright, 1963-Davidson) and the nonreductive stance (1980-Davidson, Bratman). Moreover, I agree that my approach is best seen as a variant of the belief-desire model, which implicitly takes several of the criticisms into account both of Anscombe's as well as a nonreductive point of view. For details of similarities and differences, I refer to Peijnenburg's paper. In this reply I want to concentrate on two of her points. First, to what extent is my approach, in contrast to that of Anscombe, third-person-oriented? Second, is the nonreductive argument for "unintended intentional behavior" convincing? In both cases we can focus on intriguing examples – Anscombe's shopping list and the Audi and Bratman video game.

Explaining Shopping Behavior

To be sure, I used to present my specification model from the third-person perspective. In the shopping example somebody, a detective, observes collecting behavior of someone else, the shopping man. The mistakes they can make are indeed quite different, for observing collecting behavior is quite different from the behavior itself. However, Peijnenburg's claim is that I am only engaged with the third-person perspective and the corresponding thought process, viz., that of the

In: R. Festa, A. Aliseda and J. Peijnenburg (eds.), *Cognitive Structures in Scientific Inquiry* (*Poznań Studies in the Philosophy of the Sciences and the Humanities,* vol. 84), pp. 234-236. Amsterdam/New York, NY: Rodopi, 2005.

detective, and not with the first-person perspective and the corresponding thought process, viz., that of the shopping man. Let us survey the possibilities. If I am the shopping man the detective may occasionally observe that I put a pack of flour in my trolley. His question is why I do so? Without consulting me, he may nevertheless form some hypotheses – e.g. that I want to bake an apple pie – and test them along the lines of the specification model, perhaps by consulting my wife or my notebook.

Hence, the interesting question is whether the model can also be used, perhaps with some modification, for the first-person perspective? Note first that when the detective asks me why I put the flour in the trolley, the question is third-person but the answer first-person, for I give my reasons, say an apple pie desire and a pie-needs-flour belief, hence a typical hybrid situation. Another impure case of the first-person perspective is when, also perfectly possible, I ask myself why I put the flour in the trolley. I may first check my list in order to see whether I did not make a mistake. If I see it on the list I have observed indirectly that I did put it on the list. Assuming that I remember having made the list quite consciously, the next question to myself is: why did I put it on the list? Recalling the answer in terms of an apple pie desire and some beliefs about how to make it and my home stock, I reach in this way a perfect intentional explanation for my own action in terms of my own beliefs and desires. However, it is true that in this case I consider myself from a kind of as-if third-person perspective.

Consequently, the remaining question is what a pure case of the first-person perspective amounts to. In response to the question raised to myself why I put the flour in the trolley, the belief-desire reasons may come immediately to my mind, in particular the pie-wish. In this case no further testing of the meaning components is necessary for they are self-evident to me. But this makes it neither a non-case nor a trivial case of the specification model. It would be trivial if I answered in terms of the flour-desire, that is the internal goal of the questioned action. To be sure, it is a special application of the model, which is not so much trivial but, normally, not informative. It may become informative if I experience serious memory problems or if I am cheating myself about my reasons, e.g. by replacing my unconscious wish to use flour for some peculiar activity, instead of baking an apple pie. In sum, it is perfectly possible to use the specification model to explain one's own behavior, but usually we know the answers beforehand.

The Video Game

The video game example claims to show that it is possible to perform an action intentionally, without intending to achieve its internal goal. Peijnenburg is quite right that my heuristic Principle of Intentionality (PoI, by default, actions are performed intentionally, i.e., with an external goal) presupposes that this is ruled out. I even go as far as to claim that calling some behavior an action implies that the internal goal of that action was intended (SiS, p. 104). Before I question whether the claim about the example makes sense, I give two easy, but not therefore invalid answers. First, if the claim makes sense in a particular case we may make the presupposition in PoI explicit, for example in the following plausible form: by default, internally intentionally performed actions are externally intentionally performed. Second, in particular in view of the very complicated video game story, we may readily assume that actions are *normally* internally intentionally performed. Hence, in combination with the first answer we get: by default, actions are intentionally performed, internally as well as externally. This leaves room for three kinds of exceptions: actions that are neither internally nor externally intentionally performed, actions that are internally but not externally performed intentionally, and, finally, actions that are externally, but not internally, performed intentionally. The second case is perfectly possible from my point of view, and explicitly suggested in SiS (p. 104). The first and the third are excluded as soon as we assume that when describing some behavior as an action, the actor must intend the internal goal of that action description. However, if we leave room for such actions, the third case is not only even more intriguing than the first, its conceptual possibility would also make the possibility of the first case plausible. Hence, let us look at the video game, where I have to suppose that the reader has read Peijnenburg's description of it.

In the view of Audi and Bratman it is immediately assumed that "Vincent hit target 1" is an appropriate action description. Given the peculiar construction of the game, I would think that the plausible "exclusive disjunctive" approach suggested by Peijnenburg is the beginning of the answer. The action Vincent wants to perform is 'hitting precisely one of the two targets' and that is what he achieves. That he achieves it by hitting target 1 does not imply that it makes sense to say that he performed the action of "hitting target 1," let alone that he intended to do so. A tennis player may aim at winning a match, and actually win it, say 6-2, 3-6, 6-4, without aiming at this precise score. In other words, an action description, e.g. winning, may transform into an event description entailing an action description by making it more precise than the actor had intended, e.g. winning with 6-2, 3-6, 6-4 entails winning, where only the latter was intended. Similarly, we may say that Vincent hit a target (action) by hitting target 1, an event description entailing the action description. Incidentally, the tennis example illustrates either that the video game example is much more complicated than necessary to (try to) make a point, or that I missed the intended point.

REFERENCE

Peijnenburg, J. (2000). Identity and Difference: A Hundred Years of Analytic Philosophy. *Metaphilosophy* **31** (4), 365-381.

Anne Ruth Mackor

ERKLÄREN, VERSTEHEN AND SIMULATION: RECONSIDERING THE ROLE OF EMPATHY IN THE SOCIAL SCIENCES

ABSTRACT. A basic naturalistic epistemological intuition that Theo Kuipers and I share is the idea that the differences between the natural and the social sciences do not stand in the way of co-operative, integrative, and perhaps even reductive relations between them. In several papers I have offered a teleofunctional argument against interpretationalist autonomy claims and Kuipers (2001), Chapter 6 seems to favor this type of rebuttal. However, within the last 15 years or so, there has been a revival of another kind of "verstehende," or rather "einfühlende," approach, which differs in some significant respects from the interpretationalist view. In this paper I investigate whether this so-called simulation theory might cause trouble for our naturalistic view of the relation between the natural and the social sciences.

1. Erklären, Verstehen and the Simulation Theory

A basic naturalistic epistemological intuition that Theo Kuipers and I share is the idea that the differences between the natural and the social sciences do not stand in the way of co-operative, integrative, and perhaps even reductive relations between them. Making use of Kuipers' analyses of ontological, epistemological, and methodological scientific levels and the relations between them (Kuipers 2001), in particular Chapters 3, 4 and 6, I have tried to answer the question whether his model applies to the relation between the natural and the social sciences as well.

The reason to focus on the relation between the natural and the social sciences is obvious. There are some features of psychology and the social sciences that seem to cause trouble for any reductionist model. One of the most pressing questions in philosophy of science and philosophy of mind is about folk psychology, viz. how we ascribe mental states and behavior to other agents. Traditionally, philosophers distinguish two explications of how we do this.

1. Naturalist (erklärende, positivist) philosophers claim that folk psychology is a (folk) science like other (folk) sciences. We describe, explain and predict

In: R. Festa, A. Aliseda and J. Peijnenburg (eds.), *Cognitive Structures in Scientific Inquiry* (*Poznań Studies in the Philosophy of the Sciences and the Humanities*, vol. 84), pp. 237-262. Amsterdam/New York, NY: Rodopi, 2005.

mental states and behavior in the same way as we describe, explain and predict natural scientific events: we do so by means of empirical laws and theories (e.g. Hempel, Ernest Nagel, Churchland).

2. Interpretationalist (verstehende, hermeneutic) philosophers argue that folk psychology and the social sciences that are based on it are different from the natural sciences. The way we describe, explain and predict mental, in particular intentional, states and human behavior is very different from the way we describe, explain and predict natural scientific events. We do so through interpretation of behavior in terms of social rules against the background of a "form of life" (e.g. Wittgenstein, Gadamer, Davidson).

These two views disagree, not only about the question whether the social sciences[1] are epistemologically and methodologically *different* from the natural sciences, but also with respect to the question whether they are, as a consequence of these differences, *autonomous* of the natural sciences. Many adherents of the interpretationalist view claim that they are, naturalists argue that they are not. In several papers (e.g. Mackor 1997, 1998, 1999, 2000a) I have argued against the autonomy claim of interpretationalism and Kuipers (2001), Chapter 6 seems to favor this type of rebuttal.[2]

There is no room to elaborate my argument, but roughly my strategy has been to show that the putative unique features of the social sciences that interpretationalists point at are characteristic of biology as well, and that because of these shared features, biology can bridge the gap between the natural and the social sciences. The first step is to show that in biology too, interpretation plays an important role. Function-ascriptions, like ascriptions of intentional states and behavior, demand interpretation (Millikan 1984, 1993). The next step consists in showing that although at first glance biological interpretation seems to be haunted by the same problematic features as interpretation in the social sciences (normativity, holism and indeterminacy) these features can be explained in lower-level terms. Interpretation, thus understood, does not conflict with a naturalistic view of biology. Moreover, despite the fact that it is interpretative, biology has laws, albeit laws under so-called normal conditions. Finally, restricted bio-physical identities seem possible. Therefore co-operative and even reductive relations between physics

[1] Throughout this paper, 'social sciences' should be read as 'psychology and those social sciences that make use of *folk psychological concepts and intentional explanations* to describe and explain behavior'.

[2] Thus, strictly speaking we should distinguish between naturalism as a theory about the *nature* of the social sciences and naturalism as a theory about the *relation* between the natural and the social sciences. My aim in this and earlier papers has only been to defend the latter theory.

and chemistry on the one hand and biology on the other are possible, even though these relations are more complex than those between physics and chemistry. The final step in my argument has been to show that this analysis of interpretation in biology also applies to psychology and the social sciences and that co-operative and even reductive relations between the natural and the social sciences are possible.

In this analysis, I have focused on the debate between naturalists and interpretationalists. However, within the last 15 years or so, there has been a revival of another kind of "verstehende," or rather "einfühlende," approach, which differs in some significant respects from interpretationalist views. Thus, we have to distinguish a third theory of how we understand other agents.

> 3. Simulation theorists claim that we are able to describe, explain and predict, not only the intentional states and the behavior of other agents, but also their sensations and emotions, through simulation, i.e. imaginative identification or empathy (e.g. Heal, Gordon, Goldman).[3]

The problem that has recently started to worry me, and I believe that it should bother Kuipers as well, is to what extent the simulation theory might cause trouble for our naturalist view of the *relation* between the natural and social sciences. The reason for this worry is that even if biology and psychology are on a par with respect to the role that interpretation plays, they are different as far as empathy is concerned. Although we can *interpret* non-mental biological systems, we cannot *empathize* with them.[4] Therefore, my "biological" reply to the interpretationalist view cannot hold, at least completely, as an answer to the simulation theory.

Simulation theory is a theory within developmental psychology and philosophy of mind. Its relevance for the debate in the philosophy of science has not yet been analyzed extensively.[5] In this paper I shall investigate what implications the simulation theory might have for the social sciences.

The problem I intend to explore is what role simulation plays in folk psychology-based social sciences.[6] A further question is to what extent simulation might cause trouble for a *naturalistic*, that is co-operative,

---

[3] As in the case of naturalists and interpretationalists, there are important differences between simulation theorists. I shall focus on Alvin Goldman's version and refrain from discussing their differences as much as possible.

[4] I intend this claim to be true by definition, i.e. a person simulates another person if he has *mental* states that are more congruent with another's situation than with his own situation. Note that the definition leaves open the possibility that we are able to empathize with non-human animals.

[5] See Kögler and Stueber 2000, however.

[6] In Mackor (2001) I have investigated what implications simulation theory might have for ethics, viz for our analysis of the virtues of justice and benevolence. Also see Mackor (2000b).

integrative and possibly reductionistic, account of the relation between the natural and the social sciences. This paper is mainly devoted to the first question. The second question is briefly discussed in section 7.

2. Kuipers on Verstehen

Two possible replies to the simulation theory immediately come to mind. An interpretationalist would most naturally give the first, a naturalist would give the second reply. Interpretationalists might argue that the simulation theory must be discarded since it is a revival of the mistaken Einfühlen- or Erlebnis-view. That view was fiercely attacked by, among others, Wittgenstein because of its cartesian epistemological presuppositions. In section 6.4 I shall offer a brief reply to the anti-cartesian worries that one might have about the simulation theory (also see footnote 7).

Naturalists on the other hand can argue that, although the capacity for empathy can be a very useful and reliable *heuristic* device for any social scientist, the question what role simulation plays is only a question in the context of discovery. The answer would have no direct consequences for the context of justification (see Fuller 1995, p. 19, Kögler and Stueber 2000, pp. 13-14).

I do not know whether Kuipers favors such a naturalistic answer, but it seems clear that he does not think verstehen or even einfühlen to be a problem to his approach. In a brief remark on the topic (2001, p. 101), Kuipers refers to Van Nierop (1989) and argues that verstehen or einfühlen[7] is only "a transcendental condition for the *possibility* of knowledge about human affairs".[8] It is not, or so Kuipers claims, "a necessary condition for the *acquisition* of knowledge", i.e. not "a *methodological recipe* for the

[7] Kuipers is wrong to suggest that einfühlen and verstehen are on a par. Defenders of einfühlen put emphasis on the psychological process. They argue that in order to understand the mental state someone is in, one has to "feel into" his inner feelings. Wittgenstein fiercely attacked the cartesian idea that we have infallible knowledge of our own "private and inner feelings" and that we know the mental states of others by reasoning from analogy of our own mental states. He argued that mental states have to be grasped by immersing, not into private feelings, but into the form of life of the other person, and thus by learning the public and socially shared rules that he follows. Nowadays, verstehen is usually understood in the Wittgensteinian sense of an intersubjective understanding, in the Gadamarian sense of hermeneutics, or in the Davidsonian sense of interpretation and the concepts of "feeling into and "inner feelings" are treated with suspicion. Compare section 6.4.

[8] Van Nierop himself states that "hermeneutics is a philosophical investigation into the conditions for the possibility of our interpretative way of knowing" [my translation ARM] (1989, p. 20), also see (1989, p. 19, 39, 64).

acquisition of such knowledge" (my italics, ARM). Kuipers continues and asks, rhetorically: if verstehen were a necessary condition, "who would be able to explain [in folk psychological terms, ARM] the behavior of someone like Hitler?"

It is doubtful, however, whether Van Nierop would agree with Kuipers. Van Nierop himself suggests that verstehen has more far-reaching implications than Kuipers seems to acknowledge. Taking war as an example (1989, pp. 51-52) he argues that we can only hope to understand more about it if we introduce mental notions such as "despair" and "desire." Next, he argues that if we do so, we have to realize that we could not know and describe despair and desire if:

1. these states did not have a sensory observable expression,
2. we never *ourselves experienced* despair and strong desires,
3. we were not allowed to *equate* our own experiences in relevant ways to those of the persons we study.

Thus, although having similar experiences is not sufficient for understanding the behavior and mental states of other persons (we can have experiences without understanding them and we can have experiences without understanding that others have similar experiences), such experiences do appear to be *necessary*. This seems to imply that on Van Nierop's Diltheyan view, we *must* be a little bit like Hitler if we want to understand his behavior in folk psychological terms. Thus, verstehen is not merely a transcendental condition; it seems to have methodological implications as well.

I shall not analyze Van Nierop's view in more detail however. Van Nierop explicitly states that Dilthey's verstehen is not "empathy" or "feeling into" (1989, pp. 55, 57[9]), whereas I want to investigate the implications of the simulation theory which has "empathy" and its synonyms as its key term.

In the next section I shall sketch the debate between the theory theory (a position closely related to erklären) and the simulation theory. In section 4, I shall go into more detail and discuss one of the experiments that is central to the debate: the so-called false belief task. In section 5, I briefly compare the theory theory and the simulation theory to the erklärende and verstehende positions in the philosophy of science. There I shall argue that the simulation theory seems to make at least two claims that distinguish it from erklären, and at least one claim (viz. the second) that distinguishes it from verstehen:

---

[9] Kögler and Stueber (2000, pp. 25-29) argue that Dilthey's earlier work seems closer to the simulation theory than his later work.

1. Folk psychology does not have laws.
2. The first-person point of view is important for making third-person mental attributions, viz. for doing folk psychology-based social science.

I shall argue, however, that the first claim does not really distinguish the social sciences from the natural sciences. The second claim is more problematic, both from a naturalistic and an interpretationalist point of view. I shall discuss it in section 6. A tentative conclusion is formulated in section 7.

3. Theory Theory Versus Simulation Theory[10]

The discussion that has been going on in developmental psychology and philosophy of mind over the last 15 years has two main opponents: those who adhere to the theory theory of mind and those that support the simulation theory. Their disagreement is about the question how we are able to ascribe mental states to other persons and to ourselves, i.e. how we are capable of doing folk psychology. For reasons to be discussed in the next section, the debate focuses on beliefs, in particular on the capacity to ascribe false beliefs. Both theories are meant to cover all types of mental states, however.

According to the theory theory, to be able to ascribe beliefs, one must have the concept of belief. In order to have the concept of belief, one must have a body of psychological knowledge, that we might call a (tacit) psychological *theory* (Davies and Stone 1995a, p. 3). One of the reasons to call it a theory, even though it is largely tacit, is that explanations of phenomena (implicitly) refer to *unobservable theoretical posits* (viz. mental states) that play an explanatory role, as well as to (albeit rough and ready) *laws*. On the theory theory, mental concepts are akin to natural scientific concepts, they can change and be eliminated.

Explicating the simulation theory is a more complicated task. First, note that “simulation” is an ambiguous notion. Simulation can mean “process-simulation,” but it can also mean “computer-simulation.” In the case of computer-simulation you feed theoretical posits into the simulating system and

[10] Davies and Stone (1995a) and (1995b), Carruthers and Smith (1996) and Kögler and Stueber (2000) are important anthologies about the debate between the theory theory and the simulation theory. The debate between these theories started with an experiment on chimpanzees that was set up to prove that chimpanzees are inferior mind-readers, precisely because they do not have a theory of mind but “merely” use simulation. Then simulation theorists started to argue that human beings are simulators too and the evidence was used against the theory theory (Harris 1995, p. 208).

let it "calculate over" these posits. In the case of process-simulation, the system enters into the same or at least isomorphic states as the target system. Thus, simulating a virtual fire on a computer is an example of theoretical simulation, whereas simulating a real fire in a laboratory would be an example of process-simulation.

The simulation theory claims that the person who empathizes enters into the same or at least isomorphic mental states as the person he imaginatively identifies with (Goldman 1995a, p. 85, Davies and Stone 1995a, pp. 6, p. 18-19, also see Kögler and Stueber 2000, p. 7). On this view, to have and to develop a "folk psychology" is not to have and to develop a theory, but rather a matter of having and developing a skill or a practice.[11]

The core idea of the simulation theory is that to be able to ascribe a belief to someone else, one has to be capable of *entertaining thoughts while imaginatively identifying with* this person (Davies & Stone 1995a, p. 5). Defenders of the simulation theory disagree, however, about the precise meaning of this phrase. Gordon (1995b) argues that simulation implies that you imagine *the other person* in his or her situation. You try to neutralize yourself, so to speak. On Goldman's account on the other hand, I must imagine *myself* in the situation of the other.[12] Also, Gordon and Heal seem to differ, perhaps from Goldman's, but certainly from my own view in conceptualizing imaginative identification as a purely or at least mainly *cognitive* and, one might say, *disembodied* matter. In the last paragraphs of section 4 I discuss some evidence in favor of my view that the *affective* or perhaps rather the *bodily* aspects of simulation should also be taken into account.

Another unclarity exists about the role that concepts play in simulation. Davies and Stone (1995a, p. 5) argue that on the simulation theory one need not have concepts of mental states, but Goldman (2000, p. 184) argues that in simulation "appropriate concepts are certainly needed." Some clarification of the concept of "concept" will be given in sections 6.2 and 6.3.

Although it is hard to give a uniform characterization of the simulation theory, Goldman's explication of the differences between the theory theory and the simulation theory brings us to the core of their disagreement: "ST contrasts with pure TT in its positive claim that some attribution processes

[11] On this (and other) points, the simulation theory seems to be in accord with Wittgenstein's views. See section 6.5 however.

[12] So, on Gordon's account we ask ourselves: what will *John* do when he sees a child drown, given that *he* cannot swim? On Goldman's account we ask: what would *I* do if I saw a child drown and *I* couldn't swim? Gordon (1995b, p. 53) has anti-cartesian worries about Goldman's account because the latter puts emphasis on the possession of first-person mental concepts (also see Kögler and Stueber 2000, p. 9). See Goldman (2000, pp. 179-80 and pp. 182-3). Also see section 6.

involve attempts to *mimic* the target agent, and in its negative claim that denies the use of theoretical propositions, such as scientific laws, in these attributional activities." (2000, p. 185).

4. The False Belief Task and Other Evidence

One of the topics in the debate between the simulation theory and the theory theory is the question how to explain the fact that children until about four years of age are unable to ascribe false beliefs both to other persons as well to their "former" selves. Many experiments have been done that reveal this striking failure.

In one experiment, both three and five year old children were shown a closed candy box. They were asked what they thought was inside, and all would guess that there were candies inside. After having been shown that there were pencils inside, they were asked what they had originally thought was inside. Whereas most five-year-olds would (correctly) say: "candies", most three-year-old children said that both now *and then* they thought there were pencils inside. Also, when asked to predict what other persons would say was in the candy box, five-year-olds (correctly) said "candies" but three year olds said "pencils."

Theory theorists and simulation theorists agree that the answers of the three year olds are not lies, but must be accounted for in terms of the fact that three year olds are unable to ascribe false beliefs to anyone, be it somebody else or themselves. These failures are striking since three, even two year olds have no trouble understanding that other agents have different *goals and desires* (Harris 1995, p. 212) and that other agents can be *ignorant* about facts. When it comes to divergent *beliefs and perceptions* however, they fail.

How do psychologists explain these facts? On the theory account, children are incapable of ascribing false beliefs because they lack the concept of belief. When they seem to *ascribe* a true belief to somebody, in fact they just *express* their own momentary beliefs. (They only lack the concept of belief; they do have beliefs.) This explanation is supposed to support the theory theorist's claim that the development of children's folk psychological capacities should be understood as the acquisition and refinement of concepts and laws, i.e. of a theory of mind. On a slightly different version of the theory theory, young children do have a concept of belief, but too simple a concept of belief. They have a so-called copy- or mirror-view of beliefs: what someone else believes simply mirrors the way the world is according to the children (Kögler and Stueber 2000, p. 8). I shall discuss this claim, which I consider to be the most convincing, later in this section.

Before discussing some simulationist accounts of the false belief task, it should be noticed that the theory theory makes no precise claims about how we *acquire* and *apply* concepts of mental states and that simulation theorists disagree on this point. Goldman argues that our capacity to simulate presupposes a mainly first-personal understanding of psychological concepts. Therefore he calls his view the *introspection-simulation* view (2000, p. 183). Gordon (1995b, pp. 53-4) on the other hand, argues against the view that simulation requires prior possession of mental concepts and suggests that we master psychological concepts through simulation. In section 6, in particular section 6.5, I shall return to this issue and suggest that simulation might play an important role, not only in the *attribution* of mental states to others, but also in the *acquisition* of concepts of mental states as well as in the *articulation and identification* of our own mental states.

Let us now turn to the simulation theory. On this account children do not improve a theory of mind, rather what they do is increase their imaginative flexibility.[13] One of the hard things about simulation is that you must learn to keep your own (incompatible) mental states out of the simulation process. This is something that young children find hard to do. So in false belief experiments their own beliefs enter into the simulative procedure and "keep out" or "overrule" the false belief that they should have simulated. This view fits with experimental findings that three year olds do better on false belief tasks when they are allowed to go through a story twice. Presumably this helps them in reconstructing target's mental state from memory (Goldman 2000, p. 175). Simulation theorists also argue that the questions about false beliefs are too complex for three year olds and that the real problem is one of performance, rather than competence since three year olds seem to have a passive grasp of the problem (Goldman 2000, p. 174, Perner 1995, p. 262).

The hard case for the simulation theory, however, is that these arguments do not yet explain the difference between simulation of beliefs and of desires. One would expect that their (strong) desires too would enter into the simulative procedure and disturb their imaginative identification. Although this is what we see happening to very young children (upon seeing his mother

[13] Compare Harris (1995, pp. 212-216) for a four-step explanation of this development: Step 1 (toward the end of the first year): echoing another's intentional stance toward present targets; Step 2 (toward the end of the first year, and increasingly during the second year): attributing an intentional stance toward present targets; Step 3 (three-year-olds and to some extent two-year-olds): imagining an intentional stance; Step 4 (at around four years, and systematically by five years): imagining an intentional stance toward counterfactual targets.

cry, a one or two year old may give his teddybear to comfort her[14]), by the time a child is three years old, it does understand that other agents do not have the same goals as he has, and that they do not always use the same ways to achieve goals. By that time, he is capable of ascribing desires that are different from his own, while he is still not able to ascribe divergent beliefs. So it seems that the simulation theory has to come up with an additional explanation here.

A hypothesis to this purpose says that in order to keep simulation economical, people will use their own mental states as much as possible. However, there is a clear difference between beliefs and desires in this respect. Whereas it is normally quite safe to substitute your own beliefs, at least as long as we deal with relatively "basic" beliefs and perceptions ("the apple is green", "it is raining"), this is not true with respect to desires ("I want to play with dolls, but daddy doesn't"; "daddy likes coffee, but I don't").[15] Although this solution to the problem is tempting (Perner 1995, p. 244), there is a more convincing explanation.

The theory theorist Perner (1995, pp. 245-6) argues that although three year olds can differentiate between actions according to true propositions and actions according to false propositions (i.e. they can *evaluate* propositions as true or false), they do not understand *that* they and others evaluate propositions and that people can evaluate the same proposition differently. Perner (p. 247, footnote 4) claims that experiments confirm his hypothesis; in any case his claim fits my own (casuistic and methodologically uncontrolled) observations of my son. For example, at the age of three years and three months he spontaneously commented on a picture in a children's book where a rabbit was eating an orange candle. He laughed and said: "But that is not a carrot!". When I asked, "What is it then?", he said "It's a candle!". Thus, he was clearly capable of evaluating propositions. However, when I asked: "But what does the rabbit think the candle is?", not only was he unable to answer the question, but he got confused and did not seem to understand what I was getting at.

On Perner's view therefore, three year olds do not yet have a complex enough concept of a belief, because they do not differentiate between the referent (the state of affair which is represented) and the sense (the way in which the state of affair is represented) of the representation (Perner 1995,

---

[14] Note, however, that it is most likely that the child *projects* his own desire on his mother, not because his own desire overrules his simulation of her actual desire, but because he has no inkling how else she could be comforted.

[15] Relatedly, Goldman (2000, p. 181) argues that "there may be a social premium on the communication of desire that results in greater conversational deployment of the language of desire."

p. 246; Harris 1995, p. 213). One important implication is that they cannot yet differentiate between (false) belief and pretence.

Although this would show that the concept of a belief of three-year olds is too simple, and although it seems as if individual mental concepts develop in groups of interdependent concepts (e.g. when children understand the difference between sense and reference, they will acquire both the concept of pretence and of false belief; Perner 1995, p. 264), it does not disprove the claim that simulation plays a role in the acquisition and application of these concepts. Moreover, Perner's hypothesis can also be formulated in a more simulationistic terminology. Thus, the question is: why do three-year-old children understand that others can have divergent desires and why do they understand (and enjoy) pretend play? Understanding pretend play (John pretends that the banana is a telephone) and understanding that we can have different desires (John wants a banana, I want an apple) seem to demand the ability to think counterfactually. Pretend is a true belief in another, possible, world and divergent desires refer to different states of affairs being realized in the future, i.e. possible, world. Thus, to be able to grasp pretend and divergent desires, a child must be capable of imagining *himself* in another possible world. Understanding false beliefs (John falsely believes the banana is a telephone) on the other hand seems to demand the capacity to *re-center to the perspective of another person in the real (actual or past) world*.[16] This notion of re-centering seems to fit nicely with simulationistic ideas although, obviously, a simulation theorist must come up with a detailed account of what such re-centering consists in.[17]

I conclude that the false belief task offers no conclusive evidence in favor of either TT or ST. I shall therefore discuss some further arguments in the debate. One argument against the theory theory comes from a famous experiment of Kahneman and Tversky (quoted in Davies and Stone 1995a, pp. 17-18) about Mr Crane and Mr Tees. Subjects in the experiment are told a story about Mr Crane and Mr Tees who go to the airport where they intend to catch different planes that have the same time of departure. Unfortunately their car ends up in a traffic jam and they arrive too late at the airport: both planes have left. The plane of Mr Crane, however, left in time, i.e. 30 minutes ago, whereas the plane of Mr Tees was delayed and left just 5 minutes ago. Who will be more upset? Ninety-six % of the subjects said: Mr Tees is going to be more upset. How do they know, and with so much certainty?

---

[16] I owe the terminology of real versus possible worlds to Perner (personal communication).

[17] These findings on children seem to fit with experiments on chimpanzees. They too are capable of counterfactual thinking, but probably not of re-centering to the perspective of other chimpanzees in the actual world (Tomasello 1995).

The theory theory should argue from regularities or laws. Mr Tees is going to be more upset, because "*most persons* are more upset when their plane has just left, because most persons that have come close to attaining a goal will, upon failing to meet that goal, be more upset than persons who believe that they have not come close to their goal." On the simulation-theoretical account on the other hand, we argue from simulation: Mr Tees is going to be more upset, because (I know, or at least believe, that) *I myself* would be more upset if I were in the position of Mr Tees. It looks as if simulation theory has got a point here, for will not most persons base their inference, at least in this example, on prior self-knowledge?

Another hard case for the theory theory comes from research on people with autism. Persons who are later diagnosed as autistic, are as infants poor at following the gaze of other persons and at influencing another's visual attention. Their inbuilt mechanism for establishing joint attention (important building block of full-blown empathy) does not work properly (Harris 1995, p. 215). Moreover, although autistic persons with a normal intelligence[18] fail as folk psychologists, they are capable of learning (folk) physics, chemistry and biology (e.g. physiology and neurology) as well as "normal" human beings. As Gordon (1995a, p. 70) puts it: autistic children who "… treat people and objects alike … do at least as well as normals in their comprehension of *mechanical* operations."

The problem for the theory theory is that the way that autistic people seem to learn why and when people shake hands, say thank you, become angry, etc. is exactly in the way that is suggested by the theory theory, viz by learning general (explicit) rules, by learning a *theory*. But the problem is that autistic people have trouble in *applying* these rules, i.e. in making (explicit) inferences. They do not know when the rules apply; they lack the *sensitivity* for knowing when ceteris are or are not paribus. This sensitivity might be just as much a matter of feeling as of knowing since we not only use our cognitive, but also our emotional and motivational system when we are simulating (Kögler and Stueber 2000, p. 11).

Additional evidence comes from children with Down syndrome. Although they have a lower IQ than normally intelligent autistic people of the same age, they perform significantly better on false belief tasks. Also, children with Williams syndrome, who have an average IQ of 50 and who as adults seem to be unable to undergo any of the forms of conceptual change associated with theory-learning, start to ascribe beliefs and desires to others at roughly the same age as "normal" children (Goldman 2000, p. 175). These facts all

[18] Most people with autism are mentally retarded.

suggests that there are fundamental differences between (folk) physics, chemistry and biology on the one hand and (folk) psychology on the other and thus that there is some truth in the denial of simulation theorists that normal human beings make use of laws in their folk psychological attributional activities.

Let us now look at the positive claim of simulation theory, viz. that at least some attribution processes involve attempts to *mimic* the mental states and processes of the target agent and that these states themselves, rather than theoretical posits about them, are the starting point of folk psychological practical inferences. Although simulation theory primarily focuses on cognitive states, I want to draw attention to affective and motivational aspects of our capacity for simulation (see Mackor 2001, pp. 38-42 for a more extensive overview).

For a start, it is a well-known fact that infants, soon (hours or even minutes) after they are born, are capable of imitation. When infants hear other infants cry, they'll start to cry too. This reaction fades away when infants grow older. Another famous example of early imitation is tongue protrusion. Although these are examples of behavioral imitation, it is argued that behavioral imitation has mental effects. For instance, an experiment has shown that subjects who are instructed to put on a sad face when listening to jokes find these jokes less funny than subjects who were instructed to put on a neutral face and subjects who were instructed to put on a happy face (Hoffman 2000).

Related evidence comes from physiological experiments (Levenson and Ruef 1992) that show that some of the physiological states of subjects become similar to ("resonate") the physiological states of subjects who's emotional states they are instructed to describe. This is in particular so when they have to interpret negative emotional states. The most intriguing finding is that these subjects are not only more *motivated to help* the target than subjects who's physiological states differ from the target's states, but that they are also better at *correctly describing* the mental state the target is in. Moreover, analogous research on people with autism (Althaus 2000) shows that their physiology (e.g. blood pressure, respiration, heart beat) does *not* change when they observe others.

Finally, recent research on so-called mirror neurons is interesting in this respect. Mirror neurons are a particular class of visuomotor neurons that are activated, not only when a subject *performs* a particular action, but also when the same subject *observes* the action when performed by somebody else (Gallese and Goldman 1998).

I conclude that, although evidence is certainly not conclusive, the simulation theory is a position that should be taken seriously. Therefore it is worthwhile to investigate what implications the simulation theory might have for the social sciences. *If* simulation theorists are right, there are some intricate problems to be sorted out, in particular with respect to the role of the first person perspective. Before dealing with those problems, however, I briefly compare the debate between simulation theorists and theory theorists to the naturalist and interpretationalist positions in the philosophy of science.

5. Erklären-Verstehen, Theory-Simulation[19]

The theory theory is a version of the Erklären view. According to the theory theory, folk psychology is a theory like all other folk theories, such as folk physics and folk biology. Scientific psychology and social sciences that use folk psychological notions are sciences like any other science. In particular, two implications seem to follow from the theory view. In the first place, psychology and the social sciences have *laws*, although they may be fairly rough. Second, the theory theory is a purely third-personal point of view. That is to say, it in no way implies that the seemingly direct acquaintance with my own mental states contributes to my knowledge of the mental states of other persons. The theory theory rejects the Argument from Analogy according to which I infer from my own case that others, who seem similar to myself, have similar mental states. Thus, on the theory theory the first-person point of view does not play a special role in the attribution of mental states and behavior to other persons.

The simulation theory differs from the modern interpretationalist approaches and is closer to older Einfühlen-theories such as Collingwood's and Dilthey's earlier view (cf. Heal 1995b, p. 33). In the first place, the simulation theory differs from modern versions of verstehen in that it focuses on mental states and does not say anything specific about the role that social rules and social roles play in the simulation process. Some philosophers have argued that therefore simulation theory seems particularly relevant for pre-linguistic and pre-social (universal biological) aspects of understanding that have been ignored by interpretationalists (Kögler and Stueber 2000, p. 37).[20] Second, and more importantly, although simulation theory, just as modern versions of verstehen, focuses on intentional states such as beliefs and desires, it seems particularly promising, at least more promising than verstehen, with respect to our understanding of the affective and phenomenological aspects of bodily sensations and emotions.

[19] For an extensive comparison of the two debates, see Kögler and Stueber 2000, pp. 1-61.

[20] I have doubts about this claim, but I shall not pursue it.

At the end of section 2, I have stated that the simulation theory differs from the standard naturalist approach in at least two respects. First, the simulation theory seems to imply that folk psychology is basically *casuistic*; second, the first-person point of view seems to play a crucial role in third-person attributions.

Let us deal with these claims one after the other. With respect to the first point, it is argued that we explain behavior by imagining what a particular person would do on a particular occasion. And when we do so, we take so many and so diverse factors into account that it does not seem to make sense even to try to formulate a general law afterwards, simply because these factors cannot be formulated as a "standard" ceteris paribus clause. For example, how should we "fill in" the ceteris paribus clause of the "law" that we derived from the Kahneman-Tversky example, viz. that "most persons that have come close to attaining a goal will, upon failing to meet that goal, be more upset than persons who believe that they have not come close to their goal"? Similarly, the deficiency that autistic people seem to suffer from, lends some credibility to the idea that the ceteris paribus clause is a matter of "know how" or even "feeling" and that it is difficult, if not impossible, to explicate it as "know that."

The fact, however, that folk psychologists seem to work in a casuistic manner is not a principled difference between the natural and the social sciences. Folk physicists, but also scientific physicists, especially scientists in applied sciences such as technical and medical sciences, often argue from analogy from similar cases, rather than from empirical laws (compare Thagard 1996, Chapter 5 and Barnes and Thagard 1997 on analogical reasoning). And in the natural sciences too, it is often extremely difficult to explicate scientific know how as propositional knowledge. (For instance, think of the difficulties that computer scientists encounter when they develop expert systems).

What makes the social sciences different is not that they argue from analogy, but that in doing so they argue from the first-person to the third-person case. Again, take the Kahneman-Tversky example. It is not just that we seem to argue casuistically, because then I could have argued that I believe that Mr. Tees will be more upset because, for instance, I know that my neighbor and my boss would be more upset. However, it seems (in many cases at least) that I think (and feel) that *I myself* would be more upset. Obviously, social scientists can argue from a third-person case to another third-person case. What we need to know, however, is whether they can do so, without first (whether that be in the past, or time and again) having argued from their own case to a third-person case.

The simulation theory, at least Goldman's version of it, implies that the first-person point of view plays a crucial role and therefore it differs not only from naturalism and the theory theory but also from interpretationalist views. Gordon (1995b, p. 53) and Carruthers (1996a, pp. 28-33) among others argue however, that Goldmans's view is imbued with cartesian epistemological presuppositions. In the next section I shall elaborate on the role of the first-person point of view and offer a brief negative answer to the question whether its epistemology is necessarily of a cartesian nature.

6. First- and Third-Person Conceptions of Mental States

If simulation is not merely a reliable, but an important and perhaps even necessary tool for folk psychology-based social sciences, this would seem to imply that scientists must have had mental states, not just intentional states, but also bodily sensations and emotional experiences, that were the same, or at least very similar to the mental states of the person observed. The argument is not, however, that we must have had the same attitude with respect to the same content, but we must have had the same attitude toward some content, and we must have had some attitude toward the same content. Thus, a social scientist must have a sufficiently rich sensory, cognitive, volitional and affective repertoire, and this repertoire must be sufficiently like the persons he or she studies. (Compare Vielmetter 2000, p. 96 for an example, also see Van Nierop 1989, pp. 51-2.) Moreover, this repertoire must be used (time and again) in the attribution of mental states to other persons. Such view, however, seems to conflict with Kuipers's claim about how we could understand Hitler's behavior (compare section 2), because it seems to imply that simulation is not only a *transcendental* condition; it seems to have *methodological* implications as well.

In this section I shall spell out the role that the first-person point of view might play in both the *acquisition* and the *application* of mental concepts.[21] Obviously, what I would like to hear from Kuipers is whether he agrees on the relevance of this topic for the philosophy of the social sciences, and if so, what *his* analysis of the matter would be.

6.1. *Mary and Barry*

Mary is the well-known fictitious natural scientist who has all scientific knowledge that exists about color, but who lives in a black and white room and

[21] In empirical studies of consciousness there is quite a lot of (introspectionist and phenomenological) literature on the role of the first-person point of view. See Varela and Shear (1999) for an overview. In this paper, however, I shall ignore these (partly overlapping) views and focus on the simulation theory.

thus has never actually seen any color (Jackson 1990). Naturalists such as Churchland have argued that when Mary gets out of her room for the first time, she has one extra means or medium to recognize colors, but she does not acquire new propositional knowledge. The knowledge she acquires is not "know that," but (merely) "know how." Therefore, Churchland concludes that Mary can be a professional color scientist without ever having had a first-person experience of color.

Does the same story hold for Mary's colleague Barry, the never-been-angry social scientist? Can he acquire all knowledge about anger he needs for doing social science? Can he, on the basis of purely propositional knowledge, understand anger? That is: can he describe, explain, and predict anger-involving mental states and behavior if he has never been angry in his life, i.e. if he does not know "what it is like" to be angry? And, when he becomes angry for the first time, what kind of knowledge does this new experience give him? Is it an extra means of recognizing anger in himself or in others or in both?[22]

Let us inquire what kind of theoretical third-person knowledge of anger Barry can acquire. He can learn to identify types of events that often cause anger in persons, he can learn to recognize types of (verbal and non-verbal) behavior that are typical expressions of anger, and he can learn to recognize typical effects of anger. In order to acquire this type of knowledge, he can study, among others, the social rules of the community and the physiognomy and the physiology of the subjects he studies. Moreover, his knowledge can either be lawlike or analogical. The question however, is whether this kind of third-person knowledge is sufficient to be capable of describing, explaining and predicting even the most subtle expressions of anger. Do we not need first-person "phenomenological" experience, at least to be able to know how to apply the ceteris paribus clause more reliably to the third-personal evidence?

If it is only anger that Barry is lacking, perhaps he can fill his lack of first-person experience of anger with his first-person experience of other emotions such as fear and excitement and compare this first-person experience to the stories that other people tell about what it is like to feel angry and how it affects your mental condition and your behavior. However, since anger is

---

[22] Peijnenburg and Atkinson (manuscript) express some worries about philosophical thought-experiments. The thought-experiments about Mary and Barry, however, are less esoteric than they might seem. For example, there exist types of morphine that cause people to stop caring about the pain that they feel (Carruthers 1992, p. 188). If someone takes it, he will still feel pain, but it will no longer upset him, he will not be frightened about it and will not have the desire that the pain stops. Also, it seems that people with autism recognize anger in others in exactly the same way as Barry does. For example, an autistic child knew that his father was angry when his moustache had a particular shape. The child's knowledge was not "empathic," but inferential.

considered to be a basic emotion, this will probably be difficult to do. Problems become even nastier when Barry does not have first-person experience of *any* emotion, because then he could not even make such a comparison. The question, therefore, is: does Barry need phenomenological first-person experiences of anger to *acquire* a *social scientific* concept of anger and to *apply* this concept reliably to particular cases of angry behavior?

6.2. *Substance Concepts and Conceptions*

My tentative positive answer to this question follows from an application of Millikan's analysis of substance concepts to mental concepts. In her book *On Clear and Confused Ideas* (2000, p. 2), Millikan argues that an important task of substance concepts (individuals, stuffs, natural kinds) is "… to enable us to re-identify substances through diverse media and under diverse conditions, and to enable us over time to accumulate practical skills and theoretical knowledge about these substances." The importance of the capacity for re-identification is obvious. Our knowledge of substances, and their properties, can only be of use when we are capable of recognizing the substance on different occasions (Millikan 2000, section 1.5). Only if we can re-identify substances, can we describe them and use them, either as explanans or as explanandum.

One of the reasons why we are capable of re-identification is that substances are "… things that retain their properties, hence potentials for use, over numerous encounters with them." (2000, p. 2) Therefore it is worthwhile to learn and remember that the stuff we call water (this is our *concept*) looks transparent, is thirst-quenching, tastes like water, and feels refreshing on my skin, for it is only through these (and other) *conceptions* of water, that I am capable of re-identifying, i.e. recognizing water.

Millikan's basic idea is that human beings (and animals) can acquire the same concepts as other human beings via different, even non-overlapping conceptions, conceptions being abilities to recognize substances through their properties (2000, section 1.9). Thus, on Millikan's account, Helen Keller had the same concepts of many substances (dog, piano, water, gold) and their properties as "normal" people, even though she was both blind and deaf. For Keller, touch and even more language were very important routes to achieve those concepts. Although she lacked our ordinary audio-visual conceptions, she had the same concepts, partly because and to the extent that she was capable of re-identifying the substances that the concepts refer to.

6.3. *Concepts and Conceptions of Mental States*

Millikan does not apply her theory to mental states. In this paper, however, I shall simply assume that mental states are relatively steady states of persons, and that Millikan's analysis of concepts and conceptions is applicable to mental states as well.[23]

Let us now return to Barry's case. We could say that the phenomenological first-person experience of anger in ourselves is *one conception*, one way to re-identify tokens of anger. The topic I now want to discuss is whether this first-person experience might be such a crucial conception that its absence seriously impoverishes the concept of anger that persons like Barry could ever possess.

First note that in the normal case, where I have both first- and third-person conceptions, it is logically and empirically possible that the phenomenological first-person conception that I use to re-identify anger *in myself* is not linked to the behavioral conception that I use to identify what is in fact the same feeling of anger in someone else.[24] Obviously, this would cause me to think that there are two unrelated concepts (anger-I, anger-you) and I would fail to see that our emotions are two tokens of the same type of state. There is also the opposite possibility, viz. that I falsely believe that my phenomenology and your behavior refer to the same concept (equivocation). Perhaps it seems unlikely that we make mistakes about such a basic emotion as anger. Making mistakes seems to be more plausible when we have to decide, for instance, whether an infant is bored, sad, feeling ill, or just sleepy. My son, for example, sometimes complained that he had "specks" in his leg and that they hurt (shared verbal conception). I assume that what he felt (shared first-personal conception) is that his leg had gone to sleep. I am not sure, however, because he could not describe the feeling in any more detail, and especially because he was able to stand on and walk with his leg (conflicting behavioral conception).

My suggestion is that if we lack the first-person phenomenological conception of a particular mental state altogether, and if we can only rely on the information we get from third-person conceptions (behavior, physiology, etc.), this might cause our ability to re-identify to be seriously impoverished. If so, chances of failure are likely to increase. Thus, if we have never ourselves experienced anger, if we do not have what I have called the *first-person phenomenological conception* of a bodily sensation or emotion, the chance that our concept is inaccurate, vague, or equivocal is much larger than the chance

[23] In personal communication Millikan has stated that she intended it too.

[24] It seems that people with autism fail at exactly this point. Also, one can easily imagine that such linking fails if one's mirror neurons do not work properly. Compare Williams, Whiten, Suddendorf and Perrett (2001).

that the concept of a person who does have a phenomenological conception is incorrect in these respects.[25]

6.4. *Intermezzo: Anti-Cartesian Worries*

Earlier, I said that the simulation theory might arouse the worry that it is committed to a cartesian view of the mind. Indeed, some older Einfühlen-theories started from the cartesian assumption that my own experiences are *transparently and infallibly* given to myself and that I understand the mental states of other persons *in analogy* to my own mental states (Kögler and Stueber 2000, p. 26).

To my mind however, simulation theory need not presuppose or imply such a cartesian view. In the first place, it is not in conflict with the Wittgensteinian assumption that it is only through social interaction that we acquire knowledge of our own mental states as well as those of others. Moreover, it need not give any privileged epistemological status to phenomenological consciousness. On the view that I have sketched in section 6.3, the capacity to re-identify my own mental states through my phenomenological experiences is as *opaque*, and as *fallible* as the capacity to identify your mental states through third-person conceptions, although the former might work faster and might seem to be more direct. My suggestion has only been that these fallible first-person conceptions might nevertheless play a role, both in the *acquisition* of concepts of mental states as well as in the *application* of these concepts to other agents.

6.5. *The Need for Both First-Person and Third-Person Conceptions*

In section 6.3 I have suggested that first-person phenomenological conceptions of mental states might play a role in the identification of mental states of *other persons*. In this section, I would like to consider the opposite possibility, viz. that third-person (behavioral) conceptions of mental states might similarly play a role in the proper identification of our *own* mental states. Before arguing for this view, let me briefly repeat my position with respect to the question how we *acquire* concepts (1) and how we *apply* them to *others* (2). Subsequently, I shall give a tentative answer to the question how we apply them to *ourselves* (3).

[25] Note that I am talking about states that have a phenomenology. Most philosophers argue that beliefs do not have a particular phenomenology, but I am inclined to disagree. I would argue that there is not only a conceptual, but also a phenomenological difference between being absolutely confident, being quite certain, and merely conjecturing that something is the case.

(1) In section 6.3 I have speculated that if an agent lacks either the capacity to simulate (people with autism) and/or particular mental states (Barry), his interpretation of the mental states and behavior of other agents will fail. On the view developed in this section, our capacity to simulate is partly characterized as the capacity to link our first personal conceptions of mental states to third personal behavioral conceptions.

On this view, both the capacity to have certain mental states and the capacity to simulate are developmental requirements for the *acquisition* of reliable mental concepts and thus for the possibility of having knowledge of mental states of *others*. To put it in philosophical terms, these capacities seem to be *transcendental conditions for doing social science*.

(2) A theory theorist could argue, however, that once we have acquired concepts of mental states, our own mental states and the capacity for simulation are no longer relevant. Against this view, I want to argue that we also need our own mental states and our simulative capacities in the *ongoing practice* of ascribing mental states and behavior to other persons. The neurological and physiological evidence mentioned in the last paragraphs of section 4 (the discovery of mirror neurons and the physiological experiments of Levenson and Ruef and of Althaus) suggest that the affective and physiological aspects of these mental states play an important role in this ongoing practice.

However, if having both mental states and the capacity for simulation are necessary conditions for the *ongoing practice* of ascribing mental states to others, then, philosophically speaking, one might say that they are not, as Kuipers claims, merely transcendental but also *methodological conditions for doing social science*.

(3) Finally, I want to suggest that, just as we need our own mental states plus the capacity for simulation to understand the mental states of others, we need interaction with and observation of others plus the capacity for simulation to understand *our own mental states*.[26] Mental concepts apply equally to others and to ourselves, and it is only through interaction with and observation of others that we acquire reliable concepts in the first place. Moreover, since our concepts are gradually transformed during the process of acquisition, I would argue that our own mental states become more *determinate* and are even *transformed* in this process of concept acquisition. This is certainly true of infants (obviously many of their mental states are still very indeterminate and unconscious), but it also seems to hold for adults. For example, upon seeing someone act angrily when someone else jumps the queue, I may not only

[26] It is an interesting question whether first-third person observation is sufficient or whether first-second person interaction is necessary.

understand how *he feels* and remember how *I felt* when I last got angry for the same reason, but I may also begin to realize how *I acted*, what my facial and bodily expression must have been on that occasion, and I can even improve my moral evaluation if I begin to realize how (in)appropriate his and my behavior are. Thus, behavioral observations can help us to connect first- and third-person conceptions more tightly, erase possible errors such as equivocation, and contribute to further self-knowledge.[27] It seems that the ongoing practice of ascribing mental states to others and to ourselves allows for an ongoing *refinement* of our mental concepts, and thus for an ongoing refinement of our understanding of others and of ourselves. Therefore, I would argue, finally, that having both third-person conceptions and the capacity to link them to first-person conceptions seem to be *transcendental and methodological conditions for self-knowledge.*

As a final remark, let me state that the claims made in this section are in accord with the Wittgensteinian idea that it is *through social interaction and linguistic communication* that we acquire knowledge of our own mental states in the first place. Also, it is *through social interaction and linguistic communication* with other agents that our own mental states become articulate and determinate. What distinguishes my version of the simulation theory from verstehen however is, first, that this social interaction would *basically* be a matter of simulation and even of "a bodily feeling into," rather than of purely cognitive and disembodied interpretation. Second, and more importantly, since the simulation theory is an interdisciplinary theory, developed by, among others, philosophers, developmental psychologists, ethologists and physiologists, serious efforts are being made to give a multi-leveled, more detailed, and to some extent falsifiable explication of these processes of social interaction.

[27] On the time-scale of the development of the capacity for empathy and self-knowledge, audio-visual recording techniques have occurred only very recently. They have given us an even more direct third-person view of ourselves. Many university teachers report that it was a horrible but also very educative experience when they observed themselves on video for the first time.

7. Conclusion

I have not argued for the claim *that* simulation is an important ingredient of folk psychology-based social sciences. In sections 3 and 4 I have only sketched the debate between the simulation theory and theory theory and I have mentioned some arguments in favor of the former. I have stated, however, that *if* simulation does play a role in folk psychology, we have to investigate what implications it has for the relation between the natural, especially biological, and the social sciences. In section 5 and 6 I have analyzed the two simulationist claims that I introduced at the end of section 2. Let us now see whether they might cause trouble for the kind of naturalist view of the relations between the natural and the social sciences that Kuipers and myself defend.

1. Psychology and social sciences do not have laws

In section 5 I have briefly argued that the social sciences are not the only sciences to be troubled by the absence of laws. The same claim can be made for many physical sciences, especially applied sciences that make use of analogical argumentation. If simulation theorists are right, however, the absence of laws makes standard-reduction, viz. reduction of laws impossible by definition. However, the absence of laws does not pose a threat to other kinds of co-operation or to concept-reduction.

2. The first-person point of view is important for folk psychology-based social sciences

In sections 5 and 6 I have argued that it is not the absence of laws and the role of analogy that makes the natural and the social sciences different. It is the role of the first-person point of view that makes the social sciences different from physics, chemistry, and from most biological disciplines. The way we interpret biological (non-mental) systems differs from the way we can and perhaps must simulate mental systems such as our fellow human beings.

In itself this does not imply that the simulation theory is in conflict with a naturalistic view of the relation between the natural and the social sciences. In section 6.5, however, I have argued that if my view is correct, simulation is not only required for the acquisition of mental concepts, but also for the ongoing application of these concepts. From this view it would follow that a sufficiently rich sensory, cognitive, volitional and affective repertoire of first-person experiences and the capacity for simulation are not merely *transcendental conditions for the possibility of social science*, but also *methodological conditions for the ongoing practice of social science.*

Therefore, to return to Kuipers' rhetorical question that was quoted in section 2: if simulation theory is true, we must be sufficiently like Hitler if we want to understand him in folk psychological (i.e. not *merely* in bio-pathological) terms.[28]

University of Groningen
Faculty of Law
Department Theory of Law
P.O.Box 716
9700 AS Groningen
The Netherlands
e-mail: a.r.mackor@rechten.rug.nl

REFERENCES

Althaus, M. (2000). *Visual Attention and Autonomic Adaptivity to Attention-Demanding in Children with Autistic-Type Behavioral Problems*. Ph.D. thesis University of Groningen.

Barnes, A. and P. Thagard (1997). Empathy and Analogy. *Dialogue* **36** (1997), 705-720.

Carruthers, P. (1992). *The Animals Issue*. Cambridge: Cambridge University Press.

Carruthers, P. (1996a). Simulation and Self-Knowledge: A Defence of Theory-Theory. In: Carruthers and Smith (1996), pp. 22-38.

Carruthers, P. (1996b). Autism as Mind-Blindness: An Elaboration and Partial Defence. In: Carruthers and Smith (1996), pp. 257-276.

Carruthers, P. and P.K. Smith, eds. (1996). *Theories of Theories of Mind.* Cambridge: Cambridge University Press.

Davies, M. and T. Stone, eds. (1995a). *Folk Psychology*. Oxford: Blackwell.

Davies, M. and T. Stone (1995b). Introduction. In: Davies and Stone (1995a), pp. 1-44

Davies, M. and T. Stone, eds. (1995c). *Mental simulation*. Oxford: Blackwell.

Davies, M. and T. Stone (1995d). Introduction. In: Davies and Stone (1995c), pp. 1-18.

Fuller, G. (1995). Simulation and psychological concepts. In: Davies and Stone (1995c), pp. 19-32.

Gallese, V. and A.I. Goldman (1998). Mirror Neurons and the Simulation Theory of Mind-Reading. *Trends in Cognitive Sciences* **2**, 493-501.

Goldman, A.I. (1995a). Interpretation Psychologized. In: Davies and Stone (1995a), pp. 74-99.

[28] I want to thank René van Hezewijk, Ruth Millikan, Jeanne Peijnenburg and Pauline Westerman for their helpful comments on earlier versions of this paper.

Goldman, A.I. (1995b). In Defense of the Simulation Theory. In: Davies and Stone (1995a), pp. 191-206.

Goldman, A.I. (1995c). Empathy, Mind, and Morals. In: Davies and Stone (1995c), pp. 185-208.

Goldman, A.I. (2000). The Mentalizing Folk. In: D. Sperber, *Metarepresentations*, pp. 171-196. Oxford: Oxford University Press.

Gordon, R.M. (1995a). Folk Psychology as Simulation. In: Davies and Stone (1995a), pp. 60-73.

Gordon, R.M. (1995b). Simulation without Introspection or Inference from Me to You. In: Davies and Stone (1995c), pp. 53-67.

Harris, P. (1995). From Simulation to Folk Psychology. In: Davies and Stone (1995a), pp. 207-231.

Heal, J. (1995a). Replication and functionalism. In: Davies, M. & Stone, T. eds., (1995). *Folk psychology*. Oxford: Blackwell, pp. 45-59.

Heal, J. (1995b). How to Think about Thinking. In: Davies and Stone (1995c), pp. 33-52.

Hoffman, M. (2000). *Empathy and Moral Development*. Cambridge: Cambridge University Press.

Jackson, F. (1990). Epiphenomenal Qualia. *Philosophical Quarterly* **32** (1982), 127-136. Reprinted in: W.G. Lycan, *Mind and Cognition* (Oxford: Basil Blackwell), pp. 469-477.

Kögler, H. and K. Stueber (2000). Introduction: Empathy, Simulation, and Interpretation in the Philosophy of Science. In: Kögler and Stueber (eds.), *Empathy and Agency*. (Boulder, CC: West View Press), pp. 1-61.

Kuipers, T.A.F. (2001/SiS). *Structures in Science*. Dordrecht: Kluwer.

Levenson, R.W. and A.M. Ruef (1992). Empathy: A Physiological Substrate. *Journal of Personality and Social Psychology* **63**, 234-246.

Mackor, A.R. (1997). *Meaningful and Rule-Guided Behaviour: A Naturalistic Approach*. Ph.D. thesis: University of Groningen.

Mackor, A.R. (1998). Rules Are Laws. An Argument against Holism. *Philosophical Explorations* **1**(3), 215-232.

Mackor, A.R. (1999). Natuur- en sociale wetenschappen: verscheidenheid zonder autonomie (Natural and social sciences: variety without autonomy). *Wijsgerig Perspectief* 1999/2000-2, 45-50.

Mackor, A.R. (2000a). Niet of-of, maar en-en (Not either-or, but both). *Recht der Werkelijkheid* **21**(1), 111-119.

Mackor, A.R. (2000b). De rol van intuïties, argumenten en inlevingsvermogen in de ethiek: een reactie (The role of intuitions, arguments and empathy: a reaction). *Tijdschrift voor Filosofie* **62**(4), 727-732.

Mackor, A.R. (2001). Rechtvaardigheid, barmhartigheid en empathie (Justice, benevolence, and empathy) *ANTW* (Themanummer 'Ethiek en emoties') **93**(1), 29-45.

Millikan, R.G. (1984). *Language, Thought and Other Biological Categories*. Cambridge, MA: The MIT Press.

Millikan, R.G. (1993). *White Queen Psychology and Other Essays for Alice*. Cambridge, MA: The MIT Press.

Millikan, R.G. (2000). *On Clear and Confused Ideas. An Essay about Substance Concepts*. Cambridge: Cambridge University Press.

Nierop, M., van (1989). *Denken in tweespalt - interpreteren in ambivalentie* (Thinking in discord-interpreting in ambivalence). Delft: Eburon.

Peijnenburg, J. and D. Atkinson (manuscript). Theories and Thought-Experiments in Philosophy and in Science.

Perner, J. (1995). The Many Faces of Belief: Reflections on Fodor's and the Child's Theory of Mind. *Cognition* **57**, 241-269.

Thagard, P. (1996). *Mind. Introduction to Cognitive Science*. Cambridge, MA: The MIT press.

Tomasello, M. (1995). *The Origins of Human Cognition.* Cambridge, MA: Harvard University Press.

Varela, F.J. and J. Shear eds. (1999). *The View from Within. Journal of Consciousness Studies* **6**, 2-3. Thorverton: Imprint Academic.

Vielmetter, G. (2000). The Theory of Holistic Simulation: Beyond Interpretivism and Postempiricism. In: H. Kögler and K. Stueber (eds.), *Empathy and Agency*. (Boulder, CO: West View Press), pp. 83-102.

Williams, J.H.G., A. Whiten, T. Suddendorf and D.I. Perrett (2001). Imitation, Mirror Neurons and Autism. *Neuroscience and Biobehavioral Reviews* **25**, 287-295.

Theo A. F. Kuipers

VERSTEHEN, EINFÜHLEN AND MENTAL SIMULATION
REPLY TO ANNE RUTH MACKOR

Anne Ruth Mackor introduces some very intriguing questions of methodology in the social sciences by demanding attention to the simulation theory. This theory is supposed to be an alternative to the "theory theory" about the mind, as far as our ability "to ascribe mental states to other persons and to ourselves" is concerned. Interestingly enough, the simulation theory has similarities with the old idea of "verstehen" or rather "einfühlen" as a prerequisite for doing (folk-psychology-based) social science. As Mackor reports, I have claimed in the latter connection in SiS that it is not necessary for us to assume that to explain Hitler's behavior we have to be a bit like him.

In this reply I shall first try to summarize the main claims that have been made. I then will discuss the reach of the experimental evidence which Mackor presents, and I suggest an additional perspective.

Who Claims What

To begin with, I am not so sure as Mackor is that Van Nierop is going further than claiming that "verstehen" of certain beliefs and desires is a general prerequisite for doing social science. He only claims that, for Dilthey at least, it is a transcendental condition for its very possibility, without having to play a crucial methodological role. Van Nierop (1989, p. 20) writes: "None of the three main moments which he [Dilthey] distinguishes in the interpretation process and develops in their mutual relationship: *Erlebnis* [*Experience*], *Ausdruck* [*Expression*] and *Verstehen* [*Understanding*] are genuine methodological principles. They rather form a framework of conditions for the possibility of interpretation."[1]

[1] "De drie hoofdmomenten die hij in het interpretatieproces onderscheidt en in hun onderling verband ontwikkelt: *Erlebnis*, *Ausdruck* en *Verstehen* zijn geen van drieën echte methodologische

In: R. Festa, A. Aliseda and J. Peijnenburg (eds.), *Cognitive Structures in Scientific Inquiry* (*Poznań Studies in the Philosophy of the Sciences and the Humanities*, vol. 84), pp. 263-267. Amsterdam/New York, NY: Rodopi, 2005.

Even in Van Nierop's specific example of understanding war (pp. 51-2), as summarized in Section 2 by Mackor, the points 1-3 about "despair" and "strong desire" precisely suggest not so much the requirement of having the same sensations when understanding war, but being able to imagine having them, that is, the requirement of knowing at all what it is to have such sensations. Here the occurrence of 'never' under '2' seems crucial to me. It leads to the claim of Mackor herself about Van Nierop and, indirectly, Dilthey that, according to them, we ourselves need to have *ever* experienced despair and strong desires. However, according to Mackor, these points have nevertheless some methodological implications for (folk-psychology-based) social scientists.

However, I have to concede, see Mackor's Note 7, that I certainly have been too hasty in suggesting that 'verstehen' and 'einfühlen' are on a par. I even agree that Dilthey and Van Nierop explicitly want to discard 'einfühlen', but I would like to argue that they do not succeed very well in this. Crucial terms like 'despair' and '(strong) desire' not only have cognitive but also emotive connotations. Hence, we seem to be entitled to replace the formulation of Van Nierop and Dilthey's requirement above, viz. "knowing at all what it is to have such sensations," by the requirement of "knowing at all what it is to have such cognitive, that is, verstehende, and emotive, that is, einfühlende, sensations." In other words, despite the fact that Dilthey and Van Nierop explicitly discard the emotive side, by using phrases like 'having experienced despair and strong desires', they in fact suggest that the two aspects, verstehen and einfühlen, are both relevant. Be this as it may, in view of Mackor's exposition of the simulation theory, it is clear that she will agree that the emotive aspect is at least as relevant as the cognitive aspect in understanding human behavior in folk psychological terms.

In terms of the modern simulation theory, as opposed to the theory theory, "mental simulation," that is, "imaginative identification or empathy" is a crucial ingredient in each ascription of a mental state. More specifically, on Alvin Goldman's view, "I must imagine *myself* in the situation of the other." Mackor herself does not go that far. Her methodological claim amounts to: "The argument is not, however, that we must have had the same attitude with respect to the same content, but we must have had the same attitude toward some content, and we must have had some attitude toward the same content" (see the beginning of Section 6). However, the first condition is not case-specific, and hence not methodological. The second condition, viz. "we must have had some attitude toward the same content," is case-specific, but *prima facie* rather vague. I

principes. Ze vormen veeleer een stramien van voorwaarden voor de mogelijkheid van het interpreteren."

certainly have "some attitude" to wars in general and to the Second World War and Hitler in particular. I shall discuss Mackor's more specific claim in this respect in some more detail below.

The Reach of Experimental Evidence

In Section 4 Mackor presents a number of experimental results as part of an exploration of the paper's leading question "what role simulation plays in folk psychology-based social science?" Her ultimate concern goes even further: "to what extent [might] the simulation theory [ ] cause trouble for our naturalist view of the *relation* between the natural and social sciences?" The leading question of the paper can be split into at least three questions. One, what role does simulation play in folk psychology? Sections 3 and 4 are in fact restricted to this question. Two, what role does simulation play *as a matter of fact* in folk-psychology-based social science? Three, what role *could and should* simulation play in folk-psychology-based social science? Sections 5-7 certainly deal with the third question and to some extent with the second.

In Section 4 Mackor reports a number of experiments that seem to be relevant to the debate between the theory theory (TT) and the simulation theory (ST) about the nature of folk psychology. Regarding the false-belief task, the most detailed example in Section 4, Mackor herself concludes that it "offers no conclusive evidence in favor of either TT or ST." According to Mackor the second experiment about failing to catch the plane seems to be in favor of ST. Recall that no fewer than 96% of the investigated subjects expect that Mr Crane, who arrives 30 minutes too late to catch a plane that departed according to schedule, will be less upset than Mr Tees, who arrives 5 minutes too late to catch another plane that happened to be delayed for 25 minutes. Mackor posits that according to TT the subjects predict on the basis of a statistical guess "*most persons* are [or will be] more upset when their plane has just left [, than …], "which they may even explain and predict in terms of the more general statement that most persons are more upset when failing to achieve some purpose but nearly succeeding as opposed to failing without a real chance of succeeding. However, according to ST people predict it on the basis of the "first-person perspective": "I myself would be more upset if I were in the position of Mr Tees." According to Mackor "it looks as if simulation theory has got a point here, for will not most persons base their inference, at least in this example, on prior self-knowledge?"

In my view this is a too hasty, albeit tentative, conclusion, for, as in the false-belief task, the evidence does not discriminate between the two theories. It seems plausible that in such cases some substantial percentage of the subjects responds according to ST (ST-subjects) and the remaining percentage, also substantial, according to TT (TT-subjects). Almost all members of both groups may guess, on their respective grounds, that Mr Tees will be more upset than Mr Crane, hence the experiment does not discriminate. It is likely that the percentages may vary with the kind of case, for several reasons. One reason may be, as Mackor suggests, that in case people have strong feelings about what they would do themselves in the given situation, they are more likely to behave as ST-subjects. However, the more statistical evidence is publicly known about some type of case, the more people will behave as TT-subjects. Similarly, we may expect that scientifically educated people tend more to TT-behavior than other people. Consider the paradigmatic question in The Netherlands or any other country that was occupied in 1939-1945: "What do you think that you would have done under the German occupation in the Second World War: join the resistance movement or join the collaborating party or remain passive?" One may expect that statistically well-informed people think on average that they would have been less brave than those who are not well-informed, despite the fact that in both groups relatively many people have the inclination to think *prima facie* that they would join the resistance.

In sum, the airplane experiment, like the false-belief experiment, is not very helpful for the first question, let alone for the second and the third. Let us now turn to the third question, more particularly the question of whether simulation has to play a methodological, that is, case-specific role in the correct ascription of mental states to others. As suggested, Mackor addresses the second and, even more clearly, the third question in Sections 5-7. Assuming that the first question should be answered positively, that is, assuming that the simulation theory about folk psychology is largely correct, Mackor specifically claims, in regard to the third question, at the beginning of Section 6: "… a social scientist must have a sufficiently rich sensory, cognitive, volitional and affective repertoire, and this repertoire must be sufficiently like the persons he or she studies. … Moreover, this repertoire must be used (time and again) in the attribution of mental states to other persons."

In Section 6 Mackor presents a detailed analysis of first- and third-person conceptions of mental states. In Subsection 6.5 she arrives under (2) at an underpinning of the latter, methodological, claim: "The neurological and physiological evidence mentioned in the last paragraphs of Section 4 (the discovery of mirror neurons and the physiological experiments of Levenson and Ruef and of Althaus) speak in favor of this view." Although this evidence seems to support the simulation theory as the better theory about the nature of folk psychology, I do not see why this should support the methodological claim about how "folk-psychology-based social science" has to proceed. For example, we should leave room for other possibilities, not discussed by Mackor in the present paper, but in another publication (Mackor 1997) and elaborated by myself in Ch. 6 of SiS. The case concerns the different explanations of persistent and adolescent delinquent behavior. Let me quote (SiS, p. 185) part of the summary:

> In short, persistent delinquent behavior is explained by referring to abnormal biophysical conditions leading to abnormal functional development, which under normal social conditions may lead to persistent delinquent behavior … Such an abnormal biophysical condition does not play a role in the other type of delinquent behavior. Adolescence delinquents have a perfectly normal functional development, including the (functional) tendency to choose age-specific role models. However, in the absence of classical role models and the presence of other delinquents, of a persistent or adolescence nature, they join delinquent behavior, up to the age that other role models, specific for that age, become dominant. In sum, in the case of adolescence delinquents the external social factors are crucial; they provide the abnormal factors for the specific causal explanation of the delinquent behavior.

It seems clear that the second, role model, explanation can be phrased in folk psychological terms. But for that purpose, do we ourselves need to have experience with criminal role models? It is possible that Hitler falls in a similar category. However, it is also possible that a biophysical explanation has to be given in terms of some kind of brain defect. As has become clear from recent experimental studies (see e.g. Damasio 1994), there are people who have some frontal lobe defect, at birth or later incurred by accident, which seems to be the cause of their having (almost) no emotions at all. In this case the point of a folk-psychology-based explanation is not whether we can simulate Hitler's mental condition, but whether we can imagine what would or could happen if we did not have the kind of emotions we normally have. Hence, in this case too we need not be a bit like Hitler in order to understand his behavior. In sum, folk-psychology-based social scientists not only will have to leave room for both possibilities, they can leave room for them. However, I can perfectly imagine that Mackor will be of the opinion that I am stretching the idea of folk-psychology-based social science much too far.

REFERENCES

Damasio, A.R. (1994). *Descartes' Error. Emotion, Reason, and the Human Brain.* New York: Avon Books.

Mackor, A.R. (1997). Meaningful and Rule-Guided Behaviour: A Naturalistic Approach, Ph.D. thesis: University of Groningen.

Nierop, M., van (1989). *Denken in Tweespalt* [Thinking in Discord]. Delft: Eburon.

Arno Wouters

FUNCTIONAL EXPLANATION IN BIOLOGY

ABSTRACT. This paper evaluates Kuipers' account of functional explanation in biology in view of an example of such an explanation taken from real biology. The example is the explanation of why electric fishes swim backwards (Lannoo and Lannoo 1993). Kuipers' account depicts the answer to a request for functional explanation as consisting only of statements that articulate a certain kind of consequence. It is argued that such an account fails to do justice to the main insight provided by the example explanation, namely the insight into why backwards swimming is needed by fishes that locate their food by means of an electric radar. The paper sketches an improved account that does justice to this kind of insight. It is argued that this account is consistent with and complementary to Kuipers' insight that function attributions are established by means of a process of hypothetico-deductive reasoning guided by a heuristic principle.

1. Introduction

When Hempel and Oppenheim (1948) presented the theory of explanation that became known as "the deductive-nomological model of explanation" one of the main issues, right from the start, was the question of the position of functional explanations in biology. Do such explanations conform to the proposed model and, if not, what does this mean for the scientific status of such explanations? Or for the status of the theory?

For many years now, Theo Kuipers has defended a balanced position in this debate (Kuipers 1986, Kuipers and Wiśniewski 1994, Kuipers 1996, 2001). In his view, explanation by subsumption plays an important role in the empirical sciences (especially when it comes to explaining observational laws) but there are also many sound and informative explanations in these sciences that do not satisfy this pattern. Most notable among them are intentional explanations of actions and functional explanations of biological traits. Kuipers argues that these explanations satisfy a general pattern which he calls *explanation by specification*. This pattern also applies to certain types of causal explanations, namely those explanations that select "the cause" of an event out of the entirety of factors that led to that event.

As the title of my paper indicates, I focus on Kuipers' explication of functional explanations. On Kuipers' account, a functional explanation

In: R. Festa, A. Aliseda and J. Peijnenburg (eds.), *Cognitive Structures in Scientific Inquiry* (*Poznań Studies in the Philosophy of the Sciences and the Humanities,* vol. 84), pp. 269-293. Amsterdam/New York, NY: Rodopi, 2005.

answers the question why certain organisms have a certain trait by specifying a function of that trait. A function of a trait is an effect (of the presence of that trait) that contributes to reproduction and survival. Kuipers' main claims are (1) that this reconstruction is much closer to scientific practice than reconstructions along the lines of the D-N model, and (2) that this reconstruction shows how functional explanations in biology are sound and informative, despite the fact that they do not subsume phenomena under general laws.

My main claim is that, although Kuipers' account is indeed much closer to the practice of research in functional biology than the reconstructions along the lines of the D-N model, this account neglects much of what is gained by a functional explanation. In order to account for the insights biologists gain by the kind of reasoning they call "functional explanation," Kuipers' account must be extended. I sketch the direction in which this is to be done.

The structure of this paper is as follows. In section 2 I summarize Kuipers' account. In section 3 I present an example of a functional explanation in biology, namely the explanation of Lannoo and Lannoo (1993) of why electric fishes swim backwards. In section 4 I try to reconstruct this example along the lines suggested by Kuipers. In section 5 I show that this reconstruction fails to account for an important insight gained by the explanation of Lannoo and Lannoo. In section 6 I sketch my own account of functional explanation and my own reconstruction of Lannoo and Lannoo's explanation. In section 7 I explain how Kuipers' account and mine complement each other.

2. Summary of Kuipers' Explication of Functional Explanation

Kuipers thinks of functional explanations of biological traits as statements of the form 'the function of trait *y* in organisms of kind *x* is trait *z*' in answer to a question of the form 'why do *x*-organisms have trait *y*?'. In *Structures in Science* (Kuipers 2001, hereafter referred to as SiS), he mentions four examples of such statements (SiS, p. 113):

1) the function of lungs in animals is to enable oxygen supply by breathing
2) the function of chlorophyll in plants is to enable them to perform photosynthesis
3) the heartbeat in vertebrates has the function of circulating blood through the organisms
4) the function of the fanning movement by sticklebacks is to supply the eggs with oxygen.

Kuipers calls statements of this form "specific functional statements" (in contrast with "unspecific functional statements" which state that a trait is functional without specifying what the function is).

Kuipers develops his explication of this kind of explanation in opposition to the accounts of Hempel (1959)[1] and Nagel (1961). According to Kuipers (SiS, p. 113/4), both Hempel and Nagel explicate functional explanations in terms of an "underlying argument" (Kuipers' term!)[2] which has as conclusion "*x*-organisms (must) have trait *y*".[3] In Hempel's reconstruction (as presented by Kuipers) the premises of this argument consist of lawlike statements saying that *y* is a sufficient condition for the presence of *z*, and that the presence of *z* is necessary for *x*-organisms to function adequately, together with an initial statement saying that *x*-organisms function adequately. As *y* is only a sufficient condition for the presence of *z*, the conclusion that *x*-organisms (must) have trait *y* does not follow from these premises (*x*-organisms in which *z* is brought about by other means than *y* function adequately but might lack *y*). This means that in Hempel's reconstruction the underlying argument is not valid. In Nagel's reconstruction the premises consist of a lawlike statement saying that *y* is a necessary condition for some other trait *z*, together with an initial statement saying that *z* is present in *x*-organisms. This reconstructed argument is valid.

Kuipers' own reconstruction proceeds from the intuition that functional explanations involve a valid argument but that the conclusion of that argument is different from the one in the reconstructions of Hempel and Nagel. Kuipers' reconstruction consists of two parts: (1) an analysis of the meaning of specific and unspecific functional statements, and (2) a reconstruction of the thought process by means of which a researcher produces a verified, specific functional statement that answers the original why-question (and a verified, unspecific functional statement as an interesting by-product). This thought process is hypothetico-deductive in nature.

According to Kuipers' analysis the meaning of a specific functional statement of the form 'the function of trait *y* in organisms of kind *x* is trait *z*' has three components (SiS, p. 117):

- a descriptive component: *x*-organisms have trait *y*
- a proximate causal-nomological component: trait *y* of *x*-organisms is a positive causal factor for trait *z* in the present (or selection) environment
- an ultimate causal-nomological component: trait *z* of *x*-organisms is a positive causal factor for reproduction and survival in the present (or selection) environment

[1] Kuipers mentions only the reprint of this article in Hempel (1965).

[2] It is not clear what Kuipers means by an "underlying argument" (at least not in this context). I assume that this unclarity merely reflects Hempel's and Nagel's failure to make clear what (in their view) the relation is between functional statements of the form 'the function of ... is ...' and their reconstructions of functional explanations as arguments.

[3] Actually, in Hempel's reconstruction the conclusion is a particular of the form 'at time *t* trait *z* is present in individual *i*'.

Inspired by my dissertation (Wouters 1999), I am proud to say, Kuipers distinguishes two "readings" of specific functional statements, depending on whether the environment in relation to which the function is judged is the present environment or the environment in which the trait originated (SiS, p. 116).[4]

The meaning of an unspecific functional statement of the form 'trait *y* of *x*-organisms is functional' follows naturally from the foregoing analysis of specific functional statements: there is a trait *z* such that the function of trait *y* in organisms of kind *x* is trait *z*.

The thought process that produces the specific and unspecific functional statements starts with the observation that *x*-organisms have some trait *y*. Next the question is raised why *x*-organisms have that trait. The researcher assumes that *y* has some function and starts thinking about what this function could be. If a serious candidate (*z*) has been thought of, the researcher will check whether *x*-organisms indeed have *z* and if this is the case the causal-nomological components will be tested. If the results of at least one of the tests are conclusively negative, the hypothesized specific functional statement is rejected and the researcher starts to look for another candidate function. If the results of all tests are positive, the specific functional hypothesis is accepted as true and the researcher concludes that the initial why-question is indeed answered by the now verified specific functional statement. As a side step, the researcher infers the (verified) unspecific functional statement that the trait in question is indeed functional. However, the main sequel to the acceptance of a specific functional statement is that new, related why- and how-questions are raised and pursued.

3. Example: Why Do Electric Fishes Swim Backwards?

Kuipers claims, repeatedly, that his reconstruction of functional explanations as explanations by specification (of a function) is much closer to the practice of research in biology than explications which reconstruct functional explanations as explanations by subsumption under natural laws (e.g. SiS, p. 73, 97, 115, 121/2). This is an empirical claim, which should be substantiated with a detailed discussion of examples of functional explanations taken from real science. Kuipers mentions four examples of functional explanations, but he does not work them out in any detail and he does not provide references to the scientific literature.

---

[4] Kuipers calls the latter the "selection environment." As selection might explain not only why a trait evolved but also why a trait persists this is an unfortunate choice of term.

In this section I describe a typical example of an explanation that is called "functional explanation" both by the researchers who produced the explanation and by their audience. In the next section I use this example to evaluate Kuipers' claim that his reconstruction fits the practice of biological research.

My example is the explanation of Michael and Susan Lannoo (1993) of why electric fishes swim backwards. Some 500 fish species possess the capacity to acquire information about their surroundings by means of a kind of electrical radar. These species have an electric organ that produces a stream of weak electrical discharges which radiate through the water and return to the fish. The body of these fishes is covered with a large number of electroreceptors, which are small cells sensitive to electric pulses. An object in the water can be detected because that object changes the pattern of discharges arriving at the receptors. Electric fish species belong to several unrelated taxonomic groups. Among them are the marine electric skates, the electric eels and knife fishes of South America, the elephant snout fishes of Africa, the stargazers, and certain catfishes. Quite remarkably, almost all fishes with such an active electric sense can swim backwards as easily as forwards. Why do they do so? To answer this question Lannoo and Lannoo studied the behavior of black ghost knifes (*Apteronotus albifrons*). These 95–120 mm long fishes are natives of the Amazon basin, where they hunt for zooplankton and other small prey. Lannoo and Lannoo discovered that backwards swimming is typically performed when searching and evaluating prey. Their test subjects actively search for prey by alternating forwards and backwards swimming. If the ghost knife detects a potential prey it scans it while swimming backwards until the prey is in front of it. It then catches it with a short forward lunge. This behavior is different from the behavior of animals that detect the same kind of prey by visual cues, such as the bluegill sunfish and the tiger salamander. These visual plankton hunters search for prey from a stationary position and once they have detected a prey they approach it head on. These observations support the conclusion that the ghost knifes detect their prey by electrosensoric means. This conclusion is further supported by the observation that prey is typically detected near the fish's trunk or tail and by a study of the feeding abilities of these fishes. The ghost knife hunts equally well under dark and light conditions, it takes normal prey as easily as artificially colored prey and it prefers larger to smaller prey in all circumstances. Having established that the ghost knifes search for their prey by means of electrosensory cues, the researchers continue to explain why the prey is scanned backwards. In contrast to an optical system, an electric sense lacks the ability to focus an image. As a result electric images are "blurred." "The function of scanning prey may be to pass an object across a large number of spatially separated electroreceptors in order to compensate for this limitation in image quality" (Lannoo and Lannoo 1993, p. 163). But if the fishes scanned the prey by swimming forwards they "would have the prey located near the tail and out of position for the final lunge" (Lannoo and Lannoo 1993, p. 157). Hence, the fishes swim backwards when they scan a potential prey in order to be in a favorable position to catch the prey after finishing the scan. As the authors put it: "Scanning prey for the purpose of foraging is highly dependent on backwards swimming" (Lannoo and Lannoo 1993, 163).

4. Reconstruction of the Example

In order to evaluate Kuipers' claim that his account shows how functional explanations are sound and informative, I now try to reconstruct the above explanation of Lannoo and Lannoo along the lines indicated by Kuipers. In Kuipers' view a functional explanation is an answer to a question of the form 'why do *x*-organisms have trait *y*?'. As indicated by the title of their paper, Lannoo and Lannoo seek to answer the question "why do electric fishes swim backward?". If *x* refers to electric fishes and *y* to backwards swimming, this question fits Kuipers' template.

According to Kuipers' account, in order to answer this question the researcher assumes that *y* has some effect *z* which is favorable for survival and reproduction and starts looking for such an effect. If a candidate is found the researcher will investigate whether *x*-organisms do have *z*, whether *y* is a positive causal factor for *z*, and whether *z* is indeed favorable for reproduction and survival. In our example, the researchers start to look at when backwards swimming is performed and they discover that it is characteristic of two kinds of feeding behavior, namely searching for prey and evaluating it. As it is obvious that feeding is favorable to reproduction and survival, the interpretation that the researchers seek to establish the specific functional statement 'the function of backward swimming is foraging' seems reasonable.

In Kuipers' view this specific functional statement answers the original question and the researchers will turn to new, related questions after having concluded (as a side step) that *y* indeed has a function. However, the researchers in our example in no way think of the original question as being answered by the observation that backward swimming has a role in foraging. This observation is only the beginning of an answer. The complete answer involves the verdict that food is sought and assessed by electrosensory cues, the argument that scanning is needed in order to compensate for the lack of a focusing mechanism, the argument that forwards scanning would put the fish in the wrong position, and the conclusion that scanning prey for the purpose of foraging is highly dependent on backward swimming.

How would Kuipers account for the remainder of the explanation? One possibility is to think of it as an attempt to establish a more detailed specific functional statement which specifies some intermediates between *y* (backward swimming) and *z* (feeding).[5]

It seems that there are two such detailed statements surfacing in the discussion of Lannoo and Lannoo. One ascribes the function of scanning to the swimming behavior (in some circumstances). Scanning has the further function of assessing prey, and assessing prey has a further function in feeding. Note that, as scanning can be done both backward and forward, it is doubtful whether scanning is a function of *backward swimming*. The causally relevant activity to which the function of scanning is attributed seems to be something like swimming along a potential prey. However, the trait to be explained is backward swimming. This poses a problem to Kuipers' account because according to that account a function is attributed to the trait to be explained.

The other detailed function statement attributes to the backward character of the swimming behavior the function of finishing the scan with the head near the prey. This has the further function of starting the lunge with the head near the prey, which is a positive causal factor for feeding.

Combining these two, Kuipers could view the explanation of Lannoo and Lannoo as an attempt to establish the following (complex) specific functional statement: "backward swimming has the function of scanning a potential prey (which in turn has the further function of assessing prey) in such a way that the fish ends up with the head near the prey (which in turn has the further function of starting the lunge with the head near the prey), both functions have

---

[5] Referring to Millikan (1993) and Mackor (1997), Kuipers alludes to a distinction in terms of "proximal, distal and ultimate functions" (SiS, p. 118). As far as I know, Millikan speaks of "proximal" and "distal" causes and rules (which seems appropriate) but, in contrast to Mackor and Kuipers, she does not use the term 'ultimate'. *A forteriori*, she does not use the term 'distal' as an intermediate term, referring to something between proximal and ultimate. This latter use mixes two combinations of technical terms in an unhappy way. The 'proximal'/'distal' combination originates from anatomy where these terms are used to refer to the ends of protrusions and appendices (such as wings, legs and tails): the end near the body is called 'proximal', the other end 'distal'. This distinction is relative (one may for instance speak of the proximal and the distal spots on a wing, meaning the ones nearest the body, and the ones nearest the outside, respectively). The 'proximate'/'ultimate' combination is used in the philosophy of biology to distinguish two kinds of explanation (concerned with the individual life history of an organism, or with the evolutionary history of a lineage, respectively). The classic treatment of this distinction is given by Mayr (1961). This distinction is meant to be absolute. Given these established uses, the use of 'distal' to indicate something between proximate or proximal and ultimate is confusing.

a further function in feeding." This complex statement has the structure '*y* has the function to do z_1 in such way that z_3, z_1 has the further function z_2, z_3 has the further function z_4 , both z_2 and z_4 have the further function z_5, and z_5 is a positive causal factor for *z* (reproduction and survival)'. Note that the first function statement has the form 'the function of *y* is of doing z_1 in such way that z_2'. This is different from Kuipers' 'the function of *y* is *z*'. In other words, this reconstruction introduces another subtlety beside the intermediate functions suggested by Kuipers.

5. Evaluation of Kuipers' Claims

This reconstruction (which is the best I can make along the lines plotted by Kuipers) is unsatisfactory because it ignores two main points in Lannoo and Lannoo's account. First, Lannoo and Lannoo do not merely state that (backwards) swimming has a function in both scanning and acquiring a position favorable to catch the prey. They also point out that it is *because* swimming has a function in scanning (and because the scanning is followed by a lunge) that the fish must swim backwards to acquire that favorable position. This point is left out in the above reconstruction.

Second, this reconstruction ignores the point that scanning is needed because of the physical characteristics of electrosensoric prey recognition: as an electric sense cannot be focused, scanning is the only way in which a prey can be identified. It is difficult to mold this relation (it is the electric sense which makes the scanning needed) and the reason why this relation holds (an electric sense cannot be focused) into a specific function statement of the form described by Kuipers. Actually, as I quoted above, at this point in their explanation the researchers do use a function statement, namely "the function of scanning prey may be to pass an object across a large number of spatially separated electroreceptors in order to compensate for this limitation in image quality" (Lannoo and Lannoo 1993, p. 163). However, this function statement does not have the form Kuipers requires it to have. What follows after the phrase 'in order to' is not a trait of electric fishes for which scanning is a positive causal factor but a reason why such fishes need to scan their prey.

As a result of these omissions the reconstruction above fails to show how Lannoo and Lannoo's paper is informative. The reconstruction misses their main accomplishment which is to relate the backward character of the swimming behavior of electric fishes to the fact that those fishes are *electric* fishes. Recall the title of their paper.

It is difficult to see how an account that depicts functional explanations as consisting only of statements that articulate a certain kind of consequences can account for this type of insight. At most, statements about consequences tell us how needs are solved. But in order to determine what the needs are and how such needs arise one must look beyond the consequences of the trait in question to the other traits of the organism and the environment in which it lives. In the next section I sketch a theory of functional explanation that takes this conclusion into account.

6. Sketch of an Improved Account

6.1. *The Meaning of 'Function'*

According to Kuipers' meaning analysis, functions are attributed to traits, processes and phenomena. A function of a trait, process or phenomenon (y) of an organism is another trait, process or phenomenon (z) of that organism to which y contributes and which in turn contributes to reproduction and survival. Judged from his examples, items such as lungs and chlorophyll molecules are to be considered as traits. Swimming is probably a process. Backward swimming is presumably a phenomenon (more precisely, the phenomenon to which the scanning function is attributed, is the phenomenon that swimming is often done backward).

Note that Kuipers' meaning analysis allows for many non-standard function attributions. For example, a gibbon will die if its lungs fail to breath and as a result it will be unable to move its tail. So, the lungs of gibbons are a positive causal factor for the movement of their tail. As the movement of their tail clearly is a positive factor for the survival of gibbons, it is, according to Kuipers' analysis, one of the functions of the lungs of gibbons to enable the movement of their tail. This is a strange consequence.

In order to understand functional explanation better, a distinction should be drawn between two notions of function involved in functional explanations: function as biological role and function as biological advantage (Wouters 2003). These different kinds of function pertain to different kinds of entities: biological roles apply to items (such as lungs and chlorophyll molecules), and activities (such as swimming and beating); biological advantages apply to the properties of those items and activities (such as the surface area of the lung, and the structure of the molecule that performs photosynthesis), and to properties of the organism as a whole (such as the presence of a lung, or the fact that photosynthesis is performed by means of chlorophyll). I shall use the term 'trait' to refer to the presence or character of a certain item or activity.[6] In

[6] If I am right about Kuipers' use of the terms 'trait', 'process', and 'phenomenon', I use 'item' where he would use 'trait' and I use 'trait' where he would use 'phenomenon'. So, in Kuipers'

the example of electric fishes, the swimming behavior is an activity that (during certain episodes in the life of electric fishes) has the biological role of scanning potential prey. The backward character (a property) of that behavior is a trait that has the biological advantage over forward swimming that the fish finishes the scan in a position favorable to catch the prey. The ability to scan a prey (a property of the organism) is a trait that has the advantage over the absence of such an ability that the fish gets a better impression of the form of the prey.

Attributions of *biological roles* concern the position of an item or activity in the functional organization of an organism (how an item or activity is used).[7] The functional organization of an organism is the way in which that organism manages to maintain itself and to produce offspring. Biologists explain this ability by dividing the parts and processes of the organisms they study into a number of systems (such as the circulatory system, the digestive system, and the musculoskeletal system), each of which has a number of roles (tasks) in the maintenance and reproduction of the organism. (For example, the circulatory system has the biological role to transport oxygen, carbon dioxide, nutrients, and heat through the organism. It also has an immunological role.) Each of these systems is in turn split up into a number of subsystems (for example, the circulatory system is split up into the heart, blood and blood vessels), which in turn have their own specific roles in bringing about the capacity of the encompassing system to perform its task (the heart propels the blood, the blood is the transport medium and the vessels direct the bloodstream). An attribution of a biological role situates an item in this organization of systems of subsystems with specific tasks (see Cummins 1975, 1983, Craver 2001). Consider, for example, Lannoo and Lannoo's attribution of the function (i.e. biological role) to scan the prey to the swimming behavior. This attribution owes its meaning to a tacit decomposition of the capacity of electric fishes to maintain themselves and to produce offspring into several subcapacities (tasks), one of which is feeding. Feeding is in its turn analyzed into a number of subtasks, among which are detecting potential prey, assessing potential prey and catching it. Swimming has a biological role in all these subtasks. The ability to perform the task to assess prey can be decomposed again, for example in imaging the prey and evaluating the image. The specific biological role of swimming in imaging is to pass an object across a large number of spatially separated electroreceptors (this is called scanning).

---

terms, it is traits and processes that have biological roles, whereas it is phenomena that have biological advantages.

[7] My colleagues in the Computer Science Department would probably say that a biological role of an item is its *logical* position in the maintenance and reproduction of the organism.

The notion of function as *biological advantage* refers to the biological value (utility) of a certain trait in comparison with another trait (that might replace the trait in question). The biological advantages of a trait are the abilities resulting from that trait that give the organisms with the trait better life chances than similar organisms that lack this trait (or in which this trait is replaced by another one). For example, performing the scan by swimming backward rather than forward has the biological advantage that after the scan the fish is in a better position to catch the prey. It is assumed that this better position results in greater fitness. Biological advantages are essentially comparative and relative to certain conditions. For example, in the study of Lannoo and Lannoo backward swimming is compared with forward swimming. Backward swimming is more useful than forward swimming when prey is scanned. In other situations it might be the other way round. Biological advantages of the presence or character of an item or activity are typically assessed in relation to the biological role of that item or activity. For example, the biological advantage of the swimming behavior having a backward rather than a forward character is assessed in relation to feeding: the advantage of swimming backward rather than forward is that the feeding role is carried out more effectively.

Kuipers' notion of function is in many respects similar to my notion of biological role. The main difference is that my notion of biological role is explicitly connected with both a decomposition of an organism into systems of subsytems, and with a decomposition of the capacity of an organism to maintain itself and to produce offspring into a hierarchy of subcapacities which are performed by the several systems and their subsystems. This avoids the strange function attributions allowed by Kuipers' approach ("a function of the gibbon's lung is to keep its tail moving"). What is more important, it helps us to understand the explanatory role of this kind of function attributions: by situating an item or activity in the way in which an organism is organized, attributions of biological roles provide the handle by means of which functional biologists understand their subject matter.

It seems not difficult to extend Kuipers' meaning analysis with a comparative notion of function similar to my notion of biological advantage. I should, however, warn against the possible misunderstanding that my notion of biological role corresponds to the proximate component in Kuipers' meaning analysis and my notion of biological advantage to a more distal component (or to the ultimate one). In my view, a biological advantage is not a further effect of a biological role but a different kind of thing. Advantages are comparative and the comparison is hypothetical. In order to determine what the advantage is of scanning a potential prey, the existing electric fishes are compared with hypothetical electric fishes that do not scan the prey and it is concluded that the latter would have difficulty in recognizing the fish.

6.2. *Structure of Functional Explanations*

In Kuipers' account, a functional explanation consists in the production of a specific functional statement, by means of a process of hypothetico-deductive reasoning. According to Kuipers, a functional explanation starts with a question of the form 'why do x-organisms have trait y?'. I submit that the question addressed by a functional explanation typically has the form 'why does item/activity i of x-organisms has character s_1 rather than s_2?', for example 'why do electric fishes swim in both directions (rather than forward only)?'. My point is not only that the question is comparative but also (and foremost) that the question is about the *character* of an item or activity, for example in the case of the electric fishes it is the backward character of the swimming behavior that is explained. Admittedly, there are also cases in which the question addressed by a functional explanation has the form 'why do x-organisms have/perform item/activity i?'[8], for example 'why do male sticklebacks make fanning movements in front of their nest?', but, as I shall argue, the structure of this kind of explanations is a special case of the more general structure than can be discerned in functional explanations of the character of an item or activity.

According to Kuipers, the search for a functional explanation is guided by a principle of functionality which states that if x-organisms have a certain trait, process of phenomenon y, then y is functional in the present (or selection) environment function of y. Trait, process or phenomenon y is functional if there is a trait, process or phenomenon z such that z is the function of y. The main aim of the explanation is to specify function y.

I submit that the heuristic principle that guides the search is actually more complex. It reads:

> If item/activity i of x-organisms has character s_1 and i does not have character s_2 , then there is a biological role f and there are conditions c_1 and c_2 such that:
>
> (1) conditions c_1 and c_2 apply to x-organisms;
> (2) in x-organisms item/activity i performs biological role f;
> (3) in condition c_1 it is useful to perform biological role f;
> (4) in condition c_2 biological role f is better performed if item/activity i has character s_1 than if it has character s_2.

The idea behind the conditions mentioned in clause (1) is that c_1 is a conjunction of conditions that make it useful to perform f; and c_2 is a conjunction of conditions that make it more useful to perform f by means of an

---

[8] Note that this is just a reformulation of Kuipers' question.

item/activity with character s_1 than by means of an item/activity with character s_2. The relations between c_1 and the utility of performing f (stated in (3)), and the relation between c_2 and the utility of s_1 (stated in (4)) are law-like: they are consequences of the laws of nature. The notions of "useful" and "better" are to be spelled out in terms of the fitness of the relevant organisms.

The aim of a functional explanation is:

(i) to specify f,
(ii) to specify c_1 and c_2,
(iii) to explain (3) and (4).

Let me briefly compare this version of the heuristic principle with that of Kuipers. Clause (1) is new. Clause (2) is similar to Kuipers' proximate component. Clause (3) replaces Kuipers' ultimate component. Kuipers' ultimate component states that the function is a positive causal factor for reproduction and survival in the present (or selection) environment. In my account the conditions are not necessarily environmental conditions. Quite often they are other traits of the organism. For example, in the case of the electric fishes it is the possession of an active electric sense that makes the scan useful. Clause (4) is another replacement of Kuipers' ultimate component. With regard to the aim of the explanation it will be clear that (i) is similar to Kuipers' version, and that (ii) and (iii) are additions.

Given this aim of a functional explanation the following scheme of a functional explanation will come as no surprise:

Item/activity i of x-organisms has character s_1 rather than s_2 because:
(1) conditions c_1 and c_2 apply to x-organisms (in their present environment);
(2) in x-organisms item/activity i performs biological role f;
(3) in condition c_1 it is useful to perform biological role f;
(4) in condition c_2 biological role f is performed better if item/activity i has character s_1 than if it has character s_2;
(5) explanation of (3);
(6) explanation of (4).

The explanations (5) and (6) of (3) and (4), respectively, aim to show what the law-like connection is between the conditions and the utility stated in the claims to be explained. Ultimately, the explanation should make clear how these connections relate to the laws of nature. However, in most research papers, a large part of the explanation is only tentative. Furthermore, for obvious reasons, those parts of the explanation which are obvious to the audience will, in practice, not be reported. Explanation is typically done by pointing out a biological advantage of the way things are or (which comes to the same thing) a problem that would occur if things were different.

Lannoo and Lannoo's (1993) explanation is a typical example of an explanation that conforms to this model. As related above, these researchers seek to explain why fishes with an active electric sense swim backward as easily as forward. They start by attributing a biological role to the swimming behavior (when swimming backward), namely to scan potential prey (this contributes to the capacity of the fish to assess potential prey). From there they proceed in two ways: (1) they explain why it is useful to perform this biological role, and (2) they explain the backward character of the swimming behavior. The result is that the habit of swimming backward is connected to the manner in which prey is detected. The explanation of why scanning is useful (p. 163) remains sketchy. The gist of the sketch is that in order to assess the prey, the fish needs to form an image of the prey. As the electric sense lacks a focusing mechanism, electrical images are blurred. Scanning the prey compensates for this lack of quality. The explanation of the backward character of scanning is that if the fish scanned a potential prey forward, the fish would finish the scan in an unfavorable position to catch the prey. In sum: electric fishes swim backward, because (1) those fishes detect their prey by means of an active electric sense, (2) the physical characteristics of this sense require that the potential prey is scanned, and (3) by performing the scan backward rather than forward the fish finishes the scan in a more favorable position to catch the prey.

The train of thought in the explanation can be represented as follows (the order in which the statements are presented is the order that is intuitively logical, but the clauses are numbered in such way that the connection with the general scheme becomes clear):

Electric fishes swim backward because:

(1a) electric fishes detect prey by means of an active electric sense;

(5a) if prey is detected by electro-sensoric means the image is too blurred to assess the prey, due to the lack of a focusing mechanism;

(5b) this problem is solved if the prey is scanned by swimming along it;

(3) in the condition stated in (1a) it is useful to scan potential prey (this follows from (5a, b) and the assumption that the fitness of the fish increases if its ability to assess prey improves);

(2) scanning is performed by sensing a potential prey while swimming along its length;

(1b) to catch the prey the scan is to be followed by a lunge (c_2);

(6a) if the scan is performed forward the fish ends up with the tail near the prey;

(6b) if the scan is performed backward the fish ends up with the head near the prey;

(6c) a prey is more easily caught if the fish starts the lunge with the head near the prey;

(4) under the conditions stated in (2) and (1b) it is more useful to perform the scan by swimming backward than by swimming forward (this follows from (6a,b,c) and the assumption that the fitness of the fish increases if prey is more easily caught).

We can now see how the answer to questions of the form 'why do *x*-organisms have/perform item/activity *i*?' fits into the general scheme presented above. Such questions are answered by statements corresponding to statements (1), (2), (3) and (5). That is, the answer to a question of the form 'why do *x*-

organisms have/perform item/activity i?' partially fills in the scheme of the answer to a question of the form 'why does item/activity i of x-organisms has character s_1 rather than s_2?'.

An example is Kristensen's explanation of the fanning behavior of male sticklebacks.[9] Male sticklebacks build a tubular nest. After having lured a female to lay her eggs in its nest, the male guards the nest by a complex behavior. It alternates periods of swimming around its nest with periods as long as 30 seconds in which it stays before the nest in a slanting position, head down, moving its fins in a quick regular rhythm. This latter pattern of behavior is known as "fanning behavior." In the 1940s Kristensen performed a series of experiments which showed that this behavior has the function of ventilating the nest. He showed that the eggs die if the male is removed from the nest, and also if the nest is shielded from the fanning male with a watch glass. However, if water is directed to the nest by means of a tube, the eggs survive the removal of the male, provided the water is oxygen rich but not if it is stale. Ventilation is needed because the nest is tubular: fish species which lay their eggs on leaves in running water do not need to ventilate the eggs. The scheme of the explanation is:

Male sticklebacks fan in front of their nest because:

(1) the nest of male sticklebacks is tubular;
(5) (the explanation of the connection between (1) and the need to ventilate the nest is left out);
(3) if the eggs lay in a tubular nest, the nest needs to be ventilated;
(2) the fanning serves to ventilate the nest.

The explanation of (3) (point (5)) is obvious to biologists and will therefore be left out of the explanation. The embryos in the egg need energy to develop. This energy is gained by combining carbohydrates with oxygen taken from the environment. As a result, the oxygen concentration in the nest diminishes, and the embryo will die due to lack of energy, if the oxygen is not replenished. As the nest is tubular, diffusion from the environment of the nest is too slow to replenish oxygen at the required rate. So, in order to supply the embryos with enough oxygen the nest needs to be actively ventilated.

I submit that functional explanations ultimately seek to explain the character of an item or activity. Questions such as "why do male sticklebacks fan in front of their nest?," "why do (green) plants contain chlorophyll?," and "why do land vertebrates have lungs?" are rough and/or initial formulations of complex comparative questions of the form "why do x-organisms have an item/activity with character s_1 rather than s_2?." In the case of the sticklebacks

---

[9] As I could not find the original literature, I use Tinbergen's (1976, p. 12) account of Kristensen's experiments.

the complex questions is something like "why do male sticklebacks stay near their nest and perform fanning behavior rather than leaving the nest alone after having fertilized the eggs?" Put schematically the explanation is:

> After fertilization, male sticklebacks stay near the nest and perform fanning behavior, rather than leaving the nest alone because:
>
> (1) the nest of male sticklebacks is tubular;
>
> (5) (the explanation of the connection between (1) and the need to ventilate the nest as discussed above);
>
> (3) if the eggs lay in a tubular nest, the nest needs to be ventilated;
>
> (2) the fanning serves to ventilate the nest;
>
> (6) the fanning movement results in ventilation, whereas leaving the nest alone would not;
>
> (4) it is more useful to male sticklebacks to ventilate the nest than to leave it alone (this follows from 6 and 2).

The obvious character of (4) and (6), in this case, explains the philosopher's impression that the explanation is finished if one has discovered the biological role of the fanning behavior. However, in most cases, the detailed explanation of the character of an item or activity is an important aim of research. For example, when biologists ask the question "why do plants contain chlorophyll?" (e.g. Mauzerall 1977, Seely 1977) they have in mind very specific questions about the structure and activity of chlorophyll. The chlorophyll molecule contains a porphyrin "head" and a phytol "tail." The porphyrin head is made of a tetrapyrole ring containing a magnesium atom. The research should explain why the photochemical reaction is performed by a molecule with such a structure. Why is magnesium rather than some other metal trapped in the porphyrin ring? What are the advantages of specific organic groups? Another issue concerns the question why a molecule is used that absorbs energy at the level at which chlorophyll absorbs energy (why not higher? or lower?). The statement that chlorophyll enables plants to perform photosynthesis does not provide a satisfactory answer to such questions. It is but the beginning of the explanation, not the complete explanation.

Similarly, the statement that the function of lungs in animals is to enable oxygen supply by breathing is only the beginning of an answer to the question why animals have lungs. The complete explanation should explain why an organ with the specific character that lungs have is used for respiration (rather than an organ with some other character). Why is there a special organ for respiration? Why is the surface so much enlarged? Why is the lung internal?

6.3. *Nature of Functional Explanation*

As I see it, one main aim of a philosophical account of functional explanation is to understand how the pieces of reasoning which biologists call "functional explanations" contribute to the advancement of science. It is widely acknowledged that the products of science come in three kinds: descriptions, predictions, and explanations. Philosophical theories of the nature of explanation seek to provide a general answer to the question what an explanation adds to our knowledge in addition to the descriptions included in the explanation. In other words, they seek to answer the question "what do we learn from an explanation over and above the facts cited in that explanation?" (see Salmon 1984, especially pp. 4–9).

When Hempel and Nagel discussed functional explanation, they assumed that it is the nature of explanations to show how the phenomenon to be explained is to be expected in view of the laws of nature. This is done by inferring (deductively or inductively) a description of the phenomenon to be explained from a combination of the laws of nature[10] and descriptions of conditions that apply to the phenomenon to be explained. I shall call this view of the nature of explanation the "nomic expectability view."[11] On this view an explanation presents a number of descriptions in the form of an argument. The conclusion of the argument describes the phenomenon to be explained. The premises describe the laws of nature and the relevant initial conditions. The additional knowledge gained by viewing these descriptions in the context of an argument is the insight how the phenomenon to be explained is to be expected. This insight is what explanations add to our knowledge over and above the descriptions of which they are made. It is due to this insight that explanations are explanatory.

On the nomic expectability view, functional explanations are problematic because of the so-called "problem of functional equivalents." Hempel and Nagel agree that functional explanations seek to explain the presence of a certain trait (in certain organisms). According to Hempel this phenomenon is functionally explained by showing that the trait satisfies a need; according to Nagel this phenomenon is functionally explained by showing that the presence of the trait is a necessary condition to the performance of a certain task. The conclusion that the trait is present should follow if these lawlike statements are combined with statements describing the relevant initial conditions. In Hempel's case this the statement that the relevant organisms function

---

[10] In the view of Hempel and Nagel a law is a kind of statement. If this sounds strange, replace 'laws of nature' by 'descriptions of the laws of nature'.

[11] Usually it is called "the inferential theory of explanation," but this name is somewhat confusing as it names the means (inference), rather than the aim of explanation (to provide nomic expectability).

adequately (from which it follows that all their needs are satisfied); in Nagel's case it is the statement that the relevant organisms perform a certain task. The problem is that quite often there are different (functionally equivalent) ways to satisfy a need or to fulfill a task, and hence, from the fact that a need is satisfied or a certain task is performed one may not infer the presence of a particular trait.

Hempel accepts the existence of functional equivalents and draws the conclusion that the kind of reasoning which is usually called "functional explanation" is merely heuristic. It guides the search for new descriptions but does not explain anything. Hence, Hempel's account fails to do justice to important insights gained by functional explanations (such as the insight in the relation between electro-detection and backwards swimming in the example above).

Nagel denies the existence of functional equivalents. He argues that if both the task and the conditions under which the task is to be performed are specified in detail, there remains only one way to perform that task. However, as I have argued elsewhere, Nagel's move fails because in many cases one can only exclude functional equivalents by including within the explaining law the condition that the phenomenon to be explained is present (see Wouters 1999, section 4.3.3).

In sum, on the account of the nature of explanation employed by Hempel and Nagel (the nomic expectability view), explanations show how the phenomenon to be explained is to be expected in view of the laws of nature. However, their specific accounts of functional explanation fail to show how functional explanations are explanatory in this sense. It seems that whatever we learn from a functional explanation it is not that the trait to which the function is attributed is to be expected.

Kuipers aims to show how functional explanations (and other explanations by specification) are sound and informative, despite the fact that they do not relate the presence of a trait to the laws of nature (e.g. SiS, p. 97). A functional explanation in Kuipers' reconstruction is a thought-process that consists of two or three "parts." The first part, functional specification, aims to establish a specific functional statement of the form 'the function of trait *y* of *x*-organisms is *z*'. In the second part, functional generalization, the corresponding unspecific functional statement, 'trait *y* of *x*-organisms has a function, indeed', is inferred, by way of a side step. After this, the researcher moves to new, related questions. Functional explanations are sound because of the hypothetico-deductive nature of the process of functional specification and the validity of the process of functional generalization (it is a case of existential generalization). Functional explanations are informative because they show

how a certain trait, process, or phenomenon contributes to reproduction and survival, and because they generate new research questions.

It is doubtful whether Kuipers' explication accounts for the conviction of functional biologists that they are doing *explanatory* work. Kuipers' account can easily be read as an attempt to show how so-called functional explanations (and other explanations by specification) are sound and informative, despite the fact that they are not explanatory. After all, Kuipers offers an alternative account of the *structure* of so-called explanations by specification (among which functional explanations), but he does not offer an alternative theory of the *nature* of explanation (of what it is to be explanatory). Kuipers seems to believe that the nomic expectability view captures what it is to be explanatory and that "explanations" by specification are descriptive rather than explanatory. This impression is reinforced by the way in which Kuipers draws a distinction between descriptive and explanatory research programs (SiS, pp. 6-7). Descriptive programs aim at the description of observable facts and are carried out by means of observation and experiments. Explanatory programs aim at the explanation and further prediction of the facts described. Kuipers does not state what 'explanation' means. However, his remark that "an explanatory program has a (quasi-) deductive nature" (SiS, p.7) suggests that explanations are, by definition, deductive, and hence that so-called explanations by specification are descriptive, rather than explanatory. Indeed, on p. 73 of SiS, Kuipers explicitly states that although the products of this kind of reasoning are typically called 'explanations', the programs in the context of which they are generated are of a descriptive nature and that therefore "the various patterns of explanation by specification might also be called patterns of description."

Anyway, as I have shown above, when biologists offer functional explanations they do much more than merely describe an effect that contributes to reproduction and survival. The main product of Lannoo and Lannoo's (1993) explanation of why electric fishes swim backwards is an insight into the relation between the manner in which electric fishes detect prey and the habit of swimming backwards. The statement that swimming has the biological role of scanning potential prey in order to assess it, is a first step on the road to this main product. The remainder of the explanation uses this attribution of a biological role to relate the backward character of the swimming behavior to the feeding habits. This is done, on the one hand, by relating the need to scan the prey to the fact that prey is detected by means of an active electric sense; and, on the other hand, by elucidating why swimming along a prey in order to assess it is successful, only if the swimming is done backwards. I submit that this is typical of reasoning of the kind that biologists call "functional explanation": such reasoning is explanatory because it shows how the trait to be explained fits into a fundamental structure, namely into the structure of functional interdependencies that constitute an organism (see Wouters 1999, section 8.3.4 for an elaborate account of functional interdependencies). Functional explanations show how the different traits of an organism and the environment in which its lives are functionally dependent on each other – in the words of Lannoo and Lannoo: "scanning prey for the purpose of foraging is highly dependent on backwards swimming" (Lannoo and Lannoo 1993, p. 163).

7. How Kuipers' and my Account Fit Together

Up to now, I have focused on the request for functional explanation and the answer provided to such a request. I have emphasized that the structure of the answer to a request for functional explanation (and, to a lesser extent, the structure of the question itself) is more complex than Kuipers seems to think. As a result of this, Kuipers fails to account for one of the main kinds of insights gained by functional explanations, namely the insight into how other traits of the organism are functionally dependent on the trait to be explained (for example, how locating prey by electric means is functionally dependent on being able to scan potential prey backward).

However, up to now, I have ignored another aspect of functional explanation, namely the thought process by means of which the individual statements that together answer the request for explanation are generated, evaluated and defended. It is in this area of research where Kuipers can rightly claim that his account closely fits the practice of reasoning in functional biology. As I discussed in section 3, in the example of the electric fishes, the researchers start by hypothesizing a function (more specifically, a biological role) of the swimming behavior (when it is performed backward), namely scanning a potential prey, and then provide observational and experimental evidence that this is indeed how swimming contributes to the maintenance of the organism. In the same way, they hypothesize a function (in the sense of a biological advantage) of swimming backward rather than forward (when scanning prey), namely finishing the scan with the head near the prey, and then provide theoretical evidence that this is why backward scanning increases the life chances as compared to forward scanning. Similarly, as I discussed in section 6.2, Kristensen started by hypothesizing a function (biological role) of the fanning behavior of male sticklebacks, namely ventilating the nest, and then performed experiments to show that this is how the fanning behavior contributes to the production of offspring. In these cases there are no examples of falsification of specific functional hypotheses, but this part of Kuipers' account too can easily be verified in the literature (see, among others, my examples of the snake's forked tongue (Wouters 1999, example 2.2) and the egg shell removal behavior of birds (Wouters 1999, example 3.1)).

I submit that Kuipers' account and mine are complementary in that they account for different kinds of reasoning which are both involved in functional explanation. Let me introduce this idea, by means of a quick look at the history of thinking about functional explanation.

If a 'reasoning' is defined as a sequence of statements that have a certain coherence, Hempel (1959) and Nagel (1961, 1977) attempted to account for functional explanations as a kind of reasoning, namely as valid arguments. A valid argument is a kind of reasoning in which premises are presented in support of a conclusion and the conclusion is entailed by the premises. These attempts encounter several problems. In section 6.3 above I discussed the problem of functional equivalents. Another problem is the relation between function attributions and functional explanations. Biologists think of functional explanations as explanations that employ function attributions. However, as Kuipers notes, function attributions have no explicit role in Hempel's and Nagel's reconstructions of functional explanations and it remains unclear what exactly the relation is between function attributions and functional explanations.

From Canfield (1964) onward many philosophers have argued that functional explanations in biology do not fit the pattern of explanation outlined in the deductive-nomological model of explanation, but are nevertheless genuinely explanatory. These philosophers usually abandon the idea that functional explanations are a kind or reasoning. Instead, they assume that functional explanations consist of a single function attribution in answer to a request for explanation. As Canfield puts it:

> Someone might say, 'Explain the function of the thymus', or ask, 'What is the function of the thymus?' or 'Why do animals have a thymus?' When we answer 'the function of the thymus is [such and such]' we have, it seems plain, given an explanation (Canfield 1964, p. 293).

Kuipers agrees with Canfield (and many others) that the answer to a request for a functional explanation consists of a single function attribution. This is most clear from Kuipers and Wiśniewski (1994):

> Each direct answer to a question of the form [what is the biological function of trait *y* of *x*-organisms?][12] may be regarded either as an answer to the corresponding question of the form [why do *x*-organisms have trait *y*?] or as a sentence which entails such a statement (Kuipers and Wiśniewski 1994, p. 384).

[12] I have substituted the formulae in Kuipers and Wiśniewski's quote by appropriate sentences.

However, in contrast to Canfield (and many others), Kuipers takes seriously the intuition of Hempel and Nagel that there is a reasoning process involved in functional explanation. However, unlike Hempel and Nagel, Kuipers gives a clear account of the relation between function attributions and the reasoning process: the reasoning process is the process by means of which the function attribution is established (and, as a side step, generalized).

In section 6.2, I argued that in order to do justice to the insights achieved by a functional explanation, the answer (to a request for functional explanation) itself should be seen (in line with Hempel and Nagel, and *pace* Canfield, Kuipers and many others) as consisting of several related statements, that is, as a kind of reasoning. Note that, in contrast with the accounts of Hempel and Nagel, function attributions have an explicit role in my account: functional explanations typically show that it is *because* an item or activity serves a specific biological role that the character that that item or activity actually has is more useful to the relevant organisms than the character with which it is compared. Furthermore, there is a clear relation between the argumentative reconstruction and the explanation as it is presented by the researchers: a large part of the reasoning is presented explicitly by the researchers, only the parts that are obvious to the intended audience are left out (and can be produced by the intended readers if asked).

In other words, I suggest, *pace* Canfield and Kuipers, that questions such as 'what is the function of the mammalian thymus?' and 'what is the function of the fanning behavior of male sticklebacks?' should be distinguished from questions such as 'why do mammals have a thymus?' and 'why do male sticklebacks fan their nests?'. Questions of the first kind have the form 'what is the function of item/activity *i* of *x*-organisms?'. These questions are answered by means of a function attribution (more precisely, an attribution of a biological role), for example, 'the mammalian thymus initiates the differentiation of T-lymphocytes', and 'the fanning behavior of male sticklebacks has the function to ventilate their nests'. Attributions of biological roles are the handle by means of which functional biologists approach their subject matter (see Wouters 1999, section 2.3; Craver 2001). They are applied in several types of explanations, among which are functional explanations (as discussed in section 6.2 above). Questions of the second kind ask for functional explanations. In the course of inquiry these questions are reshaped as complex comparative questions about the character of an item or activity (see, once again, section 6.2). Their answers consists of several statements, among which are attributions of biological roles and advantage articulations.

We can now see how Kuipers' account and my account are complementary. There are two kinds of reasoning involved in functional explanations. One kind of reasoning is the reasoning by means of which the individual statements that together form the answer to a request for functional explanation are established. The other kind of reasoning is the answer itself. Kuipers' account deals with the first kind of reasoning, my account with the last.

8. Conclusion

My main point has been that there is more to functional explanation than Kuipers takes into account. The main insight provided by the explanation of Lannoo and Lannoo (1993) of why electric fishes swim backward is the relation between the electric means of locating food and the backward character of the swimming behavior: the possibility of making use of an electric radar effectively is highly dependent on the habit of scanning potential prey backward. Kuipers' account neglects this kind of insight. In order to take such insights into account, Kuipers' account must be modified and extended.

First, a distinction should be drawn between two kinds of function involved in functional explanations: function as biological role and function as biological advantage. Biological roles are attributed to items and activities. Attributions of biological role inform one about the position of those items and activities in an organism's machinery. Biological advantages, on the other hand, apply to traits in comparison with other traits. Advantage articulations inform us about the consequences of the presence of that trait due to which it is more useful to the organism to have the trait in question rather than the traits with which it is compared.

Second, I have argued that the answer to a request for an explanation consists of several coherent statements (rather than the one statement that Kuipers takes into account). Functional explanations typically start with the attribution of a biological role to an item or activity and continue by explaining why, given the circumstances in which the organism lives, that role is better performed if that item/activity has the character it has, than if it had the character with which it is compared.

This shows that there are two kinds of reasoning involved in functional explanations, namely:

(1) the reasoning processes that establish the different statements of the answer to a request for explanation:

- the hypothetico-deductive processes that establishes the different function attributions involved in the explanation, with which Kuipers' account deals, and

- the inductive generalizations about the circumstances in which the organisms live; and

(2) the different statements which together form the answer to the original explanatory question, to which I have drawn attention.

For reasons of clarity, I would prefer to restrict the term 'explanation' to the answer to a request for explanation (i.e. to the statements under point (2)), and call the process by means of which this answer is established (the process in point (1)) the process of supplying support for the explanation, but perhaps this is only a matter of taste.

University of Nijmegen
Department of Philosophy
P.O. Box 9103
6500 HD Nijmegen
The Netherlands

REFERENCES

Canfield, J. (1964). Teleological Explanation in Biology. *British Journal for the Philosophy of Science* **14,** 285-95.

Craver, C.F. (2001). Role Functions, Mechanisms, and Hierarchy. *Philosophy of Science* **68,** 53-74.

Cummins, R. (1975). Functional Analysis. *The Journal of Philosophy* **72,** 741-765.

Cummins, R. (1983). *The Nature of Psychological Explanation.* Cambridge, MA: The MIT Press.

Hempel, C. G. (1959). The Logic of Functional Analysis. In: L. Gross (ed.), *Symposium on Sociological Theory*, pp. 271-287. New York: Harper and Row.

Hempel, C. G. and P. Oppenheim (1948). Studies in the Logic of Explanation. *Philosophy of Science* **15,** 135-175.

Hempel, C.G. (1965). Aspects of Scientific Explanation. In: *Aspects of Scientific Explanation*, pp. 331-496. New York: The Free Press.

Kuipers, T.A.F. (1986). Explanation by Specification. *Logique et Analyse* **116,** 509-521.

Kuipers, T.A.F. (1996). Explanation by Intentional, Functional, and Causal Specification. *Poznań Studies in the Philosophy of Science and Humanities* **47,** 209-236.

Kuipers, T.A.F. (2001/SiS). *Structures in Science: Heuristic Patterns Based on Cognitive Structures*. Dordrecht: Kluwer.

Kuipers, T.A.F. and A. Wiśniewski (1994). An Erotetic Approach to Explanation by Specification. *Erkenntnis* **40,** 377-402.

Lannoo, M.J. and S.J. Lannoo (1993). Why do Electric Fishes Swim Backwards? *Environmental Biology of Fishes* **36,** 157-165.

Mackor, A.R. (1997). *Meaningful and Rule-guided Behaviour: A Naturalistic Approach.* Ph.D. thesis: Rijksuniversiteit Groningen.

Mauzerall, D. (1977). Porphyrins, Chlorophyll, and Photosynthesis. In: A. Threbst and M. Avron, (eds.), *Photosynthesis I*, pp. 117-124. Berlin: Springer Verlag.

Mayr, E. (1961). Cause and Effect in Biology. *Science* **134,** 1501-1506.

Millikan, R.G. (1993). *White Queen Psychology and Other Essays for Alice*. Cambridge, MA: The MIT Press.

Nagel, E. (1961). *The Structure of Science*. London: Routledge and Kegan Paul.

Nagel, E. (1977). Teleology Revisited. *The Journal of Philosophy* **74,** 261-301.

Salmon, W.C. (1984). *Scientific Explanation and the Causal Structure of the World.* Princeton: Princeton University Press.

Seely, G.R. (1977). Chlorophyll in Model Systems: Clues to the Role of Chlorophyll in Photosynthesis. In: J. Barber, (ed.), *Primary Processes of Photosynthesis*, pp. 1-50. Amsterdam: Elsevier.

Tinbergen, N. (1976). *Animal Behaviour*. S.l. Time Life International.

Wouters, A.G. (1999). *Explanation Without a Cause.* Ph.D. thesis: Utrecht University. http://www.knoware.nl/users/arnow/diss/.

Wouters, A.G. (2003). Four Notions of Biological Function. *Studies in History and Philosophy of Biological and Biomedical Sciences* **34**(4), 633-668.

Theo A. F. Kuipers

FUNCTIONAL SPECIFICATION AND FISH SWIMMING BACKWARD

REPLY TO ARNO WOUTERS

Arno Wouters presents a paradigmatic critical-constructive paper. First he explains in what sense my account of functional explanation has shortcomings. He then offers an account which he advocates as an improved account, that is, a refined and extended one. A very stimulating feature of all his work is his insistence on elaborating real-(scientific)-life examples, in this case backwards swimming fishes. I shall first respond to some of the things he has missed in my account (his Sections 2-5), before evaluating his account (Section 6) separately as well as comparatively, to use some of my favorite notions elaborated in the overlapping chapters of ICR and SiS.

Specifying Why Electric Fishes Swim Backward

Wouters' general account of my approach in Section 2 is perfect. However, his treatment of the case study is somewhat problematic from my perspective. Before entering that, let me respond to some terminological points. In response to Note 2, I confirm that both Hempel and Nagel, with their distinct argumentative reconstructions of functional explanations, suggest that there is some "underlying argument" involved. In response to Note 4, I have to concede that the term 'selection environment' is technically indeed somewhat unfortunate, although in combination with 'present environment' no misunderstandings will arise. Perhaps a better combination of terms would be 'environment of origin' versus 'environment of persistence'. Finally, the term 'distal function' (Note 5) is used in 4.2 and 6.2.1 of SiS, with 'intermediate function' as an alternative. The latter is certainly to be preferred.

In regard to the case study, I appreciate Wouters' attempt at an "Explanation by Specification" (EbyS) analysis, culminating in the last paragraph of Section 4, but I am not satisfied with it, precisely because it does not deal adequately with what Wouters presents as missing points at the

In: R. Festa, A. Aliseda and J. Peijnenburg (eds.), *Cognitive Structures in Scientific Inquiry* (*Poznań Studies in the Philosophy of the Sciences and the Humanities*, vol. 84), pp. 294-298. Amsterdam/New York, NY: Rodopi, 2005.

beginning of Section 5. First, it is said to leave out "that the fish must swim backward to acquire [a] favorable position" "*because* swimming has a function in scanning." Second, according to Wouters, it "ignores the point that scanning is needed because of the physical characteristics of electrosensoric prey recognition" (which makes focusing impossible).

When reconstructing a case like this in terms of explanation by specification it is of the utmost importance to start, as far as possible, by disentangling the initial why-question, that is, why do the fish pass their potential prey backward? In my view the basic why-questions are: why do the fish *pass* their potential prey and why do they pass them *backward*? Let us start with the first question. Roughly speaking, it gets the EbyS answer: *passing* is a *positive causal factor* (henceforth, pcf) for scanning, which is a pcf for prey recognition, which is a pcf for survival. The first pcf-relation seems to be the core of Wouters' first concern. However, when this relation has been established together with the second, this raises the further task of explaining it. More specifically, the question is *how* does scanning work such that the transitive conclusion – passing is a pcf for prey recognition – becomes true? Wouters' paper teaches us that this has everything to do with the nature of electric fishes: passing many receptors is the only way in which such fishes can acquire a sufficiently high quality image. Hence, instead of ignoring the nature of scanning, Wouters' second concern, its crucial how-question comes into focus. However, none of this as yet has anything to do with the primarily surprising phenomenon, *backward* swimming. We now know that this should be interpreted as "backward *passing*," for the passing is functional for scanning, but we still don't know why it is "*backward* passing," which is the second question. This gets the EbyS answer: passing *backward* (rather than forward) is a pcf for subsequent prey catching, which is a pcf for survival.

In sum, by splitting the original why-question into two different aspects of that behavior, we get two well-structured answers in terms of functional specification, one of which generates a crucial new how-question. The answers to the first question and to the how-question generated by it, directly pertain to the two points of attention Wouters is missing according to the above quotations. Combining these answers with the answer to the second why-question, the reconstruction "relate[s] the backward character of the swimming behavior of electric fishes to the fact that those fishes are *electric* fishes," that is, the accomplishment that Wouters, at the end of Section 5, states is absent.

Wouters' Improved Analysis

To be sure, Wouters in his own analysis introduces a number of sophistications. The main one is the distinction between, on the one hand, biological roles of items and activities and, on the other hand, biological advantages of specific properties or characters of them or of the organism as a whole. In terms of roles and advantages, the passing behavior in the example above plays a role in the scanning technique of electric fishes, whereas passing backward provides an advantage relative to passing forward. Surprisingly, he calls the 'scanning-when-passing' an advantage rather than (assigning it) a role. Be this as it may, Wouters calls the presence of a certain item or activity as well as their characters *traits*, but not the items and activities themselves, nor their having a certain character. In contrast to his suggestion in Note 6, I did not presuppose such, essentially linguistic, distinctions. At several places I just added 'process/phenomenon' between brackets after 'trait', to make sure I covered everything of which it makes sense to ask for a functional explanation, including processes like photosynthesis and phenomena like the stable clutch-size of plovers.

In the first paragraph of 6.2,Wouters submits at least two claims with his analysis relative to my account of what he calls "function attribution":

> I submit that the question addressed by a functional explanation typically has the form 'why does item/activity i of x-organisms has character s_1 rather than s_2?', for example 'why do electric fishes swim in both directions (rather than forwards only)?'. My point is not only that the question is comparative but also (and foremost) that the question is about the *character* of an item or activity, for example in the case of the electric fishes it is the backward character of the swimming behavior that is explained.

Regarding his second and foremost point, I hope I have shown above convincingly that a sensible explanation by specification of backward swimming is very possible. Hence, more generally, explanation by specification can perfectly deal with characters of items and activities, and their advantages, assuming that it can deal with their comparative nature, that is, Wouters' first point. In my view, however, the latter is merely a matter of a (very important) concretization. As anticipated by Wouters, it is easy to make pcf-claims comparative, which I suggested already above with the phrase "passing backward (rather than forward)." The only thing one needs to do is to replace, and defend, probability statements of the form "$p(B/A) > p(B)$", underlying pcf-claims, by statements of the form "$p(B/A) > p(B/C)$", where C is supposed to be incompatible with (the presence of) A. When C just amounts to *non-A*, that is the absence of A, we get the weakest comparative case, which is in fact already included in the original condition, for "$p(B/A) > p(B/non\text{-}A)$"

is equivalent to "$p(B/A) > p(B)$", assuming non-zero probabilities. (See SiS, p. 122, for some further suggestions.)

In the rest of his contribution Wouters gives a detailed reconstruction of the electric fish example, and a sketch of the fanning movement of male sticklebacks in front of their nest. Although both his reconstructions and the lessons attached to them are evidently more complex than my account would be, not all complexities are enriching concretizations. However, some complexities certainly are badly needed sophistications. For example, the exclusion of unintended illustrations of my "minimal account," e.g. the Gibbon case. In such cases it becomes quite clear that, unlike me, Arno Wouters is not only a philosopher but also a biologist, and hence a plausible addressee of my simplified and idealized writing about functional explanation in biology. However, some other complexities are due to not splitting up questions. As suggested by my brief analysis of the electric fish case, and by ending my "train of thoughts" with the phrase "go to new, related why- and how-questions," my basic strategy is the splitting up of questions, rather than trying to answer several questions in a complex story.

Strengthened by some later correspondence, I subscribe to Wouters' claim that there are two important differences between our points of view. First, according to Wouters, a leading, if not *the* leading question of a functional analysis is that characters of items and activities have to be seen as advantageous solutions for problems or needs that are raised by the specific nature of them. Backward passing solves the problem that is created for *electric* fishes. Let me rephrase my analysis in this respect. The passing is functional: such fishes have to pass their prey in order to recognize it as such. Doing it in reverse is functional too: if they were to pass forwards they would end up in a poor catching position, hence, for such fishes it is advantageous to pass backwards. In general, a specific character of an activity is a solution to a problem that has arisen due to the specific nature of an item when that character is functional given that activity, and assuming that that activity is functional in view of the specific nature of the item. Hence, in my view, an EbyS analysis provides the building blocks for the answer to the type of question that interests Wouters. This brings me to the second main difference, and now I quote, with permission, from an e-mail from Wouters (June 26, 2002) in response to a draft of this reply:

> For the rest, I find your remark that your basic strategy is the splitting up of questions very illuminating. … Here lies a difference of opinion: I think that such splitting up does not do justice to the explanations given by biologists. Although in research practice complex questions are tackled by splitting them up into a number of questions, in the resulting answers to complex questions the separate questions are brought into a connection that provides more insight than the sum of the separate answers as you give them.

This I fail to see, at least in the case of electric fishes. Of course, the answers have to be put together in an appropriate way, which is usually more than mere concatenation. This may require an appeal to covering principles, such as the transitivity of causal claims in the case at hand. However, more than such connecting principles do not seem to be needed.

Adam Grobler and Andrzej Wiśniewski

EXPLANATION AND THEORY EVALUATION

ABSTRACT. It is claimed that Kuipers' approach to explanation opens the possibility for a further refinement of his own refined HD method for the evaluation of theories. One severe problem for the HD method, refined or not, is theory-ladeness. Given that experimental results are theory-laden, the comparative evaluation of alternative hypotheses is always relative to background knowledge. This difficulty can be avoided by supplementing HD considerations with the principle of inference to the best explanation. The authors sketch a program for doing this. The general idea plays on some similarities between Kuipers' account of explanation and Lipton's. The former, however, is considered more flexible than the latter, which makes it even more attractive for the purpose under consideration.

In his numerous writings Theo Kuipers promotes a revised, or refined, hypothetico-deductive (HD, for short) method of theory evaluation. The core idea, which can be viewed as an elaboration and sophistication of Lakatos' account, is that the method is not intended to serve merely as a means of error elimination. Instead, it is supposed to serve, in the first place, as a method for the comparative evaluation of theories and hypotheses in terms of their relative successes and failures. The refined HD method, so conceived, is truth-conducive in the sense that it gets closer to the truth with less and less flawed theories, rather than discarding false theories in search of a/the true one. Attractive though this may be, the HD method suffers from one serious problem. Theory-ladenness makes falsification background-knowledge relative. Even if Kuipers does acknowledge the limitations of the HD method, including those that arise from theory-ladenness, he seems to underestimate the fact that the comparative evaluation of alternative hypotheses is always relative to background knowledge. This relativity leads to a version of the Duhem problem: in the face of negative empirical results, there is always a choice whether to reject a hypothesis under test or, alternatively, to revise the present system of background knowledge so as to maintain the allegedly falsified hypothesis.

This version of the Duhem problem has never been satisfactorily solved by the most prominent proponents of the HD method. It is for this reason that Karl Popper was accused of having been an "irrational rationalist" (Newton-

In: R. Festa, A. Aliseda and J. Peijnenburg (eds.), *Cognitive Structures in Scientific Inquiry* (*Poznań Studies in the Philosophy of the Sciences and the Humanities*, vol. 84), pp. 299-310. Amsterdam/New York, NY: Rodopi, 2005.

Smith), a "contemporary irrationalist" (Stove), or, most moderately, a "conventionalist" (Brown). Lakatos, replacing background knowledge with the hard core of a scientific research program, comes quite close to a solution. Nevertheless, his conception of postponed rationality is not fully immune to Feyerabend's challenge: how long are we to wait for the decision to be made between alternatives? Watkins' and Zahar's search for the justification of Popper's basic statements with so-called 0-level statements – autopsychological reports or descriptions of noemata, respectively – represents a highly dubious switch towards internalist foundationalism. Kuipers' version, as far as the Duhem problem is concerned, seems to follow Laudan's pattern of estimating the relative problem-solving efficiency of alternative systems of theories or hypotheses. Unfortunately, in doing this, Kuipers does not take the opportunity to use some important insights of his own, which can open new prospects for the theory of scientific method.

What we have in mind here are Kuipers' ingenious remarks about explanation. He points to many aspects of scientific endeavor that are not dealt with by Hempel's law-covering account. To make up for this, Kuipers offers a novel account of explanation: explanation by specification. However, this new approach to explanation is neglected in the refined HD evaluation since, due to the symmetry between prediction and explanation, what are considered successes and failures of a theory or hypothesis are precisely the explanatory successes and failures in Hempel's defective sense of explanation. This amounts to saying that whether or not a theory or hypothesis can be used to give some explanation by specification does not contribute to its cognitive value. Such a view is dangerously close to van Fraassen's constructive empiricism. Thus, the converse seems more attractive for those who, like Kuipers and the present authors, declare their commitment to realism. Consequently, we here try to indicate how the scope of explanatory applications of a theory, in terms of Kuipers' account of explanation, is relevant for its evaluation. We discuss at some length only one kind of explanation by specification, namely explanation by causal specification, and make rather programmatic remarks concerning explanations by intentional and functional specification as characterized by Kuipers.

1. Explanation by Causal Specification

As far as explanation by causal specification is concerned, the explanation-seeking question has the form:

(1) Why did an event b occur to system a?

where b is assumed to be an *abnormal* event or factor for a. The concept of "abnormality" is not explicated in general terms; an unexpected death of a patient, a car accident, a fire, etc. are paradigmatic examples here. A question of the form (1) is then construed as:

(2) What was the cause of (abnormal) event b that occurred to system a?

A possible answer to (2) (and thus to (1)) has the form of:

(3) Event b occurred to system a due to cause x.

whereas the presupposition of (2) is:

(4) Event b occurred to system a due to some specific cause.

The meaning of a sentence of the form (3) is characterized by the following meaning postulate:

MP1: Event b occurred to system a due to cause x if and only if:

(4.1) event b occurred to system a,

(4.2) event x occurred to system a and x is an abnormal factor (event, intervention, condition) for a,

(4.3) there are factors $f_1, \ldots, f_n$ such that $f_1, \ldots, f_n$ are normal factors/conditions for a and "if x and f_1 and …. and f_n, then event b occurs to system a" is a causal law in the strict sense[1],

(4.4) x was causally effective for the occurrence of b to a.

To provide a causal explanation by specification is to formulate and verify a certain answer of the form (3) to the explanation-seeking question (1). Formulating an answer amounts to specifying a certain substitution-instance for x in (3), whereas the verification of the answer is tantamount to the verification of the corresponding substitution-instances of (4.2), (4.3) and (4.4). The term 'verification' is understood here in a very general, pragmatic sense; in particular, it does not presuppose irrevocability. Kuipers characterizes the schematic train of thought which may lead to a verified answer to the explanation-seeking question. One can show that all argumentative steps involved in such a train of thought are valid inferences, either standard or erotetic (cf. Kuipers and Wiśniewski 1994). Roughly, the process of searching for an explanation by causal specification starts with a verified hypothesis of the form "event b occurred to system a," where b is conceived as abnormal for a. Then – by the so-called principle of specific causality – the presupposition (4) of question (2) is arrived at. This presupposition is a hypothesis, however.

[1] That is, an experimental law in the sense of Nagel (1961): the factors $x, f_1, \ldots, f_n$ are space-and-time contiguous and there exists a time-asymmetry between x and b, i.e. x precedes b.

Presupposition (4) gives rise to question (2). An answer to question (2) is then proposed as a hypothesis to be tested. Of course, question (2) has many possible answers, which are substitution-instances of (3); from among them a certain one is chosen, as Kuipers puts it, "by idea." From the erotetic point of view, this step amounts to arriving at a yes-no question of the form "Is it the case that event *b* occurred to system *a* due to cause *c*?", where *c* comes "by idea." Then, on the basis of the meaning postulate MP1, one comes to a conjunctive question, the constituents of which result from (4.2), (4.3) and (4.4) by substituting *c* for *x*. Next the following question is asked:

(5) Is it the case that event *c* occurred to system *a* and *c* is an abnormal event for *a*?

If the affirmative answer to (5) is verified, the following question will be asked:

(6) Are there factors $f_1, \ldots, f_n$ such that $f_1, \ldots, f_n$ are normal factors/conditions for *a* and "if *c* and f_1 and .... and f_n, then event *b* occurs to system *a*" is a causal law in the strict sense?

If the affirmative answer to (6) is verified, the next question will be:

(7) Was *c* causally effective for the occurrence of *b* to *a*?

If the affirmative answer to (7) is verified, then – by the affirmative answers to (5) and (6) together with meaning postulate MP1 – one arrives at the following answer to (2) (and thus to (1)):

(8) Event *b* occurred to system *a* due to cause *c*.

The answer (8) is now a verified hypothesis and an explanation by causal specification. Since (8) logically entails (4), from now on (4) can be regarded as a verified hypothesis too. If, however, a negative answer to any of the questions (5), (6) or (7) is verified, or no clear results are available, the inquirer has to repeat the procedure with respect to a certain (possible) cause *d*, which, again, is taken "by idea." The process goes on until the actual specific cause is found. Nevertheless, there is no guarantee that such a cause will be found.

Among the examples of explanation by causal specification, Kuipers mentions (SiS, p. 123) the explanation of childbed fever as caused by "cadaveric matter." The well-known story of Semmelweis' discovery reported by Hempel (1966) was later retold by Lipton (1990) in a way that reinforces Kuipers' suggestion of the superiority of specific causal explanation over explanation by subsumption. In his version, Lipton argues that Semmelweis' discovery is an illustration of the way in which the principle of inference to the best explanation provides us with a better guide than the falsificationist method. Semmelweis is said to have rejected some hypotheses without even trying to falsify them, just because of their failure to give the desired explanation of the dramatic difference in mortality rates of two maternity divisions of his hospital. Among them, there was the perfectly plausible hypothesis to the effect that the membership of a higher social class, due to better nutrition, makes people more resistant to illness. This hypothesis was rejected simply because there were no considerable differences in the social composition of the two divisions. Nevertheless, the hypothesis might well have been true: the mortality rate among the members of a higher social class might have been lower than that in the rest of the population. This, however, was not even investigated just because the hypothesis under consideration appeared irrelevant to Semmelweis' explanatory endeavor. The guiding principle of Semmelweis' investigation was the search for a causally effective factor that made the difference.

In Lipton's account, explanation is an answer to a contrastive why-question, i.e. a question of the form "Why *P* rather than *Q*?" A plausible answer has to point to a factor in the causal history of *P* that has no counterpart in the causal history of non-*Q*. The concept of counterpart may be somewhat vague, but there is no need to elaborate upon it in the present context. Apparent differences notwithstanding, there are some affinities between Lipton's and Kuipers' proposals. First, the explanatory factor, let us call it *Z*, is causal. Second, in so far as the question "Why *P* rather than *Q*?" is (often but of course not always) motivated by a feeling of surprise, *P* can be considered as an event that has occurred unexpectedly as compared to the expected *Q*. Consequently, *Z* is in a sense *abnormal*, for it is precisely the factor whose occurrence has prevented the "normal" *Q* from having happened. In contrast, the shared members (up to the relation of "being a counterpart") of causal histories of *P* and *Q* can be called "normal" causal factors.

Whether or not Lipton's account of contrastive explanation and Kuipers' account of specific causal explanation are equivalent, we are not in a position to decide. Much depends on possible further explication of the concept of "abnormality." Nevertheless, the similarities between the two permit us to pursue Lipton's idea about the justificatory role of explanation, reformulated so that it can be applied to Kuipers' account. The reformulation in question is that explanatory successes and failures, in the sense of explanation by specification, count more for the purposes of theory evaluation than empirical successes and failures in the sense of the HD method, refined or not.

One may argue that, just as falsification is relative to background knowledge, so too is explanation by causal specification. This is so because the normal/abnormal distinction is pragmatic, i.e. it depends on context and, in particular, background knowledge. This, however, gives the explanatory power considerations priority over the conventional use of the HD method. As has already been stated, the HD method suffers from a version of the Duhem problem – the problem of choice, in the face of negative evidence, between the rejection of the hypothesis under test or a suitable revision of the background knowledge so that the hypothesis in question can be saved. This problem is much more easily solved when one is confronted with failures to give a specific causal explanation.

Failure of this kind suggests that the effective abnormal cause has not yet been discovered, or that there is more than one abnormal cause in operation, or that instead of an abnormal cause it is an abnormal joint occurrence of normal causes which is effective. Consequently, three different lines of research are open. The first one is rather straightforward and can be pursued as long as there is a hope of finding the cause "by idea." The two others are more complex, for they involve a hypothesis about the interaction of some causes. Such a hypothesis may go far beyond the currently accepted background knowledge, even if the causes in question are identifiable within its framework.

An explanation by causal specification may also fail when the explanation-seeking question and/or the operative questions may be sound, but unanswerable by means of a theory and/or background knowledge. For example, the theory and/or background knowledge may offer no candidate for "the cause" of the phenomenon in question (think of an empirically-oriented medieval medical doctor who tries to explain why the inhabitants of a certain village survived the "black death" epidemic whereas all the inhabitants of a village situated nearby died) or may offer no candidate to which there are no decisive objections (think of a contemporary medical doctor who observes the rapid recovery from cancer of a patient who has just visited Lourdes). In situations like these, a revision of background knowledge is needed to reopen a set of possible "ideas" for the candidate causes.

Thus a prolonged explanatory failure exerts pressure to make attempts to revise background knowledge. Indeed, assuming the account under discussion, it is plausible to claim that an explanatory failure even permits one to draw some hints about possible revisions, provided that some non-explanatory coincidences are established. The story of Semmelweis' discovery is a good example. In the maternity division with the higher mortality rate, in contrast to the other, the nursing duties were performed by medical students. This coincidence was not explanatory, however, since it was not causal. An attempted explanation "by idea" was that students dealt carelessly with patients. Investigation demonstrated the opposite. No new "idea" had come about until another coincidence was discovered. Semmelweis' colleague, doctor Koletschka, cut his finger with a scalpel and soon died of childbed fever. Before the accident, the scalpel was used in the prosectorium, where the students were regularly instructed; they attended patients only after their classes. This coincidence of two coincidences gave rise to the "idea" of transmission of a hypothetical "cadaveric matter" – the supposed cause of childbed fever – both by Koletschka's scalpel and students' hands. Clearly, Semmelweis' conclusion – that washing one's hands carefully before attending patients may help – represents a substantial revision of background knowledge.

On the other hand, even if the phenomenon in question occurred due to some specific cause and the set of conceptual possibilities offered by the theory and/or background knowledge is wide enough to offer serious candidates without substantial revisions, an attempt to provide an explanation by causal specification may fail since, in order to verify a hypothesis of the form (8), one has to answer the corresponding questions of the form (5), (6), and (7), and they are usually difficult questions. In particular, in order to answer question (6) one has to point to a certain empirical law (a question about the existence of a law can be answered only by referring to an example of an appropriate law). The required law can already belong to the theory or background knowledge, but it may also be that it yet has to be derived and/or empirically verified. Nevertheless, the theory and/or the background knowledge that we are working with may be insufficient, and may be resistant to relevant empirical extensions. Providing a successful explanation by causal specification is a difficult enterprise and therefore its success seems to present a good argument for a positive evaluation of the theory in question.

So far, we have assumed that an attempt to provide an explanation by causal specification is made by means of a single theory and the associated background knowledge. But in the case of abnormal events scientists often work with rival theories. If a given theory suggests a successful explanation by causal specification of a certain abnormal event, whereas its rival does not, one may say that the former gains superiority over the latter. Sometimes the event in question is not conceived as abnormal when viewed in the light of a rival theory, and can be explained by subsumption by means of that theory. In such cases the latter seems to gain superiority over the former.

One doubt may arise. It is stated that "an explanation by causal specification implies the possibility of providing an explanation by causal subsumption if the particular causal law is explicitly known" (SiS, pp. 125-6). This may imply that explanation by causal specification, given its heuristic value, plays a significant role in the context of discovery, but is not particularly significant in the context of justification. Once an explanation of this sort is found, it can be transformed into an explanation of the Hempelian pattern, and Hempelian-like explanatory successes are simply successes in terms of the HD method of evaluation. In such cases, however, there is a clear epistemic gain in comparison with mere subsumption, namely the identification of a causal factor. On the other hand, not every HD success is a success in giving a specific causal explanation.

Nevertheless, one may ask why an identification of a causal factor – leaving aside its heuristic value – provides us with more knowledge than the discovery of a law, whether causal or not. Or what mentioning the cause responsible for the regularity in question adds to the cognitive value of the law that expresses this regularity. Is it not the case – a positivist might ask – that the whole value of science is exhausted in discovering laws that describe regularities in nature, and that everything going beyond this is irrelevant?

Not at all. As Lakatos (1970) has pointed out, laws typically contain the *ceteris paribus* or "other things being equal" clause. Its implicit presence is responsible for all the ambiguities of falsification, since any apparently falsifying instance of a law can be explained away by an auxiliary hypothesis to the effect that a hitherto unknown factor is operating. Consequently, one can never exclude the possibility of a suitable revision of background knowledge that will transform an HD failure of the law under test into an HD success. In contrast, a failure to give a specific causal explanation is more telling, for it amounts to the lack of an identification of *the* abnormal cause operating in a test situation (and possibly suggests that the event in question occurred due to an interplay of many causes, abnormal or otherwise). Thus, a specific explanatory failure is more informative than an HD failure. On the other hand, a specific causal explanatory success provides us with more knowledge than the predictive or descriptive success of a law.

To conclude, specific causal explanatory power considerations should play an important role in theory evaluation. Hence, Kuipers' proposal to replace the principle of inference to the best explanation with the principle of inference to the best theory (ICR, p. 170), should be reconsidered in the light of his own insights concerning causal explanation. Alternatively, his definition of the "best theory" should be reconsidered so as to accommodate the present insights.

2. Other Patterns of Explanation

Apart from specific causal explanation, Kuipers considers intentional and functional explanations by specification. Their logical structure is parallel to the structure of explanation by causal specification (see SiS; see also Kuipers and Wiśniewski 1994). The introduction of other types of explanation by specification develops the prospect of going far beyond Lipton's account of inference to the best explanation. In Lipton's formulation, it is only the reference to the relevant difference in causal histories of the fact under explanation and its contrast which lends explanatory power to an answer to a why-question. Consequently, Lipton does not leave any room for non-causal explanations. This seems an unnecessary and inadequate restriction of his account. Hence, if we are right in suggesting that Kuipers' specific causal explanation is able in principle to do the job of Lipton's contrastive explanation, the other forms of explanation Kuipers considers are able to do some additional job.

One may doubt whether this additional job has anything to do with theory evaluation, for – unlike specific causal explanation – intentional and functional explanations by specification do not involve any "intentional" or "functional" law, apart from the general principles of intentionality (or rationality) and functionality (or evolution). The principles in question, however, are not law-like statements subject to evaluation, possibly in terms of their explanatory power. Rather, they are presupposed in the very concept of intentional or functional explanation, just as the principle of causality is presupposed in the concept of causal explanation. The question of what kind of theory or statements are to be evaluated by invoking successes in providing intentional and functional explanations by specification now arises.

Let us consider intentional explanation first. In this case, the explanation-seeking question has the form:

(1*) Why did agent *a* perform action *b*?

or:

(2*) What was the goal of action *b* performed by agent *a*?

A possible answer to (2*) (and thus to (1*)) has the form of:

(3*) *a* performed action *b* with the intention of approaching goal *z*,

where *z* is to be understood as an external goal, in contrast to an internal one, i.e. the one specified in the description of *b*. For example, the internal goal of "opening the window" is "having the window opened," while its possible external goal can be e.g. "letting some fresh air in."

The presupposition of (2*) is:

(4*) *a* performed *b* intentionally (with the intention of approaching a specificic external goal).

The meaning of a sentence of the form (3*) is characterized by the following postulate:

MP2: *a* performed action *b* with the intention of approaching goal *z* if and only if:

(3*.1) *a* performed action *b*,

(3*.2) *a* desired goal *z*,
(3*.3) *a* believed *b* to be useful to approach *z*,
(3*.4) the belief and desire in question were causally effective for *a*'s having had the plan to perform *b*.

As in the case of causal explanation by specification, to provide an intentional explanation is to provide an answer of the form (3*) to the explanation-seeking question (2*). Due to the structural similarity of the two patterns of explanation, the process of the search for an intentional explanation can be described similarly to that of the search for a causal explanation. Details are omitted.

In considering a related question of explaining the choice of a particular action among alternatives, Kuipers emphasizes the difference between his and the utilistic approach (SiS, pp. 110-111): the former presupposes that the specific goal of an agent is fixed beforehand, while the latter presupposes that an agent has one general goal of maximizing his expected utility so that the choice of a particular goal is a part of the agent's decision problem. Instead, Kuipers offers a generalization of the pattern of intentional specification, or a second step of intentional specification, to explain the choice of a goal in terms of the agent's approaching, as it were, a second-order goal to be attained with the goal in question. The latter is just substituted for an action in the pattern of explanation by intentional specification. Consequently, to explain why a certain goal *z* was chosen by an agent *a* is to answer the question:

(1**) Why did agent *a* choose goal *z*?

or:

(2**) What was the second-order goal *z** to be attained by *z*?

A possible answer to (2**) (and thus to (1**)) has the form of:

(3**) *a* chose goal *z* with the intention of attaining the second-order goal *z**.

Again, the presupposition of (2**) is:

(4**) *a* chose goal *z* intentionally (with the intention of approaching a specific second-order goal).

The meaning of a sentence of the form (3**) is characterized by the following postulate:

MP1: *a* performed action *b* with the intention of approaching goal *z* if and only if:

(3**.1) *a* (deliberately) chose goal *z*,
(3**.2) *a* desired goal *z**,

(3**.3) *a* believed *z* to be useful to approach *z**,

(3**.4) the belief and desire in question were causally effective in *a*'s having chosen *z*.

This flight from the utilistic approach seems quite reasonable, since the principle of maximizing one's expected utility, as a general "law" of personal behavior, is overidealized. People very rarely, if ever, perform the required calculations. Calculations may possibly be done in specific problem situations, like those in business. In such cases, however, the utility function derives from, e.g., suitable return and risk estimates, without taking into account the utilities of non-profit-oriented actions, or other actions irrelevant to the problem in question. In everyday life even crude estimations are performed only on special occasions, possibly when people ask themselves questions of the sort "Do I really want this-and-that?" Consequently, leaving much space for pragmatic considerations, as Kuipers does, seems to be the right choice.

The conventional utilistic approach presupposes just one general law, which says that people observe the principle of maximizing expected utility. Consequently, utilistic explanatory successes and failures, if they can be used at all, can be used only for the evaluation of this law. In contrast, a more flexible approach can be used to form explanations that involve claims that are more specific. Since in order to provide an explanation by intentional specification one has to verify the belief and desire claims involved, an explanatory success or failure may contribute, e.g., to the evaluation of psychological laws about, say, the preferences or inclinations of people with a certain type of personality; or to the evaluation of anthropological theories about rules of culture, taboos or prescriptions. Thus, the principle of inference to the best explanation, in the sense of intentional explanation by specification, can guide the choice of theories not only of nomothetic, but also of idiographic sciences, the latter being beyond the scope of the HD method.

On the other hand, considering the question of explaining the choice of a particular action among alternatives, Kuipers does make a limited use of the utilistic approach, albeit restricted to a two-element space of possible outcomes: attaining or not attaining the desired goal. Furthermore, in the formula for the calculation of the expected utility of an action, the cost of the action in question is taken into account. This makes room for accounting for various pragmatic factors that can be captured in the cost of an action. Even the principle of maximizing one's expected utility can in a way be re-established, if needed, by defining the cost of an action so that it covers the costs of its side effects. Kuipers' account, then, can be viewed as a generalization of the utilistic approach. And it is precisely this feature that permits the use of intentional explanation by specification in theory evaluation.

Functional explanation by specification, since it displays essentially the same structure, can also provide us with a tool for evaluating theories, e.g., of particular evolutionary scenarios. At present, the authors are not in a position to give a detailed account of the evaluative applications of Kuipers' model of functional explanation. Still, we believe that the search for such an account is a promising program in the philosophy of science.

University of Zielona Góra
Institute of Philosophy
Al. Wojska Polskiego 71A
PL-65-762 Zielona Góra
Poland
e-mail: uzgroble@theta.uj.edu.pl

Adam Mickiewicz University
Department of Psychology
ul. Szamarzewskiego 89C
PL-60-568 Poznań
Poland
andrzej.wisniewski@amu.edu.pl

REFERENCES

Bromberger, S. (1992). *On What We Know We Don't Know: Explanation, Theory, Linguistics, and How Questions Shape Them*. Chicago: University of Chicago Press.

Brown, H. (1988). *Rationality*. London: Routledge.

Hempel, C. (1966). *Philosophy of Natural Science*. Englewood Cliffs, NJ: Prentice-Hall.

Kuipers, T.A.F. (2001/SiS). *Structures in Science*. Dordrecht: Kluwer.

Kuipers, T.A.F. (2002/ICR). *From Instrumentalism to Constructive Realism*. Dordrecht: Kluwer.

Kuipers, T.A.F. and A. Wiśniewski (1994). An Erotetic Approach to Explanation by Specification. *Erkenntnis* **40**, 377-402.

Lakatos, I. (1970). Falsification and the Methodology of Scientific Research Programmes. In: I. Lakatos and A. Musgrave (eds.). *Criticism and the Growth of Knowledge*. Cambridge: Cambridge University Press.

Laudan, L. (1977). *Progress and Its Problems*. Berkeley: University of California Press.

Lipton, P. (1990). *Inference to the Best Explanation*. London: Routledge.

Nagel, E. (1961). *The Structure of Science*. London: Hartcourt.

Newton-Smith, B. (1980). *The Rationality of Science*. London: Routledge.

Stove, D. (1982). *Popper and After: Four Modern Irrationalists*. Oxford: Pergamon Press.

van Fraassen, B. (1980). *The Scientific Image*. Oxford: Clarendon Press.

Watkins, J. (1984). *Science and Scepticism*. Princeton: Princeton University Press.

Zahar, E. (1995). The Problem of Empirical Basis. In: A. O'Hear (ed.), *Karl Popper: Philosophy and Problems*. Cambridge: Cambridge University Press.

Theo A. F. Kuipers

KINDS OF EXPLANATORY SUCCESSES
REPLY TO ADAM GROBLER AND ANDRZEJ WIŚNIEWSKI

In this reply I sketch out my view of how the various types of explanation by specification could be used in the evaluation of theories and I insert my reaction to some specific points raised by Adam Grobler and Andrzej Wiśniewski at the relevant places. To be sure, the very idea of thinking about the abovementioned how-question I owe to my Polish colleagues, for which I am very grateful.

Grobler and Wiśniewski are certainly right in suggesting that my treatment of the separate and comparative evaluation of theories (SiS, Ch. 7 and 8, ICR, Ch. 5 and 6) is based on the idea of applying the HD-method in order to establish, not the truth-value of theories, but their separate and comparative merits and failures, and, indirectly, at least for realists, their comparative distance to the truth. Grobler and Wiśniewski rightly suggest, moreover, that my comparative model is a way to deal with negative empirical results that may be due to problematic background knowledge, but also, I would like to add, with negative results that seem to be straightforwardly due to the theory in question. The merits and failures of a theory(-cum-background-knowledge) can be expressed in terms of individual or general successes and problems, where the combination of general successes and individual problems (counterexamples) seems to be the paradigmatic one. More specifically, successes obtained by the HD method are always explanatory successes, and they may or may not be predictive successes. Such explanatory successes are known as DN explanations or "explanations by subsumption." My version of "inference to the best explanation" can be characterized as "inference to the best theory (in terms of successes and problems), if there is a best one, as the closest to the truth." See (Kuipers 2004) for a detailed analysis.

The decomposition model of explanations presented in Chapter 3 of SiS pertains to DN explanations, in particular, explanations of laws by theories, amounting to general successes of theories. In the introduction to that chapter I argued that the explanation of individual events is relatively uninteresting from a scientific point of view, in contrast to what most introductions to the

In: R. Festa, A. Aliseda and J. Peijnenburg (eds.), *Cognitive Structures in Scientific Inquiry* (*Poznań Studies in the Philosophy of the Sciences and the Humanities*, vol. 84), pp. 311-314. Amsterdam/New York, NY: Rodopi, 2005.

philosophy of science suggest. "In our opinion the core of explanation [in the empirical sciences] lies in the explanation of observational laws by subsumption under a theory, in short, theoretical explanation of (observational) laws. After a successful theoretical explanation of a law, we get as an extra bonus a theoretical explanation of the individual events fitting into that law" (SiS, pp. 75-6). I should have added that this point is crucial for the evaluation of models of explanation that pretend to compete with DN explanations of individual events, notably the "contrastive model" of explanation, advocated by Van Fraassen and Lipton. This model may have its value, but probably not for theoretical explanations, for there does not seem to be a plausible way to adapt it for the explanation of laws by theories. It may well be that the "contrastive model" is equivalent (as Grobler and Wiśniewski suggest) or (to some extent) competing with my (DN compatible) model of "explanation by specification," in particular, with explanation by *causal* specification, developed in Ch. 4 of SiS. The relative merits of these two models are worth investigating, but this goes beyond the scope of this reply.

Explanation by Causal Specification

For now there remains the interesting question, raised by Grobler and Wiśniewski, whether explanatory successes fitting in the model of "explanation by specification," EbyS successes for short, can play a role in the evaluation of theories. As is already clear from Grobler and Wiśniewski's concise exposition of explanation by causal specification, and in contrast to explanations by intentional and functional specification, an explanation of an abnormal or surprising individual event by specification of an abnormal causal factor implies the existence of a DN explanation of that event, using a causal law. As mentioned by Grobler and Wiśniewski, I claim in SiS (pp. 125-6) that this DN explanation is available as soon as this law is explicitly known, that is, all the relevant causal factors are known. Hence, if the causal law itself or a theory entailing this law is to be evaluated, the EbyS success implies that a straightforward individual DN success is available. That it is even an EbyS success will of course contribute to the weight assigned to that success as soon as weights have to be taken into account, for, as Grobler and Wiśniewski rightly stress, not every DN success is also an EbyS success. As Grobler and Wiśniewski point out, the case of Semmelweis's explanation of childbed fever in terms of (the abnormal factor of) cadaveric matter is a very good example of explanation by specification. Moreover, I certainly agree that the EbyS nature of this success makes it much more impressive than a mere DN success. However, even in DN terms, the alternative theories suggested by Grobler and

Wiśniewski meet DN problems, or at least DN lacunae, which the best theory did not have. Nonetheless, if the theory to be evaluated is not so much related to the causal law but to the abnormal event or factor itself, the EbyS success may well be a success that cannot be reconstructed as a special type of DN success, in which case it may certainly be taken into account as another type of success.

In sum, relative to a certain theory and a certain case, an EbyS success may either be stronger than a corresponding DN success of that theory or it may be a success of another kind, without entailing a DN success of the theory in question. In both cases EbyS successes should be taken into account in evaluation reports and, hence, in inferences to the best theory. Moreover, Grobler and Wiśniewski argue very convincingly that the search for explanations by causal specification, in one way or other related to the theory, may be very profitable for the evaluation of that theory. Of the several types of profits they indicate, I should mention that the Duhem(-Quine)-problem is more easy to tackle from this perspective. The search for abnormal factors forces us to make all relevant background assumptions explicit, for they may need revision for the special case.

Explanation by Intentional and Functional Specification

Let me turn to explanation by intentional specification, that is, the explanation of an action, a goal, or a choice among alternative actions or goals, in terms of a specific goal of which the agent assumes that it is favored by the action, goal or choice to be explained. The first question to be answered seems to be what the relation is between the specific explanation and the theory to be evaluated. In the case of the choice between actions or goals, the theory to be evaluated is likely to be the theory that specifies the particular goal that agents are supposed to achieve in such choices, be it "utility maximization" or one of its competing principles, such as "satisficing." In this case, the EbyS success is merely a DN success of that theory presented in another way, for it is apparently possible to DN explain the choice on the basis of that theory. However, in the case of an intentional explanation of an action or a goal, the specific goal put forward may well be related to the, psychological or sociological, theory to be evaluated, without being accountable as a DN success of that theory, in which case the EbyS success should be separately recorded in the theory's evaluation report.

At one point, Grobler and Wiśniewski, I am sure unwillingly, suggest by writing about my "flight from the utilistic approach" that I am reluctant to embrace that approach or its competing versions. However, the main point of my critical note about such approaches (SiS, p. 111) is that they focus on the choice between alternative actions (and goals) and have led to the neglect of a proper intentional explanation of actions (and goals) in terms of specific goals. But again, I am happy to subscribe to Grobler and Wiśniewski's suggestion that successes and failures of intentional explanations related to theories should be taken into account in the evaluation of these theories and hence in inferences to the best theory, whenever possible. As they rightly suggest, this opens extra possibilities for the evaluation of theories and hypotheses in mainly "ideographic" disciplines – in particular history, I would say.

Regarding explanation by functional specification in biology, it is plausible to count EbyS successes fitting in the particular model presented in SiS at least as successes of the theory of evolution, for that provides the crucial ultimate component in that model. Whether such successes can count as genuine DN successes is still a matter of debate around the question of the testability of the theory of evolution. Be this as it may, such an EbyS success may also be a success of a specific theory about certain kinds of organismic features or behavioral patterns or "evolutionary choices" between them. It will then at least provide an illustration of that theory, but it may amount to a straightforward DN success of it. In the case of theories regarding evolutionary choices, such as optimization theories, Looijen (2000) and Wouters (this volume) in fact suggest that the successes can be reconstrued as DN successes. Note that this is analogous to the case of choices between alternative actions or goals. Grobler and Wiśniewski report their optimism about incorporating successes of functional explanation in the evaluation of theories. The above lines concern a first attempt to actually do so.

REFERENCES

Kuipers, T. (2004). Inference to the Best Theory, Rather Than Inference to the Best Explanation. Kinds of Abduction and Induction. In: F. Stadler (ed.), *Induction and Deduction in the Sciences*, pp. 25-51. Dordrecht: Kluwer Academic Publishers.

Looijen, R. (2000). *Holism and Reductionism in Biology and Ecology. Episteme*, vol. 23. Dordrecht: Kluwer Academic Publishers.

COMPUTATIONAL APPROACHES

Jaap Kamps

THE UBIQUITY OF BACKGROUND KNOWLEDGE

ABSTRACT. Scientific discourse leaves implicit a vast amount of knowledge, assumes that this background knowledge is taken into account – even taken for granted – and treated as undisputed. In particular, the terminology in the empirical sciences is treated as antecedently understood. The background knowledge surrounding a theory is usually assumed to be true or approximately true. This is in sharp contrast with logic, which explicitly ignores underlying presuppositions and assumes uninterpreted languages. We discuss the problems that background knowledge may cause for the formalization of scientific theories. In particular, we will show how some of these problems can be addressed in the context of the computational representation of scientific theories.

1. Introduction

Background knowledge is ubiquitous in all forms of meaningful human communication. People engaged in fruitful discussion rely on a vast amount of shared background knowledge. How can we communicate if, for example, we do not have a shared understanding of the meaning of the words we utter? or make the same underlying assumptions? I vividly recall a discussion with Professor Kuipers on our common interests in artificial intelligence and philosophy of science. After much agreement, we suddenly reached an awkward difference of opinion that left me puzzled for some time. Then it turned out to be the case that Professor Kuipers was talking about the beneficial effects philosophy of science can have on artificial intelligence, and I was talking about the beneficial effects artificial intelligence can have on philosophy of science. Since these two positions are by no means incompatible, our difference of opinion was immediately resolved. This anecdote illustrates how a minor difference in the implicit presuppositions can give rise to confusion and even apparent disagreement, and moreover, how this may be resolved after the background assumptions have been made explicit. Background knowledge does not only occur in free forms of conversation, but also in more regulated discourse we make all sorts of presuppositions. This is even true for the way in which we report our findings and theories in the scientific literature. That is, even in cases where the clarity and unambiguity is

In: R. Festa, A. Aliseda and J. Peijnenburg (eds.), *Cognitive Structures in Scientific Inquiry* (*Poznań Studies in the Philosophy of the Sciences and the Humanities*, vol. 84), pp. 317-337. Amsterdam/New York, NY: Rodopi, 2005.

of principal importance, authors routinely presuppose a variety of background knowledge, for example, by the terminology that they use.

The notion of "background knowledge" is traditionally used to denote the vast amount of knowledge we take for granted when discussing a problem; this knowledge is treated as undisputed, if only for the time being and for the problem at hand (Popper 1963, p. 238). If some parts of the background knowledge are called into question, they do no longer belong to the background knowledge. As Kuipers (2001, p. 6) puts it: "It should also be stressed that, at least as a rule, observation and hence observation terms are, and remain, laden by theoretical presuppositions which are considered to belong to the so-called unproblematic background knowledge." The background knowledge is "unproblematic" in the sense that we (have to) assume that it is true or approximately true (Kuipers 2001, p. 48, p. 51). A well-known consequence of the background knowledge surrounding a theory is the Duhem-Quine thesis, i.e., the observation that we can make a theory immune for falsification by making modifications in the background knowledge (Kuipers 2001, p. 225, p. 244). This paper discusses the problems that background knowledge may cause for the formalization of scientific theories. In particular, we will address these problems in the context of computational representation of scientific theories.

Due to the fact that background knowledge is taken for granted, it will remain implicit in written expositions of a theory, and only the relevant knowledge is mentioned. The explicit treatment of underlying assumptions is one of the main reasons for the formalization of scientific theories (Suppes 1968). Of course, one may argue that the background assumptions that are left implicit are often relatively innocuous, and frequently the authors may safely assume that these implicit assumptions belong to the common knowledge of the readers. However, if our goal is to provide a version suitable for computational reasoning, this assumption is no longer valid. Computers are simply not endowed with this underlying background knowledge, and all relevant implicit assumptions need to be added explicitly. This gives rise to several problems. The first is a problem of acquisition: how to bring to light the knowledge that has been left implicit? The second is a problem of relevance: the amount of implicit background knowledge seems without an end, how to decide which part of it is relevant for the problem at hand?

The problem of background knowledge will occur in any situation where there is prior knowledge at stake, including all of the empirical sciences. Regardless of the used representation language, there will always be the question of whether one has faithfully represented the presuppositions of the domain. In order to explicate the role of background knowledge in the formalization of theories, we will situate our discussion in the context of the axiomatization in first-order logic of theories from the empirical sciences. Our experience in this area concerns the informal theories of fields like sociology rather than the mathematical theories of physics.[1]

2. Background Knowledge and Interpreted Languages

Suppose that we start out with a conventional exposition of a scientific theory, think of an article appearing in a scientific journal. A careful rational reconstruction of such a text will result in a list of statements representing the axioms of the theory, and a list of statements representing the claims or predictions of the theory. This rational reconstruction is by no means a trivial step, but we will ignore these complications and assume that, at least for some texts, it can be accomplished. As a next step, we would want to give a formal rendition of the selected statements, and thus construct an initial formal version of the theory. This initial formal theory, which we assume here to be in first-order logic, will have a number of axioms and a set of conjectures representing the statements that the theory claims to predict or explain. We can now try to find out which of these conjectures can be derived from the axioms. In particular, we can use the standard tarskian consequence relation by using standard rules of inference (see standard textbooks like Enderton 1972). As may come as no surprise, this will generally be a disappointing effort: in the (initial) formal theory many of the conjectures will not be derivable from the axioms. As is well-known, informal arguments do not straightforwardly extend to rigorous formal proofs. Admittedly, in some cases this might be due to infelicitous argumentation. Some of the informal conjectures may turn out to be false when subjected to greater scrutiny. However, more generally speaking, there are other reasons for a failure to derive some of the informal conjectures.

In particular, one may question whether the standard consequence relation is faithfully singling out the intended consequences of our theory. As Tarski (1946, pp. 121-122) put it:

[1] This is roughly based on some recent attempts to axiomatize informal sociological theory (Péli et al. 1994; Hannan 1998; Kamps and Pólos 1999). Although one may expect that the more rigorous and formal an exposition is, the more of the background assumptions have been added explicitly and that the more informal an exposition is, the greater the amount of background knowledge that is presupposed. As a consequence, one would expect that, relative to the explicitly discussed part, authors in social sciences would leave larger parts of their theories implicit than is the case in, for example, mathematical physics. However, this is only a difference in degree, and does not affect the main points of our arguments.

> Our knowledge of the things denoted by the primitive terms ... is very comprehensive and is by no means exhausted by the adopted axioms. But this knowledge is, so to speak, our private concern which does not exert the least influence on the construction of our theory. ...We disregard, as is commonly put, the meaning of the primitive terms adopted by us, and direct our attention exclusively to the form of the axioms in which these terms occur.

Now consider the empirical science theory we are axiomatizing: it contains primitive terminology that has a specific meaning – it is "antecedently understood." The standard logical consequence relation does not take into account the underlying understanding of the terminology. In other words, by using a standard consequence relation we explicitly ignore all background knowledge and assume that there are no logical relations among the atomic sentences other than those explicitly stated in the axioms. This is in sharp contrast with the discussion of unproblematic and undisputed background knowledge, which assumes that the antecedent meaning of the used terminology is taken into account – even taken for granted. The unavoidable conclusion is that the failure to derive some of the informal conjectures can be attributed to the false assumption that we are dealing with an uninterpreted language (by using a tarskian consequence relation). In particular, some of the informal conjectures might materialize into formally proven theorems, were we to use a consequence relation that takes the underlying interpretation of terminology into account.

This has some far-reaching consequences. It is simply incorrect to regard our initial formal theory as an uninterpreted first-order language, but it should be regarded as an interpreted first-order logic in which the vocabulary has a specific, fixed interpretation. A direct result of using an interpreted language is that we cannot use the standard consequence relation. This requires a non-standard consequence relation that takes into account the interpretation of the terminology in the vocabulary of the theory. That is, for every set of interpreted vocabulary we need a special consequence relation that takes the antecedent meaning of the terminology into account.[2] In order to decide whether an informal conjecture is a theorem or not, we need to use the particular consequence relation associated with the specific used interpreted language. The problem now is that the specific interpretation is left implicit in conventional discourse, and therefore the needed special consequence relation is generally unknown. We may use the standard consequence relation only if we can ensure that all relevant background knowledge is explicitly added to the theory. However, the acquisition of the relevant background knowledge is

[2] For an example of such an interpreted first-order language, see the the language of *Tarski's World* that features prominently in a textbook on logic (Barwise and Etchemendy 1992). We will later draw upon some examples from this book. An interesting discussion of logical consequence relations can be found in (Etchemendy 1990).

a far from trivial task for precisely this knowledge is taken for granted and left implicit in standard scientific discourse.

3. Logical Analysis

In order to investigate consequence relations for interpreted languages, we need to make our discussion a bit more precise. This initial formal theory in first-order logic will have a number of axioms, denoted with $\sum_{exp}$ for the explicit axioms of the (initial) theory, and a set of conjectures Γ for the statements that the theory claims to predict or explain. Now let $\models$ denote the standard (tarskian) consequence relation for an uninterpreted first-order language, and let $\models_{theory}$ denote the unknown non-standard consequence relation of the specific interpreted first-order language of our theory. We want to investigate the logical dependencies between these two possible consequence relations that can be used to determine whether a conjecture $\gamma \in \Gamma$ is derivable from the explicitly mentioned axioms $\sum_{exp}$. The four logical possibilities in Table 1 present themselves.

	$\sum_{exp} \models_{theory} \gamma$	$\sum_{exp} \not\models_{theory} \gamma$
$\sum_{exp} \models \gamma$	II.	I.
$\sum_{exp} \not\models \gamma$	II.	IV.

Table 1. Noninterpreted and Interpreted Consequences.

Let us first consider case I, $\sum_{exp} \models \gamma$ and $\sum_{exp} \not\models_{theory} \gamma$. In the case of an interpreted first-order logic, this cannot occur. The non-standard consequence relation $\models_{theory}$ will be supraclassical: all $\models$-consequences are also $\models_{theory}$-consequences.[3] Some theorems will hold irrespective of the specific interpretation of the language, that is, they will hold in any interpretation of the language (including the intended interpretation). This gives us the reassurance that we can immediately conclude that $\sum_{exp} \models_{theory} \gamma$ in case we find that $\sum_{exp} \models \gamma$ (case II).[4] As a result, if we treat an interpreted first-order language as if it were an uninterpreted language, then we can be sure that the theorems we find (using

[3] This is true for interpreted versions of classical logic. If we consider interpreted non-classical logics, the underlying consequence relation will not satisfy structural properties like monotony, and the resulting logic need not be supraclassical. This points to considerable difficulty in establishing what is implied by an interpreted nonmonotonic theory.

[4] A second result is that, by contraposition, $\sum_{exp} \not\models_{theory} \gamma$ implies $\sum_{exp} \not\models \gamma$.

the standard consequence relation $\models$) are also theorems in the interpreted language (using $\models_{\text{theory}}$)

However, as argued above, the used terminology will have antecedent meaning. Therefore, we generally expect that several of the informal conjectures will depend on the specific intended interpretation of the language. So what should we do in case a conjecture is not a $\models$-consequence of the explicit axioms, i.e., when $\sum_{\text{exp}} \not\models \gamma$? One option is case IV, the informal conjecture is no theorem, i.e., when also $\sum_{\text{exp}} \not\models_{\text{theory}} \gamma$. We will return to case IV below. The remaining option is case III, the informal conjecture is a theorem when we take the interpretation of the language into account, that is when $\sum_{\text{exp}} \not\models \gamma$ and $\sum_{\text{exp}} \models_{\text{theory}} \gamma$. This is the crucial case for here it would be an important failure to ignore the (implicit) interpretation of the language – we would falsely judge a theorem as a false conjecture. What can we do to prevent this?

The obvious way out is to find a way to ensure that all the relevant implicit background knowledge is explicitly added to the formal theory. Of course, if we would have an axiomatization of all underlying background knowledge, call this set $\sum_{\text{imp}}$, then there would be no more implicit relations between atomic sentences, and we could use the standard consequence relation. In case the background knowledge is first-order expressible (which we may assume in case of an interpreted first-order language) and finitely axiomatizable, we have that

$$\textstyle\sum_{\text{exp}} \models_{\text{theory}} \gamma \text{ if and only if } \sum_{\text{exp}} \cup \sum_{\text{imp}} \models \gamma$$

Under these conditions, we can reduce the question of how to use the unknown non-standard consequence relation, to the question of how to make the relevant part of the implicit background knowledge explicit. The situation we are interested in can now be reformulated as:

$$\textstyle\sum_{\text{exp}} \not\models \gamma \text{ and } \sum_{\text{exp}} \cup \sum_{\text{imp}} \models \gamma$$

with $\sum_{\text{imp}}$ being the unknown set of implicit background knowledge. Our goal is now to make relevant parts of $\sum_{\text{imp}}$ explicit.[5]

We can push our analysis even further by considering this situation in terms of formal semantics. A first observation is that $\sum_{\text{exp}} \not\models \gamma$ implies that there must exist models $\mathcal{M}$ such that $\mathcal{M} \models \sum_{\text{exp}} \cup \{\neg\gamma\}$. In fact, constructing such a model would be one of the straightforward ways of proving that $\sum_{\text{exp}} \not\models \gamma$. Moreover, there is nothing magical about the construction of these models for it involves only the explicitly known axioms and the conjecture, and the standard consequence relation – a simple algorithm suffices for constructing

[5] Note that, even in case the total background knowledge is not first-order expressible or not finitely axiomatizable, some parts of it may still be.

these models (as we will illustrate in the next section). Each of these models represents a counterexample against the derivation of the conjecture under the assumption that the language is uninterpreted. In our case, however, the conjecture would become derivable in case we would succeed in explicitly adding the background knowledge that enforces the interpreted language, that is $\sum_{\text{exp}} \cup \sum_{\text{imp}} \models \gamma$. A second observation is that all the models that are counterexamples (in the uninterpreted case) must be violating the implicit background knowledge. That is, for all these models $\mathcal{M}$, it must be the case that $\mathcal{M} \not\models \sum_{\text{imp}}$ (since $\mathcal{M} \models \sum_{\text{exp}} \cup \{\neg \gamma\}$ and $\mathcal{M} \not\models \sum_{\text{exp}} \cup \sum_{\text{imp}} \cup \{\neg \gamma\}$). The models that are counterexamples in the uninterpreted case are 'witnesses' of the implicit background knowledge that we need to add explicitly to the axiomatization. Therefore, finding such models can allow us to come to grips with the implicit background knowledge. Consider what happens when we inspect such a model: it necessarily conflicts with some part of our implicit background knowledge on the domain of the theory. To a human observer these models appear strange or extraordinary in some respects. This will prompt us to formulate appropriate axioms that will prevent these models from occurring – axioms that make part of the implicit background knowledge explicit (i.e., some elements of $\sum_{\text{imp}}$). Since these background axioms are based on the specific models that we have examined, this need not be a one-step approach. Further testing may reveal different counterexamples, giving rise to more of the background knowledge being made explicit.

This will, in general, not lead to the axiomatization of all underlying background knowledge. This is hardly unfortunate, since there seems to be no end to the underlying background knowledge. Attempting an axiomatization of all the background knowledge that is taken for granted is at least impractical, if not impossible. We propose to use the informal conjectures for determining which parts of the background knowledge are relevant for the question at hand. That is, we want to use it as a sufficient condition for relevance: if some implicit background knowledge is used for deriving one of the informal conjectures, this is a tell-tale sign for its relevance. In this case there are obvious benefits to making these particular background assumptions explicit, for example, they can become part of future discussion. Note that we do not think that this is a necessary condition, there may be other reasons for including parts of the background knowledge. Also, in a later stage one may want to extend the set of conjectures we want to explain, which may require more of the background knowledge to be explicitly added to the theory.

It is well known that, from a logical point of view, one can always find some additional assumptions that will make a conjecture derivable (Quine 1953, p. 43). So it is a legitimate concern if we are able to distinguish false conjectures from informal conjectures that can be made derivable by explicating background knowledge (case IV in Table 1 we delayed discussing above). That is, how can we identify false conjectures, i.e., conjectures for which we have that $\sum_{exp} \not\models_{theory} \gamma$? Inspection of the models that are counterexamples provides an easy safe-guard against this. In this case there will, again, be models $\mathcal{M}$ such that $\mathcal{M} \models \sum_{exp} \cup \{\neg \gamma\}$. However, since in this case $\sum_{exp} \cup \sum_{imp} \not\models \gamma$, some of these counterexamples will be in perfect harmony with all the background knowledge that we would take for granted, i.e., $\mathcal{M} \models \sum_{exp} \cup \sum_{imp} \cup \{\neg \gamma\}$. Inspection of these models will reveal a genuine counterexample – an intended model of the theory in which the conjecture fails – proving that the informal conjecture does not hold. We may only rebut a potential counterexample by relying on unproblematic background knowledge. Otherwise, we must conclude that $\sum_{exp} \not\models_{theory} \gamma$.

4. Applications and Computational Support

Our discussion up to this point has been rather abstract. However, as we will show in this section, our analysis above can be directly applied to concrete situations. In particular, we will show how this can immediately be supported by standard tools from automated reasoning. The formalization of an empirical science theory is typically using an interpreted language, with the interpretation being enforced by the implicit background knowledge that is taken for granted. Barwise and Etchemendy (1992) introduce the interpreted first-order language of Tarski's world with an associated computer program that visualizes this blocks world. The vocabulary of this first-order language contains constants (a through f plus n1, n2, ...), predicates (unary predicates: Tet, Cube, Dodec, Small, Medium, Large; binary predicates: =, Smaller, Larger, LeftOf, RightOf, BackOf, FrontOf; and a tertiary predicate Between), and no functions. The predicates have a fixed interpretation in the associated computer program, for example an object cannot be both a cube and a tetrahedron. The fixed interpretation assigned by Tarski's world is one that is "reasonably consistent with the corresponding English verb phrase" (Barwise and Etchemendy 1992, p. 11). The authors assume that readers share common background knowledge on names of these predicates, and that the program's interpretation is consistent with it. Although the predicates have a very precise meaning, the authors do not give the axioms that are assumed to hold. Instead, they invite the reader to experiment with the program, and get acquainted with their meaning by trial and error – not unlike in ordinary language acquisition. We can use some examples from (Barwise and Etchemendy 1992) in order to illustrate the strategy for elucidating implicit background knowledge discussed in the previous section.

4.1. *Interpreted Consequence*

We are particularly interested in arguments which depend on the fact that the language of Tarski's world is an interpreted language. In this case, the (formal) proofs do strictly depend on the interpretation as given in the program, and would not hold for arbitrary interpretations of the predicates. If we would substitute the used predicate symbols with fresh ones having the same associated arity, the arguments would not hold. An exercise which relies on the specific interpretation of the predicates is (Barwise and Etchemendy 1992, Problem 5-30, p. 143).

> Is $\exists x$ [FrontOf(c, x) $\wedge$ Cube(x)] a consequence of
>
> A1. $\forall x$ [Cube(x) $\vee$ (Tet(x) $\wedge$ Small(x))]
>
> A2. $\exists x$ [Large(x) $\wedge$ BackOf (x, c)]

According to the instructor's manual this is indeed the case in Tarski's world (Eberle 1993). It will be impossible to build a world that is a counterexample using the Tarski's World program. Is it also valid for arbitrary interpretations of the predicates? To answer this question, we can use standard tools from automated reasoning, like automated theorem prover OTTER (McCune 1994b) and automated model generator MACE (McCune 1994a). The answer turns out to be negative: theorem prover OTTER fails to find a proof, and model generator MACE has no trouble in finding counterexamples. These counterexamples are models of the premises in which the conjecture is false (the first model on universe {0,1} is shown in Table 2).[6]

Finding this model proves that the argument does not hold using a standard consequence relation, in symbols,

{A1,A2} $\nvDash$ $\exists x$ [FrontOf(c, x) $\wedge$ Cube(x)}

However, the argument should hold when we respect the interpretation of the predicates. According to our above discussion, the counterexample in Table 2 must conflict with the interpretation of the predicates in the Tarski's world

[6] For example, by invoking MACE with the options '`-n2 -p -m100`' (see for details McCune 1994a). MACE generates 16 models on {0,1} in less than a second. A formal model consists of a universe (here the two elements {0,1}) and a mapping between the non-logical symbols (here constant c; unary predicates Cube, Tet, Small, and Large; and binary relations BackOf and FrontOf) and elements of the universe. For example, consider the model in Table 2: here, the constant symbol c is interpreted as object 0, and predicate symbol Cube is interpreted as both Cube(O) and Cube(l) are true. That is, all objects in the universe of this model are cubes making the sentence $\forall x$ [Cube(x) $\vee$ (Tet(x) $\wedge$ Small(x))], the first premise, is indeed true in this model.

program. Moreover, this model must also be in conflict with the ordinary language meaning of the corresponding English phrases. That is, anybody with some proficiency in English should find this model in violation of his or her background knowledge of the domain. Our expectation is that, when confronted with this model, a person is able to articulate *why* this model should not be allowed to occur.

c	
	0

Cube	
0	T
1	T

Tet	
0	F
1	F

Small	
0	F
1	F

BackOf	0	1
0	T	F
1	F	F

FrontOf	0	1
0	F	F
1	F	F

Large	
0	T
1	F

Table 2. Counterexample I.

Upon inspecting the model in Table 2, we immediately note a strange feature: BackOf(0, 0) is true, the object 0 is in back of itself. This is not in accordance with the normal English interpretation of this predicate, and we decide to spell out this background knowledge explicitly:

B1. $\forall x\,[\neg\,$BackOf$(x,x)]$

After adding this background assumption explicitly to the premises, the model in Table 2 will no longer be a model of the theory. We can now test anew if we can now formally derive the conclusion. Notice that this need not be the case for there may exist different counterexamples. Indeed, theorem prover OTTER still fails to find a proof, and model generator MACE is able to construct further counterexamples (now 8 on {0,1}, the first is shown in Table 3).

c	
	1

Cube	
0	T
1	T

Tet	
0	F
1	F

Small	
0	F
1	F

BackOf	0	1
0	F	T
1	F	F

FrontOf	0	1
0	F	F
1	F	F

Large	
0	T
1	F

Table 3. Counterexample II.

There must still be more background knowledge at stake. Inspecting the model in Table 3, there seems to be no problem with the interpretation of each predicate independently. However, some natural relations between the predicates are not properly taken into account. In the model BackOf(0,1) is true while at the same time FrontOf(1, 0) is false. This conflict with our background understanding of these predicates. Our intuitions say that these two predicates are inversely related, so we decide to explicitly add this relation between FrontOf and BackOf:

B2. $\forall\, x, y$ [FrontOf(x, y) $\leftrightarrow$ BackOf(y, x)]

Have we now added all relevant background knowledge? We test again using theorem prover OTTER, but still fail to find a proof. Yet again, the model generator MACE is able to construct further counterexamples (still 4 on {0,1}, the first is shown in Table 4).

c	0

Cube	
0	T
1	F

Tet	
0	F
1	T

Small	
0	F
1	T

BackOf	0	1
0	F	F
1	T	F

FrontOf	0	1
0	F	T
1	F	F

Large	
0	F
1	T

Table 4. Counterexample III.

Again, we examine this new counterexample to verify whether it is an intended model of this domain. Inspection reveals that this model is not conform the normal English interpretation of the predicates: both Small(l) and Large(l) are true, the same object is both small and large. We do not want to exclude that an object is neither small or large (as object 0 in Table 4) for, after all, there might be medium sized objects. An object being both small and large at the same time, however, conflicts with our implicit understanding of these two predicates, and we decide to add a further background assumption explicitly to the theory:

B3. $\forall\, x \neg$ [Small(x) $\wedge$ Large(x)]

Did we now make all relevant background knowledge explicit? At last, the answer is positive: theorem prover OTTER finds a proof using the two premises and two of the background assumptions (B2 and B3).[7] The proof constructed

[7] That is, we may decide to relax the first background assumption Bl again because it is not necessary for this argument. Notice that this implies that the model in Table 2 is also violating the other background assumptions, otherwise this counterexample would still disprove the argument (in this case B2).

by OTTER is a clause-based resolution proof. Paraphrasing this formal proof, we find that OTTER derives from premise A2 and background assumption B2, that c is in front of a large object; and using premise A1 and background assumption B3, that this large object must be a cube. That is, we can now formally derive that c is in front of a cube, in symbols,

$$\{\text{A1, A2, B2, B3}\} \vDash \exists x\, [\mathsf{FrontOf}(\mathsf{c}, x) \wedge \mathsf{Cube}(x)\}$$

The answer to this problem is, indeed, positive. We have now proved that the argument is valid when respecting the (implicit) interpretations of Tarski's world, in symbols,

$$\{\text{A1, A2}\} \vDash_{\mathsf{TW}} \exists x\, [\mathsf{FrontOf}(\mathsf{c}, x) \wedge \mathsf{Cube}(x)\}.$$

4.2. *Interpreted Non-consequence*

An exercise with the same premises as above is (Barwise and Etchemendy 1992, Problem 5-31, p. 143).

> Assume the following premises:
>
> A1. $\forall x\, [\mathsf{Cube}(x) \vee (\mathsf{Tet}(x) \wedge \mathsf{Small}(x))]$
>
> A2. $\exists x\, [\mathsf{Large}(x) \wedge \mathsf{BackOf}(x, \mathsf{c})]$
>
> Does it follow that $\neg \exists x\, [\mathsf{Small}(x) \wedge \mathsf{BackOf}(x, \mathsf{c})]$?

We are asked to establish whether this argument is valid when respecting the interpretation of predicates. According to the instructor's manual this is not the case in Tarski's world (Eberle 1993). Since we established above that B1, B2, and B3 are part of the implicit background knowledge, we will start with the set of explicit premises {A1, A2, B1, B2, B3}. As expected, theorem prover OTTER fails to derive the conjecture from this set of premises. We resort to model generator MACE in order to find models of the premises in which the conjecture is false, that is, in which there exists a small object in the back of object c. Model generator MACE fails to find any model of cardinality 2, but produces 24 models of cardinality 3 (the first of them is reproduced in Table 5).

c	
	1

Cube	
0	T
1	T
2	T

Tet	
0	F
1	F
2	F

Small	
0	F
1	F
2	T

BackOf	0	1	2
0	F	T	F
1	F	F	F
2	F	T	F

FrontOf	0	1	2
0	F	F	F
1	T	F	T
2	F	F	F

Large	
0	T
1	F
2	F

Table 5. Counterexample IV.

Can we rebut this model by mobilizing part of the background knowledge? Examining the model in Table 5, we have a configuration of one cube c that is placed in front of two other cubes, one of which is large (as required by premise A2), and the other small (refuting the conjecture). The interpretation of the predicates in this model is conform our implicit background knowledge. We can confirm this by replicating a corresponding world using the Tarski's World program. We must conclude that this model is a genuine counterexample disproving the conjecture. That is, we have proved that the argument does not hold when respecting the interpretations of Tarski's world, in symbols,

$$\{A1, A2\} \not\models_{TW} \neg \exists x\, [\mathsf{Small}(x) \wedge \mathsf{BackOf}(x, \mathsf{c})].$$

These simple examples demonstrate the necessity of taking implicit background knowledge into account in languages that have antecedent meaning. Moreover, they illustrate how automated reasoning tools can assist in the acquisition of implicit background knowledge by constructing the models that are in conflict with our understanding of the domain. This has also proved to be crucial in more substantial applications: uncovering implicit theoretical presuppositions is one of the main problems in the reconstruction of informal sociological theories (Kamps and Pólos 1999; Kamps 1999). A word of warning is in place for it is important not to underestimate the general complexity of this task. The use of automated tools is subject to important limitations, both in principle (first-order logic is not decidable), as in practice (time, memory, CPU-power). The above examples are well within these limits: none of the successful or failed proof attempts or model searches lasted more than a single second. It is interesting to note that the models conflicting with our implicit understanding of the domain are particularly difficult to find by hand. Since we ourselves possess the underlying background knowledge, we have a natural tendency to focus our attention towards the intended models of the theory. Computer programs, not endowed with this underlying understanding, are not hindered with such a bias.

5. Discussion and Conclusions

Scientific theories about empirical phenomena are human constructions. Formal theories do interface, on the one hand, with empirical reality, leading to all the familiar problems of confirmation, falsification, and truth approximation. On the other hand and less frequently discussed, formal theories also interface with human conceptions and theoretical intuitions. It is well-known that in the empirical sciences, terms "have a clear meaning independent of the theory and they retain this meaning within the context of the theory" (Kuipers 2001, p. 44). The terminology is "antecedently understood" (Hempel 1966, p. 75). One of the main reasons for the formalization of scientific theories is to bring out the meaning of concepts in an explicit fashion (Suppes 1968). Making the underlying background knowledge explicit contributes to our understanding of the theory by avoiding ambiguity.[8] This may work even if there is no full consensus on the meaning of terminology (as is rarely the case in the social sciences). In case of partial consensus, researchers would still agree that some background axioms should hold. At the same time, making the underlying background knowledge explicit is a highly non-trivial task. This is immediately clear once we realize that much of our background knowledge is tacit knowledge (Polanyi 1958). This means that, even though we are carriers of implicit background knowledge, the articulation of it may be beyond our own control. That raises the question whether we can ever be sure that all relevant background knowledge has been made explicit. If we must assume, as Polanyi does, that part of our background knowledge will always remain implicit, this has some important methodological implications.

The general conclusion is a call for caution when discussing what are the sets of consequences or models of a formal theory. If we cannot be sure that all background knowledge is explicitly added to the theory, we must anticipate that we can only derive part of its consequences, and that the set of formal models of the theory contains models that are conflicting with our intuitions. In

---

[8] Arguably, it is more acceptable to revise implicit background knowledge, than to retract some explicit statements of a theory. That does not imply that explicit background knowledge cannot evolve over time. In fact, formalization is known to trigger the further development of terminology. Even in the simplified setting of Tarski's World the background knowledge may change over time: in the new version of the program, the interpretation of the Between predicate has changed (Barwise and Etchemendy 1999).

particular, this may interfere with attempts to compare formal theories by their sets of consequences or models, as in approaches to truthlikeness (Kuipers 2000). If we are to compare theories by the statements they imply, we must take into account that we are systematically underestimating the set of consequences. Hence, in general, a statement approach to truthlikeness will miss out some of the successes and failures of a theory. If we compare theories by their models, we must take into account that we are systematically overestimating the set of models. Thus a semantic approach to truthlikeness will find, in general, more successes and failures than warranted by the theory. Keeping this in mind, there is even more reason to pursue efforts to make relevant parts of the implicit background knowledge explicit. After all, if our implicit background knowledge is (approximately) true then adding this explicitly to a formal theory should bring us even closer to the truth.[9]

Background knowledge not only affects the hypothetical problem of enumerating deductively closed sets of consequences or all the models of a theory. Even apart from the question of implicit background knowledge, comparing all consequences or models of a theory is already infeasible in practice for these sets are generally infinite. As a result, theory comparison is relativized to a particular set of key predictions (such as the comparison of electrodynamic theories shown in [Kuipers 2001, Table 8.2, p. 236]). Precisely in such a setting we would want to avoid falsely discarding conjectures by not fully taking into account the meaning of terminology. In our discussion of the axiomatization in formal logic above, we have focused on this case by assuming that a specific conjecture is at stake. We have shown how formal semantics may be used to avoid the unjustified rejection of a conjecture. Just as logic provides a formal notion of proof, it also provides a formal notion of refutation. A formal refutation of a conjecture is a formal model (in the logical sense) of the premises in which the conjecture does not hold. In the context of implicit background knowledge, a formal refutation need not correspond to an empirical refutation (only models that respect the terminology may correspond to an empirical possibility). If we are able to construct a formal model refuting a conjecture, we can inspect the model to verify whether the refutation is

[9] Technically this will be somewhat more involved, since it points out an asymmetry between the statement and models view on a theory. The "extra" models of the theory are necessarily all in conflict with the implicit background knowledge (i.e., these are all nonintented or nonsensical models). Thus, we will approximate truth in terms of models. However, even if the background knowledge is (approximately) true, the "missed" consequences of the theory may contain both interesting statements and nonsensical ones. As a result, explicitly adding background knowledge may increase both the number of successes and failures of the theory (in terms of statements). Only in case the explicit axioms of the theory are also (approximately) true, the "missed" consequences will also be all (approximately) true. Then, we will also approximate truth in terms of consequences of the theory.

warranted. If this is not the case, inspecting the model immediately suggests which background knowledge needs to be added explicitly to the theory. Recall that much of our underlying knowledge is tacit, however, this need not prevent us from identifying models that are in conflict with it. Identifying such a model makes us aware of our tacit understanding, and can provide crucial help in its articulation.[10] This results in an interesting interplay between conjectures, proofs, and refutations. Although the antecedent meaning of terminology is a principal feature of the empirical sciences, we may also have background knowledge on terminology in non-empirical fields like mathematics and philosophy. In fact, our discussion shows some remarkable similarities with discussions on mathematical discovery (Pólya 1945; Lakatos 1976).[11] The main difference is that in the non-empirical sciences, we have the luxury of being able to stipulate that the concept as characterized in a theory is the 'real' concept.

It is known for long that logical axiomatization can contribute to theory development in the empirical sciences (Woodger 1937; Kyburg 1968). In recent years, this has resulted in the formalization of a number of sociological theories (Péli et al. 1994; Hannan 1998; Kamps and Pólos 1999). Having this is mind, we find it difficult to agree with the remark that a logical axiomatization or so-called statement approach is not very useful and very difficult (compared with a semantic approach). Kuipers (2001, p. 319) has it that

> the statement approach is certainly more difficult for specific reconstructions. ... Happily enough, not all interesting theoretical questions need logical treatment. ... Given our intention to be as useful as possible for actual scientific research we will restrict our attention to the structuralist approach.

We do not disagree on the merits of semantic approaches. There are many examples of fruitful axiomatization in the structuralist approach (Balzer et al. 1987, 2000). We also immediately admit that, more generally, a semantic approach has specific advantages over a statement approach. To mention just a few, a semantical approach immediately suggests itself for establishing the consistency of a theory or domain, or for disproving a conjecture. However, we disagree on the decision to de-emphasize logical axiomatizations. Generally speaking, a statement approach also has specific merits, think of

---

[10] It is important to bear in mind that tacit knowledge can be made explicit, and that doing so has contributed to the theoretical development of various fields (Polanyi 1958). This does, however, require significant effort, and it will be impossible to formalize all the tacit knowledge in a particular field – a point with which we concur.

[11] Our discussion of models conflicting with the implicit antecedent meaning of terminology shows resemblance with Lakatos' monster-barring heuristic (dealing with doughnut-shaped or picture-frame polyhedra discussed in [Pólya 1954, p. 42] and [Lakatos 1976, p. 19]).

establishing the inconsistency of a theory. For certain cases a statement approach is intuitively more appropriate. It is of interest to analyze reasons that might explain this discrepancy between these views on formal theorizing.[12]

Perhaps a difference in appreciation of the statement and semantical approaches is rooted on the difference between the sciences. Our experience in logical reconstruction has focused on informal theories in sociology, whereas the structuralist approach is based on reconstructions of mathematical physics (Sneed 1971), although later also applied to various other fields (Balzer et al. 1987), including sociology (Manhart 1994). The axiomatization of a highly mathematical theory would also require the axiomatization of the used mathematical techniques. This is a highly nontrivial task in case of the advanced, quantitative mathematics used in mathematical physics. This view is consistent with the axiomatization of one of the rare mathematical theories in sociology, a mathematical model of social groups (Simon 1952). The resulting axiomatization is almost completely concerned with the differential equations used in the mathematical model (Kyburg 1968, Ch. 12). The structuralist approach, in contrast, allows for freely using all kinds of useful mathematics, allowing the reconstruction to focus on the theory at hand without first having to axiomatize various mathematical theories. This will make reconstructions certainly easier in case of advanced mathematical theories such as in theoretical physics. However, the mathematical finesse of physics is not a rule in the empirical sciences. In fact, in fields like sociology, mathematical theories are even rare, and the standard discourse is in natural language. At least for non-mathematical theories in the empirical sciences, the statement approach to formalization seems a viable option.

It is important to note that the flexibility of the structuralist approach does not come without a price. Since the standard mathematical vernacular is only partially formal, it requires substantial mathematical background knowledge usually shaped by years of mathematical training. If the goal is to provide a computational implementation of a theory, we are again confronted with the fact that computers lack the mathematical background knowledge. It is unclear to what extent a structuralist formalization renders our theories in a form that

---

[12] Some have argued that there is some form of resentment against logical empiricism (Friedman 1991, 1999). There may be some truth in this, e.g., the structuralist approach is also sometimes referred to as the "non-statement view" (Stegmüller 1973). Needless to say, the field of logic has changed dramatically since the days of positivism. As Hintikka (1998, p. 304) writes: "[W]hen the sharpest philosophers of science realized that a study of 'the logical syntax of the language of science' was not enough, they resorted to set theory for their conceptualizations. Ironically some misguided philosophers of science have continued to seek salvation in set theory long after the development of logical semantics and systematic model theory." However interesting such arguments may be from a historical point of view, we will restrict ourselves here to substantial reasons.

can be interpreted by a computer.[13] Standard mathematics is usually too informal to allow for constructing formal proofs. Formal logic, in contrast, provides the needed rigorousness. A formalization using the so-called statement approach immediately allows for computational implementation. In fact, the automation of logical reasoning is one of the oldest applications of artificial intelligence (Newell and Simon 1956; Beth 1958). Current implementations of automated reasoning programs are powerful tools that can support the formal reconstruction of theories in various ways (Kamps 1998, 1999). By using such tools, the construction of a logical axiomatization need not be more difficult than a structuralist reconstruction. One of the reasons why logical axiomatization is considered to be difficult, is because manually deriving theorems using a particular formal proof system can be painstaking and prone to errors. Unlike humans, computers are well-equipped for performing tedious tasks like proof checking or proof finding in a formal proof system. In fact, the detailed rigorousness is precisely what makes a logical axiomatization suitable for computational reasoning. In sum, using these programs can greatly facilitate the process of reconstructing scientific theories in formal logic. Moreover, if the aim is to provide a computational representation of theories, an axiomatization produced by the statement approach is still an attractive alternative.

In our experience, both a statement approach and a semantical approach have their respective merits. Since these merits do not coincide, it is of particular interest to investigate ways that can exploit both views. In this light, it is important to note that most logics come with both a proof theory and a formal semantics. This allows us to view our theory as either a set of statements and a set of models, depending on which point of view is better suited for the question at hand. For example, for proving a particular conjecture we can use syntactic proof theory and for disproving a particular conjecture we can use semantic model theory. That is, the "pragmatic choice" between a "statement approach" and a "semantic approach" as discussed in (Kuipers 2001, p. 319) need not be made: there is no reason why we cannot have the best of both worlds. Our earlier discussion on background knowledge is an illustrative example of how we can exploit a semantical view on the statement approach. One can only hope that such considerations may ultimately lead to a reconciliation of the two approaches.

[13] Although Kuipers (2001, p. 302) writes: "it will become quite clear ... that the structuralist analysis of theories can almost directly be used for the computational representation of theories." This is far from obvious to me, in fact, it seems to require pencil, paper, and a philosophy professor in order to operate a structuralist representation.

ACKNOWLEDGMENTS

This research was supported by the Netherlands Organization for Scientific Research (NWO, grant # 400-20-036).

University of Amsterdam
Institute for Logic, Language and Computation
Nieuwe Achtergracht 166
NL-1018WV Amsterdam
The Netherlands
e-mail: kamps@illc.uva.nl

REFERENCES

Balzer, W., C. U. Moulines and J. D. Sneed (1987). *An Architectonic for Science: The Structuralist Program. Synthese Library*, vol. 186. Dordrecht: D. Reidel Publishing Company.

Balzer, W., C. U. Moulines and J. D. Sneed, eds. (2000). *Structuralist Knowledge Representation: Paradigmatic Examples. Poznań Studies in the Philosophy of the Sciences and the Humanities*, vol. 75. Amsterdam: Rodopi.

Barwise, J. and J. Etchemendy (1992). *The Language of First-Order Logic.* Revised and expanded third edition. Stanford, CA: CSLI Publications.

Barwise, J. and J. Etchemendy (1999). *Language, Proof and Logic.* New York NY: Seven Bridges Press. Stanford, CA: CSLI Publications.

Beth, E. W. (1958). On Machines which Prove Theorems. *Simon Stevin: Wis- en Natuurkundig Tijdschrift* **32**, 49-60.

Eberle, R. (1993). Instructor's Manual to Accompany Jon Barwise and John Etchemendy's *The Language of First-Order Logic*. Stanford University: CSLI Publications.

Enderton, H. B. (1972). *A Mathematical Introduction to Logic.* New York: Academic Press.

Etchemendy, J. (1990). *The Concept of Logical Consequence.* Cambridge, MA: Harvard University Press.

Friedman, M. (1991). The Re-Evaluation of Logical Positivism. *The Journal of Philosophy* **88**, 505-519.

Friedman, M. (1999). *Reconsidering Logical Positivism.* Cambridge: Cambridge University Press.

Hannan, M. T. (1998). Rethinking Age Dependence in Organizational Mortality: Logical Formalizations. *American Journal of Sociology* **104**, 126-164.

Hempel, C. G. (1966). *Philosophy of Natural Science. Foundations of Philosophy Series.* Englewood Cliffs, NJ: Prentice-Hall.

Hintikka, J. (1998). Truth Definitions, Skolem Functions and Axiomatic Set Theory. *The Bulletin of Symbolic Logic* **4**, 303-337.

Kamps, J. (1998). Formal Theory Building Using Automated Reasoning Tools. In: A. G. Cohn, L. K. Schubert, and S. C. Shapiro (eds.), *Principles of Knowledge Representation and Reasoning: Proceedings of the Sixth International Conference* (KR'98), pp. 478-487. San Francisco, CA: Morgan Kaufmann Publishers.

Kamps, J. (1999). On Criteria for Formal Theory Building: Applying Logic and Automated Reasoning Tools to the Social Sciences. In: J. Hendler and D. Subramanian (eds.), *Proceedings of the Sixteenth National Conference on Artificial Intelligence (AAAI-99)*, pp. 285-290. Menlo Park, CA: AAAI Press/The MIT Press.

Kamps, J. and L. Pólos (1999). Reducing Uncertainty: A Formal Theory of Organizations in Action. *American Journal of Sociology* **104**, 1776-1812.

Kuipers, T.A.F. (2000/ICR). *From Instrumentalism to Constructive Realism: On Some Relations between Confirmation, Empirical Progress, and Truth Approximation. Synthese Library*, vol. 287. Dordrecht: Kluwer Academic Publishers.

Kuipers, T.A.F. (2001/SiS). *Structures in Science: Heuristic Patterns Based on Cognitive Structures. Synthese Library*, vol. 301. Dordrecht: Kluwer Academic Publishers.

Kyburg, H.E., Jr. (1968). *Philosophy of Science: A Formal Approach.* New York: The Macmillan Company.

Lakatos, I. (1976). *Proofs and Refutations: The Logic of Mathematical Discovery.* Cambridge: Cambridge University Press.

Manhart, K. (1994). Strukturalistische Theorienkonzeption in den Sozialwissenschaften. *Zeitschrift fiir Soziologie* **23**, 111-128.

McCune, W. (1994a). A Davis-Putnam Program and Its Application to Finite First-Order Model Search: Quasigroup Existence Problems. Technical report, Argonne IL: Argonne National Laboratory. DRAFT.

McCune, W. (1994b). OTTER: Reference manual and guide. Technical Report ANL-94/6, Argonne IL: Argonne National Laboratory.

Newell, A. and H. A. Simon (1956). The Logic Theory Machine. *IRE Transactions on Information Theory*, IT-2(3), 61-79.

Péli, G., J. Bruggeman, M. Masuch and B. Ó Nualláin (1994). A Logical Approach to Formalizing Organizational Ecology. *American Sociological Review* **59**, 571-593.

Polanyi, M. (1958). *Personal Knowledge: Towards a Post-Critical Philosophy.* Chicago IL: University of Chicago Press.

Pólya, G. (1945). *How to Solve It: A New Aspect of Mathematical Method.* Princeton, NJ: Princeton University Press.

Pólya, G. (1954). *Induction and Analogy in Mathematics. Mathematics and Plausible Reasoning*, vol I. Princeton New Jersey: Princeton University Press.

Popper, K.R. (1963). *Conjectures and Refutations: The Growth of Scientific Knowledge.* London: Routledge and Kegan.

Quine, W.V.O. (1953). *From a Logical Point of View.* Cambridge, MA: Harvard University Press.

Simon, H. A. (1952). A Formal Model of Interaction in Social Groups. *American Sociological Review* **17**, 202-211.

Sneed, J. D. (1971). *The Logical Structure of Mathematical Physics.* Dordrecht: D. Reidel Publishing Company.

Stegmüller, W. (1973). *Logische Analyse der Struktur ausgereifter physikalischer Theorien, 'Non-statement view' von Theorien. Probleme und Resultate der Wissenschaftstheorie und Analytischen Philosophie*: Band II *Theorie und Erfahrung*. Berlin: Teil D. Springer-Verlag.

Suppes, P. (1968). The Desirability of Formalization in Science. *Journal of Philosophy* **65**(20), 651-664.

Tarski, A. (1946). *Introduction to Logic and to the Methodology of Deductive Sciences.* Second revised edition. New York: Oxford University Press.

Woodger, J. H. (1937). *The Axiomatic Method in Biology.* Cambridge: Cambridge University Press.

Theo A. F. Kuipers

BACKGROUND KNOWLEDGE AND THE STRUCTURALIST APPROACH

REPLY TO JAAP KAMPS

In ICR and SiS I have emphasized the nature of the long-term dynamics of science. For example, in SiS (p. 38) I wrote:

> However, the [two-level] picture [of theoretical and observational terms] hides the long-term dynamics. When a proper theory is accepted as (approximately) true, it usually enables the establishment of criteria for the determination of its theoretical terms. In this way it becomes an observation theory, and the corresponding theoretical level transforms into a higher observational level, enabling new observations and hence the establishment of new observational laws, requiring new, 'deeper' theories to explain them.

In fact, this long-term dynamics leads to the growth (and occasional repair) of the "unproblematic" background knowledge. As far as scientists are aware of a specific increase in this respect, it concerns *explicit* background knowledge, which can be taken into account when questions of implication and hence falsification or confirmation are concerned. However, as Jaap Kamps argues first in general and then by way of a very nice "Tarski world" example, implicit background knowledge is something we have to excavate and computational means can be very helpful for that purpose.

I find his exposition very elegant and convincing, so I will only deal with some points raised in the last section where he presents his conclusions and discusses them. The first point deals with the question whether making true background knowledge explicit is a form of truth approximation. The second point concerns the advantages and disadvantages of the statement and the structuralist approach.

Truth Approximation by Adding True Background Knowledge

In the second paragraph of Section 5 and Note 9 Kamps discusses the effect of adding true background knowledge for a model as well as a statement or, more specifically, a consequence approach to truth approximation. From additional

In: R. Festa, A. Aliseda and J. Peijnenburg (eds.), *Cognitive Structures in Scientific Inquiry* (*Poznań Studies in the Philosophy of the Sciences and the Humanities*, vol. 84), pp. 338-342. Amsterdam/New York, NY: Rodopi, 2005.

correspondence it became clear that for the first he has primarily the basic definition of ICR in mind and for the second Popper's original definition. It turns out to be interesting to elaborate Kamps' main points about them in some detail.

According to the (basic) model definition, 'ψ is at least as close to the truth as ϕ' iff:

all correct models of ϕ are (correct) models of ψ

all incorrect models of ψ are (incorrect) models of ϕ

where a model is correct iff it is a model of the truth, that is, the strongest true theory about the intended domain within a given vocabulary.

Popper's consequence definition has a similar form, in brief:

all true consequences of ϕ are (true) consequences of ψ

all untrue consequences of ψ are (untrue) consequences of ϕ

It is not difficult to prove that, whereas the first consequence clause is equivalent to the second model clause, the second consequence clause is essentially stronger than the first model clause. For further details, among other things, about underlying intuitions, see ICR Section 8.1.

Now let us see what adding background knowledge amounts to, according to the two definitions. Adding β to a theory ϕ, results of course in a stronger theory, $\psi = \phi \& \beta$. Since all consequences of ϕ are consequences of $\phi \& \beta$, or equivalently, all models of $\phi \& \beta$ are models of ϕ, we get, in line with the abovementioned equivalence, that the second model clause as well as the first consequence clause are automatically satisfied.

If β is true, $\phi \& \beta$ will drop only incorrect models of ϕ, hence the first model clause is satisfied, hence $\phi \& \beta$ is at least as close to the truth as ϕ, and, we may add, as a rule, closer to the truth, according to a plausible extra condition. On the other hand, as Kamps rightly hints upon in Note 9, if β is true, $\phi \& \beta$ may not only have extra true consequences, but also extra untrue ones. Hence, the second consequence clause is not guaranteed. Therefore, $\phi \& \beta$ need not be as close to the truth as ϕ, let alone closer to the truth. Of course, if ϕ is also true, $\phi \& \beta$ has only true extra consequences compared to ϕ (and β). In this case Popper's definition even guarantees truth approximation, and the model definition does too.

Kamps presents the diverging conclusions, evidently assuming as a condition of adequacy for a definition of 'closer to the truth' that adding true background knowledge should always leave us as close to the truth and, as a rule, bring us closer to it. In other words, excavating background knowledge should, if true, be functional for truth approximation.

More generally, I would like to submit as a general condition of adequacy for a 'content definition' (to use the apt expression of Zwart (1988/2001) for the type of definitions we are discussing now) of 'closer to the truth' that adding some true statement (or its model equivalent; dropping incorrect models) should be functional for truth approximation in the indicated sense. From the above it follows that the model definition satisfies this general condition, whereas Popper's definition fails to do so.

I am particularly eager to point to this condition for the following reason. The famous impossibility theorem against Popper's definition, independently proved by Miller and Tichý, typically assumes that "the truth" is complete, due to having one intended model. In that case, a false theory cannot be closer to the truth than another theory according to Popper's definition (see ICR Section 8.1 for a detailed reconstruction). In view of my general belief that the truths that one looks for in theoretically oriented empirical sciences are incomplete , one might say that the impossibility theorem should not be that impressive, for it applies only in an extreme, atypical case. However, the general condition of adequacy proposed above for content definitions, in the line of Kamps' discussion of background knowledge, provides an argument against Popper's definition and in favor of the model definition that also applies to paradigmatic cases of theory improvement: adding true statements about the domain of interest should never be counterproductive for truth approximation but, as a rule, productive.

The important question remains whether the refined definition of truth approximation presented in ICR (Ch. 10, p. 250), being a likeness definition in the sense of Zwart, satisfies the general condition of adequacy. Since the refined second model clause is a weakening of the corresponding basic one, it is again automatically satisfied for any added statement. However, the refined first model clause is a strengthening of the corresponding basic one. For this reason, the refined one need not always be satisfied when a true statement is added. More specifically, adding a true statement β to ϕ (leading to $\phi\,\&\beta$) does not exclude the possibility that all incorrect models of ϕ get lost that could be the "intermediate" model, required by the refined first clause, between a given incorrect model of ϕ and a given intended model, both being no model of β. It is a question for further research whether this should be seen as a really problematic aspect of the refined definition, as Kamps probably thinks, or whether the discussed condition of adequacy is too ambitious or even undesirable for likeness theories in general.

The Logical Versus the Structuralist Approach

Regarding the advantages of the structuralist approach, I have certainly overstated my claim in one respect, viz. "that the structuralist analysis of theories can be used almost directly for the computational representation of theories" (SiS, p. 302). It is rightly criticized by Kamps in Note 13, though perhaps for the wrong reason. It is not so much that you need a "pencil, paper, and a philosophy professor," but that for computational purposes you need some kind of syntactically tractable transcription of all the relevant set-theoretical aspects. My suggestion that this is already (almost) always possible is far from the truth. However, in many cases such a transcription is indeed possible, as in the case of many sociological theories, but also in Kamps' nice example of Tarski's world. See also Balzer and Moulines (2000, p. 9), for general optimism with respect to implementing set-theoretic representations in AI.

Apart from the computational claim then, I would insist on the general claim that for many purposes the structuralist representation is less complicated than a logical one. To begin with, above we have seen that it is possible to give a logical definition of 'closer to the truth' – in fact three equivalent versions are possible: the given version purely in terms of models, a complicated version in terms of consequences, and a dual version, combining the first clauses of the two definitions discussed above. However, the core of all three versions can easily be reproduced completely in set-theoretic terms (ICR, pp. 184-6), viz. by replacing models by structures of a certain type and conceiving theories as sets of structures and, when desired, their (set of) consequences as (the set of) supersets of these sets. As a matter of fact, I invented that definition by starting, in 1982, to think in the latter way (ICR, pp. 150-3), more specifically, the one purely in terms of theories as sets of structures (corresponding to the purely model formulation), and hence first avoiding all complications in surveying and comparing the sets of true and false consequences of theories.

Another nice example of structuralist representation is suggested by Kamps' own illustration. If Tarski's world were not designed for didactic logical purposes, but to illustrate the nature of empirical theories, assuming that some serious empirical law would be involved, see below, the set-theoretic representation would be superior, not in principle, but in (non-computational) practice, for a couple of reasons. As a matter of fact, it is an elementary exercise to give the set-theoretic representation (see SiS, Ch. 12) of what then would plausibly be called "Tarski Worlds," for example Kamps' counterexamples I to IV. A Tarski World is a set-theoretic structure of a certain type, e.g. with one base set (domain) and a number of unary, binary and ternary relations, satisfying a number of analytic or semantic axioms and a number of synthetic or substantial axioms. For example, let *D* indicate a domain of objects and *C* the subset of cubes, *T* the subset of tetrahedrons, *S* (*L*) the subset of small (large) objects. Analytic (background) axiom B3 amounts to $S \cap L = \varnothing$ and synthetic axiom A1 amounts to $D - C \cup (T \cap S) = \varnothing$. Of course, these clauses can be transcribed in first-order claims, as Kamps has done, but for many purposes the former are just simpler than the latter. Unfortunately, A1 is not at all like an empirical law, but within the present boundaries one might think of a condition to the effect that a cube cannot be positioned on top of a tetrahedron. Although this is conceptually possible, we may assume that it would fall to the ground. It would be instructive to transform the example into a more serious example of similar structures of a physical theory. To be sure, for Kamps' computational purposes, some syntactic redescription is required. For that purpose one should first look for a first-order redescription, for if that is possible, as in Kamps' case, programs like OTTER and MACE can be used.

Let me close by referring to the contributions of Zwart, Van Benthem, Burger and Heidema, and my replies in the companion volume. Among other things, the comparison of the logical versus the structuralist approach is discussed as well as the desirability of an "alternative model theory" that is more suitable for the structuralist approach.

REFERENCES

Balzer, W. and U.C. Moulines. (2000). Introduction. In: W. Balzer, J. Sneed, C. U. Moulines (eds.), *Structuralist Knowledge Representation*, pp. 5-17. Amsterdam/Atlanta: Rodopi.

Zwart, S. (1998/2001). *Approach to The Truth. Verisimilitude and Truthlikeness*. Dissertation: Groningen. Amsterdam: ILLC-Dissertation-Series-1998-02. Revised version: *Refined Verisimilitude*, *Synthese Library*, vol. 307. Dordrecht: Kluwer Academic Publishers.

Alexander P.M. van den Bosch

STRUCTURES IN NEUROPHARMACOLOGY

ABSTRACT. This paper explores structuralism as a way to model theories from scientific practice. As a case study I analyzed a theory about the dynamics of the basal ganglia, a part of the brain that is involved in Parkinson's disease. After introducing the case study I explore how to structurally represent qualitative assumptions about disease, intervention and dynamical systems in general. I further explicate the structure of the basal ganglia theory in detail, how it explains Parkinson's disease and how it implies treatments. I close with a consideration of how a structuralist representation could be useful in practice to explore and develop theories with the aid of a computer.

1. Introduction

As a case study of the application of structuralism – as put forward in Kuipers (2000, 2001) – to scientific practice, I analyzed an example practice in neuropharmacology (Van den Bosch 2001a, 2001b). This paper presents a structuralist analysis of a theory in the research for drugs to treat Parkinson's disease. The question I address in this paper is, how can one understand the structure of the dopamine theory of Parkinson's disease? In answering this question I explicate how this theory explains the effect of known treatments for Parkinson's disease.

The next section presents a short introduction to neuropharmacology and the dopamine theory of Parkinson's disease. To analyze the structure of the dopamine theory, in section 3 I discuss the structuralist approach to represent theories in general, and how it can represent theories about dynamical systems in particular. Then, in section 4, I formally represent a theory about the basal ganglia – a brain cell group studied in neuropharmacology – as a qualitative equation that imposes conditions on possible models of the phenomena that the theory explains, and demonstrate how the theory implies and predicts treatments for Parkinson's disease. I argue that the structuralist approach is not only instrumental in explicating the structure of a scientific theory, but may also be able to aid scientific practise by using a computer program that can effectively infer predictions from a theory, based on its structuralist representation. I end the paper with a brief conclusion in section 5.

In: R. Festa, A. Aliseda and J. Peijnenburg (eds.), *Cognitive Structures in Scientific Inquiry* (*Poznań Studies in the Philosophy of the Sciences and the Humanities*, vol. 84), pp. 343-359. Amsterdam/New York, NY: Rodopi, 2005.

2. Neuropharmacology

In this section I describe the field of neuropharmacology in general and drug research related to Parkinson's disease in particular. One aim of drug research in neuropharmacology is to find a way to intervene in neurophysiological and neurochemical processes such that pathological properties or symptoms are suppressed, or desired properties are induced (Vos 1991). Those unwanted properties are determined and discovered in numerous ways. The history of pharmacology and medicine is rich with serendipitous cases where a patient with a particular disease comes into contact with a compound that enhances his condition, hence providing a clue about the disease mechanism. A systematic study involves comparison of properties of pathological processes of patients with those of control subjects. In some cases, such as in Parkinson's disease, a cause of disease symptoms can be traced back to different concentrations of a single neurotransmitter compound.

Neural disorders have their origin in shifts in delicate balances of neurochemicals, which can be caused by *e.g.* cell damage or degeneration. The plasticity of the brain is sufficient to restore imbalances, *e.g.* by increasing the sensitivity to a particular neurotransmitter. But when it fails, *e.g.* when a substance is depleted almost completely as in the case of *dopamine* in Parkinson's disease, a severe neurological disorder results.

Fundamental research in neuropharmacology investigates the processes of the brain and how drugs interact with those processes. One research tool employed is building models of neurochemical and neurophysiological processes that aim to fit data acquired by laboratory studies on animal models. In the Pharmacy Department of Groningen University this is done by employing electrophysiological methods and microdialysis to track nerve signals.

A nerve propagates a signal by conducting an electric pulse called an action potential. This signal initiates the release of transmitter chemicals at the terminals of the cell that affect receptors of nearby nerve cells that may further propagate a signal. The placement of an electrode in the brain can aloow one to monitor the electrical activity. The release of transmitters can be measured by means of a microdialysis probe. This probe can also be used to release chemicals locally and measure the effect *in vivo*. At the Pharmacy Department of Groningen University the function of neurophysiological pathways is studied using these two techniques.

Specific studies of the functional relation between several variables together contribute to an understanding the function of a brain area, or cell groups called *nuclei*. To describe these neural circuits, box and arrow models are drawn showing positive and negative influence relations (Timmerman 1992). These models are further tested for their correctness and used to explain and predict the functioning of the system. Newly developed drug compounds play a bootstrap role in this research: they are used to revise and refine the model and the experiments conducted, while the model is in turn used to understand their effect. A drug that works very selectively for one particular type of pathway can be used to further explore the function of that pathway. The data acquired may then serve to refine the model, so that the effects of the new drug can be explained and predicted.

A group of subcortical nuclei called the basal ganglia are being studied in Groningen (Timmerman *et al.* 1998). These nuclei play an important role in the control of voluntary behavior. In the case of Parkinson's disease a part of them, called the *substantia nigra pars compacta* (*SNC*), decays due to an unknown cause. The *SNC* is a supplier of an important neurotransmitter called *dopamine*, which is postulated to serve a modulating function. It is thought to maintain a delicate balance in influencing signals from the *cortex*. To understand this balance a schematic model is used to represent neural activity in the basal ganglia in Parkinson's disease, see Figure 1, which is a schematic representation of neural activity in the basal ganglia in Parkinson's disease, as postulated in studies by Timmerman (1992, p. 18). An arrow in the diagram is a neural pathway, consisting of a bundle of individual nerve cells. A box is a nucleus, or clustering of nerve cells. Increased inhibition induced by receptors sensitive to the transmitter *GABA* of the external segment of the *globus pallidus* (*GPe*) leads *e.g.* to disinhibition of the *subthalamic nucleus* (*STN*). In turn, this provides increased excitatory drive to the internal segment of the globus pallidus (*GPi*) and *substantia nigra reticulata* (*SNR*), therefore leading to increased thalamic inhibition. This is reinforced by reduced inhibitory input to the *SNR/GPi*. These effects are postulated to result in a strong inhibition of *brainstem* neurons. *D*1 and *D*2 are two different types of receptors, postulated to react excitatory and inhibitory, respectively, to *dopamine* (*DA*).

The model ascribes a dual function to *dopamine*. It enforces the direct path from the *striatum* to the *SNR/GPi* while it inhibits the indirect path, via the *GPe* and *STN*. This balance maintains an inhibition of both the *brainstem* and the *thalamus*. Yet when *dopamine* is nearly depleted, the balance becomes disrupted, resulting in a strong increase of the activation of an area called the *SNR/Gpi* (see Figure 1). This hyper-activation causes strong inhibition of *brainstem* neurons and is correlated with some of the major symptoms of Parkinson's disease.

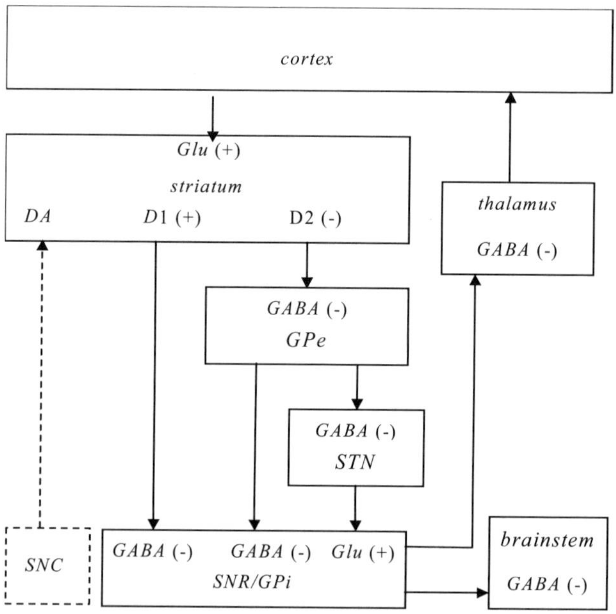

Fig. 1. Diagram model of the basal ganglia

Most of the traditional research on Parkinson's disease is focused on restoring levels of *dopamine*. This compound cannot be administered as an oral drug because it does not pass the so-called blood-brain barrier. Yet it was discovered that *L-dopa*, which metabolizes in the brain to *dopamine*, can pass this barrier. Administering regular doses of *L-dopa* is currently the most successful therapy for dealing with Parkinson symptoms.

Administering *L-dopa* also causes *dopamine* levels in other parts of the body to increase. However, this higher concentration of *dopamine* in the blood causes nausea as a side effect due to stimulation of *dopamine* receptors elsewhere in the body. And after three to five years of use the therapeutic effect declines drastically. Further research is investigating the use of highly selective *dopamine* receptor agonists, compounds that interact only with particular *dopamine* receptors. The *dopamine* receptors on the direct route from the *striatum* to the *SNR/GPi* were discovered to be mainly of another type (*D*1) than that on the indirect route (*D*2) via the *GPe*. Both receptors can be stimulated by *dopamine*, but with different effects. *D*1 receptor stimulation with *dopa-*

mine has an exciting effect on a cell, while stimulation of the *D*2 receptor with *dopamine* inhibits the cell. Clinical studies are being conducted to investigate the therapeutic effects of using different compounds that differ in selectivity to both the *D*1 and *D*2 receptors. These studies show that the use of only a selective *D*1 agonist, a compound that stimulates *D*1 but not *D*2 receptors, is not successful.

The model in Figure 1 is used to understand the effect of selective compounds. However, in the literature opinions about these kinds of models vary rather widely. Some people use them extensively to understand and theorize about physiological phenomena, while others are wary of using them because they are too simple, do not respect the subtlety of the data, and are therefore not realistic. An article in the movement disorder literature states:

> On the one hand, efficient models have to be simple, but simple models can provide only part of the reality and are thus bound to be wrong (for example, current basal ganglia model) ... On the other hand, an elaborated model that would embody all the complexities of a given reality [...] is doomed to be useless. (Parent and Cicchetti 1998)

The practical problem of the diagram model is that it is informally represented. Its consequences are infered by tracking the boxes and arrows. The general basal ganglia model is already fairly elaborate. A more realistic picture would have to be substantially larger, including more transmitters, peptides, small interactions and feedback loops. Including these would cloud the bird's eye view, drowning it in the complexity of all the consequences of the model.

The following section describes in general terms a part of the reasoning involved with such models, introducing the use of qualitative equations to represent them. These allow for systematic and computational exploration of their consequences and have the potential to aid in both the understanding and the testing of the models, but also to explore them for suggestions that might lead to new drugs.

3. Structures

The first question I address in this paper is, how can we understand the structure of the *DA* theory of Parkinson's disease? And secondly, how does it explain the effect of known treatments? In this section I introduce a structuralist analyses of theories.

The structuralist approach in the philosophy of science characterizes a theory according to its models, conceived as structures (Kuipers 2000, 2001). A structure, in this context, is usually represented as an ordered set of variables, functions and constants. A structure is called a model of a theory if the theory, seen as a proposition about that structure, is true.

The core of a theory consists of a set of models M which is a subset of all conceptually possible models M_P given the vocabulary of the theory. M_P minus M is the set of models that the theory excludes and is called the empirical content of a theory. It contains all the potential falsifiers of the theory. Given a domain D of application of the theory it is assumed that there is a subset of M_P that contains the empirically possible models of that domain. A weak empirical claim states that all empirically possible models are models of the theory. A strong claim also asserts that they are equal.

For the purpose of this exposition I will characterize a theory in terms of its vocabulary of variables V, the quantity spaces Q of those variables (a quantity space of a variable defines the range and type of values of a variable), and conditions C on the values of those variables. These conditions C determine the set of models of the theory as a subset of all possible coceptual models based on V and Q.

I further draw a distinction between a theory T, which is basically a set of definitions about relations between variables in V, and a hypothesis H, which is a statement that asserts that the properties of phenomena in a domain D can be characterized by the vocabulary V and by the models of theory T.

Definition 1. *Theory*. The ordered set $\langle V, Q, C\rangle$ containing variables V, quantity spaces Q and conditions C, represents a theory. The theory determines an ordered set $\langle M_P, M_T\rangle$ that contains the conceptually possible models M_P, given V and Q, and the models of the theory M_T, given the conditions C on V.

Definition 2. *Hypothesis*. The ordered set $\langle V, Q, C, D\rangle$ represents a hypothesis where a theory is applied to a domain D. The hypothesis determines the ordered set $\langle M_P, M_T, M_E\rangle$ that contains the conceptually possible models M_P of a domain D given possible descriptions by variables V and quantity spaces Q; the models M_T of the theory of the domain given conditions C on variables V; and the empirically possible models M_E of the phenomena of domain D. The hypothesis asserts that the set of empirically possible models M_E is a subset of, or equal to, the set of models M_T of the theory.

A model of a phenomenon in a domain is a structure that represents certain aspects of that phenomenon in terms of a set of interpreted variables with particular quantities. The structures that are possible according to the conditions C from a theory are called the models M_T of that theory. The conceptually possible models M_P are the set of all the models that are possible if you combine all possible variables from V with all their possible quantities from Q.

The relation between the conceptually possible models M_P, the models of the domain M_E and the models M_T of a theory in a hypothesis can be graphically represented as in Figure 2.

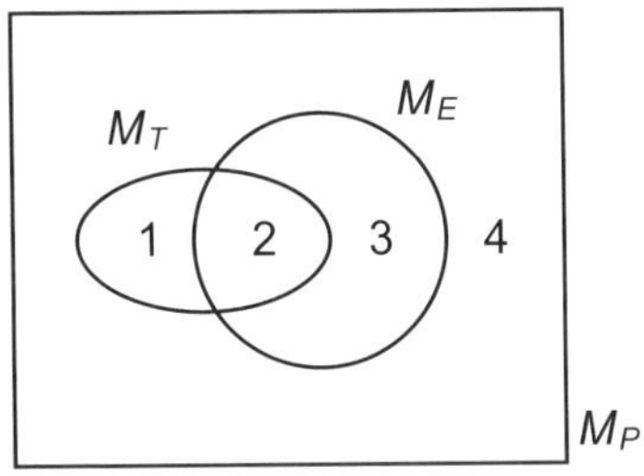

Fig. 2.: Models M_T of a hypothesis and empirically possible models M_E of the phenomena of a domain, both part of the conceptually possible models M_P

The different intersections represent subsets of structures that constitute either a success, an anomaly, or a problem for the theory. The goal of explanation is to find a hypothesis such that a better hypothesis has fewer problems (subset 1) or anomalies (subset 3) than a competitor (cf. Kuipers 2000, p. 150).

Subset	M_T	M_E	
1	1	0	Explanatory problem
2	1	1	Empirical success, confirming instance
3	0	1	Empirical anomaly, counter example
4	0	0	Explanatory success

Table 1. Subsets of conceptually possible models M_P of a domain

To understand the theory of Parkinson's disease we can understand it as a hypothesis about the dynamical behavior of the brain. The theory asserts what kind of states and behaviors are possible. The sets V and Q describe the known structural properties of the brain, and the conditions in C describe the assumed functional relations between those properties. *A* variable x of a structure is related to variable y if there is a functional condition in C, such that $y=f(x)$.

Disease and Intervention

To understand the research problems in pharmacology we need to extend our vocabulary. Pharmaceutical research is not only interested in how to explain observations of a pathological biological system. It also aims to know how to treat it, and why a treatment works. For this we can introduce two extra subsets of M_P, the models of a biological system that is influenced by a (drug) intervention, M_I, and the models of phenomena that we wish to cause, the set M_W, see Figure 3.

Given a set of conceptually possible models of the behavior of a biological system a set of drug interventions can be assumed to cause behaviors represented by the set M_I, while the set M_W represents the set of desired behaviors.

Let M_E represent the empirically possible behaviors of a living organism with a given biological structure. Hence if the assumptions are correct M_I should be a subset of M_E.

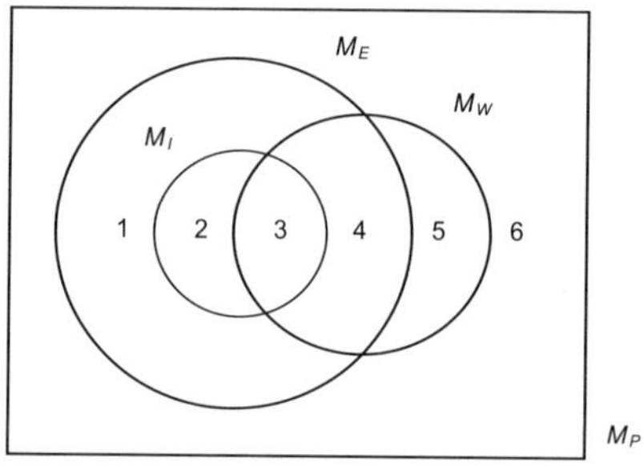

Fig. 3. Empirically possible models M_E of a biological system, wished for models M_W, and models M_I of a system that is influenced by an intervention, all part of conceptually possible models M_P of a biological system

In Figure 3 subset 1 denotes undesired behavior that is not treated by known interventions. Subset 2 contains unsuccessfully treated system behavior and unwanted side effects of a partially successful drug treatment, while subset 3 denotes behavior that is successfully treated. Subset 4 may correspond to health, given that M_W denotes health. Subset 5 can contain a behavior that is not possible given the biological structure of the organism, but can still be desired. Subset 6 equals the periphery of both possibility and interest.

Subset	M_E	M_I	M_W	
1	1	0	0	Disease, untreated by known interventions
2	1	1	0	Disease, treated with side effects
3	1	1	1	Successfully treated
4	1	0	1	Health
5	0	0	1	Desired, but not empirically possible
6	0	0	0	Periphery of interest and possibility

Table 2. Subsets of conceptually possible models M_P of a biological system

These three sets define the main goals of neuropharmacology. It is a goal to describe and explain M_E, what kinds of values of variables describing the brain and behavior of the organism are empirically possible, and why. It is also a goal to determine what states and behaviors M_W constitute health, or are desired for other reasons. And finally, what kind of drug or other medical interventions cause those desired behaviors M_I.

Dynamical Systems

In neurobiology the function of the brain is described and explained as a complex dynamical system. In physics, the tool for modeling a dynamical system is the use of differential equations. Variables represent properties of the system, the values of which can change over time. By defining the specific relations between those variables, those values can be predicted, given an initial state of the system.

Empirical studies of both the brain and behavior in Parkinson research result in many quantitative data, correlating variables of the activation frequency of nuclei and neural pathways and local concentrations of different kinds of neurotransmitters. Yet those relations are not sufficiently known to define them as a quantitative equation. The relation is only known qualitatively. Many results of empirical studies of the brain amount to conclusions, such as, if the value of this variable changes in this direction, the change of the value of that variable in that direction is statistically significant. In this way the theory that explains Parkinson's disease can explain why the activation of the *thalamus* decreases, when the concentration of *DA* in the *striatum* significantly decreases.

While these results are insufficient to define a model with the aid of an ordinary differential equation, they can be represented by a more abstract qualitative equation, *cf.* B. Kuipers (1994). In a qualitative equation the possible values of the variables in V are constrained by formulas in C. Conditions in C can consist of conditions corresponding to additions, multiplications, negations, derivatives, and incompletely known functions specified only as being part of a monotonicity class.

The last category is relevant for our case. We can know about a function f between two variables $v_1(t)$ and $v_2(t)$, $v_1(t) = f(v_2(t))$, that f belongs to either M^+, the class of monotonically increasing functions, or M^-, the class of monotonically decreasing functions. That is, for every $f \in M^+, f' > 0$, and for every $f \in M^-, f' < 0$ over the domain of the function. These classes can be generalized to multivariate functions so that *e.g.* M^{+-} is the class of functions $v_1(t) = f(v_2(t), v_3(t))$, such that $\partial f / \partial v_2 > 0$ and $\partial f / \partial v_3 < 0$.

The conditions C in a qualitative equation define which qualitative states and behaviors are possible. So C amounts to a theory about a system. We can define the qualitative state of a system at a given point in time, or on an interval between two give points in time.

Definition 3. *Qualitative state.* The qualitative state (QS) of a system described by variables V at point in time t_i is an ordered set of individual qualitative values (QV) at a certain point in time, or time interval from t_i, to t_{i+1}:

$$\mathrm{QS}(V, t_i) = \langle \mathrm{QV}(v_1, t_i), \dots, \mathrm{QV}(v_m, t_i) \rangle$$

$$\mathrm{QS}(V, t_{\mathrm{i}}, t_{\mathrm{i+1}}) = \langle \mathrm{QV}(v_1, t_{\mathrm{i}}, t_{\mathrm{i+1}}), \ldots, \mathrm{QV}(v_{\mathrm{m}}, t_{\mathrm{i}}, t_{\mathrm{i+1}})\rangle$$

The qualitative behavior of a system can now be defined as an ordered set of qualitative states:

Definition 4. *Qualitative behavior.* The qualitative behavior of a system with variables V on time interval $[t_0 < \ldots < t_{\mathrm{n}}]$ is a sequence of qualitative states:

$$\mathrm{QB}(V) = \langle \mathrm{QS}(V, t_0), \mathrm{QS}(V, t_0, t_1), \mathrm{QS}(V, t_1), \ldots, \mathrm{QS}(V, t_{\mathrm{n}})\rangle$$

The possible states and behaviors of a system can be seen as models of the qualitative equation. Benjamin Kuipers developed a computer program called QSIM that can generate such models (B. Kuipers 1994). It takes as input a qualitative equation and an initial qualitative state description and produces a tree of possible state sequences. This can be seen as:

$$\mathrm{QSIM}(\langle V,Q,C\rangle, \mathrm{QS}(t_0)) = M$$

such that M is an ordered set $\langle S, B\rangle$, where S is a set of all possible qualitative states and B is a set of all possible qualitative behaviors, *i.e.* totally ordered sets of qualitative states consistent with C, *cf.* Schults and B. Kuipers (1997).

In the next section I use the qualitative equation representation to explicate the structure of the dopamine theory of Parkinson's disease, and how it explains the function of known treatments.

4. Structures in Neuropharmacology

Neurobiologists study the processes of the brain, *e.g.* by recording values of activation frequencies and concentrations of neurotransmitters in different locations of the brains of guinea pigs, Wistar rats, or monkeys. When the values of two variables v_1 and v_2 are consistent with a monotonic function in all trials of an experiment, a correlation could be proposed.

This is a simple style of descriptive induction: the variables are monotonically related in the sample, so they are monotonically related in all brains, of the sample organism or even in the human brain. It becomes an explanation if a hypothesis is formed about what processes underlie the variables acting in that way.

In Parkinson research it is observed that the increase of symptoms is correlated with a substantial decrease of the availability of the neurotransmitter *DA*, which is due to a decay of the *substantia nigra pars compacta* (*SNC*). The model of the basal ganglia aims to explain why the decrease of *DA* can lead to these symptoms, by explaining why the activation of the *SNR* increases as a result of this decrease.

I shall now reconstruct this explanation by first representing the theory of the basal ganglia with the aid of qualitative equations. These equations serve as a hypothesis from which it can be deduced that, given a decrease of *DA*, an increase of the *SNR* activation is a consequence. I also show how the activity of known treatments can be explained and how such explicit models can be used to infer possible new interventions.

Theory of the Basal Ganglia

The basal ganglia theory is a qualitative theory about a system, so we can represent it as a qualitative equation. In the basal ganglia theory there are two basic variables describing firing rate (f) of nerve cells in a cell group, nuclei or pathway, and the amount (a) of a particular neurotransmitter released in the vicinity of a cell group, nuclei or neural pathway. The qualitative equation $y = M^+(x)$ abbreviates $y = f(x)$ and $f \in M^+$ and is used to state that the change of values of y over time is monotonically related to the change of value of x. It is a matter of debate whether this relation represents a causal direction from x to y, for discussion see Iwasaki and Simon (1994).

I represent the model of the basal ganglia as depicted in Figure 1, which was used by Timmerman (1992). While this model could be further extended to include other influences, such as those of the compounds *substance P* and *encephalin*, the simpler model suffices for my analysis of the observed practice. The notation *x*-to-*y* in the cell groups denotes the neural pathway from cell group *x* to cell group *y*. I further abbreviate *SNR/Gpi* to *SNR*, since it is functionally the same. So we can define the basal ganglia theory as follows:

Definition 5. *Basal ganglia theory.* $T_{BG} : \langle V, Q, C \rangle$ is an ordered set such that:

1. **Variables in V**

   - Cell groups *G*, containing nuclei and neural pathways

     G: {*striatum*, *GPe*, *STN*, *SNR*, *thalamus*, *brainstem*, *cortex*-to-*striatum*, *SNC*-to-*striatum*, *striatum-D*1-to-*SNR*, *striatum-D*2-to-*GPe*, *GPe*-to-*SNR*, *GPe*-to-*STN*, *STN*-to-*SNR*, *SNR*-to-*thalamus*, *SNR*-to-*brainstem*}

   - Set of neurotransmitters *N*: {*Glu*, *DA*, *GABA*}

   - The firing rate $f(g)$ of cell group g is a value of quantity space F
     $f: G \rightarrow F$

   - Amount $a(n, g)$ of neurotransmitter n in cell group g is a value of A
     $a: N \times G \rightarrow A$

2. Quantity spaces in *Q*

- Boundaries of firing rates *F*: {0, MAX}
- Boundaries of amounts *A*: {0, MAX}

3. Conditions in *C* on:

- Firing rates of nuclei in the basal ganglia

c.1 $f(striatum) = M^{+}\ (a(Glu, striatum))$
c.2 $f(GPe) = M^{-}\ (a(GABA, GPe))$
c.3 $f(STN) = M^{-}\ (a(GABA, STN))$
c.4 $f(SNR) = M^{-+}(a(GABA, SNR), a(Glu, SNR))$
c.5 $f(thalamus) = M^{-}\ (a(GABA, thalamus))$
c.6 $f(brainstem) = M^{-}\ (a(GABA, brainstem))$

- Firing rates of neural pathways between nuclei

c.7 $f(cortex\text{-to-}striatum) = M^{+}\ (f(cortex))$
c.8 $f(SNC\text{-to-}striatum) = M^{+}\ (f(SNC))$
c.9 $f(striatum\text{-}D1\text{-to-}SNR/GPi) = M^{++}(f(striatum), a(DA, striatum))$
c.10 $f(striatum\text{-}D2\text{-to-}GPe) = M^{+-}\ (f(striatum), a(DA, striatum))$
c.11 $f(GPe\text{-to-}SNR) = M^{+}\ (f(GPe))$
c.12 $f(GPe\text{-to-}STN) = M^{+}\ (f(GPe))$
c.13 $f(STN\text{-to-}SNR) = M^{+}\ (f(STN))$
c.14 $f(SNR\text{-to-}thalamus) = M^{+}\ (f(SNR))$
c.15 $f(SNR\text{-to-}brainstem) = M^{+}\ (f(SNR))$

- Amounts of released neurotransmitters in nuclei

c.16 $a(DA, striatum) = M^{+}\ (f(SNC\text{-to-}striatum))$
c.17 $a(Glu, striatum) = M^{+}\ (f(cortex\text{-to-}striatum))$
c.18 $a(GABA, GPe) = M^{+}\ (f(striatum\text{-}D2\text{-to-}GPe))$
c.19 $a(GABA, STN) = M^{+}\ (f(GPE\text{-to-}STN))$
c.20 $a(GABA, SNR) = M^{++}(f(striatum\text{-}D1\text{-to-}SNR), f(GPe\text{-to-}SNR))$
c.21 $a(Glu, SNR) = M^{+}\ (f(STN\text{-to-}SNR))$
c.22 $a(GABA, thalamus) = M^{+}\ (f(SNR\text{-to-}thalamus))$
c.23 $a(GABA, brainstem) = M^{+}\ (f(SNR\text{-to-}brainstem))$

- Metabolism of *dopamine*

c.24 $a(DA, \mathrm{x}) = a(L\text{-}dopa\ , x) \times \text{Enzyme-ratio}$
c.25 $\text{Enzyme-ratio} = a(AADC, x)\ /\ a(MAO\text{-}B, x)$

I have included assumptions about the metabolism of *dopamine* as part of the theory of the basal ganglia. The availability of *dopamine* outside the dopaminergic cell terminal dependents on the activation of the cell by the neural pathway from the *SNC*, see *c*.24 where location *x* is the *SNC*. But *DA* can only be released by the vesicles of the terminal if the precursor *L-dopa* and the enzyme *AADC* is available. The enzyme *MAO-B* breaks down the excess of *dopamine* to *DOPAC,* see *c*.25.

Explanation of Parkinson's Disease

The theory of the basal ganglia can be applied to explain observations in Parkinson's disease research. The hypothesis of the basal ganglia states that the empirically possible states *E* of the basal ganglia, given the empirical study of the basal ganglia *D*, are part of the theoretically possible states *M*.

Definition 6. *Basal ganglia hypothesis.* $H_{BG} : \langle V, Q, C, D\rangle$ represents a hypothesis about the basal ganglia brain structure where *V*, *Q*, *C* are part of the T_{BG} and *D* is the set of instances of the basal ganglia, the domain of application of the theory.

We saw that the symptoms of Parkinson's disease are assumed to be caused by an increase of activation of the *SNR*, which in turn is explained by a steep decrease of *DA* in the *striatum* due to the decay of dopaminergic nerve cells from the *SNC*. One question in this chain, how the observed decrease of *DA* causes the assumed increase of *SNR* activation, is explained by the theory about the basal ganglia. This proposition can be deduced from the basal ganglia theory by programs like QSIM (B. Kuipers 1994).

In the following example proof I reduce the values of the variables to just their qualitative direction, abstracting from time and qualitative magnitude. From $y = f(x)$ where $f \in M^+$ we know that *x* and *y* both increase or decrease together, while if $f \in M^-$, *y* increases when *x* decreases, and vice versa. If $z = f(x, y)$ and $f \in M^{++}$, the direction of change of *z* is unknown if *x* increases and *y* decreases, since we do not know their magnitude, *cf.* Table 3. This is similar for $f \in M^{+-}$, when both variables increase or decrease in value

$y \setminus x$	inc	std	dec
inc	inc	inc	?
std	inc	std	dec
dec	?	dec	dec

Table 3. Derivative values for *z* if $z = f(x,y)$ and $f \in M^{++}$

As background assumptions we assume that the amount of *dopamine* in the *striatum* decreases and the firing rate of the *striatum* is steady. I use the notation $v = qdir$ as shorthand for $\text{QV}(v, t) = \langle y, qdir\rangle$, abstracting from time and qualitative value.

Theorem 1. $H_{BG} \cup B$: $\{a(DA, striatum) = \text{dec}, f(striatum) = \text{std}\} \vdash$
P: $\{f(SNR) = \text{inc}\}$

Proof: As a proof I deduce the conclusion P from the premises B by applying the conditions C from the basal ganglia hypothesis H_{BG}.

$a(DA, striatum)$ = dec $\wedge$ $f(striatum)$ = std
$\Rightarrow f(striatum\text{-}D1\text{-to-}SNR)$ = dec $\wedge$ $f(striatum\text{-}D2\text{-to-}GPe)$ = inc (c.9, c.10)
    $f(striatum\text{-}D2\text{-to-}GPe)$ = inc
    $\Rightarrow a(GABA, GPe)$ = inc (c.18)
    $\Rightarrow f(GPe)$ = dec (c.2)
    $\Rightarrow f(GPe\text{-to-}SNR)$ = dec $\wedge$ $f(GPe\text{-to-}STN)$ = dec (c.11, c.12)
        $f(GPe\text{-to-}STN)$ = dec
        $\Rightarrow a(GABA, STN)$ = dec (c.19)
        $\Rightarrow f(STN)$ = inc (c.3)
        $\Rightarrow f(STN\text{-to-}SNR)$ = inc (c.13)
        $\Rightarrow a(Glu, SNR)$ = inc (c.21)
        $f(GPe\text{-to-}SNR)$ = dec $\wedge$ $f(striatum\text{-}D1\text{-to-}SNR)$ = dec
        $\Rightarrow a(GABA, SNR)$ = dec (c.20)
$a(Glu, SNR)$ = inc $\wedge$ $a(GABA, SNR)$ = dec
$\Rightarrow f(SNR)$ = inc (c.4) (Q.E.D)

Deducing Treatments

I now first introduce a new set in my terminology. Besides to a hypothesis H, background assumptions B, and propositions P that are explained or need to be explained, we also have a set of interventions I. This set contains propositions that describe a property of the world, usually a value of a particular variable, that can be set by a manipulation. All consequences of that manipulation hold for all the structures in the set M_I.

A theory can explain why a particular intervention has a particular consequence. With H_{BG} we have a hypothesis that explains the symptoms of Parkinson's disease by linking them to the observed decrease of *DA*. The hypothesis also explains the function of metabolites like *L-dopa*, *MAO-B* and *AADC*. These metabolites can serve as an artificial intervention by changing their concentration with the aid of a drug. Parkinson drugs all serve to increase the amount of *dopamine* which, according to the theory, would decrease the activation of the *SNR*, reducing the behavioral symptoms. In the theorems below I demonstrate how the basal ganglia hypothesis explains the activity of known drug interventions for Parkinson's disease. All these drugs aim to influence the amount of *dopamine*, so I first pose the following theorem:

Theorem 2. $H_{BG} \cup B$: $\{f(striatum) = \text{std}\} \vdash$
P: $\{a(DA, striatum) = \text{inc} \rightarrow f(SNR) = \text{dec}\}$

From H_{BG} it can be deduced according to Theorem 2 that an increase of *DA* implies a decrease of the firing rate of the *SNR* output nuclei of the basal ganglia. The proof follows similar lines to the proof of Theorem 1.

Theorem 3 states that an increase of *L-dopa* in the *striatum* will increase *DA* in the *striatum*, which is a consequence of *c*.24, and given that the enzyme ratio does not increase.

Theorem 3. $H_{BG} \cup$ *I*: {*a*(*L-dopa* , *striatum*) = inc} $\vdash$
P: {*a*(*DA*, *striatum*) = inc}

But to increase *L-dopa* by a drug intervention, which is taken up in the bloodstream, means that *L-dopa* is increased in the entire body, causing side effects. A decrease of the amount of *AADC* in the periphery by also administering an inhibitor that cannot cross the blood-brain barrier, will cause *DA* to increase in the brain, but to be relatively steady in the periphery. Next, Theorem 4 is a consequence of *c*.24 and *c*.25, given the assumption that the amount of *MAO-B* does not increase in the periphery.

Theorem 4. $H_{BG} \cup$ *I*: {*a*(*L-dopa*, body) = inc, *a*(*AADC*, periphery) = dec} $\vdash$
P: {*a*(*DA*, *striatum*) = inc, *a*(*DA*, periphery) = ?}

By *c*.24 and *c*.25 one can also prove Theorem 5, which states that decreasing the enzyme that breaks up *DA* will increase the amount of *DA*, assuming that both the amount of *AADC* and *L-dopa* in the *striatum* do not increase:

Theorem 5. $H_{BG} \cup$ *I*: {*a*(*MAO-B*, *striatum*) = dec} $\vdash$
P: {*a*(*DA*, *striatum*) = inc}

The function and activity of these treatments can be explained by the theory of the basal ganglia, but another question is whether the hypothesis is true. That is, are all the states that are possible in the emperical domain also states allowed by the theory?

A structuralist description of qualitative theories such as the basal ganglia model can also be useful in the research practice itself. The problem of the basal ganglia model, as noted in Section 2, is that it is too simple to be real and becomes too complex to work with were it to be extended to incorporate all details.

The advantage of a structuralist description is that you can add more kinds of details, while you can still easily explore predictions by making use of a computer program like QSIM which easily compute the consequences for the variables you are interested in. I have explored a number of computable predictions of different effects on the *SNR* after intervening in the direct and indirect pathways of the basal ganglia with selective dopaminergic agonists (van den Bosch 2001). Comparing these kinds of predictions with laboratory observations could in principle result in more detailed and accurate models of biological structures, such as the basal ganglia.

So, in summary, a stucturalist analysis can explicate theories from the studied practice of neuropharmacology. Moreover, the task of exploring predictions from these kinds of theories could in principle be aided by both a structuralist representation and a computer program that can reason about that representation.

5. Conclusion

In neuropharmacology the basal ganglia area in the brain is studied in drug research for Parkinson's disease. The theory of the basal ganglia consists of qualitative relations between variables of chemical and electrical neural activity in nuclei. This theory can be represented by a set of qualitative conditions on variables that describe the brain. In the structuralist approach this theory can be defined by its models, based on the set of conditions on conceptually possible models defined by a set of variables and possible values. The structuralist representation can in this case be used to both explicate a theory and possibly aid research because it enables a computational investigation of the theory's consequences.

University of Groningen
Faculty of Philosophy
Oude Boteringestraat 52
9712 GL Groningen
The Netherlands

REFERENCES

Bosch, A.P.M., van den (2001a). Logic of Drug Discovery – A Descriptive Model of a Practice in Neuropharmacology. *Proceedings of the Fourth conference on Discovery Science. Springer Lecture Notes in Artificial Intelligence* **2226**, 476-481.

Bosch, A.P.M., van den (2001b). *Rationality in Discovery – A Study of Logic, Cognition, Computation and Neuropharmacology.* Ph.D. thesis: Groningen. Amsterdam: Institute for Logic Language and Information.

Bosch, A.P.M., van den (1999). Inference to the Best Manipulation: A Case Study of Qualitative Reasoning in Neuropharmacy. *Foundations of Science* **4** (4), 483-495.

Iwasaki, Y. and H.A. Simon (1994). Causality and Model Abstraction. *Artificial Intelligence* **67**(1), 143-194.

Kuipers, B. (1994). *Qualitative Reasoning, Modeling and Simulation with Incomplete Knowledge.* Cambridge, MA: The MIT Press.

Kuipers, T.A.F. (2000/ICR). *From Instrumentalism to Constructive Realism.* Dordrecht: Kluwer Academic Press.

Kuipers, T.A.F. (2001/SiS). *Structures in Science.* Dordrecht: Kluwer Academic Press.

Parent, A. and F. Cicchetti (1998). The Current Model of Basal Ganglia Organization under Scrutiny. *Movement Disorders* **13**(2), 199-202.

Timmerman, W. (1992). *Dopaminergic Receptor Agents and the Basal Ganglia: Pharmacological Properties and Interactions with the GABA-Ergic System.* Ph.D. thesis: Groningen University.

Shults, B. and B. Kuipers (1997). Proving Properties of Continuous Systems: Qualitative Simulation and Temporal Logic. *Artificial Intelligence* **92**, 91-129.

Timmerman, W., F. Westerhof, T. van der Wal and B.C. Westerink (1998). Striatal Dopamine-Glutamate Interactions Reflected in Substantia Nigra Reticulata Firing. *Neuroreport* **9**, 3829-3836.

Vos, R. (1991). *Drugs Looking for Diseases. Innovative Drug Research and the Development of the Beta Blockers and the Calcium Antagonists.* Dordrecht: Kluwer Academic Press.

Theo A. F. Kuipers

STRUCTURES FOR COMPUTATIONAL ASSISTANCE IN DRUG DESIGN

REPLY TO ALEXANDER VAN DEN BOSCH

The title of Alexander van den Bosch's contribution is a nice allusion to the title of SiS. However, it not only deals with structures in the more specific sense of the structuralist approach as characterized in Ch. 12, it also deals with two other topics that are presented in SiS, viz. design research (Ch. 10) and computational approaches (Ch. 11). Van den Bosch explicitly deals with design research, notably drug design. Design research is normally (almost) neglected by philosophers of science, but as Van den Bosch's paper nicely illustrates, although (modern) design research is strongly related to nomological research, it makes very much sense to distinguish it from the latter, not only in goal but also in method, despite the fact that both types of research can be represented in set-theoretic terms. Moreover, Van den Bosch also indicates in his paper the way in which computational means can be used in drug design research when described in these terms, of course, with modest pretensions. Here he refers to some impressive computational studies which others from time to time attribute to me. Incorrectly, unfortunately, for they are the work of my namesake Benjamin Kuipers (no relation).

In this reply I confine myself to two related points of terminological criticism dealing with nomological research. In both cases it not only seems conceptually important in theory, but also in practice I frequently meet people who, like myself and Van den Bosch, are not always aware of some important distinctions that can and should be made.

Epistemological and Methodological Categories

In Table 1 and Figure 2 Van den Bosch categorizes the four types of conceptually possible models that are generated by the comparison of the models allowed by a theory and those that are, as a matter of *unknown* fact, empirically or nomically possible. Unfortunately, he uses the terminology that I find, apart from a specific point (see below) more appropriate for categorizing empiri-

In: R. Festa, A. Aliseda and J. Peijnenburg (eds.), *Cognitive Structures in Scientific Inquiry* (*Poznań Studies in the Philosophy of the Sciences and the Humanities*, vol. 84), pp. 360-363. Amsterdam/New York, NY: Rodopi, 2005.

cally established results. Because these (and only these) categories are methodologically useful I call them the methodological categories, as distinct from the epistemological categories (ICR, p. 150 versus p. 158), corresponding to Van den Bosch's Figure 2. So, let me insert in his Table 1 my favorite epistemological terminology between brackets, where the first inserted possibility refers to my (1992, p. 303) and the second to my (ICR, p. 150):

Subset	M_T	M_E	
1	1	0	Explanatory problem (explanatory/external mistake)
2	1	1	Empirical success, confirming instance (instantial/internal match)
3	0	1	Empirical anomaly, counterexample (instantial/internal mistake)
4	0	0	Explanatory success (explanatory/external match)

Table 1. Subsets of conceptually possible models M_P of a domain (the numbered subsets in the first column refer to Figure 2 of Van den Bosch's paper)

Hence, instead of the "problem/success terminology," which I find more appropriate for methodological purposes, I prefer for (abstract) epistemological characterization the "mistake/match terminology." Regarding the two suggested *sub*categorizations, viz. "explanatory/instantial" (1992) versus "external/internal" (2000), I have no strong preferences.

The background to the main preference is the following. As soon as we become methodologically realistic, and no longer suppose that we dispose of the set of empirical or nomic possibilities (M_E), we have to base our judgements on realized (and investigated) (types of) possibilities at a certain moment (R) and the empirical regularities based on them. The latter essentially arise by inductive generalization on the basis of R. Their conjunction, which is the strongest established empirical regularity, will be indicated by S. In view of the fact that Van den Bosch explicitly speaks of "descriptive induction" at the beginning of Section 4, it may well be that he assumes in fact that S may be equated with M_E. Under certain conditions this may be reasonable, though not without the risk of being incomplete (M_E may still be a proper subset of S) or incorrect. The assumption that the data are correct in the sense that the characterizations of R and the inductive jumps leading to S are correct amounts to the claim that R is subset of M_E, and that the latter is a subset of S.

Be this as it may, as a long as we assume that R is a proper subset of S, with, if correct, M_E as an unknown set in between, we get again four categories, now methodological ones, see Figure 1.

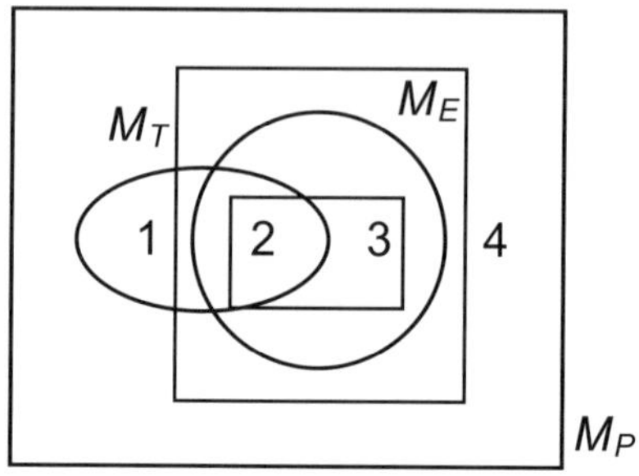

Fig. 1 (adapted from *Fig.* 2 of Van den Bosch's paper): Models M_T of a hypothesis and empirically possible models M_E of the phenomena of a domain, both part of the conceptually possible models M_P. The small rectangle indicates R, the large one S.

In our Table 2 we list first the "problem/success" names as used in (1992, p. 307) and then the first ones from ICR (p. 158), that is, the ones mentioned above, but with the qualification 'established', abbreviated by 'est'.

Subset	M_T	$R \subseteq M_E \subseteq$	S	
$1 = M_T - S$	1	0	0	Explanatory problem/ est. external mistake
$2 = M_T \cap R$	1	1	1	Instantial success/ est. internal match (example)
$3 = R - M_T$	0	1	1	Instantial problem/ est. internal mistake (counterexample)
$4 = M_P - S \cup M_T$	0	0	0	Explanatory success/ est. external match

Table 2. Subsets of conceptually possible models M_P of a domain, relative to data R/S (the first column refers to the adapted version of *Fig.* 2 of Van den Bosch's paper, i.e., our *Fig.* 1)

In this way we obtain a clear distinction between epistemological and methodological categories. Of course, I do not bother about these terms as such, but about the distinction. Note that Van den Bosch talks about "empirical" successes and problems, whereas I used the qualification "instantial," but this difference is not very important.

Confirming Instances

From the foregoing it follows that one problem with Van den Bosch's terminology of 'empirical success' and 'confirming instance' is that it could better be used for the members of $M_T \cap R$ instead of those of $M_T \cap M_E$. However, my main criticism of this terminology and, for that matter, of my 1992 terminology of 'instantial success', is that the category $M_T \cap R$ not only covers proper successes, but also realized possibilities that are merely compatible with *T*. For this reason I add to the phrase 'est. internal match' in the table on p. 158 of ICR, besides the term 'example', the phrase: *individual success* or *neutral instance*, where the former could of course also have been called 'positive instance'. This distinction is also already made in the so-called evaluation matrix (ICR, pp. 117-9; SiS, pp. 235-7, p. 307), in terms of positive and neutral instances, besides negative instances (or counterexamples), with the corresponding refinement of the notion of "being more successful."

A simple example of the crucial distinction is the fact that the hypothesis "all ravens are black" has only one type of counterexample (non-black ravens), but two types of individual successes, that is, not only black ravens, but also non-black non-ravens, and one type of neutral case: black non-ravens. The latter are merely compatible with the hypothesis, that is, the hypothesis has nothing to offer, neither when you start with something black, nor when you start with a non-raven. For a detailed analysis, see ICR, Ch. 2 and 3; see, however, also the contribution of Maher and my reply, both in the companion volume. For the moment I conclude that we should already refine our concepts and diagrams corresponding to the epistemological categories by introducing (hypothetical) proper subsets of M_T and M_E with respect to which *T*, resp. the true theory (i.e., the one characterizing M_E) has nothing to offer. This would automatically generate the suggested refinement of the methodological category of 'established internal match'. Refined diagrams for both types of categories are still missing. They will easily get complicated, in particular the methodological ones, so the challenge is to make them nevertheless as appealing as possible. For the epistemological point of departure it may be useful to start from a diagram in SiS (p. 281), drawn for a similar problem, viz. bringing 'irrelevant properties' into the picture of design research.

REFERENCE

Kuipers, T. (2002). Beauty, a Road to The Truth. *Synthese* **131** (3), 291-328.

Paul Thagard

WHY IS BEAUTY A ROAD TO THE TRUTH?

ABSTRACT. This paper discusses Theo Kuipers' account of beauty and truth. It challenges Kuipers' psychological account of how scientists come to appreciate beautiful theories, as well as his attempt to justify the use of aesthetic criteria on the basis of a "meta-induction." I propose an alternative psychological/philosophical account based on emotional coherence.

1. Introduction

In a recent article, Theo Kuipers (2002) offers an account of the relation between beauty, empirical success, and truth. Building on his impressive work on the nature of truth approximation (Kuipers 2000), he provides a "naturalistic-cum-formal" analysis that supports the contention of McAllister (1996) that aesthetic criteria are useful for scientific progress and truth approximation. I agree with this contention, but will challenge Kuipers' psychological account of how scientists come to appreciate beautiful theories, as well as his attempt to justify the use of aesthetic criteria on the basis of a "meta-induction." I propose an alternative psychological/philosophical account based on emotional coherence (Thagard 2000).

2. Kuipers on Beauty and Truth

According to Kuipers, the truth is beautiful in the sense that it has features that we have come to experience as emotionally positive due to the mere-exposure effect. This effect is a robust finding in experimental psychology that an increasing number of presentations of the same item tends to increase the affective appreciation of the item. Kuipers introduces the mere-exposure effect because it suggests that the human mind does a kind of affective induction in addition to the more familiar cognitive kind. Kuipers proposes that scientists do a kind of affective induction that leads them to react with positive emotions to recurring features of science that are not conceptually connected with empirical success, for example simplicity, symmetry, and visualizability.

In: R. Festa, A. Aliseda and J. Peijnenburg (eds.), *Cognitive Structures in Scientific Inquiry* (*Poznań Studies in the Philosophy of the Sciences and the Humanities*, vol. 84), pp. 365-370. Amsterdam/New York, NY: Rodopi, 2005.

Assuming that there is indeed a correlation between such features and empirical success, the philosopher of science can then do a "cognitive meta-induction" that justifies scientists' affective inductions on the grounds that beauty really does correlate with truth. On this view, scientists acquire the tendency to find beautiful theories that possess features such as simplicity and symmetry on the basis of exposure to previous successful theories that had such features. Moreover, the acquisition is legitimate because, by the cognitive meta-induction, such features really do correlate with experimental success, which is an objective feature of theories. Kuipers not only tries to argue that the empirical success of theories signals their approximation to truth, but also that the correlating *non*-empirical features directly signal approximation to truth. Hence it is reasonable that scientists let themselves be guided by non-empirical features as well as empirical success.

I do not want to challenge Kuipers account of truth approximation, which strikes me as the most sophisticated currently available, but I see several problems with the way he connects beauty and truth. First, note that the mere-exposure effect is very different psychologically from affective induction. When mere exposure leads me to like something, the structure of the episode is: exposure to *X* → increased liking of *X*. In contrast, affective induction has a structure something like: *X* goes with *Y* and *Y* is liked → increased liking of *X*. Affective induction requires exposure to two features, e.g. simplicity and empirical success, whereas the mere-exposure effect does not require any such correlation. Hence the mere-exposure effect is logically and psychologically irrelevant to affective induction. I would not be surprised if human thinking does in fact use something like affective induction, but Kuipers needs to find empirical support for this kind of thinking from experiments other than those that support the existence of the mere-exposure effective.

Second, evidence is needed to support the claim that the positive emotional attitude toward simplicity and symmetry that many scientists exhibit is acquired by affective induction. Does scientific education really involve juxtaposition of aesthetic features and empirical success in ways that could lead budding scientists to acquire the emotional appreciation of simplicity and symmetry? In the first place, do scientists have an antecedent positive emotional attitude toward empirical success that would provide the basis of the affective induction that aesthetic features are good? I conjecture that science students acquire the tendency to find some theories beautiful through a partly innate and partly acquired ability to recognize coherence; the next section defends an emotional coherence account of aesthetic judgments in science. If this account is correct, then scientists acquire aesthetic attitudes by means different from affective induction.

Third, I am less confident than Kuipers about the connection between empirical success and truth. Even if there is a legitimate meta-induction connecting beauty and empirical success, it remains to be shown that there is a connection between empirical success and truth. On Kuipers view, the connection is direct, by virtue of the definition of approximate truth and the theorem that if *Y* is closer to the truth than *X*, then *Y* is at least as empirically successful as *X*. I agree that in general empirical success is a sign of truth, but it is hard to make the connection directly, since we have no independent way of establishing truth. This is concealed in Kuipers' framework because he identifies *the truth* as the strongest true theory rather than as how the world really is. In order to conclude that empirical success is a guide to how the world really is, we need to bring in other aspects of science such as its technological applicability, the substantial degree of agreement among scientists, and the largely cumulative nature of scientific development (Thagard 1988, ch. 8). In the past few hundred years, we have learned that empirical success is a much better guide to truth than other determinants of belief such as a priori reflection and divine inspiration, but it might have been otherwise. Hence the connection between empirical success and truth is just as much in need of argument as the connection between beauty and truth. The argument cannot be a cognitive meta-induction, because we have no way of identifying what is true. Rather, the form of argument is theoretical: we can infer that science acquires true theories because that is the best explanation of its technological success and largely cumulative development.

3. Beauty as Emotional Coherence

I will now sketch a different picture of the role of beauty in scientific inference. My most recent book develops a theory of emotional coherence that is used to explain how judgments of beauty arise (Thagard 2000, ch. 6). The theory extends a general theory of coherence as constraint satisfaction: when people make inferences, they do so in a way that maximizes coherence by maximizing the satisfaction of multiple positive and negative constraints among representations. The kind of inference most relevant to scientific thinking is explanatory coherence, in which the representations are of evidence and hypotheses, the positive constraints are based on explanation relations between hypotheses and evidence, and the negative constraints are based on relations of contradiction or competition between hypotheses. When scientists choose between competing theories, they do so by accepting those hypotheses that are part of the maximally coherent account. Various algorithms are available for maximizing coherence, including psychologically plausible algorithms using artificial neural networks.

The theory of emotional coherence postulates that human thinking is a process that involves affective as well as cognitive constraints and that both kinds of constraint satisfaction are intimately related. Representations acquire *valences,* which constitute their emotional content, in addition to their degrees of acceptability. For example, your concept of *beer* involves in part a valence that represents whether or not you like beer. Propositional representations such as "Beer is good for you" also have a valence, as is evident in the different emotional reactions that might be given to this proposition from avid beer drinkers as opposed to those of teetotalers. From the perspective of emotional coherence theory, beauty is not a property of individual representations, but is a "metacoherence" property that arises as the result of a general assessment of coherence. A feeling of happiness emerges when most constraints are satisfied in a person's unconscious processing of cognitive and affective constraints, whereas feelings of sadness and anxiety can emerge when constraints are not satisfied. In particular, scientists find a theory beautiful when it is highly coherent with the evidence and with their other beliefs. Such coherence is largely a matter of empirical success, in that many of the constraints on a theory concern the data which it is intended to explain. But simplicity is intrinsically part of the coherence calculation, since the constraints that tie hypotheses with evidence are stronger if the explanations involve fewer hypotheses (see Thagard 1992, for a full exposition). Moreover, symmetry, which is another one of the aesthetic factors mentioned by Kuipers, is also a matter of coherence, of an analogical sort. Symmetry is a matter of having multiple parts of a theory or other set of representations that are analogous to each other (Thagard 2000, p. 203). For example, a face is symmetrical to the extent that the left side is analogous to the right side. Like explanatory inference, analogical thinking can be thought of in terms of satisfaction of multiple constraints (Holyoak and Thagard 1995).

In contrast to Kuipers, who views simplicity, symmetry, and analogy as problematic because they are nonempirical, I see them as an integral part of the coherence-based inferences about whether to accept or reject a theory. Beauty is the feeling that emerges to consciousness when a theory is very strongly coherent with respect to explaining the evidence *and* being consistent with other beliefs *and* possessing simplicity, symmetry, and other kinds of analogies. Psychologically, the beauty of a theory does not arise from affective inductions connecting aesthetic features with empirical success, but rather from the coherence of the theory that intrinsically includes those features.

4. Assessment

I have offered an alternative to Kuipers' psychological and philosophical explanations of why beauty is a road to the truth. Whose explanations are more plausible? First consider the competing psychological explanations of how scientists come to experience some theories as beautiful because of aesthetic features such as simplicity and symmetry.

Kuipers: Scientists come to like such aesthetic features because of a psychological mechanism of aesthetic induction akin to the mere exposure effect.

Thagard: Scientists find theories with such features beautiful because of their contribution to coherence which is inherently pleasurable.

There is currently little experimental evidence to enable us to discriminate directly between these two explanations; I have already argued that aesthetic induction is a very different process from the mere-exposure effect, so the considerable psychological evidence for the latter does not support the general plausibility of the former.

My main reason for preferring the emotional-coherence explanation of the pleasurable nature of simplicity and symmetry is that it derives scientific beauty from the same kind of psychological mechanism that produces intellectual pleasure in other domains, such as art, music, and mathematics. Aesthetic theorists such as Collingwood and Hutcheson, as well as mathematicians such as Hardy, have described beauty as deriving from unity, harmony, and coherence. Emotional coherence provides a unified (i.e. more beautiful!) explanation of scientific judgments of beauty, because it describes the same mechanism at work in science as in art and mathematics. Kuipers could well maintain that aesthetic induction on particular features operates in these other domains as well, which might serve to explain emotional preferences for particular kinds of art or mathematics. But aesthetic induction does not explain the general appreciation of beauty deriving from an overall appreciation of a work of art, a mathematical construction, or a scientific theory. In contrast, the theory of emotional coherence provides a specific computational mechanism by which positive feelings can emerge from global judgments of coherence, including ones that incorporate simplicity and symmetry.

I also think that the emotional-coherence account provides a better basis for the philosophical issue of justifying scientists' use of aesthetic judgments than Kuipers inductive account. Here are the two positions:

Kuipers: Scientists' use of aesthetic criteria such as simplicity and symmetry is justified by the cognitive meta-induction that these features correlate with empirical success and truth.

Thagard: Scientists' use of aesthetic criteria is justified more indirectly by the fact that they are integral to the coherence assessments that promote the largely cumulative development of theories, many of which are technologically successful.

I prefer the indirect strategy because it does not require the accumulation, by practicing scientists or by philosophers combing the history of science, of a large body of instances of correlations between aesthetic features and truth. It is also immune to the likely existence of counterexamples in the form of cases where theories that turned out to be false were initially adopted in part on the basis of aesthetic criteria. Judgments of scientific beauty, like all inductive reasoning, are highly fallible. My indirect method of justifying explanatory coherence assessment as scientific method does not assume that it always or even usually works, as meta-induction requires. Scientific reasoning, based on explanatory coherence and including judgments of beauty, is justified because it is sometimes successful and there is no other method that is anywhere near as successful in finding out how the world really is. Beauty is a road to truth, but the road can be a winding one.

In conclusion, I applaud Theo Kuipers for his development of elegant and plausible accounts of scientific reasoning and approximation to truth, and for his noble attempt to extend these accounts to explain the role of aesthetic judgments in science. But I have argued that the role of beauty in science is more fruitfully understood from the non-inductive perspective of emotional coherence.

University of Waterloo
Philosophy Department
Waterloo, Ontario ON N2L 3G1
Canada

REFERENCES

Holyoak, K.J. and P. Thagard (1995). *Mental Leaps: Analogy in Creative Thought.* Cambridge,MA: The MIT Press/Bradford Books.

Kuipers, T. (2000/ICR). *From Instrumentalism to Constructive Realism*. Dordrecht: Kluwer.

Kuipers, T. (2002). Beauty, a Road to the Truth. *Synthese* **131** (3), 291-328.

McAllister, J. W. (1996). *Beauty and Revolution in Science*. Ithaca, NY: Cornell University Press.

Thagard, P. (1988). *Computational Philosophy of Science*. Cambridge, MA: The MIT Press/BradfordBooks.

Thagard, P. (1992). *Conceptual Revolutions*. Princeton: Princeton University Press.

Thagard, P. (2000). *Coherence in Thought and Action*. Cambridge, MA: The MIT Press.

Theo A. F. Kuipers

AESTHETIC INDUCTION VERSUS COHERENCE
REPLY TO PAUL THAGARD

Paul Thagard's brief contribution deserves a long reply, but I confine myself here to some basic issues. I start with some concessions relative to SiS regarding simplicity and analogy, followed by rebutting Thagard's general and specific reserves about my recent naturalistic-cum-formal inductive account of the relation between beauty and truth. Finally, I raise some doubts about the exhaustiveness of his coherence account of that relation and its supposed incompatibility with my account.

Aesthetic Induction, Empirical Success, and Truth Approximation

Let me start by reporting some new considerations that are relevant to Thagard's contribution. In SiS I went as far as to claim that simplicity should only play a role in case of equal success (SiS, p. 238, and Section 11.2) and for analogy I saw no role at all (SiS, p. 297). Contrary to my previous beliefs, at the time of completion of SiS, very much stimulated by reading McAllister (1996), I was beginning to understand that there might be a relation between truth and simplicity, and, more recently, stimulated by a discussion with Thagard when he visited Groningen on the occasion of Alexander van den Bosch's promotion, even one between truth and analogy. Hence, in the light of my recent article on beauty and truth (Kuipers 2002), I have to qualify these claims in SiS.

Since "simplicity" figures, at least in certain periods of certain disciplines, in the prevailing aesthetic canon, to use McAllister's nice phrase, it has cognitive merits related to empirical success and even to truth approximation, which scientists favoring the dominant theory may value more than some empirical successes of a new theory that are failures of the old one. Repairs may well come to grips with these failures. Similarly, as McAllister (1996) also illustrates and my article implicitly justifies, "analogy" may also be seen

In: R. Festa, A. Aliseda and J. Peijnenburg (eds.), *Cognitive Structures in Scientific Inquiry* (*Poznań Studies in the Philosophy of the Sciences and the Humanities*, vol. 84), pp. 371-374. Amsterdam/New York, NY: Rodopi, 2005.

as a nonempirical feature of certain theories that may play a cognitively justified role. Certainly, the relative weight assigned to such features should take into account that these features are based on "meta-induction," that is, induction of a recurring nonempirical feature correlating with empirical success, whereas general empirical successes are based on "object-induction," induction of a regularity about (the behavior of) a certain kind of objects. Although object-inductions are not very trustworthy, they are certainly more trustworthy than meta-inductions.

To be sure, the "uniform" notion of being "empirically more successful," as presented in ICR and SiS, leaving no room for empirical failures compensated by more impressive empirical successes, can be extended to the more general notion of "more successful," taking also "nonemprical" successes and failures uniformly into account. However, as explained in Section 6 of my article on beauty and truth, the interesting cases of nonempirical considerations come into the picture when they point in another direction than the empirical considerations. This would require a combined definition of 'more successfulness' taking relative weights of different kinds of considerations into account. Depending on one's weights, to use an example suggested to me by Thagard, one may then value the phlogiston theory or even the oxygen theory as less successful than the classical theory, according to which there are only four substances, viz., air, earth, fire, and water, because this theory is much simpler than the two famous competing theories.

I am happy to agree with Thagard's claim that my view of the relation between beauty and empirical success needs new experimental and historical evidence, although I would not say that the well established "mere-exposure effect" is irrelevant. In the article I argue that the aesthetic induction may be a variant of the mere-exposure effect, more precisely, a concretization, provisionally called a qualified-exposure effect. In line with its naturalized approach, I suggest at the end a number of experiments with normal and toy pieces of art and with scientific examples to establish the conditions and limitations of the effect. Moreover, further evidence for the varying character of the aesthetic canon when different phases or different research programs of the same discipline or of different disciplines are compared would strengthen the basic ideas around aesthetic induction as such and its diagnosis as a variant of the mere-exposure effect. Finally, as I also stress in my reply to Miller, in the companion volume, my refined claim about aesthetic induction can be falsified: determine a nonempirical feature which happens to accompany all increasingly successful theories in a certain area from a certain stage on and which is not generally considered beautiful by the relevant scientists. To be sure, the common interesting point of our diverging views is, of course, that both suggest (comparative) experiments and possible pieces of historical evidence (see below), a rare but welcome aspect of primarily philosophical theories.

Apparently I did not convince Thagard by arguing in ICR (p. 162) that there is a direct connection between empirical success and truth, and that we do not need his detour, as I explained in SiS (p. 298). The crucial point seems to be that I identify *the truth* as the strongest true theory (given a domain and a vocabulary) "rather than as how the world really is." Here Thagard is transgressing the boundaries of my kind of constructive realism and enters some kind of essentialist realism. In the introductory chapter to this volume I summarize my direct argument for a relation between truth and empirical success. In my reply to Hans Mooij in the other volume I try to specify my metaphysical position in some more detail. Since Thagard's truth does not exist in my view, his detour argument, that empirical success is a sign of truth, essentially pertains to my non-essentialist kind of truth(s), like my direct argument.

Emotional Coherence

Let me now turn to Thagard's theory of beauty as an aspect of emotional coherence. According to him, "scientists find a theory beautiful when it is highly coherent with the evidence and with their other beliefs," where simplicity, symmetry and analogy (of which symmetry is a special case) are intrinsically part of the coherence calculation. In SiS (Section 11.2), I argue in general against Thagard's "unstratified" theory of explanatory coherence (and its implementation in the ECHO program), in favor of the stratified priority of explanatory superiority (implemented by the evaluation matrix EM), by using a meta-application of simplicity considerations. I show that both are equally successful in accounting for all historical choices provided and "prepared" by Thagard himself, whereas ECHO is much more complicated than EM. (See my reply to Vreeswijk.) In other words, Thagard's coherence theory asks for historical cases in which explanatory superiority is sacrificed to simplicity, which would go against the stratified view.

Thagard associates the beauty of theories with all kinds of coherence. Hence, incoherent aspects of theories should be seen as ugly. Thagard (2000, pp. 199-200) argues in general that symmetry is aesthetically appreciated for its contribution to coherence, and asymmetry is ugly due to its incoherence. He mentions the symmetry of (most) human faces, as opposed to the asymmetry of a misshapen face. This type of example is interesting for two reasons. First, after habituation to a misshapen face, e.g. of a movie star, we may come to find it very beautiful. Second, we are used to pictures of the arrangement of organs in the human body, including all kinds of asymmetries, and many of us will find the composition very beautiful, not least for these asymmetries. Hence, an overall coherence account of beauty is difficult to combine with the fact that at least certain people appreciate incoherencies, including scientists. The biologist Stephen Gould, for example, stresses in an interview (Kayzer 2000) that he, in contrast to the physicist Steven Weinberg, counts diversity, unrepeatable contingencies and irregularities among the sources of his ultimate aesthetic satisfaction. Gould mentions as examples of great aesthetic satisfaction the diversity of a certain species of land snails, called cerions (p. 32), and the incoherencies in the revolutions of earth and moon, which make it impossible to design a coherent calendar (p. 29). Ironically enough, Weinberg (Kayzer 2000, p. 78; see also Weinberg 1993, p. 119) mentions the gravedigger scene in Shakespeare's Hamlet as a surprising intermezzo in a logical sequence of events, which, according to Weinberg, illustrates the fact that in the arts there are even higher aesthetic phenomena than in science. Hence, Gould's claim and examples seem to be incompatible with an overall coherence view of beauty in science, and Weinberg's example at least suggests that coherence cannot be the only source of aesthetic appreciation in the arts, which makes it difficult to understand why there would be no experiences of beautiful incoherencies in science.

In the last part of his contribution Thagard gives a very clear statement of our diverging psychological and philosophical explanations of why beauty is a road to the truth. However, from the above it will be clear that I am not yet converted to his view. But I would also like to stress that they may be less incompatible than Thagard suggests. First, as to the psychological side, overall coherence might well be a feature that in certain disciplines and at certain stages can belong to the "aesthetic canon" as the result of aesthetic induction. Second, as to the philosophical side, I have already indicated that Thagard's supposed indirect connection between beauty and the essentialist truth, that is, the truth about how the world really is, boils down to a connection between beauty and constructive truths, for which connection there is a direct argument which, as a matter of fact, has not been disputed by Thagard.

REFERENCES

Kayzer, W. (2000). *Het Boek over de Schoonheid en de Troost.* Amsterdam: Contact.

Kuipers, T. (2002). Beauty, a Road to The Truth. *Synthese* **131** (3), 291-328.

McAllister, J. (1996). *Beauty and Revolution in Science.* Ithaca, NY: Cornell University Press.

Thagard, P. (2000). *Coherence in Thought and Action.* Cambridge, MA: The MIT press.

Weinberg, S. (1993). *Dreams of a Final Theory.* London: Vintage.

Gerard A. W. Vreeswijk

DIRECT CONNECTIONISTIC METHODS FOR SCIENTIFIC THEORY FORMATION

ABSTRACT. Thagard's theory of explanatory coherence (TEC) is a conceptual and computational framework that is used to show how new scientific theories can be judged to be superior to previous ones. In *Structures in Science* (SiS), Kuipers criticizes TEC as a model that does not faithfully reflect scientific practice. This article tries to explain the machinery behind TEC, and tries to indicate where TEC falls short (conceptually speaking) and where it can be improved. The main idea proposed in this article is not to *derive* a coherence network from the input (à la TEC), but to construct a coherence network right from the input itself.

> "I'm all for a bad story and incoherent quests (wait a minute... no I'm not)."
> (Diablo 2 Review, Rob Pecknold for www.mastergamer.com. Rating: Average.)

1. Introduction

Did you know that complex connectionistic (neural-network) computations are still done by hand? For "only" 45 minutes? If you did not, then please consult Kuipers' *Structures in Science* (SiS), Ch. 11, Sec. 2.3, p. 313. In that section, Kuipers takes pains to show his readers why the principle of explanatory superiority (PES) is conceptually simpler and more to the point than the theory of explanatory coherence (TEC), a theory proposed by the Canadian philosopher of science Paul Thagard (1994). Kuipers does so by simulating the computations of both PES and TEC by hand.

In this article I do not so much want to discuss Kuiper's PES, but rather Thagard's TEC. TEC is about coherence, and coherence is an important if not central notion in the philosophy of science. Arguments for coherence stem from mainstream epistemology, where it is called *coherentism.* The basic idea of coherentism is that *all* beliefs are justified inferentially, that there are no basic foundational beliefs, and that justification works both ways (Everitt and Fisher 1995). There are various forms of coherentism, and several coherentists have explored in some detail the ways in which coherentism can be developed.

In: R. Festa, A. Aliseda and J. Peijnenburg (eds.), *Cognitive Structures in Scientific Inquiry* (*Poznań Studies in the Philosophy of the Sciences and the Humanities*, vol. 84), pp. 375-403. Amsterdam/New York, NY: Rodopi, 2005.

One way to understand the nature of coherence – one way that is particularly relevant to philosophers of science – is to think of coherence as inference to the best explanation based on a background system of beliefs (Keith Lehrer 1992). Thus, coherence and philosophy of science are intimately related.

A recent offspring of coherentism in the philosophy of science is TEC (Thagard 1994). The surplus value of TEC, compared to other theories of coherence, is that it is supported by a computer program, ECHO, that is able to compute the coherence of formalized scientific theories. Although philosophers of science are familiar with computational approaches to cognitive processes (cf. Shrager and Langley 1990; Darden 1997), and although epistemologists are familiar with coherence (cf. Dancy and Sosa 1992; Everitt and Fisher 1995), TEC's computational approach is exceptional in philosophy of science. Thagard's work on TEC did not remain unnoticed and gave rise to many discussions within the scientific philosophers' community (Thagard 1989).

Thagard's work on TEC did not pass unnoticed by Theo Kuipers either. In Chapter 11 of *Structures in Science*, entitled "Computational philosophy of science," Kuipers discusses TEC. In fact, he severely criticizes it. Kuipers maintains that TEC (and ECHO) use "a non-transparent updating process, which may nevertheless lead, as a rule, to an ambiguous conclusion." (Cf. SiS, Ch. 11, Sec. 2.1.2.) Kuipers further explains why his evaluation matrix (EM) is a more simple and transparent approach to the evaluation of scientific theories, arguing that TEC uses unnecessary complicated connectionistic techniques to compare the explanatory power of two competing scientific theories. According to Kuipers, two competing theories can be compared just as well with the more simpler and transparent EM. The EM simply enumerates the successes, failures and lacunas of both theories and then compares them on the basis of an aggregated performance measure. (For further details, the reader is referred to SiS.) Broadly speaking, Kuipers maintains that the architecture of theory selection of TEC "is on the wrong track."

This article tries to explain the machinery behind TEC, and tries to indicate where TEC falls short (conceptually speaking) and where it can be improved. Although I am a proponent (some would say follower) of TEC, this article does not try to defend it. Neither does it try to explain where I think that Kuipers goes wrong in his criticism on TEC.

Although the author of this article works in a computer science department, and although the paper sometimes has a relatively high formula density, the implications for philosophy of science are immanent and direct. This will be further explained in the summary at the end of this article.

2. TEC

Thagard's theory of explanatory coherence, TEC, is a conceptual and computational framework that is used to show how new scientific theories can be judged to be superior to previous ones. This section explains how TEC works. For the motivations *behind* TEC, I refer to Thagard's exceptionally well-written monograph *Conceptual Revolutions* (1994). See also (2000).

The essentials of TEC are implemented in ECHO. ECHO is a computer program that uses propositions, contradictions, explanations, data elements and analogies as input. ECHO was implemented done by Thagard (in Lisp) and Donaldson (in Java). Propositions are represented by atomic identifiers that correspond to evidence, hypotheses, and other logical statements. Pieces of evidence usually start with an E, and hypotheses usually start with an H.

An example of this format is the following, in which Thagard represented essential statements of two competing theories of combustion, viz. Stahl's (1723 et seq.) phlogiston theory of combustion and Lavoisier's (1772 et seq.) oxygen theory of combustion.

Example 2.1. (Competing theories of combustion). Below is the input given to ECHO to represent Lavoisier's argument in his 1783 polemic against phlogiston. These propositions do not capture Lavoisier's arguments completely, but do recapitulate its major points.

proposition El	In combustion, heat and light are given off.
proposition E2	Inflammability is transmittable from one body to another.
proposition E3	Combustion only occurs in the presence of pure air.
proposition E4	Increase in weight of burned body is weight of absorbed air.
proposition E5	Metals undergo calcination.
proposition E6	In calcination, bodies increase weight.
proposition E7	In calcination, volume of air diminishes.
proposition E8	In reduction, effervescence appears.
proposition OH1	Pure air contains oxygen principle.
proposition OH2	Pure air contains matter of fire and heat MFH.
proposition OH3	In combustion, oxygen from air combines with the burning body.
proposition OH4	Oxygen has weight.
proposition OH5	In calcination, metals add oxygen to become calxes.
proposition OH6	In reduction, oxygen is given off.
proposition PH1	Combustible bodies contain phlogiston.
proposition PH2	Combustible bodies contain matter of heat.
proposition PH3	In combustion, phlogiston is given off.
proposition PH4	Phlogiston can pass from one body to another.
proposition PH5	Metals contain phlogiston.
proposition PH6	In calcination, phlogiston is given off.

```
explain OH1 OH2 OH3 El     explain OH1OH5 E7          contradict PH3 OH3
explain OH1 OH3 E3         explain OH1OH6 E8          contradict PH6 OH5
explain OH1 OH3 OH4 E4     explain PH1 PH2  PH3 El
explain OH1 OH5 E5         explain PH1 PH3  PH4 E2
explain OH1 OH4 OH5 E6     explain PH5 PH6  E5

data   El   E2   E3   E4   E5   E6   E7   E8
```

For example, OH1, OH2, and OH3 together explain El, because the heat and light in a combustion can be explained by assuming that the oxygen in the air combines with the burning body.

ECHO'S task is to investigate which propositions cohere, and which propositions incohere, on the basis of the input given. ECHO's outcome for the above input, for example, is that there is more coherence between 0-type hypotheses and the evidence supplied, than between P-type hypotheses and the evidence supplied. According to TEC, this would suggest that the oxygen theory of combustion is superior to the phlogiston theory of combustion. (End of example.)

Again, I refer to Thagard's work on TEC for further motivation (Thagard 1989, 1994; Thagard and Millgram 1995; Thagard et al. 1997). In later publications, (e.g. Verbeurgt and Thagard 1998; Thagard 2000), TEC is generalized to a more comprehensive theory of coherence, in which an expression of the form $P_1, \ldots, P_m \rightarrow Q$ is no longer viewed exclusively as an explanation, but more generally as some form of "soft" implication. When TEC is mentioned in this paper, we refer to this more general type of coherence.

The next few sections describe how TEC works and how the corresponding computer program, ECHO, computes the coherence between propositions.

2.1. *Coherence Networks*

Computing the coherence between propositions is a three-step process.

1. Derive a coherence network from the input given (propositions, contradictions, explanations, data elements and analogies).
2. Initialize the coherence network.
3. Maximize the global coherence of the network.

After global coherence has been maximized, propositions possess an activation value. Propositions with similar activation values are likely to cohere, and propositions with different activation values are likely to incohere. The propositions with high activation values are usually the ones that are accepted.

I will first explain the notion of a coherence network, and then explain how such a network is derived in ECHO.

Definition 2.1. (Coherence)

1. Coherence is a symmetric, real-valued relation between two propositions, that ranges from 1 (absolute coherence) to -1 (absolute incoherence).
   - If P and Q cohere with degree 0.57 we write $P \sim_{0.57} Q$.
   - If P and Q incohere with degree 0.23 (or cohere with degree -0.23, which is the same) we write $P \sim_{-0.23} Q$.
2. A *coherence network* is a graph with weighted and undirected edges, such that the nodes correspond to propositions, and the edges correspond to a (fixed) coherence relation.
3. Propositions may possess different *activation values.* Activation ranges from 1 (accepted, believed) to -1 (rejected, disbelieved). E.g., $\text{ACT}(P) = 1/2$, or $\text{ACT}(Q) = -3/4$. The value 0 expresses indifference. The activation values of nodes in a coherence network may vary.
4. The degree in which an incoherence relation between two propositions is satisfied, is expressed by the product of the activation values of both propositions, and the weight of the link that connects them. This is sometimes called *local coherence.*

   For example, if

   $\text{ACT}(P) = 1/2$, $P \sim_{-2/3} Q$, and $\text{ACT}(Q) = -1$,

   then the local coherence is equal to $1/2 \times (-2/3) \times -1 = 1/3$. Don't make the mistake of confusing the (local) coherence between two propositions (1/3) with the weight of the link that connects them (-2/3).
5. The (global) *coherence* of a network is the sum of the local coherence values. Global coherence is also named *harmony,* or *goodness-of-fit.*
6. – An *optimal solution* is an assignment of activation values that maximizes global coherence.
   – A *perfect solution* is an assignment of activation values such that every (in)coherence relation is maximally fulfilled.

   It is easy to verify that perfect solutions imply extreme activation values (i.e., activation of each node is either 1 or -1). Further, it is easy to verify that optimal solutions always exist, and that perfect solutions do not always exist. If a perfect solution exists, it is optimal.

Further observations:

a. Coherence can be a local matter, or it can refer to the entire constellation of propositions. Accordingly, items (3) and (4) concern local coherence, while items (5) and (6) concern global coherence.
b. The notion "incoherence" is intended to mean more than just that two propositions do not cohere: to incohere is to *resist* holding together.

c. The global coherency of a network is a *non-standardized measure of coherence:* largernetworks usually possess a higher coherency than smaller ones, simply because they have more links.
d. A standardized measure of coherence could be

$$\frac{\text{global coherence}}{\text{coherence of an optimal solution}}$$

Thus, the best solution would always have measure one. But this measure is difficult to obtain, since the value of the optimal solution is generally not known. (Since an optimal solution is generally not known.)
e. Another standardized measure of coherence could be

$$\frac{\text{global coherence}}{\text{coherence of a perfect solution}}$$

This one is easy to compute because the coherence of a perfect solution is always equal to the sum of the absolute values of the weights of the links in the corresponding graph. The ratio does not necessarily indicate the closeness to the optimal solution as the previous measure would, but it does have the property that the higher the ratio, the closer the solution is to optimal. It thus gives a size-independent measure of coherence.
f. The above definition does not tell us how to compute the coherence of individual propositions (within the network), nor does it tell us how to compute, or define, the coherence of a subset of propositions in a network. There are two reasons for doing so. First, there are different ways in which the coherency of subsets may be defined, but none of them is satisfactory. A second (and more pragmatic) reason for not trying to define coherency for subsets is that TEC works equally well without such a concept.

Example 2.2. If $C = \{P,Q,R\}$ is a coherence network with links

$$P \sim_{0.98} Q \sim_{0.54} R \sim_{-0.97} P$$

and p, q, and r are the activation values of P, Q, and R, then

$$global\_coherence(\mathrm{C}) = 0.98\mathrm{pq} + 0.54\mathrm{gr} - 0.97\mathrm{rp} \quad (1)$$

Here are some examples for several values of p, q, and r:

p	0.00	1.00	-1.00	1.00	1.00	0.98	1.00	1.00
q	0.00	1.00	-1.00	1.00	1.00	1.00	0.98	1.00
r	0.00	1.00	-1.00	0.00	-1.00	-1.00	-1.00	-0.98
global coherence	0.00	0.55	0.55	0.98	1.41	1.37	1.40	1.40

For example, the combination $(p, q, r) = (1.00, 1.00, -0.98)$ yields a relatively high global coherence of 1.40.

2.2. *Deriving a Coherence Network*

TEC uses the following principles to derive a coherence network from the input given.

1. *Implication.* Each implication $P_1, \ldots, P_m \rightarrow Q$ increases the coherence between
   (a) P_i and Q,, for each i with $1 \leq i \leq m$.
   (b) P_i and P_j, for each i and j with $1 \leq i < j \leq m$.
   In both cases, the additional strength in coherence is inversely proportional to the number of co-formulas in the antecedent of the rule.

   For example, if $P, Q,, R \rightarrow S$, and $\in$ is the standard excitation value, then $P\sim_{\in} S$, $Q\sim_{\in} S$, $R\sim_{\in} S$ (1a). Further, for P, Q,, and R the number of co-formulas in the antecedent is 2, so that $P\sim_{\in/2} Q$, $P\sim_{\in/2} R$, and $Q\sim_{\in/2} R$(1b). If $P,Q \rightarrow T$ as well, for instance, then $P\sim_{\in/2} Q$ raises to $P\sim_{\in/2+\in} Q$.
2. *Analogy.* An *analogy* is formed by two implications $P_1 \rightarrow Q_1$, $P_2 \rightarrow Q_2$, together with an explicit statement that P_1 is analogous to P_2, and Q_1 is analogous to Q_2. Each analogy $(P_1 \rightarrow Q_1, P_2 \rightarrow Q_2)$ strengthens the coherence between
   – P_1 and P_2
   – Q_2 and Q_2
3. *Contradiction.* Each contradiction between two propositions diminishes the coherence between them.
4. *Competition.* Two propositions *compete* if they occur in the antecedent of two different rules with similar consequents, but do not occur in the same rule. Each form of competition between two propositions diminishes the coherence between them.

   For example, if P, $Q \rightarrow R$ and Q, $S \rightarrow R$, then P and S compete, since they both explain R but do not occur in the same rule.
5. *Data.* Propositions that are represented as data (because they are observed, for example) cohere with the special proposition true.

For simplicity's sake, a number of minor details have been left out here. For example: the implication principle officially works with a simplicity factor, α, which is in practice always set to 1. For the details, cf. (Thagard 1994). Table 1 describes how to derive a coherence network from logical data.

2.3. *Initializing a Coherence Network*

ECHO's next step is to initialize the coherence network by assigning to every proposition an activation value (Table 2). The value 0.01 can be considered as a seed that initially gives all propositions some benefit of the doubt. The rest of their activation, then, must be obtained from other propositions. The activation of true is clamped to 1 throughout the process. Thus, the special proposition true is completely accepted, and remains accepted throughout the entire process.

PROCEDURE derive network

1. Create nodes for all propositions, plus a node for the proposition true.
2. Increase the degree of coherence between all propositions that are coherent according to the implication and analogy principles by a standard amount, say 0.04.[a] (Take into account that weights are additive, so that if more than one principle applies, the weights sum.)
3. Set the degree of coherence between true and data propositions to a small positive value, say 0.05.
4. Decrease the degree of coherence between all propositions that are incoherent according to the contradiction and competition principles by a standard amount, say 0.06.

[a] The numbers are more or less arbitrary and are determined from experience.

Table 1. Deriving (setting up) a coherence network

PROCEDURE initialize network

1. Set the activation of true to 1, and of all other propositions to a small positive value, say 0.01.

Table 2. Initializing a coherence network

2.4 *Harmonizing a Coherence Network*

A network is usually incoherent after the initialization phase. ECHO's third step, then, is to make the network as coherent as possible. This is done by easing the "logical tension" that exists among the different propositions. The situation might be seen as a three-dimensional graph, where links between nodes are spiral springs between wooden balls. Some springs are shorter than others. A short spring between two nodes means that the two nodes are coherent. A long spring between two nodes means that the two nodes are incoherent. Pulling two coherent nodes apart costs energy, and putting two incoherent nodes together costs energy as well. Since the network consists of multiple springs, it may happen that two incoherent nodes are brought together by other nodes in the network, because the two incoherent nodes both belong to the same coherent clique. Conversely, it may happen that two coherent nodes are pulled apart because they belong to two different groups that are incoherent. Thus, certain configurations of the nodes cause more tension in the network than other configurations. The least strenuous configuration is obtained simply by releasing the network, i.e., by letting it loose, so that all nodes assume a position that optimally contribute to the greatest possible decrease of tension in the network.

To harmonize the network, it is run in cycles to synchronously update all units using the following equation:

$$\mathrm{ACT}(p)_{new} := \mathrm{ACT}(p)(1 - \theta) + \begin{cases} \mathrm{NET}(p)(\max - \mathrm{ACT}(p)) \text{ if NET} > 0 \\ \mathrm{NET}(p)(\mathrm{ACT}(p) - \min) \text{ otherwise} \end{cases} \quad (2)$$

where ACT(p) is the activation value of p, θ is a so-called decay factor, max is the maximum activation (usually 1), min is the minimum activation (usually -1), and NET(p) is the net input to p:

$$\mathrm{NET}(p) =_{Def} \sum_{\text{all neighbours } q \text{ of } p} \mathrm{ACT}(q)\omega_{pq}$$

where ω_{pq} is the strength, or weight, of the link that connects p to q in the coherence network. Formula (2) is taken from (McClelland and Rumelhart 1989). More about the why and how of this update formula is given in the next section.

If this is done for the input displayed in Example 2.1, ECHO produces the following output:

accepted propositions		*rejected propositions*	
true	1.0	PH4	-0.44132495
OH1	0.91564536	PH6	-0.71097136
OH3	0.8557134	PH5	-0.71097136
OH5	0.82189536	PH2	-0.79307806
OH2	0.79902226	PH1	-0.8158864
OH4	0.686075	PH3	-0.8158864
E3	0.60112447		
E7	0.59894043		
E4	0.59236825		
E6	0.5908484		
E8	0.5758307		
OH6	0.48836628		
E5	0.48127842		
El	0.45618105		
E2	0.21289238		

(Source: `http://cogsci.uwaterloo.ca/JavaECHO/echoApplet.html`.)

Since hypotheses of the oxygen type are accepted and hypotheses of the phlogiston-type are rejected, ECHO suggests that the oxygen theory of combustion is superior to the phlogiston theory of combustion.

3. Problems with TEC

TEC is an important and attractive account of coherence that has withstood the test of severe criticism. Several objections to Thagard's proposal were made (Thagard 1989), and Thagard replied to all of them in a clear, cogent and convincing manner (Thagard 1989, 1994).

Nevertheless, I maintain that TEC still has some problems. These problems are not fatal, and do not in any way compromise TEC's basic principles. Nevertheless, none of them is mentioned, or suggested, by Thagard in the problem section, while they are relevant enough to be discussed. Here are the problems:

I. The use of the update formula (2) above is not well motivated in the main exposition of TEC (Thagard 1994), nor is it well-motivated in related work on coherence-as-constraint-satisfaction (Thagard 1989; Thagard and Millgram 1995; Thagard et al. 1997; Verbeurgt and Thagard 1998). What pattern of convergence does (2) imply? How does it relate to, say, local hill-climbing techniques, known from traditional differential calculus?

II. The principles for deriving a coherence network from logical input (Sec. 2.2) are, in large measure, empirically determined rather than being theoretically underpinned. Can such empirical justifications be scientifically defended?

III. In TEC, the coherence network is *derived* from the logical data available. This makes Thagard's notion of coherence an indirect one. Would it be possible to construct a coherence network right out from the logical data themselves? (And, if so, how?)

IV. In TEC, propositions are sentences without structure. But we often need more expressive languages to make our statements. Would it be possible to extend the idea of coherence to more expressive languages, such as the language of propositional logic, or the language of first-order logic? If so, how?

V. TEC settles for a global optimum. However, it is always possible that a global optimum is established by other network configurations as well, especially if the network is harmonized without decay. What is the meaning of the existence of different optimal network configurations, and how does this influence the overall acceptance of propositions?

Problems I-V will be discussed in turn below. Sometimes I present solutions; at other times I suggest approaches that might lead to a solution.

Problem I. Formula (2) is neither explained nor motivated in (Thagard 1989, 1994; Verbeurgt and Thagard 1998). Thagard and Verbeurgt refer to McClelland and Rumelhart (1989), but do not explain what (2) does. Below, I explain what (2) does, and argue that it is not necessarily the most obvious choice for updating all nodes in a coherence network.

Apparently, (2) is a pseudo-addition of ACT(p)(1 - θ) and the net input to p, NET(p). That is, apparently, (2) is

$$\text{ACT}(p)_{new} := \text{ACT}(p) \oplus \text{NET}(p),$$

where $\oplus$ is a kind of addition on the interval [-1, 1] such that normal properties of addition hold [$x \oplus 0 = x$ and $x \oplus y = y \oplus x$ and $(x \oplus y) \oplus z = x \oplus (y \oplus z)$], with 1 behaving as ∞ and -1 behaving as $-\infty$ [$x \oplus 1 = 1$ and $x \oplus -1 = -1$, and $-1 \leq x \oplus y \leq 1$]. In (2), Thagard uses

$$x \oplus y =_{Def} \begin{cases} x + y(1 - x) & \text{if } x \geq 0 \\ x + y(x+1) & \text{otherwise} \end{cases} \tag{3}$$

but the simpler

$$x \oplus y =_{Def} (x + y)/(1 + xy) \tag{4}$$

could have been used just as well. Experiments support this observation. Experiments also support the observation that (4) leads faster to solutions than (3). This is one point.

Another point is that the use of (3) or (4) is not self-evident. It is also possible to compute the next value of ACT(p) by gradient ascent, for example:

$$\begin{aligned} \text{ACT}(p)_{new} \quad &:= \min\{\max\{\text{ACT}(p) + \eta \frac{\vartheta global\_coherence(C)}{\vartheta p}, -1\}, 1\} \\ &= \min\{\max\{\text{ACT}(p) + \eta\, \text{NET}(p), -1\}, 1\} \end{aligned} \tag{5}$$

where "min" the "max" ensure that ACT(p) remains between -1 and 1, and η is a constant, sometimes referred to as the *learning rate.* Experiments suggest that (5), with $\eta = 0.5$, leads faster to solutions than (4) and that (4) leads faster to solutions than (2) or (3). So why not use gradient ascent?

Another problem is that the use of decay values is questionable. In *Conceptual Revolutions,* we read:

> … ECHO automatically increases the value of a decay parameter in proportion to the ratio of unexplained evidence to explained evidence (...) (Thagard 1994, p. 80)
>
> … Another important parameter of the system is decay rate, represented by θ .... We can term this the *skepticism* of the system, since the higher it is, the more excitation from data will be needed to activate hypotheses. If skepticism is very high, then *no* hypothesis will be activated. (p. 81)
>
> … θ is a decay parameter that decrements each unit at every cycle (p. 100)
>
> … greater decay values tend to compress asymptotic activation values towards 0 (p. 101)

All of the above is true, but the problem is that a positive decay value causes more than moderate activation values: when I ran my own implementation of ECHO (in Perl), the results indeed suggest that large, or at least positive, decay values tend to compress asymptotic activation values towards 0, and that small decay values tend to compress asymptotic activation values towards the boundaries of [-1, 1]. But the same experiments also suggest that the best coherency is reached for $\theta = 0$, and *not* for $\theta > 0$. Thus, if attaining moderate activation values is the most important objective, then the decay should be indeed positive. If optimizing coherency is the most important objective, however, then there should be no decay at all. Since TEC is aimed at optimizing coherence, it should always be the case that $\theta = 0$.

Problem II. In section 4.1.2, Thagard (1994) goes to some lengths to justify the principles of explanatory coherence (Sec. 2.2 above). Section 4.1.2 thus forms the theoretical justification for these principles. His argumentation is convincing and seems to be correct.

To me, section 4.1.2 also shows, however, that almost *every* principle can be supported by a plausible justification, as long as it is not too far-fetched, and its advocate is able to "sell" it. The latter is not a problem in Thagard's case, because Thagard's writing style is cogent and convincing. But this means that additional principles can be introduced at will, as long as the supporting argumentation is good. For example, why not introduce a "Principle of Conjunction," saying that *"P* and *Q"* coheres with *"P"* and "*Q*"? One possible answer would be that ECHO's language doesn't allow conjunctions. But why, then, opt for implications (explain) and contradictions (contradict) over other connectives, such as negation, conjunction, or disjunction? Why not opt for negation, accompanied by a "Principle of Negation," saying that "not *P*" incoheres with "*P*" ?

Thagard, however, also uses *empirical* arguments to justify the principles of explanatory coherence. This is most manifest in section 4.1.3 of *Conceptual Revolutions* (1994), where Thagard explains that certain earlier principles were abandoned because "they lack interesting scientific applications," or "do little to illuminate actual scientific cases." Further, new principles (such as competition) were adopted "to cover cases" that initially did not come out right in the first version of ECHO (Thagard 1994, in a footnote on p. 66). Thus, it seems that the principles for deriving a coherence network from logical input, i.e., the *principles of explanatory coherence* are, to an important degree, empirically determined rather than being theoretically underpinned.

Let me first state that I have no problem with an empirical justification of coherence principles. If ECHO works better with certain parameter settings than with other parameter settings, then why not use the better parameter settings? In particular, if ECHO gives better outcomes when the principle of competition is incorporated, then why not use the principle of competition?

Still, the danger of using empirically justified principles is that any principle may function as a candidate-coherence principle, since the only criterion that counts (empirically speaking) is performance. As long as a principle helps ECHO to produce the right outcomes, it may be selected as a principle of explanatory coherence. This does not seem to be right and opens the door to arbitrary principles.

Another problem is that one (fixed) principle can be translated in a number of different ways. Even *if* the introduction of additional theoretical principles is taken for granted, we still have to determine how these principles are translated into coherence relations between nodes. For example, the contradiction principle (p. 6) states that each contradiction between two propositions diminishes the coherence between them. The default parameters of ECHO, then, for diminishing the weight between two propositions, is 0.06. So if the weight between P and Q was 0.78, say, then a contradiction between P and Q would diminish the coherence between P and Q to 0.78 - 0.06 = 0.72. But why -0.06 and not, say, -0.05 or -0.07? In Thagard (1994), it is explained that different parameter settings lead to essentially the same outcomes, qualitatively speaking, except for the ratio between standard excitation (0.04) and standard inhibition (-0.06) of weights. If this ratio is ill-chosen, then either too many or too few nodes will be activated. But then why set the standard excitation for all positive coherence relations (implication, analogy) to the same value (0.04)? Similarly, why set the standard inhibition for all negative coherence relations (contradiction, competition) to the same value (-0.06)? These choices indicate that there are numerous degrees of freedom in the translation of coherence principles, with the unpleasant consequence that the derivation of a coherence network becomes a relatively arbitrary process.

The problem of determining which principles are important in TEC and which are not, seems to be a metaphysical one: we try to capture reality, but all we do is devise principles about how we *think* about reality. To me, such principles depend on the metaphysical preferences of their creator and are therefore arbitrary. So it seems that we have to abandon our ideal of having five or six core principles of TEC.

4. Direct Coherence

Problem III. In TEC, the coherence network is *derived* from the logical data available. This problem is related to Problem II (p. 9). This section offers a possible answer to both problems.

An important property of Thagard's notion of coherence is that it is a *derived* one, in the sense that the various coherency and incoherency relations among propositions are derived from the logical data available (such as rules, analogies, contradictions, and competing explanations). A derived notion of coherence works well in most cases, as it more or less reflects the logical relation between the various propositions. A disadvantage of a derived notion of coherence, however, is that it is indirect. The problem with an indirect notion of coherence is that its maximization does not necessarily maximize the coherence between the logical concepts themselves. In this way, Thagard's notion of coherence becomes a secondary, or artificial notion of coherence, which must be derived from existing rules and propositions. Indirectness is not raised as a point of criticism in (Thagard 1989), by the way.

The goal of this section is to come to a more direct notion of coherence - direct in the sense that we are aiming at a notion of coherence that already resides in the logical input itself. To explain how this might be achieved, we use a network flow metaphor, based on analogies between network flows and propagation of "truth" through rules of inference. If some analogies appear somewhat constructed and artificial, then please bear in mind that they are meant in the first place to help.

The idea behind the flow metaphor is to "pump" (infer) a "truth serum" (validity) from one or more "sources" (observations) through a "network of pipelines" (rules) to one or more "sinks" (unobserved propositions). If a pipe or node is saturated, the serum cannot pass and the flow must find its way through other channels. The flow metaphor offers a convenient analogy, but comes with a few problems. The first problem appears when the network is saturated. In that case, no pipe has extra capacity, so that a computer implementation of reason-as-flow-net works keeps sending back and forth superfluous "truth serum," *unless* the programmer has ensured that the computer program keeps track of which channels already have been tried and which not. Another more serious problem is to select where to "drain off" truth serum that turns out to be superfluous. In a real physical network consisting of pipes and T-joints, the source must eventually take back all the flow that cannot be handled by the network. Thus, in normal situations the network is supposed to be watertight, so that the surplus of flow will be sent back to the source eventually. But here the situation is different. Once it has been observed that a certain amount of flow cannot be handled by the network whatsoever, one or more rules must be selected to drain off the extra flow. To this end, the idea is put "safety valves" on the selected rules, so that the surplus amount of flow can "leak" through those vents. (Feel free to smile if you find the analogy somewhat labored and artificial.)

Which rules must leak? From an epistemological point of view, the standpoint is that perception is more direct and, hence, more reliable than weakly supported rules that are obtained indirectly through inductive reasoning. Thus, according to this point of view, the degree of belief of propositions obtained through perception must be respected more than products of inductive reasoning, viz. rules. Since a logical (and hence artificial) network permits us to introduce leaks everywhere (it is just a matter of programming), we can "jab" leaks at weak rules, or weakly supported rules, while leaving the stronger epistemic beliefs (such as observations and deductive rules) untouched. In this way propositions obtained through perception are prioritized at the expense of weak rules that are obtained indirectly through inductive reasoning.

4.1. *Basic Concepts*

To carry out the above ideas, we need three basic concepts, namely, degree of belief (DOB), degree of support (DOS) and activation (ACT). We begin with the DOB. Some (but not all) propositions possess a fixed *degree of belief,* DOB $\in [0,1]$. A degree of belief ascribes an inherent degree of belief to a proposition, due to observation, or due to the fact that the proposition in question is input.

All propositions possess a variable *activation value,* ACT $\in [0, 1]$, that is initially set to

$$ACT(x) = \begin{cases} DOB(x) & \text{if DOB}(x) \text{ exists,} \\ 0 & \text{otherwise.} \end{cases} \tag{6}$$

with TEC, the point of departure in defining a direct notion of coherence is a collection of logical rules of inference and propositions. Only this time we allow for weighted implications.

Example 4.1. Consider

rule		DOB	*proposition*	DOB	ACT
rule1:	$g, \neg h, k$ -(0.94)→ $\neg a$	1.00	a	0.89	0.89
rule2:	$\neg e, f, \neg d, b$ -(0.78)→ $\neg a$	1.00	$\neg a$		0.00
rule3:	$a, \neg b$ -(0.92)→ c	1.00	b		0.00
rule4:	d -(0.89)→ c	1.00	$\neg b$	0.87	0.87
rule5:	$\neg a, e$ -(0.87)→ k	1.00	c		0.00
rule6:	$d, \neg e$ -(0.93)→ k	1.00	$\neg c$		0.00
rule7:	$\neg g$ -(0.98)→ k	1.00	d	0.56	0.56
⋮			$\neg d$		0.00
			⋮		

The fact that 0.94 > 0.78, suggests that g, ¬h, and k imply ¬a with more certainty than ¬e, f, ¬d, and b imply ¬a. We stop with the example for now, but continue with it in a moment.

The driving force behind establishing direct coherence is that activation values are usually "wrong" and must be adjusted. To see why, we introduce the notion of derived activation, or *degree of support*, DOS $\in [0, 1]$. Propositions as well as rules are supported. Support for a proposition cannot be computed directly, but must be computed via rule support.

Definition 4.1. (Support) 1. Let r = "$a_1, \ldots, a_n$ -(s)→ a". The support that r gives to a, or the support that a receives through r, is the minimum activation of the elements in the antecedent, times the implication strength s of that rule:

$$\text{DOS}(r) =_{Def} s * \min\{\text{ACT}(a_1), \ldots, \text{ACT}(a_n)\} \tag{7}$$

2. The (accumulated) support of a proposition a is the sieve-sum of the support of all rules that support a:

$$\text{DOS}(a) =_{Def} \oplus \{\text{DOS}(r) \mid r \text{ is a rule for } a\} \tag{8}$$

The sieve-sum is defined by $x \oplus y =_{Def} x + y - xy$. This sum behaves like ordinary addition (it is commutative and associative, for example) except that if $0 \leq x, y \leq 1$, then $0 \leq x \oplus y \leq 1$.

Definition 4.1(1) is based on the principle that the support that a rule gives to its consequent (in this case, a) is determined by the weakest element. Definition 4.1(2) is based on the idea that the support for one proposition from multiple sources, accrue.

We now continue our running example. The support that Rule 3 gives to c can be computed, because the DOBS, and hence the ACTs, of all elements of the antecedent of Rule 3 are known:

$$\text{DOS(“Rule3”)} = s * \min\{\text{ACT}(a), \text{ACT}(\neg b)\}$$
$$= s * \min\{\text{DOB}(a), \text{DOB}(\neg b)\}$$
$$= 0.92 * \min\{0.89, 0.87\}$$
$$= 0.80$$

Similarly with Rule 4:

$$\text{DOS (“Rule4”)} = 0.56 * 0.89$$
$$= 0.50$$

Because *c* is supported by Rule 3 and Rule 4, *c*'s support is

$$\text{DOS}(c) = \text{DOS (“Rule 3”)} \oplus \text{DOS (“Rule 4”)}$$
$$= 0.8 + 0.5 - 0.8 * 0{,}5$$
$$= 0.9$$

Now let us suppose that ACT(c) = 0.00 at the time we were computing *c*'s support. In that case, DOS(c) $\neq$ ACT(c). This difference indicates an incoherence between *c*'s activation proper (ACT), and what the rules of inference say that *c*'s activation should be (DOS). It is our task to "smooth out" the differences between ACT and DOS, with the prospect that eliminating the difference at one node almost always introduces differences at other nodes.

There are several ways in which the difference between support and activation can be lessened. We consider two of them, viz. (forward) propagation ("prop") and back-propagation ("backprop").

Propagation. According to the first approach we assume that all activation values are "wrong" and must be modified to the support (derived activation) that has been derived from the (old) activation values. We call this method "prop," since the (old) activation values propagate through the rules forward to compute the new activation values. Thus, with "prop" we would add 0.9 to ACT(c) to obtain DOS(c) = 0.9. This is a relatively straightforward computation.

Back-propagation. Another way to look at support is to say that the derived support values (rather than the activation values) are "wrong" because they are computed on the basis of "wrong" activation values. Here, the approach is to go back in the rules to modify the activation of predecessors. We call this method "backprop," because activation propagates backward through rules to compute new activation values. Thus, with "backprop," we reduce one of the activation values of one or more elements of one of the antecedents of Rule 3 and Rule 4, to reduce *c*'s support to 0.0. In the running example, DOS(c) = 0.00 might be achieved by choosing to set DOB(d) = 0.00 and by setting either DOB(a) = 0.00 or DOB($\neg b$) = 0.00.

Back-propagation is more complicated because we must choose which rules, which antecedents of those rules, and which elements of those antecedents, must be modified. Thus, normal propagation is straightforward, while back-propagation is more difficult. If DOS(c) ≠ ACT(c), there are two cases to consider.

1. DOS(c) < ACT(c). In this case, we will have to "boost" one or more rules that support c. The choice between boosting one rule or boosting more rules depends on what you want. Almost always, you would like to increase the difference among rules concerning throughput of conclusive force. In that case select the best rule and increase its throughput, provided that this rule is able to compensate for the difference
   |DOS(c) - ACT(c)| of itself. (If not, then also improve the second-best rule, up to and including the nth-best rule, if necessary.) The other possibility is that we would like to establish the opposite, namely, to level out the difference among rules. In that case we give all rules a bit extra. The definition of "best" rule may vary. It can be defined as the rule with the greatest throughput, capacity (strength), DOB, ACT, or a combination of these factors. This is entirely up to the designer of the network. How a rule's throughput, or activation, may be increased is explained in the next paragraph.

2. DOS(c) > ACT(c). In this case, we will have to "temper" one or more supporters of c. Here too we have the choice of modifying one or more rules, depending on whether or not we would like to increase the difference in rule support.

Eq. 7 above indicates that a rule's throughput is determined by the element of the antecedent that has the lowest activation value. Therefore, to change a rule's throughput it usually suffices to change the activation value of one element of the antecedent, namely, the element that has the lowest activation value. If this does not produce the desired effect, then change the one-but-smallest, up to and including the nth-but-smallest element of the antecedent, if necessary. Alternatively, it is also possible to uniformly decrease or increase all elements of the antecedent. Which modification method you use depends on what you are after. If a rule's throughput must be increased and you would like to enlarge the difference among activation values, then increase the activation value of all elements in the antecedent. Otherwise, increase the activation value only of the element of the antecedent with the smallest activation value, and leave all other elements in the antecedent untouched. If a rule's throughput must be decreased and you'd like to enlarge the difference among activation values, then decrease the activation value only of the element of the antecedent with the smallest activation value, and leave all other elements in the antecedent untouched. Else, decrease the activation value of all elements in the antecedent. (Note the reversed order.) See also Table 3.

Your choice is to…	*Boost rule*	*Temper rule*
…increase difference in rule activation	Boost the entire antecedent	Temper the minimum element of the antecedent only
…level out difference in rule activation	Boost the minimum element of the antecedent only	Temper the entire antecedent

Table 3. Changing activation values in back-propagation.

Additional constraints. Principles of logical inference suggest a number of additional constraints.

A. An additional constraint could be that

support ≥ activation

for each node in the network. The idea underlying this constraint is that support is considered as a *facilitator* of activation, in the sense that activation exists by the grace of support. (Just as physical activity [movement, light, sound] exists by the grace of energy resources [fuel, electicity].) The difference

slack $=_{Def}$ support – activation

represents the "leakage" (remainder, or residue) of conclusive force from the supporting rules.

B. Another plausible constraint is that deductive rules of inference, i.e., rules with an implicational strength equal to 1, are not allowed to "leak." Thus, this constraint amounts to *slack* = 0 for deductive rules.

C. A refinement of (B) is to require that weak rules may be compromised more than strong rules. An alternative is to require that rules with a low DOB may be compromised more than rules with a high DOB.

These constraints are meant as implementation options. Listing them does not imply that they are written on a biblical stone or that they must be followed unconditionally!

4.2. *Knonet*

KNONET is an implementation of the above ideas on direct coherence, with the following design choices:

1. Activation is adjusted with the mean of node support and what is indicated by back-propagation. Thus, we simply take the average of "prop" and "backprop".
2. Back-propagation is done such that it increases differences in activation that might exist among nodes. (See Table 3.)
3. We permit situations in which activation is strictly greater than support (cf. point A above).
4. The burden to compensate differences between DOS and ACT lies with rules that are believed relatively less (regardless of their strength). Thus, rules with a low DOB are permitted to "leak" more than rules with a high DOB.
5. With respect to strength, all rules are considered equal when it comes to compensating the difference between DOS and ACT (cf. points B, C above). We merely look at the DOB.

As an example, we translate Example 2.1 to KNONET. Although the translation is simple, it preserves all essentials of the original example:

i. Every contradiction "*X* contradicts *Y*" is replaced by two rules, viz. X -(1.00)$\rightarrow \neg Y$ and Y -(1.00)$\rightarrow \neg X$.
ii. Because explanations and contradictions are considered self-evident, we give them a DOB of 1.00.
iii. Because evidence is considered indisputable, we give each piece of evidence a DOB of 1.00.
iv. Since we do not know how strictly the rules must be interpreted, we give each rule a strength of 0.90.
v. As described above, all claims receive an activation value. If they have a DOB, the activation value is equal to the DOB, otherwise it is 0.00.

In this way, Example 2.1 changes into

```
# evidence      # rules                      # contradictions
e1 1.0          oh1 oh2 oh3 0.9 e1 1.0       ~ph3 1.0 oh3 1.0
e2 1.0          ph1 ph2 ph3 0.9 e1 1.0       ~oh3 1.0 ph3 1.0
e3 1.0          ph1 ph3 ph4 0.9 e2 1.0       ~ph6 1.0 oh5 1.0
e4 1.0          oh1 oh3 0.9 e3 1.0           ~oh5 1.0 ph6 1.0
e5 1.0          oh1 oh3 oh4 0.9 e4 1.0
e6 1.0          oh1 oh5 0.9 e5 1.0
e7 1.0          ph5 ph6 0.9 e5 1.0
e8 1.0          oh1 oh4 oh5 0.9 e6 1.0
                oh1 oh5 0.9 e7 1.0
                oh1 oh6 0.9 e8 1.0
```

If KNONET is applied to the present case 30 times, it produces:

oh1	1.00	oh6	0.98	~e2	0.50	~e1	0.50	
e2	1.00	ph4	0.96	~e4	0.50	~e3	0.50	
e4	1.00	oh5	0.96	~ph4	0.50	~e5	0.50	
e6	1.00	oh2	0.90	~ph2	0.50	~e7	0.50	
e8	1.00	ph1	0.90	~oh6	0.50	~oh1	0.50	
e1	1.00	~ph6	0.86	~oh4	0.50	~ph1	0.50	
oh4	1.00	oh3	0.82	~oh2	0.50	~oh3	0.28	
e7	1.00	~ph3	0.77	~e8	0.50	ph3	0.25	
e5	1.00	~ph5	0.50	~e6	0.50	ph6	0.13	
e3	1.00	ph2	0.50	ph5	0.50	~oh5	0.05,	g_err 2.25

The global error is defined by

$$g\_err = \sqrt{(\mathrm{ACT}(n_1) - \mathrm{DOS}(n_1))^2 + (\mathrm{ACT}(n_2) - DOS(n_2))^2 + \ldots}$$

where n_1, n_2,… are nodes. Since the global error becomes smaller if the global difference in activation and derived activation (support) becomes smaller, it is a respectable measure of the incoherence of the entire network.

Let us consider e_1 as an example, where e_1 is the evidence that, in combustion, heat and light are given off. This proposition is supported by two rules, viz.

$$oh_1\ oh_2\ oh_3 \ {+}(0.90) \rightarrow e_1$$
$$ph_1\ ph_2\ ph_3 \ {+}(0.90) \rightarrow e_1$$

If we write behind every proposition in the antecedent its activation value (at a specific point in the iteration process), we obtain

$$oh_1\ [\mathbf{1.00}]\ oh_2\ [\mathbf{1.00}]\ oh_3\ [\mathbf{0.88}] \ {+}(0.90) \rightarrow e_1\ ph_1$$
$$[\mathbf{1.00}]\ ph_2\ [\mathbf{0.50}]\ ph_3\ [\mathbf{0.67}] \quad {+}(0.90) \rightarrow e_1$$

0.86 e_1

With this information, and the information that every rule has a DOB of 1.00, we can compute the support that every rule gives to its consequent. For the first rule this is min{1.00, 1.00, 0.88} * 0.90 = 0.79. For the second it is min{1.00, 0, 50, 0.67} * 0.90 = 0.30:

$$oh_1\ [1.00]\ oh_2\ [1.00]\ oh_3\ [0.88] \ {+}(0.90) \rightarrow \mathbf{0{,}79}\ e_1$$
$$ph_1\ [1.00]\ ph_2\ [0.50]\ ph_3\ [0.67] \ {+}(0.90) \rightarrow \mathbf{0{,}30}\ e_1$$

0.86 e_1 [1,00]

The next step is to accumulate (with $\oplus$) the rule support of e_1 to the total support of e_1.

$$\begin{array}{r} oh_1\ [1.00]\ oh_2\ [1.00]\ oh_3\ [0.88]\ \leftarrow\!\!\!\!\!+ 0.90) \rightarrow\ 0{,}79\ e_1 \\ ph_1\ [1.00]\ ph_2\ [0.50]\ ph_3\ [0.67]\ \leftarrow\!\!\!\!\!+ 0.90) \rightarrow\ 0{,}30\ e_1 \\ \hline \mathbf{0.86}\ e_1\ [1{,}00] \end{array}$$

Behind e_1 we have written its activation. Thus, the activation proper is

$\mathrm{ACT}(e_1) = \mathrm{DOB}(e_1) = 1.00$ (since it is evidence) while the derived activation, or DOS, is

$\mathrm{DOS}(e_1) = 0.86$. Thus, the local error lerr at e_0 is

$$\mathrm{Lerr}(e_0) = \sqrt{(0.86 - 1.00)^2} = |\ 0.86 - 1.00\ | = 0.14 \quad (9)$$

All local errors are relatively low, since the proposition tables are made after the network has gone through a number of cycles in which the global error decreased.

5. Direct Propositional Coherency

Problem IV. TEC deals with atomic propositions. But in making scientific statements, or any type of statements for that matter, we often need languages that are more expressive. Would it not be possible to extend the idea of coherence to more expressive languages, such as the language of propositional logic, or the language of first-order logic? If so, how?

One possible answer to this question is to replace TEC's language by a slightly more expressive (formal) language. An obvious candidate here is the language of propositional logic. Such a language enables us to formulate additional principles that express the coherence between logical formulas and their constituents. In this way, $P \wedge Q$ would cohere with P, $P \wedge Q$ would cohere with Q, and so forth:

$P \wedge Q \sim P, \qquad P \vee Q \sim P, \qquad \neg P \nsim P,$

$P \wedge Q \sim Q, \qquad P \vee Q \sim Q$

This approach produces a number of new problems: are all coherence relations treated equally? For example, are P and $P \wedge Q$ as coherent as P and $P \vee Q$ are? How is the material implication $P \supset Q$ to be interpreted? One possible answer to this question is to treat $P \supset Q$ as $\neg P \vee Q$ so that $P \supset Q$ is incoherent with P

but coherent with Q. In this way, we could extend the principles of coherence (§2.2) as follows:

- *Negation.* Each negation $\neg P$ diminishes the coherence between $\neg P$ and P.
- *Conjunction.* Each conjunction $P \wedge Q$ strengthens the coherence between $P \wedge Q$ and P and $P \wedge Q$ and Q.
- *Disjunction.* Each disjunction $P \vee Q$ strengthens the coherence between $P \vee Q$ and P and $P \vee Q$ and Q.

Implication, would then already be covered by Principle 1 above. I have not run experiments on the basis of these additional principles, but their implementation of them seems straightforward. Whether they reflect Thagard's idea on coherence is another matter.

5.1. *Continuous Truth-Values*

Another approach to propositional coherency, and one that I have tested experimentally, is to use continuous truth-values, i.e., truth-values that range from 0 to 1. To determine the coherency of a set of propositional formulas, we create a network with nodes that correspond to subformulas of all formulas.

Example 5.1. Suppose we would like to investigate the coherency of

$$C = \{P, (\neg P) \vee Q, \neg Q\}$$

Intuitively, C's coherency should be low, since it is inconsistent. Create nodes for all subformulas:

node	*subformula*	*node*	*subformula*
P	P	S	$\neg Q$
Q	Q	T	$R \vee Q$
R	$\neg P$		

The number of nodes of the network thus obtained, depends linearly on the length of the input: there are as many nodes as there are subformulas, and one can prove that the number of subformulas depends linearly on the length of a formula. Thus, setting up coherence networks for large sets of propositional formulas is computationally feasible. (End of Example.)

At this point, the network consists of *triples* and *pairs*. Triples for binary connectives ($\wedge$, $\vee$, and $\supset$), and pairs for unary connectives ($\neg$). An example of a $\wedge$-triple is (U, V, W), with $W = U \wedge V$. In this case we say that W is a *parent* of U and V, and that U and V are the *children* of W. Every parent has either one or two children, depending on the connective. A child can have arbitrarily many parents.

Example 5.2. Consider the formula $(\neg P) \wedge ((P \vee P) \supset P)$, with subformulas $Q = \neg P$, $R = P \vee P$, $S = R \supset P$, and $T = Q \wedge S$. Then P is a child of many nodes, viz. Q, R, and S. (End of Example.)

Given a propositional coherence network, we change the Boolean variables to real numbers from 0 to 1 and redefine the logical operators as follows:

$$\begin{aligned} \neg P &= 1 - P \\ P \wedge Q &= PQ \\ P \vee Q &= P + Q - PQ \\ P \supset Q &= \min\{Q/P, 1\} \end{aligned} \tag{10}$$

This extension of connectives from discrete to continuous values is sometimes referred to as the *Goguen*-extension of truth-functional connectives. There are more extensions of connectives (min/max, Lukasiewicz, Kleene-Dienes, Zadeh, Reichenbach, Weber-family, Hamacher-family, Yager-family), but a disadvantage of some of these alternatives is that they are algebraically more complex than the Goguen-type of extension (Zadeh, Reichenbach), or else are less suitable for optimization of coherency (min/max). Kruse *et al.* (1994) contains a clear and concise overview of real-valued logical connectives. The Goguen extension of logical connectives is almost exclusively used in the realm of fuzzy logics, but I hasten to add that computing the coherence of propositional formulas is still remote from fuzzy logic.

Like Thagard's coherence networks, nodes have activation values. But since the language of propositional logic is more expressive than the language of TEC, it is no longer necessary to have activation values ranging from -1 to 1. Instead, it suffices to set the bounds at 0 and 1. Disbelief in a formula P can now be expressed as $\text{ACT}(\neg P) = 1$, rather than $\text{ACT}(P) = -1$, thanks to the enhanced expressiveness of the language.

The next step, then, is to update the network in cycles, by updating triples and pairs synchronously. This is done by trying to make every triple and pair more coherent. For example, the pair (U, V) with $V = \neg U$ is optimally coherent if $\text{ACT}(V) = 1 - \text{ACT}(U)$. This can be verified by the reader for discrete truth values $U = 0$ and $U = 1$. An example of extreme *in*coherence would be $U = V = 1$, or $U = V = 0$. Often $\text{ACT}(V) \neq 1 - \text{ACT}(U)$. In the *xy*-plane, optimal coherent pairs lie on the line $y = 1 - x$. To make (U, V) into a coherent pair (U', V') such that the distance between (U, V) and (U', V') is minimal, we have to choose a point on the line $y = 1 - x$ that is close to (U, V). This point is

$$(U', V') = \frac{1}{2}(U - V, V - U) + (1, 1).$$

It can be verified that $\text{ACT}(V') = 1 - \text{ACT}(U')$. Thus, merely taking (U, V) into consideration, (U, V) should change into (U', V'), or at least move in the direction of (U', V'), to maximize coherency.

Similarly, to increase the coherency of the triple (U, V, W) with $W = U \wedge V$, we should look at triples (U', V', W') such that

1. The distance between (U, V, W) and (U', V', W') is minimal.
2. (U', V', W') is optimally coherent, where an $\wedge$-triple (U', V', W') is considered optimally
   coherent if $\text{ACT}(U')\text{ACT}(V') = \text{ACT}(W')$. (Cf. Equation 10.)

Geometrically, these two constraints can be fulfilled by drawing a line l through (U, V, W) perpendicular to the $z = xy$ surface in R_3. The triple (U', V', W'), then, is where l pierces the $z = xy$ surface. Algebraically, (U', V', W') can be determined less easily. There are two approaches. The first is to formulate an equation of all lines l perpendicular to the $z = xy$ surface, and then investigate which of these lines meet (U, V, W). Another approach is to express the distance between (U, V, W) and an arbitrary point (x, y, xy) on the $z = xy$ surface, and then to minimize on the distance by taking derivatives. Neither approach works, because they produce polynomials of degree ≥ 5, for which no general solution exists (Galois). What I did in my computer experiments, was simply to approximate (U', V', W') with the Gauss-Newton method [4]. Depending on the desired accuracy, this generally took about 5-15 iterations on average.

Similarly, disjunctive triples, i.e. triples of the form (U, V, W) with $W = U \vee V$, are moved in the direction of the $z = x + y - xy$ surface, and implication-triples are moved in the direction of the $z = \min\{y/x, 1\}$ surface (Equation 10 above).

A (final) problem with modifying pairs and triples is that one node can be a member of several triples. For example, a node can be the parent of two children, but can itself be a child of seven different parents. Such a node takes part in eight different relations: one for its children, and seven for its parents. This is a problem, because children and parents might send conflicting values, so that coherency cannot be achieved. The approach I took in the computer experiments was simply to take the average of all inputs and use this as the incoming update value.

The *local error* at each triple is defined as the distance between the triple and the corrected triple (i.e, the triple for which the truth-condition would hold). The *global error, E,* is defined as the sum of the squares of the local errors. (Which brings the problem into the realm of *least-mean squares* optimization problems.)

Global coherency, then, is considered to increase if the global error decreases. We could quantify global coherency as $1/(1 + E)$, or as $exp(-E)$, but I do not know if that is common practice. (Cf. Hertz et al. 1991; Kröse and van der Smagt 1993; Haykin 1994.) If $E = 0$ we have found activation values that satisfy all logical constraints in the network. This does mean that the network is optimally coherent. (It does not mean, however, that the activation values correspond to a propositional model that satisfies the input, for some input-formulas may be activated at values < 1.)

Assessing the significance of single global coherency values of random networks is hard. Not from a computational point of view, but from a quantitative point of view. Apart from a global error of $E = 0$ (maximum coherence) cases in which $E > 0$ say little about the quality of the outcome since the minimum value of E is generally unknown. I therefore have chosen to test the performance of the propositional coherence algorithm against GSAT. GSAT is a simple but renowned algorithm for testing the satisfiability of propositional formulas, and is famous for the speed with which it finds models for large satisfiable propositional formulas (Trick 1996; Hoos and Stützle 2000; Selman et al. 1992).

5.2. *Propositional Satisfiability*

The algorithm that implements direct propositional coherency can be used to verify whether a propositional formula, or a set of propositional formulas, is satisfiable.

If φ is a formula of which we would like to know whether it is satisfiable, we proceed as above with φ's activation clamped to 1. Then harmonize the network (no decay) and compute the truth-value of φ on the basis of the activation values of nodes that correspond with atomic propositions in the stabilized network. If truth-value(φ) = 1, then stop, since φ is apparently satisfiable. If not, then scramble the network and restart. Give up after max_tries restarts.

The algorithm that implements direct propositional coherency is written in Perl, and is able to solve random 150-variable, 645-clause 3SAT instances (when a solution exists) in about 2two minutes on a Pentium Pro. Of course this isn't competitive with GSAT (Trick 1996; Hoos and Stützle 2000; Selman et al. 1992). However, the code is not optimized, and the approach is promising enough to be investigated further.

In connection with propositional satisfiability, the following problem is important and touches upon the general credentials of TEC.

Problem V. In TEC, the network converges to a specific state in which all nodes assume a particular activation value. The problem is that this state need not be unique. It is always possible that, after a restart, the network will reach the same optimum with different activation values.

Example 5.3. If we compute the coherency of

input = {P contradicts Q}

as described in TEC, a network is created with nodes P and Q and a link $P\sim_{0.06}Q$. This network can settle in two states: (ACT(P), ACT(Q)) = $(a, -a)$ and (ACT(P), ACT(Q)) = $(-a, a)$, where $a \in (0,1]$ and depends on the value of the decay parameter. Both states correspond to a global coherency that is equal to the optimal global coherency, which is $0.06a^2$. (End of Example.)

It is perhaps helpful to draw an analogy with the concept of validity in propositional logic. In propositional logic we would say that $\varphi_1, \ldots, \varphi_n \models \varphi$ is valid if all models that satisfy $\varphi_1, \ldots, \varphi_n$, satisfy φ as well. Not just one model but *all* of them. Similarly, in the theory of coherence it would make sense to say that φ is implied by $\varphi_1, \ldots, \varphi_n$ if the acceptance of $\varphi_1, \ldots, \varphi_n$ implies the acceptance of φ, – *not for one configuration of optimal activation values, but for all configurations of optimal activation values.* Likewise, it would be more in line with common sense to say that $T = \{\Psi_1, \ldots, \Psi_K\}$ is a coherent scientific theory if and only if T is accepted in all possible configurations of optimal activation values – not just one. In this way, TEC would reject scientific theories that are accepted in one state of the network, but (partially) rejected in another state of the network (i.e., another "state of the world"). This feature would contribute to TEC as a plausible model of epistemic coherence.

6. Summary

The objective of this article was to explain the machinery behind TEC and to suggest improvements to it. I also hope that this article has taken away some of Kuipers' skepticism about TEC, and that it has removed one of his objections to "computational coherentism," namely, that it makes use of an obscure and ambiguous connectionistic update mechanism to achieve its results.

Here is a summary of possible improvements:

1. Experiments have shown that simple gradient ascent (Eq. 5) leads faster to solutions than ECHO's update mechanism (Eq. 4). Thus, use gradient ascent instead of Rumelhart's update formula.
2. To make accurate scientific statements, languages are needed that are more expressive than the language of TEC. One step in the direction of more expressive languages is to allow the conjunction, disjunction and negation of sentences. The language of TEC can be extended to the language of propositional logic, including additional coherence principles that express the relation between propositions and their subformulas.

3. Propositional coherency can not only be computed by means of the indirect method of TEC, but also directly, by means of minimizing the incoherency of truth-values that exist between composite propositions and their immediate subformulas.
4. Direct propositional coherency is closely related to propositional satisfiability. The results in this paper suggest that algorithms to harmonize propositional coherence networks can also be used to find models for propositional formulas.

A number of problems that Kuipers raised against explanatory coherentism have remained untouched here. An example of such a problem is brought forward by the important observation that TEC is result-oriented rather than process-oriented. Thus, TEC does not foster the ambition to model the actual scientific process itself. I recommend the reader to consult *Structures in Science* to obtain an impression of problems that go well beyond the alleged obscurity of connectionism.

I hope that one of Kuipers' students, or any student for that matter, implements Kuipers' evaluation matrix to compare it with competing evaluation methods, in particular TEC. In this way, a comparison between Thagard's TEC and Kuipers' EM would come down to testing it against a database of formalized cases such as displayed in Example 2.1. Another pleasant side-effect would be that Theo Kuipers would be relieved from doing manual computations that last 45 minutes or longer.

ACKNOWLEDGEMENTS

Many thanks to Theo Kuipers for creating an extraordinary pleasant and stimulating research environment during my stay in Groningen. Many thanks to Atocha Aliseda Llera for her help in making this article more consistent.

Utrecht University
Dept. of Computer and Information Sciences
PO Box 80.089, 3508 TB Utrecht.
email: gv@cs.uu.nl

REFERENCES

Dancy, J. and E. Sosa, eds. (1992). *A Companion to Epistemology. Blackwell Companions to Philosophy Series*. Oxford: Blackwell Ltd.

Darden, L. (1997). Recent Work in Computational Scientific Discovery. In: M. Shafto and P.Langley (eds.), *Proc. Of the 19th Ann. Conf. Of the Cognitive Science Society*, pp. 161-166, Mahwah, NJ: Lawrence Erlbaum.

Everitt, N. and A. Fisher (1995). *Modern Epistemology: A New Introduction.* McGraw-Hill.

Haykin, S. (1994). *Neural Networks: A Comprehensive Foundation.* Macmillan.

Hertz, J.A., A. Krogh, and R.G. Palmer (1991). *Introduction to the Theory of Neural Computation.* Redwood City, CA: Addison-Wesley Publishing Company.

Hoadley, C.M., M. Ranney and P. Schank. (1994). WanderECHO: A Connectionist Simulation of Limited Coherence. In: A. Ran and K. Eiselt (eds.), *Proc. Of the 16th Ann. Conf. Of the Cognitive Science Society*, pp. 421-426. Hillsdale, NJ: Erlbaum.

Hoos, T. and H.H. Stützle (2000). SATLIB: An Online Resource for Research on SAT. In: I. Gent, H. van Maaren, and T. Walsh (eds.), *SAT 2000.* IOS Press.

Kröse, B.J.A. and P.P. van der Smagt. (1993). *An Introduction to Neural Networks.* Fifth edition. University of Amsterdam.

Kruse, R., J. Gebhardt, and F. Klawonn. (1994). *Foundations of Fuzzy Systems.* Chichester, England: J. Wiley and Sons.

Lehrer, K. (1992). Coherentism. In: Dancy and Sosa (1992), pp. 67-70.

McClelland, J.L. and D.E. Rumelhart (1989). *Explorations in Parallel Distributed Processing.* Cambridge, MA: The MIT Press.

Selman, B., H. Levesque, and D. Mitchell (1992). A New Method for Solving Hard Satisfiability Problems. In: *Proc. of the Tenth National Conf. on Artificial Intelligence* (AAAI-92), pp. 440-446. San Jose, CA.

Shrager, J. and P. Langley (1990). *Computational Models of Scientific Discovery and Theory Formation.* San Mateo, CA: Morgan Kaufmann.

Thagard, P. (1989). Explanatory Coherence. *Behavioral and Brain Sciences* **12**, 435-467.

Thagard, P. (1994). *Conceptual Revolutions.* Princeton: Princeton University Press, 1992. Italian translation published by Guerini e Associati.

Thagard, P. (2000). *Coherence in Thought and Action.* Cambridge, MA: The MIT-Press.

Thagard, P. and E. Millgram (1995). Inference to the Best Plan: A Coherence Theory of Decision. In: A. Ram and D.B. Leake (eds.), *Goal-Driven Learning,* pp. 439-454. Cambridge, MA: The MIT-Press.

Thagard, P., C. Eliasmith, P. Rusnock, and C.P. Shelley (1997). Knowledge and Coherence. In R. Elio (ed.), *Common Sense, Reasoning, and Rationality.* Oxford: Oxford University Press.

Trick, M.A. (1996). Second DIMACS Challenge Test Problems, vol. 26 *DIMACS Series in Discrete Mathematics and Computer Science*, pp. 653-657. American Mathematical Society.

Verbeurgt, K. and P. Thagard (1998). Coherence as Constraint Satisfaction. *Cognitive Science* **22**, 1-24.

Theo A. F. Kuipers

COHERENCE

REPLY TO GERARD VREESWIJK

In a way, Gerard Vreeswijk's contribution could better be seen as a contribution to a Volume in Debate with Paul Thagard, so a reply by Paul Thagard would be more interesting than one from me. In particular for Vreeswijk himself, I hope that Thagard will reply in some way or other. Be that as it may, I am pleased that the present volume stimulated Vreeswijk to design a new connectionist method that claims to evaluate theories in a way that improves on the method advocated by Thagard in terms of his theory of explanatory coherence (TEC), implemented in ECHO. Of course, the plausible question for me is whether Vreeswijk's version of TEC, which I will indicate by TEC-V, and his implementation in the program KNONET escapes the main criticisms that I raised in SiS against TEC/ECHO by comparing that combination with my simple principle of the Priority of Explanatory Coherence (PES), "implemented" by the even more simple comparative Evaluation Matrix (EM). In this reply I will first deal with this question, followed by some remarks about the prospects for the computational implementation of PES/EM.

Comparing TEC/ECHO, TEC-V/KNONET, and PES/EM

Let me start by specifying Vreeswijk's opening paragraph which, incidentally, reflects his typical straightforward style of debate. In SiS I report (p. 313) that it took me forty-five minutes to calculate by hand two cases of theory comparison, indeed relatively very complicated ones, viz. Copernicus versus Ptolemy and Newton versus Descartes, by applying PES/EM on the two cases as propositionally structured by Nowak and Thagard (1992). As Vreeswijk wrongly suggests, I did not recalculate by hand their computational application of TEC/ECHO to these cases. It is all the more true that forty-five minutes is a long time, but since it indicates the time of a computation by (head and) hand, it nowadays means that an appropriate computer program might do it in a split second. Hence, what I did must be computationally very simple indeed.

In: R. Festa, A. Aliseda and J. Peijnenburg (eds.), *Cognitive Structures in Scientific Inquiry* (*Poznań Studies in the Philosophy of the Sciences and the Humanities*, vol. 84), pp. 404-406. Amsterdam/New York, NY: Rodopi, 2005.

My points of criticism were in fact two related points. One, "ECHO-selection" is a non-transparent updating process (p. 306). Two, as long as you can achieve the same results in a much more simple way, you should prefer that way (p. 310). Of course, the claim that PES/EM is "much more simple" than TEC/ECHO should be judged on the basis of a hypothetical computer program implementing EM. My additional claim was that all historical examples of the products (not the processes) of theory selection reproduced by Thagard and his colleagues could be reproduced by PES/EM. My main worry about the non-transparency was that considerations of explanatory success and simplicity are intermingled by TEC/ECHO, whereas they are clearly separated in the PES/EM approach. In my reply to Thagard I make clear that I have in principle liberalized my separation claim, leaving room for weighted roles of (desired and undesired) empirical and nonempirical features. But first there should be a proof that it is needed. That is, the following challenge formulated in SiS (p. 313) should first be met:

> In general, the challenge of new cases is that they may lead to strong counter-examples of the claim that the EM-method reproduces the historical choices: the EM-method might prescribe the opposite choice. If there are such cases, our stratified model is descriptively inadequate, i.e., even with respect to the simulation of products.

It is highly questionable whether the only (appealing, hypothetical) example suggested to me by Thagard (see my reply to him) viz. the classical theory of air, earth, fire, and water, has really ever been found more successful, in a generalized, weighted sense, than the phlogiston theory or even the oxygen theory (after their conception, of course). Unfortunately, Vreeswijk does not provide such cases, either.

One of the main things Vreeswijk argues is that ECHO's crucial update formula (2) can better be replaced by the "gradient ascent" formula (5). Not, however, for reasons of greater clarity, but for reasons of greater computational speed. Moreover, although his direct connectionist coherence approach in Sections 4 and 5 certainly has some plausibility, in terms of the transparency of the resulting calculations it is obviously much less effective than PES/EM.

In sum, as long as there are no clear historical cases going against PES/EM, I take it that there is no need for indirect or direct coherence approaches to theory selection. However, I should concede that if such cases were to be produced, PES/EM is in trouble and the computational coherence approaches of Thagard and Vreeswijk may well be the proper answer.

Implementing PES/EM and the Need for Justifying Normative Selection Algorithms

At the end of his paper Vreeswijk expresses the hope that somebody will implement PES/EM in order to compare it with TEC(-V). I am happy to relate that Alexander van den Bosch is far advanced with this project and is preparing a paper entitled "Explanatory coherence and the evaluation matrix." One important problem to overcome is that PES/EM, as it is formulated in SiS, compares just two theories, whereas TEC in fact compares all pairs of subsets of relevant propositions.

For the moment I would like to confine myself to stressing a point that Van den Bosch suggested to me about the paper by Vreeswijk. Although Vreeswijk is not very clear about this, it seems clear that he has only normative pretensions, in contrast to Thagard, who mainly has historical pretensions, not only regarding resulting selections, but also processes of selection. However – and this is Van den Bosch's basic point – in contrast to my PES/EM approach, which is rooted in the theory of empirical progress and truth approximation as developed in ICR, Vreeswijk still has to come up with some justification of his constraints, for otherwise you obtain an efficient but non-effective means, for the goal to be served is not specified. That is, one may concede that his constraints are very efficient, in the sense that they can easily be applied computationally. They may also be effective means to achieve some cognitive goal, but it is still not clear with respect to which goal they are effective. If such a goal could be identified, however, it would represent a convincing justification of Vreeswijk's constraints.

REFERENCE

Nowak, G. and P. Thagard (1992). Copernicus, Ptolemy, and Explanatory Coherence. In: R. Giere (ed.), *Cognitive Models of Science*, pp. 274-309. Minneapolis: The University of Minnesota Press.

THEORIES AND STRUCTURES

Emma Ruttkamp

OVERDETERMINATION OF THEORIES BY EMPIRICAL MODELS: A REALIST INTERPRETATION OF EMPIRICAL CHOICES

ABSTRACT. A model-theoretic realist account of science places linguistic systems and their corresponding non-linguistic structures at different stages or different levels of abstraction of the scientific process. Apart from the obvious problem of underdetermination of theories by data, philosophers of science are also faced with the inverse (and very real) problem of overdetermination of theories by their empirical models, which is what this article will focus on. I acknowledge the contingency of the factors determining the nature – and choice – of a certain model at a certain time, but in my terms, this is a matter about which we can talk and whose structure we can formalise. In this article a mechanism for tracing "empirical choices" and their particularized observational-theoretical entanglements will be offered in the form of Yoav Shoham's version of non-monotonic logic. Such an analysis of the structure of scientific theories may clarify the motivations underlying choices in favor of certain empirical models (and not others) in a way that shows that "disentangling" theoretical and observation terms is more deeply model-specific than theory-specific. This kind of analysis offers a method for getting an articulable grip on the overdetermination of theories by their models – implied by empirical equivalence – which Kuipers' structuralist analysis of the structure of theories does not offer.

1. Introduction

Almost all projects that aim at demarcating the "purely" observational (in the sense of so-called "raw sense data") from the theoretical are beset with certain difficulties which are invariably the result of two major issues. On the one hand, these difficulties arise as a result of the nature of the links postulated to exist between these two kinds of entity and the languages with which they are described, and, on the other hand, the difficulties are caused by the nature of the set of "intended applications" of a theory, especially in terms of the existence of more than one "empirical model" as the "real" domain of reference of the terms of theories. I claim here that a model-theoretic realist analysis of the structure of scientific theories may clarify the motivations underlying choices in favor of certain empirical models (and not others) in the above context of demarcation in a way that shows that "disentangling" theoretical and observation terms is more profoundly model-specific than theory-specific. A mechanism for tracing

In: R. Festa, A. Aliseda and J. Peijnenburg (eds.), *Cognitive Structures in Scientific Inquiry* (*Poznań Studies in the Philosophy of the Sciences and the Humanities*, vol. 84), pp. 409-436. Amsterdam/New York, NY: Rodopi, 2005.

"empirical choices" and their particularized observational-theoretical entanglements is offered in the form of Shoham's version of non-monotonic logic.

A model-theoretic realist account (see Ruttkamp 1999, and Ruttkamp 2002) of science places linguistic systems and their corresponding non-linguistic structures at different stages or different levels of abstraction of the scientific process. The philosophy of science literature offers two main approaches to the structure of scientific knowledge analyzed in terms of theories and their models: the "statement" and the "nonstatement" approaches. The statement depiction of scientific theories is cast in terms of an analysis of scientific knowledge as embodied by theories formulated in some (appropriate first-order) symbolic language with certain observational links of correspondence to reality. Defenders of the nonstatement approach (such as Suppes, the structuralists including Theo Kuipers, Beth and Suppe), in their turn, place more emphasis on the (mathematical) structures satisfying the sentences of some scientific theory in the Tarskian sense, than they do on the language in which the particular theory is formulated.

A model-theoretic realism retains the notion of a scientific theory as a (deductively closed) set of sentences (usually formulated in some first-order language), while simultaneously emphasizing the interpretative and referential role of the conceptual (i.a. mathematical) models of these theories. Rather than looking to typical statement approaches's notions of correspondence rules, or bridge principles to address observational-theoretical translations and referential questions concerning terms in theories, a model-theoretic approach acknowledges the re-interpretability of the language(s) in which theories are formulated and so turns to mathematical models of theories as the crucial links in the interpretative and referential chain of science. Merely "presenting" the theory "in terms of" its mathematical structures (or the set-theoretical predicates representing the class of these structures), which is typical of the so-called nonstatement accounts of theories, is not considered sufficient, since these accounts seem to eliminate – or at least de-prioritize – the possibility of addressing within a realist context the nature and role of general terms and laws – expressed in some appropriate formal language – in science. Model-theoretically speaking, this is unacceptable, since the links between the terms of scientific theories (as linguistic entities) and their interpretations in the various models of these theories in this context are taken to regulate the whole referential process, since such links offer particularized theoretical/ observation distinctions.

Advocates of the structuralist program take $\langle M_P, M\rangle = K$ (Moulines 1991, p. 319; Balzer, Moulines, Sneed 1987, pp. 36*ff.*) to be the (conceptual) "theory-core" of a particular theory. The core *K* plus the class of intended applications, call it I, form the simplest set-theoretic structure that may serve as a logical reconstruction of an empirical theory. Sneed's answer to the questions surrounding the question of theoreticity is roughly close to the criterion that Kuipers (2001, Chapter 12) uses to denote epistemological stratification, i.e. a criterion refering to the theory in which the concept under discussion appears. Kuipers (2001, Chapter 12) offers a more simple formulation than Sneed's for a general distinction between two kinds of "non-logico-mathematical" terms in relation to a statement S, but here I shall explain the more general formulation of so-called T-theoreticalness as Stegmüller (1979, p. 116) sets it out, following Sneed.

Stegmüller (p. 116) summarizes this criterion as follows: "... a quantity *f* is theoretical relative to a theory T iff the values of *f* are calculated in a T-dependent manner". Stegmüller (pp. 117-118) stresses the pragmatic implications of Sneed's criterion when he remarks that it may be viewed as a "... partial explication of the phrase 'meaning as use'." The structuralist emphasis on the *use* of laws determining the latter's empirical extensions fits in with the default framework for choices of empirical models, sketched in the following sections. The consequence of the application of this "T"-criterion to the structure M (i.e. to the structure representing the so-called "fundamental" laws, which holds for every application of the relevant theory) is a "decomposition" (p. 118) of M, as follows: the class M_P is the class of possible models of the "full conceptual apparatus". (In most cases M will only form a small subset of M_P.) Removing all theoretical components from M_P, leaves us with the set M_{PP} of partial potential models.

This further class of partial potential models M_{PP} is obtained by taking the elements of M_P and for each of them forming what we could call – following Kuipers (2001, Chapter 12) – an "observational reduct." Recall that a "reduct" in model-theoretic terms is created by leaving out of the language and its interpretations some of the relations and functions originally contained in these entities. In the structuralist case it is relations, functions, and constants which correspond to T-theoretical terms that are left out to define such a reduct.

In Kuipers' terms this comes down to the fact that within the class of partial potential models lies the class πM of the observational reducts of the structures in the class of actual models, M. Also in the class M_{PP} lies I, the class of intended applications. The empirical claim associated with a certain theory then, is that I is a subset of πM. The question to be asked within the context of this article is whether this implies that the structuralist theoretic/observational distinction might be as naive as the positivist one, in the sense that they do not relativize their reduct to particular applications of M. Surely more than one reduct exists, both of the class of potential models and of the class of actual models, depending on both the real system under consideration and the nature of the classes M_P and M, since non-isomorphic models may have isomorphic empirical substructures – so the structuralist reduct projections may be many-to-one – without any harm done either to (moderate) realist ideals or to theory-observation disentanglements.

An obvious motivation (on which both realists and anti-realists would surely agree) for empirical theory construction is the (successful) application, in one way or the other, of that (empirical) theory. That is why it is not completely correct to claim that we know what an empirical theory looks like if we know its core. We also need some information on the nature of its intended applications. Structurally speaking, then, if we take I as the set of intended applications of a given empirical theory identified by a specific given K, we have to know the nature of the elements of I, as well as the extension of I. Note again that cores of theories and the applications of theories together – i.e. M_P, M, and I – are the "material" out of which empirical claims may be formulated.

Now, the elements of I are taken – by the structuralists – to be not "simply the 'real things', independent of any conceptualisation, to which the theory is supposed to apply" (Moulines 1991, p. 319)[1], but rather systems, which are nothing other than structures that present us with ways of "... conceptually carving up reality in pieces and putting these pieces in certain relationships" (ibid., p. 320). Thus, we can take a system, s, to be a structure of the form $<A_1, \ldots, A_m, R_1, \ldots, R_n>$. Sneed (1994, p. 196) points out that I should be seen as the "totality" of potential data for which the theory in question is supposed to account. I agree, and model-theoretically speaking "real systems" are just such structures (i.e. elements of I). These structures are represented in model-theoretic terms as empirical conceptualizations of data – more about this in the following section.

Determining the identity of I for a given theory is something to which, structuralists stress, there is *no purely semantic answer*. Any kind of approach to this issue has to be preceded by what they term "pragmatic-diachronic considerations" (Moulines 1991, p. 321), because of the fact that for every given theory core, K, there has to exist a *scientific community* that will *use* (in Stegmüller's sense mentioned above) the theory identified by the core in "real life." Because I is dependent on the scientific community within which the theory under consideration has been constructed or will be applied, the structuralists refer to the class of intended applications as a "genidentical" (p. 322) entity. It is this kind of scientific community-relativity (or rather disciplinary matrix-relativity) plus the constant being-in-motion of science that I claim non-monotonic logic can rationally represent in a model-theoretic account of science – see Section 4.

Recall that in Kuipers' terms, modifications aiming at better – or stricter – definitions of I are made to the mathematical structure M in terms of the structuralist notion of T-theoretical-ness, so-called "constraints," and "special laws." I shall discuss below a new non-classical method of analysing choices concerning the members of the class I at specific times, which is adequate for the purposes of establishing the continuance of science from a realist point of view, and which also focuses on certain subsets of the class M. Before I explain this

[1] In my terms, the elements of I would be *representations* of systems of the "real things."

further I shall briefly outline what I mean by a "model-theoretic" account of scientific theories (see also Ruttkamp 1999).

In what follows I shall first briefly offer a sketch of the framework of a model-theoretic account of science. The next section focuses on the problem of overdetermination of theories by empirical models, or, as I refer to it sometimes, the problem of "empirical proliferation." Thereafter I offer a model-theoretic non-monotonic default model for dealing with the problem of empirical model choice. Finally I make a few comments on the implications for realism of the semantic use of models in analyses of scientific theories and show the relations between model-theoretic and constructive realism.

2. A Model-Theoretic View of Science

As mentioned above, in model-theoretic terms both the linguistic and the non-linguistic aspects of scientific knowledge and its expression(s) are woven into an articulated referential chain. In such an account, models of theories are defined in the usual Tarski sense. The method of ("empirical") verification of each of these models (i.e. how well do each of them reflect the system in the real world?), is decided by the specific nature of the specific model in question, *as well as* by the nature of the specific real system in question. Hence (see Figure 1) I claim that if the phenomena in some real system and the experimental data concerned with those phenomena are logically reconstructed in terms of a mathematical structure – call it an "empirical" model – the relation of empirical adequacy then becomes – close to Van Fraassen's depiction – a relation which is an isomorphism from the empirical model into some empirical reduct of the relevant model of the theory in question.

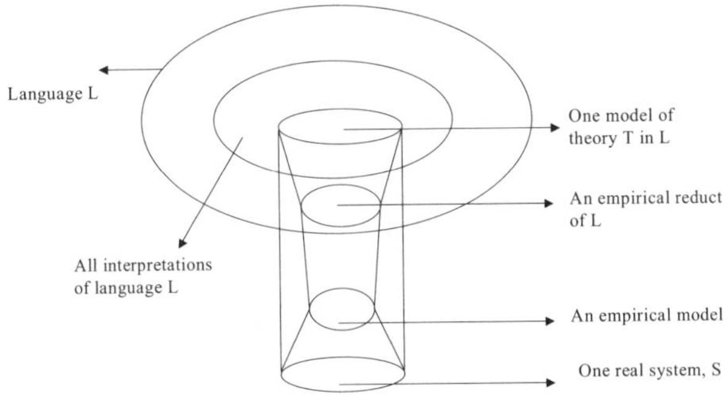

Fig. 1. A model-theoretic account of science I

Consider what it really means to formulate a model of a particular theory. A model of a theory sees to it that every predicate of the language of the theory has a definitive extension in the underlying domain of the model. Now, focusing on a particular real system at issue in the context of applying a theory, which in turn implies a specific empirical set-up in terms of the measurable quantities of that particular real system, it makes sense to concentrate only on the predicates in the mathematical model of the theory under consideration that may be termed "empirical" predicates. This is how in my context an empirical reduct is formulated. Recall that a "reduct" in model-theoretic terms is created by leaving out of the language and its interpretations some of the relations and functions originally contained in these entities. This kind of structure thus has the same domain as the model in question but contains only the extensions of the empirical predicates of the model. Notice that these extensions may be infinite since they still are the full extensions of the predicates in question.

Now, as sketched above, from the experimental activities carried out in relation to the real system on which we are focusing, a conceptualization of the results of these activities, i.e. of the data resulting from certain interactions with this system, may be formulated. This (mathematical) conceptualization of data is refered to as an empirical model. Then, if it is the case that there exists some relation of reference between our original theory and the real system we are considering, we may then find that there is a one-to-one embedding function from the empirical model into the empirical reduct in question. Why? The empirical model contains finite extensions of the empirical predicates at issue in the empirical reduct, since only a finite number of observations can be made at a certain time.

To summarize: the interpretative model interprets all terms in the appropriate relevant language and satisfies the theory at issue. In the empirical reduct are interpreted only the terms called "empirical" in the particular relevant context of application or empirical situation. Think of this substructure of the interpretative model as representing the set of all atomic sentences expressible in the particular empirical terminology true in the model. An empirical model – still a mathematical structure – can be represented as a finite subset of these sentences, and contains empirical data formulated in the relevant language of the theory. See Figure 2 for the example following below.

Say we take Newtonian mechanics as our theory. Take our solar system as a model, *M*, of the theory. Take one empirical reduct of this model, call it E_{Red}, a substructure of *M*, containing (only) events, that is, four-tuples (x, y, z, t) pinpointing the position(s) of Mars on its elliptical orbit. Notice that we acknowledge that the elliptical form of the orbit is an approximation, since we assume for now that the sun is heavier than any of the other planets and that we exclude predicates concerning forces, accelerations, and other so-called theoretical predicates – such as mass – which are not the "direct" result of observations in this case[2]. This subset E_{Red} then is the set of all points (x, y, z, t) lying mathematically on the elliptical orbit of Mars. Should we now consider the empirical models that resulted from observations of countless astronomers through the ages, we would find empirical models $E_{\text{emp}}{}^{i}$, $\forall i \in N$ all isomorphically embedded in our empirical reduct E_{Red} (assuming for our purposes here that Mars's orbit has not shifted for any reason). Thus we find that the conceptual four-tuples we get (at a certain time) from observing the positions of Mars in space and time, that is, the elements of some empirical model E_{emp}, are amongst the elements of E_{Red}, that is, the four-tuples (x, y, z, t) showing us the position of Mars at various time instances.[3]

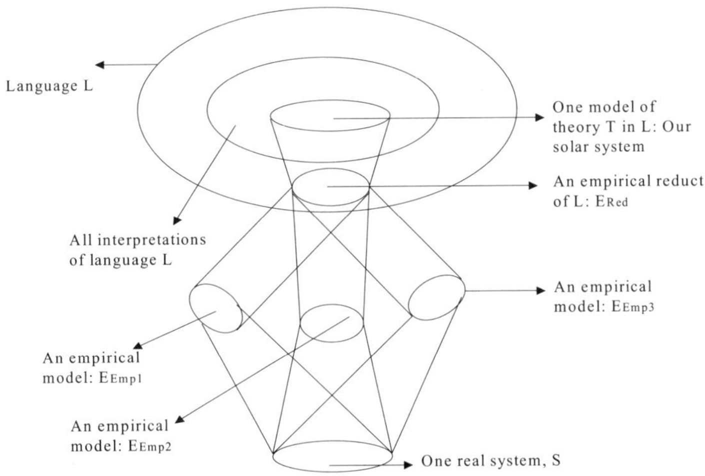

Fig. 2. A model-theoretic account of science (Newton's theory)

In terms of theory-observation distinctions in this context, notice the following: the requirement for a set of *c*-rules (or "postulates") to connect theoretical terms to their observational counterparts was supposed by some to be the tool for actualizing the positivist dream of rooting out all forms of pseudoscience, but, in a sense, turned into the biggest enemy of the positivist

[2] Note that this distinction between so-called "theoretical" and "empirical" predicates is model-specific rather than unique or absolute.

[3] Note that in this case, the embedding function simply is the identity function, mapping elements of E_{emp} onto elements of E_{Red}.

ideal. Briefly, the reason for this is that it is impossible – given all of the above – to find one clear, unambiguous method in which to draw the observational/theoretical distinction, mainly because of, on the one hand, the spurious nature of the positivist definition of *c*-rules; but also, on the other hand, because of the fluid nature of scientific knowledge.

In Chapter 2 of *Structures in Science* (Kuipers 2001), Kuipers comments on the problems concerning theory-observation distinctions. He writes (p. 37): "The law-distinction [i.e. the distinction between experimental or observational laws and proper theories (my insert)] forms a crucial construction principle for the hierarchy of knowledge and therefore an important heuristic factor in the dynamics of knowledge development." Obviously this distinction is closely related to theory-observation distinctions (as he also points out). In model-theoretic terms it can also be shown – focusing on models rather than theories as units of construction – that theory-observation distinctions are constructive of different levels of knowledge. This notion of the "multi-level-ness" of science also reminds of the structuralist notion of theory-nets built up in terms of T-theoretical distinctions.

Model-theoretically the prominent issue in formulating a realism containing both linguistic and non-linguistic systems may be viewed in terms of reconciling intensive and extensive definitions of terms in theories (intensive definitions are linguistic descriptions, while extensive definitions are listings of cases). The formulation of a theory in terms of some appropriate first-order language offers no more than an intensive definition of the terms in theories concerned, i.e. theories are systematic descriptions of the defining attributes of terms in theories in such a way that the "basic terms of the theory are 'implicitly defined' by the postulates of the theory" in Nagel's terms (Nagel 1961, p. 91). Against the notion of a "fully articulated scientific theory" (p. 91) having "embedded in it an abstract calculus that constitutes the skeletal [deductive] structure of the theory" and thus the conviction that connotations of terms in theories are irrelevant to this bare deductive skeleton, in a model-theoretic context, however, the connotations of the terms in theories are important in so far as they are relevant to the interpretation of the deductive elaboration of the postulates of the theory. In this sense it is, though, still the case – in a typical statement way – that the "fundamental assumptions of the theory formulate nothing but an abstract relational structure" (p. 91) since the terms in theories are not "tied down to definite observational [situations] by way of a fixed set of experimental procedures" (p. 89) and are thus general enough for these terms to be applicable to "diverse areas" (p. 89) in the empirical sense.

The role of the connotations of terms in theories becomes most evident at the level of the (conceptual) models interpreting these terms, since here the connotations of these terms serve to present the first referential links of these terms by making more precise or particular their general intensive definitions by interpreting them in such a way that the sentences of the definitions come out true. The denotation or extension of at least some of the terms in theories, i.e. the classes of all the individual cases to which the terms in theories in question apply, is given by the notion of empirical models isomorphically embedded into some empirical reduct of some mathematical model of the theory concerned. Model-theoretically thus, "rules of correspondence" (and thus, extensive definitions of some terms in theories) are given by the reduction functions fashioning empirical reducts from models, and also by empirical models and the isomorphic relations between such models and empirical reducts. Note that in this context the distinction between so-called "theoretical" and "empirical" predicates is model-specific rather than unique or absolute, which points towards a changeable – although traceable – model-specific interpretation of theory-observation "entanglements."

Notice that non-isomorphic models may have isomorphic empirical substructures. Also, theories are interpreted by many different models – think of the difficulties involved in pinning down standard models of theories. Moreover, theories, as well as their models, are also further referentially linked to many empirical reducts. In other words the theory/observation distinction cannot be a unique one, but must, of necessity, be model-specific first, but also empirical reduct-specific. This should not lead to conclusions of rampant relativism, however, since these distinctions can all be precisely defined and articulated in terms of model theory such that theory-observation distinctions are actually accepted as contingent on particular theory-model-empirical reduct-interpretative links.

Nagel (1961) offers one of the most well-known distinctions between so-called "experimental laws" and "proper theories." In his sense experimental laws contain only so-called observational terms, while the purpose of the formulation of proper theories is to explain experimental laws by the theoretical terms they introduce. However, Kuipers (2001, Chapter 2) points out the equally well-known fact – stated above – that this distinction is far from a clear cut or neat division. Kuipers (p. 3) claims that the so-called "law-distinction" should be viewed on the basis of "… a theory-relative explication of theoretical and observation terms … [This] suggests a disentanglement of the so-called theory-ladenness of observations. In particular, an observation may not only be laden by a theory, if unladen by it, it may nevertheless be relevant for it, and even be guided by it." The above analysis implies that Kuipers's specification of theory-relativeness (typical of structuralists) is too weak to embody the full complexity of theory-observation distinctions, since these distinctions concern only T-theoretical-ness. Obviously, pointing out the theory-relativity of these distinctions is a step in right direction, but it does not take into account – or perhaps, can at least not fully account for – the potentially changing (semantic) relations between models, empirical reducts, and empirical models.

In general the structuralist and Hempelian accounts of theoretical-observational distinctions terms were taken simply as a new kind of interpretation of the old two-level distinction between the theoretical and observational levels. Kuipers (p. 38) claims rather that these accounts – perhaps especially that of Sneed's – point to a new multi-level distinction between these kinds of terms. He (p. 38) explains that in terms of the long-term dynamics of science, if some proper theory is accepted as "approximately true" it is usually possible to set up criteria for the determination of its theoretical terms. Then, he claims, as soon as the theoretical terms are identified the proper theory "becomes" (p. 38) an observation theory, and "the corresponding theoretical level transforms into a higher observational level, enabling new observations and hence the establishment of new observational laws, asking for new, 'deeper' theories to explain them" (p. 38).[4]

I find Kuipers' remarks concerning a multi-level interpretation of science insightful, and view them, as mentioned already, as related to the structuralist notions of specializations and theory nets. In my terms the theoretical terms in a proper theory will be "identified" as soon as an interpretation of the theory is formulated in terms of some model. The proper theory "becomes an observational theory" when some reducing function has "reduced" the relevant model to an empirical reduct (substructure) containing only "observational" terms (in that particular context). Notice again that the reducing function is changable in the sense of "reducing" the same model to different empirical reducts. Recall here that the set I of intended applications is not a "Platonic entity" but "an open class frequently originating through gradual expansion from a paradigmatic original class" (Stegmüller 1979, p. 116). This shows that the evolution of "corresponding theoretical levels" into "higher observational levels" is further complicated by the ever growing class of empirical models (intended applications in structuralist terms) the elements of which (may) contain different entities and relations available as possible referents of terms in a specific theory. The following section focuses on a way to articulate decisions made for a particular relation of empirical adequacy at a particular time.

More precisely, in the second half of this article I show how a model-theoretic account of scientific theories, augmented, at the level of empirical reducts, by the machinery of non-monotonic logic, may enable us to express reference relations between theories and empirical (observational) models in the face of theory

---

[4] This also recalls Patrick Suppes' (1967) hierarchy of theories and models – he articulates the empirical relation between a (conceptual) model (of a given theory or class of systems) and a system in reality as a highly articulated, composite relation, with an articulation that depends on the experimental or observational situation in question.

change in general, and multiple model choice in particular. Rather than focusing only on progress in terms of a gradings of truth and success, I want to focus on the choices made when one is faced with more than one empirical model and the motivations for these choices. Finding a way to trace these motivations and link them with the formulation of models of theories might help to refine the relations between target sets and their approximations, in Kuipers' sense (or between the "actual" and the "nomic"; Kuipers 2001, Chapter 8), and so, in the end, might also have something to add to our conception of scientific progress.

3. The Problem of Empirical Proliferation

My answer when confronted with questions concerning model choice has usually been that these are about very particular concerns that will depend on the particular intentions of a particular scientific community at a particular time – notice the echoes of the structuralist concerns regarding the limits of the mechanisms of pure semantics to present these intentional choices. Although I still claim this to be the case, I have always been dissatisfied with the – at least apparent – informal character of such an answer. In this context, I want to consider with you the possibility of introducing into the wide empirical equivalence debate, concentrated on issues concerning overdetermination of theories by data, the non-monotonic mechanism of default reasoning, refined into a model-theoretic non-monotonic logic (based on the logic of Yoav Shoham) offering a formal method to rank models.

In terms of what I call "temporary knowledge" we need at least to consider the following questions: Where in the process of science would we find these particular pockets of temporary knowledge? In what sense exactly may scientific knowledge be temporary? How does such knowledge affect our final judgments about the nature of scientific progress?

Briefly, in answer to these questions: where do we find such pockets of temporary knowledge? We find such knowledge everywhere in the process of science, obviously, since we know that even the "best" theory at a certain time might in all probability be refuted at some point in the future. However we find the most extreme form of it at the level of the process of science where empirical adequacy is determined, that is, in my terms, the level at which we are considering so-called "empirical reducts" and their relations to so-called "empirical models."

The sense in which I mean this knowledge to be "temporary"is the one in which we make choices for certain models (and so sometimes for certain theories) at certain times. The context of this discussion is that of empirical equivalence in Van Fraassen's sense of the notion: he (1980, p. 67) explains that if for every model M of theory T there is a model M' of T' such that all empirical substructures of M are isomorphic to empirical substructures of M', then T' is *empirically at least as strong as* T. Earlier Van Fraassen (1976, p. 631) wrote that "Theories T and T' [each being as least as strong as the other in the above sense] are *empirically equivalent* exactly if neither is empirically stronger than the other. In that case ... each is empirically adequate if and only if the other is."

But what is the status of the models or empirical reducts – or even the relations of empirical adequacy – we do not choose at a specific time, then? The knowledge or information about the particular empirical model(s) in question that they carry, certainly still *is* knowledge, is it not? Well, yes and no. What we need is a formal mechanism by which we can depict our choices, the motivations for our choices, and the change of both of these, should there be a change of context within which we are applying some theory. We choose to work with a certain model or empirical reduct at a certain time, but we may always change our minds and make a different choice, which might imply a change in the set of knowledge claims (and the meta-tracings of reference links and theory-observation distinctions) our theory is offering, and this is where non-monotonic logic in the form of default reasoning comes in, as I explain below.

Related to this, as far as the nature of scientific progress is concerned, my (multi-level) view is the following. Theories change very slowly, conceptual models more quickly, and empirical reducts, and the empirical databases (the accumulation of empirical data via observations and experiments) they depict, change the quickest. The general theory of relativity was formulated by Einstein (and Hilbert) in 1915. For more than 80 years now physicists have been constructing literally dozens of different types of models – all models of precisely the same theory – to fit both experimental and observational data about the space-time structure of the real universe *and* certain paradigmatic preferences. Now, in this sense, I agree with Kuhn that neither the content of science nor any system in reality should be claimed to be "uniquely exemplified" by scientific theories from the viewpoint of studies of "finished scientific achievements." And, therefore, one has to accept the open-endedness (see Section 1 again) of theories as a permanent feature of the total process of science. Notice, though, that this open-endedness to me is represented by the ebb and flow of the models (including their empirical reducts) of the theory which ensures the continuity of science at least at a formal (meta) level of analysis.

Hence I imply that issues of theory succession or reduction are often, for long periods of time better – or at a finer level of analysis – interpreted as issues of model succession or reduction, and that this implies that certain aspects of our knowledge are more temporary than others. I claim that the terms of an already established theory can be said to be "about" an ongoing potential of entities in some system of reality to give reference to some objects and relations in *any* model of that theory. The actualization of this potential requires human action in the sense of finding and finally articulating "satisfying" referential relations between systems in reality and certain empirical aspects (reducts) of models of the theory. And it is the nature of these referential relations that will be the topic of the rest of this article.

Let us now focus on what I term "empirical proliferation." In a sense this is the reverse of the traditional scenario of the underdetermination of theories by data. In the philosophy of science the issue of the underdetermination of theories by data is the original problem of explaining – and perhaps justifying – the existence of empirically equivalent, yet incompatible, scientific theories. In the history of science instances of such theories are quite common – think of the various ways in which an electromagnetic field has been described, from Faraday through Einstein to Feynman.[5] In the context of underdetermination of theories by data, the bottom line thus is that empirical data are too incomplete to determine uniquely any one theory.

Turning now to the flipside of underdetermination, notice that we interpret "empirical equivalence" in the traditional (Van Fraassen-ian) way – i.e. theories are empirically equivalent just in case they have the same class of empirical consequences. Also bear in mind that contact between scientists and real systems that result in scientific data is relative to the state of scientific knowledge and of technological development at the time, as well as the research tradition or disciplinary matrix in which scientists work at that given time. Scientific knowledge is amendable and even defeasible, because of its contingent and particularized links with the reality it describes (and explains). Recall that according to Van Fraassen (1976, p. 631) a theory is empirically adequate if "all appearances are isomorphic to empirical substructures in at least one of its models." This view leads the way to the model-theoretic interpretation of empirical equivalence according to which theories with the same empirical reducts, or at least some empirical models, are empirically equivalent. These definitions point to the reverse case of traditional underdetermination of theories by data, i.e. a specifically model-theoretic interpretation of traditional underdetermination – i.e. underdetermination of data by theories. This article focuses on this very important (and different) aspect of traditional empirical equivalence.

---

[5] More precisely, traditionally the nature of underdetermination has been understood in terms of two kinds of relations between the "real world" and scientific theories. The first kind is taken to exist between phenomena (or whole systems) in reality and the observation terms of theories, while the second kind of relation is said to exist between sets of protocol sentences (formed from the observation terms and expressing data) and possible theories incorporating or explaining such a set of protocol sentences – that is, the existence of incompatible but empirically equivalent theories.

In general scientific theories, depicted as syntactic (linguistic) entities that need to be interpreted to be given semantic meaning and reference, are not able to uniquely capture their semantic content. In terms of theory application, within a model-theoretic context, two sets of relations are conducive to empirical proliferation: the set of relations between the terms in some theory and their extensions in its various models; and the set of relations between the terms of models (or of only one model) – via an empirical reduct (or empirical reducts) of that (those) model(s) – and the objects and relations of some real system (or systems) conceptualized in one or many empirical models.

Retaining the notion of scientific theories as linguistic expressions at the "top" level of science solves the problems regarding the justification of the existence of many (conceptual) models as interpretations of any one theory by the simple (formal) fact of the incompleteness of formal languages. Thus the possibility of a given scientific theory being interpreted in more than one mathematical model (structure) is natural in a very basic sense in model-theoretic terms. The second proliferation of relations between models and their empirical reducts and between these and empirical models may also turn out to be less counterintuitive than might be supposed at first glance, if it is understood that the possibility of articulating a chain of reference is *not* jeopardized under such circumstances.

Recall now that in model-theoretic realist terms, theories are empirically adequate if and only if they are true in certain models, some of the empirical reducts of which may conceptually encompass the empirical data of the relevant real system. In this sense the first step of the model-theoretic way to confront the model-theoretic overdetermination implied by either the choice of a model for interpreting a particular theory, or the choice of a model in which to embed certain empirical data, is to keep in mind the following structural fact regarding the scientific process. The choice of empirical reduct has to be such that it has embedded in it (an isomorphic copy of) some empirical model in which certain "observation" sentences are true. However, simultaneously, the mathematical model of which this empirical reduct is a substructure must be one that also "makes" or "keeps" true the sentences in the language of the theory that is shown to be empirically adequate.

This characteristic of a model-theoretic analysis of scientific realism ensures that tracing theory-model-reality links – even if presenting a rather complicated undertaking – is still articulable. Simultaneously, however, this also shows the complexity of theory-model-data links. In what follows I claim in particular that an application of non-monotonic default logic to situations of overdetermination of theories by models and data may enable us to formalize and get a grip on this complexity in terms of a particular kind of preferential ranking of these models. My claim is further that this ordering induces an ordering both of empirical reducts and models of theories themselves, and may ultimately even result in a ranking of theories.

4. Empirical Proliferation on a Model-Theoretic "Default" Model

The context of looking to non-monotonic reasoning as a possibility of rationalizing model choice is that of abduction.[6] Simply put, in the face of overdetermination of theories by empirically equivalent models, we are faced with a situation analogous to inference to the best explanation, since we have a "theory" but have to choose under certain particular contingent circumstances, out of many options one empirical reduct – and first a model – via which it (i.e. the theory) is linked to a particular empirical model and so to a particular system in reality. Kuipers (1999, p. 307) states that abduction is "the search for an acceptable explanatory hypothesis for a surprising or anomalous (individual or general) observational fact." The fact that our knowledge at the level of empirical models is finite and incomplete and therefore changeable does *not*, however, imply that we cannot discover *some* rational aspects of the kind of abductive reasoning required in this context.

Yoav Shoham (1988, p. 80) points out that in certain issues regarding incomplete information, we should concentrate on distinguishing between the meaning of sentences on the one hand, and our reasons for adopting that particular meaning and no other, on the other. The latter will naturally be outside the domain of the system of logic within which we are working at the time. I agree and acknowledge the contingency of the factors determining the nature – and choice – of a certain model at a certain time. But in my terms this is a matter to be articulated or pinpointed via the empirical models of the theory (about the construction of which admittedly not much can be said external to some particular context of application of the theory in question). Once confronted with more than one empirical model though, I claim we may make use of Shoham's kind of extra-logical motivations to rank these empirical models in a certain order.

Formalizing this is a rather complex task. One way in which to do so might be to take all existing possibilities present at a certain time into account, and summarizing the reasons for picking a certain empirical model – and so a particular empirical reduct of a certain model – at a certain time in such a way that the existence of other models – and other empirical reducts – is not denied, but simply, for a certain period of time, put on hold, as it were. A method for doing

[6] Heidema and Burger (forthcoming), p.1 note Paul's (1993) remark that abduction is often related to conjecture; diagnosis, induction, inference to the best explanation, hypothesis formulation, disambiguation, and pattern recognition.

this is offered to us by the nature of non-monotonic logic in general. In particular for our purposes here Shoham's model-theoretic non-monotonic logic is preferable, since it offers a fairly simple way of ranking models, which perhaps is not as adequately possible in other versions of non-monotonic logic.[7]

The general idea behind Shoham's reasoning that I find has some appeal in our context is that it is necessary sometimes to take "decisions" in our reasoning, while ignoring some information that is potentially relevant, but at the same time accepting or expecting to "pay the price of having to retract some of the conclusions in the face of contradicting evidence" (1988 p. 80). The trick is to have some rational way of keeping track of these retractions. Traditionally, logic is concerned with cautious and conservative reasoning. It finds its natural home in mathematics, the theorems of which are immune to fashion and the passage of time. But life in general and science in particular need more than mathematics – we need common sense and contextualization. This involves the capacity to cope with situations in which one lacks sufficient information for one's decisions to be logically determined, so that one has to try to distinguish between possibilities that are more plausible (i.e. "normal") and those that are less plausible at a given time.

Shoham (1988), pp. 71-72 sets out his non-monotonic scheme as follows:

> The meaning of a formula in classical logic is the set of interpretations that satisfy it, or its set of *models*[8] ... One gets a non-monotonic logic by changing the rules of the game, and accepting only a subset of those models, those that are 'preferable' in a certain respect (these preferred models are sometimes called 'minimal models' ...). The reason this transition makes the logic non-monotonic is as follows. In classical logic $A \rightarrow C$ if C is true in all the models of A. Since all the models of $A \wedge B$ are also models of A, it follows that $A \wedge B \rightarrow C$, and hence that the logic is monotonic. In the new scheme we have that $A \rightarrow C$ if C is true in all *preferred* models of A, but $A \wedge B$ may have preferred models that are not preferred models of A. In fact, the class of preferred models of $A \wedge B$ and the class of preferred models of A may be completely disjoint! Many different preference criteria are possible, all resulting in different non-monotonic logics. The trick is to identify the preference criterion that is appropriate for a given purpose.

In other words, inference from uncertain laws is non-monotonic since additional knowledge may make previously derived consequences underivable (Schurz 1995, p. 287).

The process of making informed guesses on the basis of a mixture of definite knowledge and default rules is called defeasible reasoning. The word "defeasible"

---

[7] For instance: Clark's (1978) predicate completion, Reiter's (1980) default logic, McDermott and Doyle's (1980) non-monotonic logic, McCarthy's (1981) circumscription, or McDermott's (1982) non-monotonic logic II. See also Ginsberg (1987), Kraus, Lehmann and Magidor (1990), and Shoham (1987).

[8] Where 'interpretation' means "truth assignment for [propositional calculus], a first-order interpretation for [first-order predicate calculus], and a <Kripke interpretation, world>-pair for modal logic." (Shoham 1988, pp. 71-72)

reflects the fact that our guess may turn out to be wrong, in other words that the default rule may be "defeated" by exceptional circumstances, or a change of circumstances caused by a change in the content of our knowledge. Defeasible inferences are inherently non-monotonic, since amending our system of knowledge might change our conclusions.

As an example of the need to go beyond the irrefutable logical consequences of one's definite information, consider a simple physical light-fan system.[9] Say we take an ordinary two-valued propositional language with atoms p and q, where p: the light is on, and q: the fan is on. p can be T/F (1/0) or q can be T/F (1/0) such that the four possible states of the system are depicted by the set $\mathbf{W} = \{11,10,01,00\}$ (where a specific valuation depicts a specific state of a system). Say, now, that we determine theoretically that it is the case that $p \vee q$, this reduces the frame of our language to $\{11, 10, 01\}$. Then we – or some of us at least – discover say, in reality, that we can see whether the light is on, but are too far away to see or hear whether the fan is on. Thus we have limited knowledge about the system. Now suppose the system is really in state 11, i.e. that the light and the fan are both on. We will know only that the light is on, i.e. that p is the case, not that both components are on, i.e. not that p and q are both the case. Our definite knowledge suffices to cut our current frame of states down even more to the frame consisting of the models of p, i.e. Mod(p) = $\{11, 10\}$. So far, so good. Where's the problem?

Suppose we urgently need to know what the state of the system is, because state 10 is an unwanted state for whatever reason. This implies that we want to cut down the frame Mod(p) = $\{11, 10\}$ to a frame with just one element in it. We need to go beyond our definite (although incomplete) knowledge, but without making blind guesses. How can we do this in a reasoned way? We can use a default rule such as "Experience and descriptions of the system have shown that when the light is on, the fan is normally on too" to make the informed guess that the state is actually 11.

Exactly how do default rules justify cutting down the set of models of our definite knowledge though? Or rather, what would we be willing to regard as a default rule? After all, not every rule of thumb can be taken seriously as a default rule. The standard representation of "meta"-information – motivating choices scientists make at given times (in our case), and distinct from "sentential" information about aspects of real systems – is as a relation on the set of states – or possible worlds – (of a system).[10] (In the context of our example, the possible

---

[9] This example is borrowed from discussions with Willem Labuschagne from the Department of Computer Science at Otago University, Dunedin, New Zealand.

[10] There are two approaches to ordering possible worlds: by using numbers, or without using numbers. The best known numerical ways are those using *fuzzy sets* or using *probabilities*. Neither of these would give us the kind of formal mechanism I am looking for in the current context.

worlds are just the states of the system, namely $\mathbf{W} = \{11, 10, 01, 00\}$.) In the case of the minimal model semantics related to non-monotonic logics this relation is a preference relation and is depicted as a "total pre-order," which is a reflexive, transitive relation capable of effecting comparisons between arbitrary elements. Intuitively, such relations are thought of as allocating states to levels of normality, or preference.

Shoham (1988) requires that a default rule should be expressible as such an ordering on possible worlds (or models). He focuses on using non-numerical default rules, such as the rule "11 is more normal than 10, which in turn is more normal than 01 and 00" as the basis for "informed guesswork". All we require is that the rule arranges the states of the system in levels, with the most normal states occupying the lowest level, then the next most normal states, and so on, until the least normal, least typical, least likely states are put into the top level. The given rule yields the ordering:

01 00

10

11

Now we can choose between the two models of p in our previous example, because 11 is below 10. Our choice reflects not merely our definite knowledge that p is the case, but also our default knowledge that 11 is a more preferred state of the system than 10 is (by the default rule stated above). (See the Appendix for formal definitions.)

In summary, default rules may be used to justify defeasible reasoning as follows: order the possible states of the system from bottom to top in levels representing decreasing preference; given definite knowledge α, look at the states in Mod(α) – the set of all models of α; pick out the states in Mod(α) that are *minimal*, i.e. lowest in the ordering; then any sentence true in each of these minimal models of α may be regarded as plausible, i.e. as a good guess. So, whereas α classically entails β, i.e. $\alpha \rightarrow \beta$, when among ALL the models of α no counterexample to β can be found, α defeasibly entails β when among all the most PREFERRED models of α no counterexample to β can be found.

Note though that a default rule is not an absolute guarantee. Our informed guess may turn out to be wrong. Normally if Tweety is a bird then Tweety is able to fly. But exceptional circumstances may defeat the default rule. Tweety may be a penguin or an ostrich. Tweety may be in Sylvester's tummy. Abnormal states or a change in the content of the body of knowledge concerning a certain situation can sometimes occur. That is why, after all, in such cases we call our reasoning "defeasible."

Now, back to the context of science, given all of the above, the possibility of after-the-fact semantic reconstructions of reference links from theories to some real systems formulated with the help of model theory and non-monotonic logic offers a way to get us out of at least some of the apparent difficulties implied by overdetermination and empirical equivalence in the model-theoretic way as follows. In the scientific context I claim that a default rule containing at least the following two conditions – or orderings – might be useful. The first condition induces an ordering or ranking of empirical models in terms of precision or accuracy. This condition has to do with the highest quality of data and the finest level of technology. For now, I am considering cases here where we have to choose among different equivalent empirical models, all of which may be embedded into the same reduct, or at least empirical reducts of the same type. The second condition that I would include in my default rule is also more often concerned with a choice of empirical reduct, together with a choice of empirical model, since here the condition implies a ranking of empirical models are preferred that may induce a ranking of empirical reducts. Here the rule states that empirical models are preferred that can be embedded into empirical reducts of a type that contains a larger class of empirical terms from the theory than others.

The second condition has two noteworthy implications. First it shows how such a ranking distinguishes between weaker and stronger links between theories and reality, since a theory that is model-theoretically linked to an empirical model embedded into an empirical reduct containing a larger class of empirical terms than others, may be said to be more effectively "about" some real system than would otherwise be the case. Also, in terms of the progress of science it might be preferable to have a mechanism for justifying the inclusion of previously exogenous factors as endogenous ones in a particular model of a theory. This becomes possible if we enlarge the type of empirical reducts.

If we combine both these conditions together in one default rule, we may find that the resulting rankings of empirical models induce rankings of empirical reducts, which might induce rankings of models themselves, which may ultimately induce rankings of theories. Let us look at a simple example, again in terms of our light-fan system. Theory: $p \vee q \equiv \mathrm{T}$

- Empirical situation: Only the light can be observed
  This implies that
    - p: empirical term
    - q: theoretical term

Models of T	Empirical Reducts	Empirical models
11	1-	1-
10	1-	
01	0-	

- The observation of the light in an on position cancels the empirical reduct 0-, which in turn cancels the model 01
- Our choice of empirical model thus induces the following ordering of empirical reducts:

0-
1-

and the following ordering of models:

11 10

- This changes our theory to $T' \equiv p$
- Suppose the empirical situation is enhanced by developments in technology and we can observe that whenever the light is on the fan is off. Then our frames of models become

Models of T'	Empirical Reducts	Empirical Models
11 10	11 10	10

- The result of our observations now is that the empirical model "cancels" the empirical reduct 11, and this, in turn "cancels" the model 11
- Our new enhanced empirical model now induces the following ordering of empirical reducts:

11
10

and the following ordering of models:

11
10

- This changes our theory to $T'' \equiv p \wedge \neg q$

Recall that, given my view of scientific progress, generally theories change much more slowly than models. Specifically, theory changes usually occur only when the possibility of changing and modifying the models of the theory concerned has been exhausted, which confirms the continuity of scientific knowledge. This view may be viewed as a different kind of "multi-level" view than the one that Kuipers (2001, Chapter 2) advocates. The difference in terms of a model-specific notion of truth and a notion of approximate truth is not important here, what is important is the acceptance of the fact that science's processes are realized at different levels.

Returning to the conclusions I draw from the above, I claim that non-monotonic default rules and consequent rankings enable us to reduce the available

– or possible – choices of models, empirical reducts, and empirical models. This kind of analysis offers a method for gaining an articulable grip on empirical equivalence of any kind. The mechanism of non-monotonic logic fulfils what Kuipers (1999, p. 307) calls the "main abduction task," i.e. "the instrumentalist task of theory revision aiming at an empirically more successful theory, relative to the available data, but not necessarily compatible with them," although this is done here mostly through revision – or change – of relations of empirical adequacy, implying possible revision of choices concerning empirical models, empirical reducts and (conceptual) models. Although the above application of non-monotonic logic starts at a finer level of analysis than is usually the case in non-monotonic contexts (where we simply look at rankings of the states – models – of the system in question), the model-theoretic structuring of relations between models, empirical reducts, and empirical models makes possible the kind of "carrying over" of rankings that I have set out above.

Notice that relations of empirical adequacy are thus temporary and contextual, as Laudan and Leplin (1991) also concluded in their 1991 article *Empirical equivalence and underdetermination*. Science progresses fastest at the level of empirical models, but continuity is ensured by the fact that these models remain conceptualizations of observations, even if these observations are also contextual. The point of a model-theoretic realism is exactly that, instead of offering simply one intended model of "reality," a theory is depicted as a way of constructing or specifying a collection of alternative models, each of which may represent, explain, and predict different aspects of the same real systems (or different ones) via the same or different empirical reducts isomorphically linked to the same or different empirical models.

Above we have mostly concentrated on cases of empirical equivalence in terms of model-theoretic overdetermination. What – in terms of realist concerns – about underdetermination in the traditional (Laudan/Leplin) sense? (I.e., different theories, same empirical model.) In this sense – in a realist context – a scientist can "know" – or at least determine – that she is working with the "same phenomenon", even if using "different" theories or "different" models, because of the possibility of analyses that a model-theoretic realism offers of the different empirical links between different empirical models of different (conceptual) models of (perhaps) different theories. Detailed analyses of these empirical links will reveal common factors on the reality side of the link (e.g. light blobs observed through different telescopes by different people at different times indicating – by careful analyses – a common factor called "Neptune") which entails the "same phenomenon." And, moreover, cases where the same empirical model is embedded in different empirical reducts also show the continuity of science at the empirical level. Kepler took Brahe's precise empirical observations, i.e. the empirical data forming the empirical model of the theories in terms of celestial spheres that Brahe worked with, and fitted these data – i.e. Brahe's empirical model – into his (Kepler's) theory in terms of elliptical orbits. Applying non-monotonic logic within a model-theoretic context also may help to minimize traditional underdetermination of theories by models and data within a context of scientific progress, since it leads to choices of more accurate, more encompassing (empirical models and so) empirical reducts, and in certain cases it may even help to eliminate certain models or, ultimately, even theories.

5. Conclusion

Thus, even in the face of the fact that our fallible sensory experience and the finiteness of experimental data at a given time indicate that our knowledge of reality at such a time is limited, contextual, and temporary, we *can* rationally discuss the choices we make concerning so-called "empirically equivalent" models *and* keep track of changing theory-observation distinctions. It might be then possible, after all – contrary to Popper – to give some kind of rational motivation for the so-called "creative" leap that we make from data to theories. Kuipers (2001, Chapter 10) also comments that "... discovery, contrary to traditional opinion in philosophy of science, is accessible for methodological analysis ..." (p. 287), although he chooses to show this by his distinction between different kinds of research programs and explores relations between discovery, evaluation and revision by means of computational philosophy of science mechanisms. A non-monotonic logical analysis of empirical model choice admittedly does not "simulate" the "processes in the minds or brains of scientists" Kuipers (2001, p. 290), but rather it makes sense of the motivations underlying certain of these scientists's actions, based on the status and development of the knowledge claims they make.

I do not necessarily agree with Kuipers' claim (2001, p. 201) that "the realist ultimately aims to approach the strongest true hypotheses, if any, i.e. the theoretical-cum-observational truth about the subject matter". Perhaps this may be said to be the case for a certain kind of realist. A realist with a more sophisticated, moderate view of science and its processes ultimately aims at establishing reference relations between terms in theories and entities in real systems and is content with acknowledging that questions of truth are contextual and temporary matters. Questions of truth cannot be settled before questions of reference are settled. Accepting this will go a long way towards accepting the contingent and defeasible nature of science without harming the (realist) status of scientific theories in any important way. Recall also my emphasis on the re-interpretability of the language of science, or of theories in particular, and then it will be clear that claiming *model-theoretic reference* is sufficient to establish *some* form of realism, since in this *referential semantic* sense it *can* be shown that unobervables "exist" in real systems (i.e. terms in theories might after all be shown to refer to them). The contextually empirical terms refer directly, and the contextually theoretical terms indirectly, "by implication," via their conceptual and logical links to the empirical terms established by the theory. Some philosophers might be scornful about this kind of "weak" realism, while actually this realism is "weak" only because "strong" means traditional metaphysical realism. "Weak" means non-absolutist, and in that sense model-theoretic realism (supported by a non-monotonic semantics) is much stronger and more flexible than typical metaphysical scientific realism.

In general, then, I conclude that scientific theories may indeed say something about reality, but it is not possible when faced with an *uninterpreted* theory and possibilities of overdetermination of the theory by both data and models to determine or claim that it will definitely or uniquely be applicable to a certain aspect of reality and to no other. The model-theoretic notion of articulated reference and truth, augmented by non-monotonic mechanisms to get a grip on empirical overdetermination, may render the process of science expressible to rather finer and more accessible detail than may be possible on other accounts of science. When reference is traced via model-theoretic relations between theories, models, and data, and extra-logical default rules are used to formally order our choices in a rationally responsible way, Quine's inscrutability of reference becomes an even vaguer notion than before. Hence reference – at least in this sense – does not appear to be indeterminate after all. Secondly, this implies that the content of the meta-verification procedures for the processes of science cannot be given uniquely, but is rather a result of the context-specific actions and constructions of human scientists. In other words, theory-observation distinctions – or the definition of *c*-rules – remain somewhat less precise than one might wish in a positivist sense, but overall at least these distinctions remain articulable in the model-theoretic sense – which is more important for the success of a realist quest.

It might be that a model-theoretic realism aided by a non-monotonic ranking of models (empirical reducts and empirical models) offers, at least partly, some response to Laudan and Leplin's (1991) concerns about the "collapse" of epistemology into semantics in terms of traditional underdetermination and empirical equivalence issues, taken almost as two sides of the same coin. Non-monotonic default rules are extra-logical and are determined by the state of knowledge of a system at a particular time (i.e. "the agent knows that the light is on"). The new perspective on the consequence (entailment) relation that non-monotonic semantics offers might thus present us with a different way of looking at Laudan and Leplin's (1991) claim that evidential support for a theory should not be identified with the empirical consequences of the theory.

To conclude this article I review a model-theoretic realism according to the five questions Kuipers asks in the beginning of *From Instrumentalism to Constructive Realism* (2000, Chapter 1, pp. 3*ff*) in order to show the common features and the differences between such an approach and that of Kuipers' constructive realism. The first question is "Does a world that is independent of human beings exist?" I agree with Kuipers that a positive answer to this question – especially in a philosophy of science and a realist context – interprets the question as "does a non-conceptualized natural world that is independent of human beings exist?" Both constructive realism and model-theoretic realism answer positively to the latter, and it is granted that the nomic version of this form of ontological realism is stronger than the actual one, since in such a case it is not only a particular actual possibility that is conceptualized, but rather many nomic ones.

The second question (the first of four epistemological ones) is "Can we claim to possess true claims to knowledge about the natural world?" (p. 3). Again I agree to interpret this question as asking whether "we can have good reasons for assuming that certain claims, by definition phrased in a certain vocabulary, about the natural world are true in some objective sense, while others are false" (p. 4). A supporter of model-theoretic realism will answer positively, but will qualify "some objective sense" as a methodological sense – i.e. the model-theoretic way to "trace" references to entities and relations in real systems – since such a supporter will believe in the actual contingency of such links. Thus both model-theoretic realism and constructive realism are forms of epistemological realism.

The third question Kuipers poses is "Can we claim to possess true claims to knowledge about the natural world beyond what is observable?" (Kuipers 2000, p. 4). Again, this should be interpreted as Kuipers (p. 4) claims, as asking whether more than observational knowledge is possible. Here I think Van Fraassen is correct in believing that the point in this context is not whether theoretical terms refer or not, or whether proper theories are true or false, as Kuipers (p. 5) points out. It is true that the point is rather whether theories are empirically adequate – or, in Kuipers' sense observationally true. The model-theoretic point, though, is that determining empirical adequacy is important since it is the *final* step in articulating the referential link between terms in theories and entities and relations in real systems. Determining empirical adequacy is not only *not* all that matters (as defenders of Van Fraassen's view claim), but also cannot be done – at least in a realist context – without certain preceding steps in terms of the construction of models interpreting the language in which theories are formulated (set out in Section 4).

The fourth question is "Can we claim to possess true claims to knowledge about the natural world beyond (what is observable and) reference claims concerning theoretical terms?" (p. 6). Here I classify model-theoretic realism with

Cartwright and Hacking's referential realism, since an advocate of the former will also claim that "entity and attribute terms are intended to refer, and frequently we have good reasons to assume that they do or do not refer" (p. 6), although I do not support the metaphysical form of realism that Cartwright seems to favor in her later writings (e.g. Cartwright, 1989, 1994).

The final question that Kuipers considers is "Does there exist a correct or ideal conceptualization of the natural world?" (p. 7). My answer is no, and so is Kuipers'. Given the contingency and defeasible nature of our knowledge claims, linked as they are to disciplinary matrices and everything this entails, no other answer is possible. I agree with Kuipers that

> [v]ocabularies are constructed by the human mind, guided by previous results. ... one set of terms may fit better than another, in the sense that it produces, perhaps in cooperation with other related vocabularies, more ... interesting truths about the domain than another. The fruitfulness of alternative possibilities will usually be comparable, at least in a practical sense ... . There is however no reason to assume that there comes an end to the improvement of vocabularies (p. 8).

My point here is that representing a real system from a different perspective – i.e. linking some theory model-theoretically to a different empirical model than before – can augment the content of our knowledge claims regarding that system, but are not necessarily an "improvement" on the claims generated by the first linkage, although in both cases we can speak of "contextual" truth or truth of the theory in the particular chosen model.

6. APPENDIX: Formal Definitions

***Definition* 6.1.** Let G be any set. A relation $R \subseteq G \times G$ is a *total preorder* on G iff

- R is *reflexive* on G (i.e. for every $x \in G$, $(x, x) \in R$), and
- R is *transitive* (i.e. if $(x, y) \in R$ and $(y, z) \in R$, then $(x, z) \in R$), and
- R is *total* on G (i.e. for every $x \in G$ and $y \in G$, either $(x, y) \in R$ or else $(y, x) \in R$.

***Definition* 6.2.** Let L be a propositional language over some finite set **A** of atoms. Let **W** be the set of all local valuations of L (i.e. functions from **A** to {T, F}). A *ranked finite model* of L is a triple M = (G, R, V) such that

- G is a finite set of possible worlds,
- R is a total preorder on G, and
- V is a labelling function from G to **W**.

By a *default* model of L we understand a ranked finite model (G, R, V) in which G = **W**, R is a total preorder on **W**, and V is the identity function (i.e. V(w) = w for all w∈**W**).

***Definition* 6.3.** Suppose that L is a propositional language over a finite set **A** of atoms, and that M = (G, R, V) is a ranked finite model of L. Given a sentence α of L and a possible world $x \in$ G, the following rules determine whether M satisfies α at x:

- if α is an atom in **A**, then M satisfies α at x iff the valuation V(x) assigns to α the truth value T;
- if α is ¬β then M satisfies α at x iff M does not satisfy β at x;
- if α is β∧γ then M satisfies α at x iff M satisfies both β and γ at x;
- if α is β∨γ then M satisfies α at x iff M satisfies β at x or γ at x;
- if α is β→γ then M satisfies α at x iff M satisfies ¬β at x or satisfies γ at x;
- if α is β↔γ then M satisfies α at x iff M satisfies both β and γ at x or satisfies neither at x.

***Definition* 6.4.** Suppose L is a propositional language over a finite set **A** of atoms, and that M = (G, R, V) is a ranked finite model of L. Let α and β be any sentences of L. The sentence α *defeasibly entails* β iff M satisfies β at every possible world x such that

- M satisfies α at x, and
- x is minimal amongst the worlds satisfying α, i.e. there is no possible world y of M such that α is satisfied at y and $(y, x) \in$ R and $(x, y) \notin$ R.

University of South Africa
Department of Political Sciences and Philosophy
Discipline of Philosophy
PO Box 392, 0003 Pretoria
South Africa
e-mail: ruttkeb@unisa.ac.za

REFERENCES

Balzer, W., C.U. Moulines and J.D. Sneed (1987). *An Architectonic for Science - The Structuralist Programme*. Dordrecht: D. Reidel.

Cartwright, N. (1989). *Nature's Capacities and Their Measurement*. Oxford: Clarendon Press.

Cartwright, N. (1994). Is Natural Science Natural Enough? A Reply to Philip Allport. *Synthese* **94** (2), 291-301.

Clark, K.L. (1978). Negation as Failure. In: G. Hervé and J. Miller (eds.) (1978) *Logic and Data Bases* (Symposium on Logic and Data Bases, Centre d'études et de recherches de Toulouse), pp. 293-322. New York: Plenum Press.

Ginsberg, M.L., ed. (1987) *Readings in Nonmonotonic Reasoning*. California: Morgan Kaufman.

Heidema, J. and I. Burger (forthcoming.) *Degrees of Abductive Boldness*.

Kraus, S., D. Lehmann and M. Magidor (1990). Non-Monotonic Reasoning, Preferential Models and Cumulative Logics. *Artificial Inteligence* **44**, 167 - 207.

Kuipers, T.A.F. (1999). Abduction Aiming at Empirical Progress or Even Truth. *Foundations of Science* **4** (3), 307-323.

Kuipers, T.A.F. (2000/ICR). *From Instrumentalism to Constructive Realism. On Some Relations Between Confirmation, Empirical Progress, and Truth Approximation. Synthese Library*, vol. 287. Dordrecht: Kluwer Academic Publishers.

Kuipers, T.A.F. (2001/SiS). *Structures in Science. Heuristic Patterns Based on Cognitive Patterns. An Advanced Textbook in Neo-Classical Philosophy of Science. Synthese Library*, vol. 301. Dordrecht: Kluwer Academic Publishers.

Laudan, L. and J. Leplin (1991). Empirical Equivalence and Underdetermination. *The Journal of Philosophy* **88** (9), 449-472.

McCarthy, J.M. (1981). Circumscription – A Form of Non-Monotonic Reasoning. Reprinted in: B.L. Webber and N.J. Nilsson (eds.), *Readings in Artificial Intelligence*, pp. 466 - 472. California: Tioga Publishing Company.

McDermott, D.V. (1982). A Temporal Logic for Reasoning about Processes and Plans. *Cognitive Science* **2** (3), 101 - 155.

McDermott, D.V. and J. Doyle (1980). Non-Monotonic Logic. *Artificial Intelligence* **13**, 41 - 72.

Moulines, C.U. (1991). Pragmatics in the Structuralist View of Science, in: G. Schurz and G.J.W. Dorn. (eds.), *Advances in Scientific Philosophy. Essays in Honour of Paul Weingartner*, pp. 313-326. Amsterdam: Rodopi.

Nagel, E. (1961). *The Structure of Science*. London: Routledge & Kegan Paul.

Paul, G. (1993). Approaches to Abductive Reasoning: An Overview. *Artificial Intelligence Review* **7**, 109-152.

Reiter, R. (1980). A Logic for Default Reasoning. *Artificial Intelligence* **13**, 81 - 132.

Ruttkamp, E.B. (1999). Semantic Approaches in the Philosophy of Science. *South African Journal of Philosophy* (Special issue on philosophy of science) **18** (2), 100 - 148.

Ruttkamp, E.B (2002). *A Model-Theoretic Realist Interpretation of Science*. Dordrecht: Kluwer Academic Publishers.

Schurz, G. (1995). Theories and Their Applications - A Case of Nonmonotonic Reasoning. In: W.E. Herfel, W. Krajewski, I. Niiniluoto and R. Wójcicki (eds). *Theories and Models in Scientific Processes. Poznań Studies in the Philosophy of the Sciences and the Humanities*, vol. 44, pp. 269-294. Amsterdam: Rodopi.

Shoham, Y. (1987). A Semantical Approach to Nonmonotonic Logics. In: *Proceedings: Logics in Computer Science*, 275 - 279.

Shoham, Y. (1988). *Reasoning about Change: Time and Causation from the Standpoint of Artificial Intelligence*. Cambridge, MA: The MIT Press.

Sneed, J. D. (1994). Structural Explanation. In: P. Humphreys (ed). *Patrick Suppes: Scientific Philosopher*; vol. 2: Philosophy of Physics, Theory Structure, and Measurement Theory, pp. 195-216. Dordrecht: Kluwer Academic Publishers.

Stegmüller, W. (1979). The Structuralist View: Survey, Recent Developments and Answers to Some Criticisms. In: J. Hintikka (ed.) (1979). *The Logic and Epistemology of Scientific Change*. (*Acta Philosophica Fennica* **30**. Amsterdam: North-Holland Publishing Company), pp. 113 - 129.

Suppes, P. (1967). What is a Scientific Theory? In: S. Morgenbesser (ed). *Philosophy of Science Today*. (New York: Basic Books), pp. 55 - 67.

Van Fraassen, B.C. (1976). To Save the Phenomena. *The Journal of Philosophy* **73** (18), 623-632.

Van Fraassen, B. C. (1980). *The Scientific Image*. Oxford: Oxford University Press.

Theo A.F. Kuipers

OVERDETERMINATION AND REFERENCE

REPLY TO EMMA RUTTKAMP

A couple of papers deal with the two (almost entirely) overlapping chapters of ICR (5, 6) and SiS (7, 8) and one or more chapters from either ICR or SiS. However, only the paper by Emma Ruttkamp mainly deals with the topics of other chapters from ICR *and* SiS. Her main aim is to defend a kind of realism, called model-theoretic realism, that can make sense of the problem of *over*determination of theories by empirical data, using nonmonotonic ways of reasoning. Instead of going into details about her widely encompassing and intriguing approach, I would like to elaborate on two points that are directly related to her main themes, viz. the problem of overdetermination and the problem of reference of theoretical terms.

Underdetermination by Overdetermination

In Section 3 Ruttkamp suggests most of the time that the problem of overdetermination of theories by data is strongly related to the distinction between observational and theoretical terms, the O/T distinction, and the changing semantic relations between models, empirical reducts, and empirical models. However, in Note 6 she gives a formulation that makes clear that this problem is already present without the O/T distinction and without changing semantic relations. I would like to call attention to this basic version of the problem within my own framework in ICR. I will explain that, besides the traditional problem of underdetermination, due to theoretical terms that leave room for observationally equivalent theories, there is a more basic problem of determination operative in scientific research, a kind that can partly be conceived as a problem of overdetermination. In my ICR framework (see Section 7.3.2) the data are represented by $R(t)$, the set of realized possibilities up to t, i.e., *the accepted instances*, and by $S(t)$, *the strongest accepted law*, based on $R(t)$, where both are formulated within a previously chosen observational vocabulary. These data by

In: R. Festa, A. Aliseda and J. Peijnenburg (eds.), *Cognitive Structures in Scientific Inquiry* (*Poznań Studies in the Philosophy of the Sciences and the Humanities*, vol. 84), pp. 437-439. Amsterdam/New York, NY: Rodopi, 2005.

no means determine a theory, let alone the strongest true (observational) theory *T*, corresponding to the set of nomic possibilities. Even if we restrict the attention to theories that are compatible with *R*(*t*) and *S*(*t*), that is, theories that can be represented as both a superset of *R*(*t*) and a subset of *S*(*t*), there will be, as a rule, many other theories besides *T*. Although by enlarging *R*(*t*) and hence narrowing down *S*(*t*) we zoom in on *T* in a two-sided way, normally speaking *T* remains underdetermined. However, *R*(*t*) or, more precisely, the theory with *R*(*t*) as its set of models, assuming that such a theory can be formulated, entails all the remaining theories "between *R*(*t*) and *S*(*t*)," including *T* and many more. As a matter of fact this holds for any subset and even any member of *R*(*t*). That is, after performing an experiment we can give a complete description of the realized physical possibility (relative to the observational vocabulary), which entails very many theories, including *T* itself. I am happy to agree with Ruttkamp's Note 6 that this is, in a sense, a problem of overdetermination.

A Problem of Reference

In her concluding section, following the five questions I put forward in the introductory chapter of ICR, it becomes particularly clear that Ruttkamp's model-theoretic realism and my constructive realism are close relatives. The main difference seems to lie in our view of reference. Although she does not criticize my analysis in ICR in detail, it is clear that she favors an epistemological kind of reference, whereas my basic analysis is semantic and metaphysical. Since I came to realize after closing ICR that I leave an important problem concerning reference open there in Ch. 9, I would like to take the opportunity to formulate this problem briefly. It will certainly suggest that the contrast with Ruttkamp's approach of reference be investigated further.

Let me start by quoting the most relevant summarizing claim in the concluding Chapter 13 of ICR (pp. 325-6):

> Now we arrive at a highly idealized picture of (new) research, in which we make the main metaphysical assumptions explicit. The scientist assumes the existence of two unconceptualized natural worlds, THE ACTUAL WORLD and THE NOMIC WORLD. THE ACTUAL WORLD includes its history, and its future, and is at least partially made by humans, among others, by scientists who perform experiments. THE NOMIC WORLD on the other hand, exists independently of human beings. It encompasses THE ACTUAL WORLD, and is to be studied via that world. Studying THE ACTUAL and THE NOMIC WORLD requires conceptualizing them.

The specific topic of reference (and ontology) is summarized on p. 329:

> Recall that in CHAPTER 9 we have defined 'reference' primarily in a 'domain and vocabulary' relative way, viz., in terms of the nomic truth generated by them and THE NOMIC WORLD, according to the Nomic Postulate. For attribute terms, the crucial question

> was whether the nomic truth is constrained by them; for entity terms, it was whether they occur as a domain-set of referring attribute terms. But we also suggested the possibility of basing on these definitions an absolute definition, viz., whether the term refers in at least one 'domain and vocabulary' combination. Note that the link with the nomic truth assures that reference may just be a potential matter, not (yet) actual, in the sense that the relevant nomic possibilities need not (yet) have been realized. In other words, terms always refer to THE NOMIC WORLD if they refer at all, and they may or may not refer to THE ACTUAL WORLD.
>
> The corresponding ontology is roughly given by: entities and attributes exist as far as the corresponding terms refer. Note that the definitions are such that attributes only exist as far as there are entities having the attribute. Note also that, since reference is defined in terms of the nomic truth, there are again two kinds of existence, actual and potential.
>
> To be sure, speaking of reference to, and existence in, THE NOMIC but not ACTUAL WORLD, is a way of speaking that has its risks. The more cautious way of speaking is to systematically talk about *potential* reference and existence.

As said, after closing ICR I came to understand that there is a problem with this way of dealing with reference. Whether a combination of an entity term and an attribute term refer, using a set of these (potential) entities as one of its domain-sets, will, in a context in which truth *approximation* is taken seriously, basically depend on whether something like these entities exists to which something like this attribute may or may not apply. However, what is "something like" in such a context? When do we say that there is nothing like that type of entity and that type of attribute, even apart from our probable lack of the epistemological means to apply the relevant terms? Maybe we should just take a formal point of view. As soon as the theoretical vocabulary introduces an entity and an attribute term they are supposed to be coupled to a combination of entities and an attribute "that are around" in the intended domain of application and that are not yet taken care of by the observational vocabulary. Of course, when more options are possible a choice will have to be made. I would like to conclude by conceding that these informal remarks still leave much to be desired.

Robert L. Causey

WHAT IS STRUCTURE?

ABSTRACT. In *Structures in Science*, Theo A. F. Kuipers presents a detailed analysis of reductive, including microreductive, explanations. One goal of a microreduction is to explain the laws governing a structured object in terms of laws about its parts, plus a description of its structure. Kuipers refers to structures in his book, and uses the idea of a "structure representation function," but does not characterize the relevant concept of structure. To characterize microreductions fully, we need an adequate characterization of the relevant sense of "structure." After discussing examples, I present general analyses of bonds and of structured wholes. My analyses apply from physics to the social sciences, the latter illustrated by a hypothetical robotic social structure. Since Kuipers' philosophical position appears to be generally compatible with my own, I do not critique of any part of his work. Instead, this article is intended to fill in a gap in his presentation.

1. Introduction

Theo A. F. Kuipers presents rich and detailed analyses of many aspects of scientific knowledge and explanation in his book, *Structures in Science* (Kuipers 2001; hereafter referred to as SiS). The scope of the book is so large that it is impossible to discuss adequately any major part of it in a short article. Moreover, since Kuipers' philosophical position appears to be generally compatible with my own, I do not undertake a critique of any part of his work. Instead, I hope to fill in a gap in his presentation. I shall therefore limit this contribution to an issue that has concerned me for many years, and which has also been a gap in my own work. My discussion assumes familiarity with Kuipers' book, and general familiarity with the related philosophical and scientific literature.

In Chapter 5, "Reduction and Correlation of Concepts," Kuipers presents a detailed, semi-formal analysis of reductive explanations. Explanations of this form often play the central role in inter-theoretical reductions, i.e., major scientific advances in which a theory pertaining to one domain of research is explained in terms of a theory pertaining to another domain of research. Kuipers' book mentions a number of examples of such reductions and includes many literature references. Much of the discussion of reduction that is found in

In: R. Festa, A. Aliseda and J. Peijnenburg (eds.), *Cognitive Structures in Scientific Inquiry* (*Poznań Studies in the Philosophy of the Sciences and the Humanities*, vol. 84), pp. 441-462. Amsterdam/New York, NY: Rodopi, 2005.

the philosophical literature is concerned with the logical and ontological status of the "connecting sentences" which relate the terms of a reduced theory to those of a reducing theory. This is also true of my own past work. *Unity of Science* (Causey 1977) contains extensive discussions of thing-identity connecting sentences and attribute-identity connecting sentences in reductive explanations. The subsequent critical discussions of this book focused on issues related to these types of inter-theoretical connections, especially what I had written about attribute-identities. Kuipers' is also largely concerned with inter-theoretical connections and discusses them in the light of more recent analyses involving supervenience and other ideas.

In the present article I shall not address the general issues about connecting sentences in a reduction. Instead, I shall direct attention to a key aspect of microreductions, the role of descriptions of structure. As is well known, a microreductive explanation applies to an integrated whole composed of parts. One goal of a microreduction is to explain the laws governing the whole in terms of laws governing the parts, plus a description of the structure of the whole, and perhaps some other information. General adequacy conditions for microreductions are presented in great detail in *Unity of Science*, yet this book leaves the concept of "description of the structure" rather vague (see pp. 60–61). Kuipers also refers to structures in his book, and Section 5.2 makes use of the idea of a "structure representation function." Yet, I find no definition or general characterization of functions of this kind. Now it might be thought that characterizing "structure," in the sense required for microreductions, is not a very significant philosophical problem. In fact, I believe that the contrary is the case, and that we cannot fully characterize microreductive explanations without an adequate characterization of the relevant sense of "structure." In this article I develop an analysis of this concept of structure. Of course, the word "structure" is used in many ways. For example, a mathematical structure can simply be an abstract set together with specified types of relations and functions defined on this set. This is a useful concept, but too general for my purposes. I am concerned with what I called *structured wholes* in *Unity of Science*. A structured whole (SW) is an object that exists in the real world and is composed of parts. As is well known, there is a large literature on mereology, which is concerned with parts and wholes. I shall not review this literature here because I have not found it helpful in my quest to characterize SW's. Instead, let us begin with some examples and work from there.

2. Some Motivational Examples

In order to motivate the explication of structured whole (SW), I shall briefly discuss a few familiar examples drawn from the natural sciences and everyday life. Among the most familiar types of SW's considered in the natural sciences are the various types of molecular structures. Usually one is concerned with a type of molecule rather than a particular molecule. In order to describe a type of molecule it is at least necessary to mention the types of atoms composing it and the spatial configuration of these atoms. For instance, a methane molecule has a carbon atom surrounded by four hydrogen atoms in such a way that the carbon atom can be considered to be in the center of a regular tetrahedron with a hydrogen atom located at each vertex. To be considered a molecule, a configuration of atoms must exhibit some reasonable degree of stability. Stability under a range of environmental conditions suggests internal forces holding the atoms of a molecule together in their characteristic configuration. As molecular theory developed during the Nineteenth Century, it became customary to represent these forces rather abstractly as chemical bonds. Eventually, different types of bonds were distinguished, for instance, single, double, and triple bonds, represented by one, two, or three dashes, respectively, in molecular diagrams. In the Twentieth Century additional types of chemical bonds were distinguished. In order to describe a molecular structure it is not sufficient to mention the atoms and their configuration in space. One must instead list the various types of atoms in the molecule, together with the bonds between these atoms. An elaborate general theory of chemical bonds now exists. This theory is based on quantum mechanics, and it allows one to derive many other attributes of a molecule from a description of its structure in terms of its atoms and their bonding arrangement. In principle, the spatial configuration of the atoms in a molecule should be derivable from this type of description plus the general theory of chemical bonds. Yet, not all structures have spatial configurations; at least not in the sense of physical space. For instance, a social structure may be described in terms of a relatively stable configuration of types of actions performed by individual agents or institutions in certain roles. So, instead of referring to "spatial configuration," I shall use the term, *stable configuration*, when discussion SW's. This concept will be refined in later parts of this article. For now we can say that a description of the *stable configuration* is an *explanatory consequence of the description of the structure of the molecule, in terms of parts plus bonds,* rather than an essential part of the description of this structure. This idea will be generalized.

It can be seen that many other types of SW's are correctly described in terms of their parts and how these parts are bonded. In the case of a particular

SW we must describe its particular parts; in the case of a type of SW we must describe the types or kinds of parts it has. Consider a type of brick wall that is constructed from bricks of uniform type and size, which are mortared together in a particular repeating pattern with a particular type of mortar. We can describe this type of wall by describing the type of bricks in it, the type of mortar used, and the way each brick is mortared to each of its neighboring bricks. Consider any two neighboring bricks in the wall. By describing the type of mortar between them and exactly how this mortar is placed between them (e.g., a certain amount of mortar placed between adjacent ends of the two bricks), we are describing the type of bond between these two bricks. Changing the type of bricks, the type of mortar, or the way adjacent bricks are mortared together will produce a different type of SW (or no SW at all).[1]

A structured whole can have moving parts. For example, a bicycle is an SW. Also the solar system is an SW. In this case the stable configuration is described in terms of the orbits of the various planets and their satellites, and the functions which describe the positions and velocities of these bodies at various times. The bonds between these bodies can be described in terms of the various gravitational and inertial forces affecting them in such a way as to maintain a stable configuration of the entire solar system. There are many kinds of stable configurations with moveable parts. Consider a chain. The separate links are the parts; the bond between any two adjacent links consists in the state of their being linked in the way they are. The exact spatial arrangement of the chain is variable within limits. If we examine two adjacent links, there will be some range of possible positions they can have with respect to each other without their linkage breaking or without producing substantial distortion of either link. Suppose that these two links are labeled *a* and *b*, and to simplify the discussion, suppose that *a* is fixed in space. Then the range of possible positions of *b* will be limited by the fact that it is bonded to *a*. We can call this limitation *a restriction on the degree of freedom of b with respect to a.* Now consider the entire chain. Each link has its degree of freedom somewhat restricted with respect to other links. This produces a range of possible positions that can be reached by the entire chain. This range of possible positions can be considered the configuration of the parts of the chain.

The bonds of any finitely determined SW can be broken or destroyed if the structure is exposed to sufficiently strong stresses. This is certainly true of the examples just discussed. However, in a sufficiently benign environment these SW's will be stable without any significant interaction with the environment.

[1] There are more complex types of structures. For instance, in some SW's some parts form substructure SW's which are in turn parts of the larger SW. These and other kinds of complications should not require any essential modifications of the analyses presented in this article.

Not all SW's have this feature. Consider a protozoan, such as an amoeba. It has a complex structure with internal parts such as mitochondria and nuclei. But its stability as a structure depends on exchanges of materials and energy with its external environment (Parker 1982, pp. 1406-1407). Similar interactions with external environments are found in multicellular plants and animals, and in social structures. My explication of SW's will be sufficiently general to include SW's whose stability requires environmental interactions. Before stating it, we should consider a few more examples of SW's and some non-SW's.

In Section 6 I shall briefly discuss how a container of gas is described in the kinetic theory of gases. The examples in this paragraph and the next may help to prepare for the later discussion of the kinetic theory. Suppose that we have several light, round, rubber toy balloons inflated with air. In a still room, each of these balloons would, if unsupported, slowly fall to the floor. Suppose, however, that a number of streams of air are directed towards the center of the room above the floor from several different strategically placed blowers. Suppose that a clump of several balloons is positioned above the floor in the region of the room where the air streams converge. The balloons are in no way attached to each other, but each one is either barely touching one or more neighboring balloons, or is close by and not touching. Finally, suppose that the balloons and the airstreams are so arranged and balanced that the clump of balloons remains suspended above the floor in a fixed configuration. This is an improbable, but not impossible, state of affairs. Label this suspended clump of balloons *B*, and consider the airstreams and all else to be the external environment *E* of *B*. *B* is a (relatively) stable configuration of balloons. Yet, the only forces maintaining this configuration are the external force of gravity, the small forces of buoyancy, and the forces produced by the air streams. (We can assume that there are no frictional forces between the balloons. In fact, they may not even touch each other.) Thus, the configuration of balloons is maintained entirely by external causes, and there are no internal bonds in *B*. We can say that *B* is an example of an externally constrained configuration of objects, and I doubt that anyone would consider it to be a structured whole.

Now consider a bunch of marbles *Marb*, which are held together in a certain configuration because they are tightly wrapped in a sealed plastic bag *Bag*. First suppose that *Bag* is considered to be part of the external environment of *Marb*. Then *Marb* is similar to the example of the balloons, and *Marb* is not an SW. Now consider *Marb* together with *Bag* to be one object (which I denote *Marb* ⊕ *Bag*), and consider the external environment to consist of everything outside of *Bag*. Since *Bag* is tightly wrapped around *Marb*, there are internal strains in *Bag* which transfer forces to the marbles adjacent to *Bag*, which in turn transfer forces to the other marbles in *Marb*. All of these forces are produced internally in *Marb* ⊕ *Bag*, and they bond the marbles and the bag together into a fixed configuration. Thus, *Marb* ⊕ *Bag* is a structured whole. Let us say that the specification of the *boundary* of an object distinguishes the surface and inside of the object from its external environment. This example illustrates that one must precisely specify the boundary of an object before one can make a definite decision whether it is an SW. Specifying boundaries of objects is related to the way in which a theory classifies the kinds of elements in its domain. The construction of a classification system sometimes requires making somewhat conventional distinctions. Some use of convention is also to be expected in specifying boundaries of objects.

3. Configurations, Constraints, and Bonds

So far I have been using the term *bond* in a vague and intuitive way. My explication of SW's is intended to be very general. Because of this generality, it is impossible to use a very precise definition of *bond*. Yet, I believe that the general idea of a bond can be described adequately for our purposes. In order to fashion this description, I shall now develop the general explication relative to a scientific theory. I use some of the terminology in Causey (1977).

Let us suppose that we have a scientific theory **T** that consists of a set of laws about the attributes (and behavior) of the things in some domain, *Bas*. The things in *Bas* may themselves be SW's and thus may be decomposable into smaller parts under certain conditions. However, it is assumed that the laws of **T** describe attributes of these things under conditions such that these things are integral units. Thus, from the point-of-view of **T**, the elements of *Bas* are basic (indecomposable) elements. **T** will be formulated with the help of some background logic, and it will also make use of a set, Voc(**T**) of nonlogical predicates. Some of the predicates in Voc(**T**) will denote kinds of things in *Bas*, and some will denote attributes (properties, relations, and quantities) of these things. If **T** refers to particular things in *Bas*, it will be assumed that Voc(**T**) is augmented with proper names for these particular things. The various things in *Bas* will exhibit different attributes under different conditions, for example, an atom may be at rest or it may be moving under different environmental conditions. Thus, Voc(**T**) must also contain predicates that enable us to describe various, relevant environmental conditions in which the things in *Bas* can exist. It is important to realize that in any normal scientific theory the predicates in Voc(**T**) that denote kinds of things in *Bas* are predicates that make no reference to environmental conditions. For instance, *hydrogen atom*, *horse*, NaCl–*crystal*, *human being* all refer to kinds of things without referring to any environmental conditions. It should also be noticed that what is considered to be the set of relevant environmental conditions depends on the theory **T** and its ontology. The economic conditions in Namibia will not be relevant if **T** is the atomic theory and *Bas* consists of atoms.

Suppose now that **T** and *Bas* satisfy the general conditions above. Consider an arbitrary element of *Bas*. Depending on its external environmental conditions, it may be more or less constrained. For example, imagine a small elastic particle trapped in an elastic box. It can bounce around within the box, but it will be assumed to be incapable of penetrating through the walls of this box. The movements of this particle are constrained within a particular region of space. Yet other attributes of the particle may not be constrained. For example, at least in classical mechanics, there will be no limit on its kinetic energy; it may be at rest, or it may be bouncing around with an extremely high velocity. An analogous example is this: a person's movements may be constrained by locking him (or her) in a prison cell, yet this person may be allowed the freedom to sing or not to sing while in the cell. In general, if **T** is fairly well developed, it will be able to specify, either deterministically or probabilistically, the various attributes or ranges of attributes which an element of *Bas* will have under specified environmental conditions. Some of these attributes may be lawfully correlated with others, so it is customary to pick out a set of independent attributes in terms of which to specify the *state* of an element of *Bas*. For example, in classical mechanics, the state of a particle is specified by giving its three position coordinates and its three momentum coordinates. A set of independent attributes used to specify the state of an arbitrary element of *Bas* is a set of *state coordinates* or *state dimensions*. These attributes may be either qualitative or quantitative, and they may assume a finite or infinite number of degrees. Thus a classical particle may in principle assume an infinite number of positions along an x-axis, a y-axis and a z-axis. The particle in the box, however, is constrained to a restricted subset of all possible position values. Let p be an arbitrary element of *Bas* and let E describe some arbitrary set of environmental conditions within the scope of **T**. Let $\mathbf{s} = < s_1, \ldots, s_k >$ represent the various state coordinates of **T** expressed as a state vector. Suppose that p is under conditions E. Then we will assume that, for each s_i, **T** can specify the range of possible s_i-values which p can have under E. Thus, if no environmental conditions are specified at all, then **T** can specify the total range of possible s_i-values that can be reached by an arbitrary element of *Bas*.

The examples in the previous section indicate that an SW has a stable configuration of parts that is determined by bonding relations. It is therefore important to be able to describe configurations. In the physical sciences, configurations are often described in spatial terms. For example, in mechanics a configuration of particles at a particular time can be described by giving for each particle its x, y, and z coordinates. Note that this is not a complete description of the state of the set of particles, since the state also includes the momenta of the particles. Thus, in describing a configuration one usually uses only a proper subset of the set of state coordinates. This proper subset may consist of spatial coordinates, but it need not be spatial. It may instead be quite abstract, for instance, it may consist of possible dimensions of behavior that an animal might exhibit. I do not know any characterization of the general types of state attributes that can be used in descriptions of configurations. This is an issue for further investigation. In general, the attributes used will depend on the theory **T** and on the general category of configuration under consideration. Returning to **T**, I shall assume the following: Among the state coordinates, $\mathbf{s} = < s_1, \ldots, s_k >$, a certain subvector, $\mathbf{c} = < c_1, \ldots, c_n >$ is specified. These c_i are the *configuration coordinates* (*dimensions*) of *Bas*. The set of all possible c_i-values that can be reached by an element p of *Bas* is the *degree of freedom of p along the coordinate* (or *dimension*) c_i. The "*position*" (understood abstractly) of an element, at a particular time, is given by specifying a vector **c** that truly applies to this element at the time. The *configuration space* of **T** is the set of all possible values of **c** corresponding to the degrees of freedom of all of the kinds of elements in *Bas*.

Let $P = \{p_1, \ldots, p_m\}$ be a finite set of elements of *Bas*. At time t we specify the relative position of each p_i with respect to the others. Relative positions are specified in terms of the configuration coordinates introduced in the previous paragraph. If these relative positions are stable during a time interval, then P maintains a *stable configuration* during this interval. This does not mean that P is stable in any absolute sense of configuration. It means that the configuration of the elements of P with respect to each other is stable during the time interval. In addition, in this context, "stable" does not mean constant or invariant. Recall the example of the chain. Its links are not fixed with respect to each other; they can move within certain limits. Yet, we want to say that the chain has a stable configuration. In general, I shall say that the elements of P have a *stable configuration*, or that P has a *stable configuration*, over a time interval if and only if the relative configuration positions of these elements remain within specified ranges during this time interval.

Now recall the balloon example. The clump of balloons has a stable configuration in physical space, but the stability of this configuration is maintained by external forces. The clump of balloons is not an SW. We still need to examine the concept of bonding. Most bonds appear to be binary, between two objects, so I shall first consider the case of two elements of *Bas*, a and b, which are possibly of the same kind of basic element of **T**. Suppose that *a* is within a specified environment, *E*, and there are no other objects in this environment. This is, of course, an idealization, but it is the kind of idealization that is commonly used in theoretical science. Under these conditions, I shall say that *a* is free in *E*, or that *a* is in the free state in *E*. When *a* is free in *E*, it will have a certain degree of freedom along each of the configuration coordinates. This will determine a set F_a of possible vector values of these coordinates. I shall say that F_a is *the degree of freedom of a under E*. Similarly, F_b denotes the degree of freedom of *b* under *E*. Now assume that both *a* and *b* are simultaneously in environment *E*. If they do not interact in any way, then they would each still have the degrees of freedom, F_a and F_b. If this happens, I say that there is no restriction on their *relative* degrees of freedom.

Now suppose that there is a restriction on the relative degrees of freedom of *a* and *b* in *E*. This restriction may only be temporary. For example, suppose that *b* is a star and *a* is a spacecraft traveling through space initially in a straight line with an initial constant velocity. The spacecraft may approach *b* in such a way that it passes by *b* without crashing into it or getting trapped in an orbit around *b*. In this kind of situation, *a* continues on in space past *b*, but the path of *a* is bent by their mutual gravitational attraction (see, for instance, Goldstein 1950, pp. 65–66). If this happens, I say that the *relative degree of freedom of a with respect to b is constrained or restricted*. When this occurs, the relative degree of freedom of *b* with respect to *a* is also constrained. Yet, these two objects do not have a stable configuration because their relative configuration positions are not stable over the time interval under consideration (i.e., the entire time of flight of *a*, which might be extremely lengthy). Assuming that no other states, and no forces other than gravity, are involved, I say that *a* and *b* are not bonded in this example. From this example, it should be clear that a stable configuration is required for a bond.

We are now in a position to characterize bonds. More precisely, I shall state the conditions for the existence of a binary bond, and then discuss these conditions. I continue to assume that we have a theory **T** about a domain of things, *Bas*. The language of **T** is used to describe environmental conditions, as well as being used in the statements of laws about the things in *Bas* and their attributes. It is assumed that the reader is familiar with the features of deductive-nomological derivations, and their limitations. In spite of these limitations, I believe that good, causal explanations within well-developed theories can be formulated in the form of deductive-nomological derivations. Thus, when I mention "causal explanation," it will be assumed that such an explanation can, in principle, be formulated in deductive-nomological form within the theory **T**. Of course, in order for the explanation to be reliable and acceptable, the theory must have empirical support. If **T** has unsupported hypotheses, the "explanations" are only possible explanations. Additional details are in Causey (1977, Chapter 2), and of course in Kuipers' SiS (Chapter 3). To simplify the presentation, the following condition is stated for a particular bonding relation between particular elements. It can be generalized in a straightforward way to a kind of bonding relation between kinds of elements.

BB: Existence condition for a binary bond. Let *a*, *b* be distinct elements of *Bas* associated with theory **T**. Let *E* be a description of the environmental conditions external to *a* and *b*. Then, *a is bonded to b in E during a time interval* if and only if all of the following hold.

B1. The relative degree of freedom of *a* and *b* is constrained during the time interval.

B2. There is a causal explanation (which we may or may not know) of the relative constraint mentioned in B1. This explanation makes essential reference to attributes of *a* and *b*, makes essential use of general laws of **T**, and may use the description *E* as a boundary condition.

B3. The explanation mentioned in B2 does not refer to any elements of *Bas* other than *a* and *b*, except possibly for certain environmental conditions described in the following paragraphs.

In order for *a* and *b* to be bonded, it is not necessary to have a restriction on every state of either one. For instance, if *a* can exist in different colors, we *normally* would not require a restriction on its colors to be an essential part of a bonding relation.[2] When we speak of bonds we presuppose some relevant configuration coordinates of the bonded objects. This is presupposed in B1. In order for a bond to exist, or for us to hypothesize that a bond exists, it is not necessary for us to know how to construct the appropriate causal explanation in B2. It is only necessary that such an explanation could, in principle, be given. Thus, when we assert the existence of a bond, we are at least tacitly assuming that such an explanation is possible. I believe that the Nineteenth Century chemists who hypothesized chemical bonds made such tacit assumptions.

Condition B3 requires that the causal explanation not refer to any elements of *Bas* other than *a* and *b*. This condition is included as part of the existence condition for *binary* bonds. For contrast, suppose that the relative degree of freedom of *a* and *b* is constrained only when some third object *c* is present. Also, suppose that the explanation of the *a*–*b* constraint makes essential

[2] There could be exceptions. If *a* and *b* are socially bonded chameleonic creatures (see Section 5 below), and part of their behavioral states include their changeable colors, then color might be a configuration coordinate that plays a role in their bonding relation.

reference to *a*, *b*, and *c*, and their attributes. In other words, the presence of *c* is a *necessary condition* for the constraint between *a* and *b*, according to the relevant theory of these objects. In this kind of situation we can distinguish two kinds of cases: the relative degree of freedom of *c* with respect to *a* and *b* is also constrained, or it is not. In the former case, it is natural to say that we have a tertiary bond between all three objects. In the latter case, which seems unlikely to occur, it is not clear what to say. I shall adopt the convention that this latter case is not a case of tertiary bonding, but rather that it is a rare situation in which the presence of *c* is simply considered to be a part of the environmental conditions affecting *a* and *b*. In the realm of social structures, it is conceivable that there is a *ménage à trois* that is stable and constrains all three people only because of interactions between all three, and is such that no two of the persons would stay together without the third. This would be an example of a tertiary bond. The existence condition can easily be extended to bonds between four or more objects in a similar way.

This distinction between binary and tertiary bonds requires some additional clarification. Consider a hypothetical structure, *a–b–c*. In order for there to exist an *a–b* bond, we would expect that *a* and *b* both need to be in certain states. Suppose that *b* must be in some state S_b. For instance, if *b* is a person, S_b might be some kind of psychological state. If *b* is an atom, S_b might be a state of its outer electron shell. If *b* is a supporting cable in a suspension bridge, S_b would probably include features such as its tensile strength, elasticity, mass, etc. Now, this relevant state S_b required of *b* in an *a–b* bond, may not be stable without having *c*, or some surrogate for *c*, bonded to *b*. Now we distinguish two cases: (i) Without an object of kind *c*, state S_b *could not exist*, according to **T**. (ii) State S_b could exist, according to **T**, through other, surrogate means. The other means could be the presence of objects different in kind from *c*, or they could simply be some environmental conditions that put *b* into state S_b. The phrase in the previous paragraph that the presence of *c* is a *necessary condition* for the constraint between *a* and *b* is to be understood as (i). Thus, to have a tertiary bond, (i) must hold, and the description of the bond must make *essential reference* to *a*, *b*, and *c*. Similar remarks apply to more complex multiple bonding relationships.

BB is only an existence condition for a binary bond; it does not provide a criterion of identity for distinguishing types of bonds. However, different *types, or kinds, of bonds between a and b* can be distinguished by different types of relative constraints mentioned in B1. For instance, under some conditions, *a* and *b* may be bonded in a very strong and restricted way, and under other conditions they may be bonded loosely and weakly. Naturally, we would also expect that these different kinds of bonds would have different B2 explanations. It should be observed that conditions B1 and B2 refer to a constraint in the relative degrees of freedom of *a* and *b*. This is to be understood as symmetric; i.e., if *a* is constrained relative to *b*, then *b* is constrained relative to *a*. As a result, when we are referring to one type of bond, there is no difference between saying that *a* is bonded to *b* or saying that *b* is bonded to *a*. The bonding relation is symmetric. Again, the symmetry refers to the *existence* of the bonding relationship, not to the particular types of constraints. For instance, in a *Master-Slave bond*, both the Master and the Slave are bonded to each other. Moreover, each of them is constrained by the existence of this bond, although the nature and degree of these constraints are different. Abraham Lincoln wrote, "As our case is new so we must think anew and act anew. We must disenthrall ourselves and then we shall save our country." The institution of slavery enthralled both the Masters and the Slaves. It is assumed that the language of **T** is adequate to define the different kinds of possible bonds, by using descriptions of the relative constraints corresponding to different kinds of bonds.

Condition **BB** says nothing about the strength of the bonding relation. Some bonds are strong and others weak, and there may be different ways of measuring bond strengths. For example, we may be interested in the resistance of a bond to acids, to heat, to physical bending, or to stretching forces. In general, the strength of a bond (however measured) will depend not only on attributes of *a* and *b*, but also on the environmental conditions *E*. The strength of an *a–b* bond may also be affected by other nearby objects in the environment *E*. Condition B3 only says that the explanation of the bonding does not require reference to any other elements of *Bas* (except in the special, and seemingly unlikely, case previously discussed). This means that we can explain the *existence* of an *a–b* bond without referring to other elements of *Bas*, except in very special environmental situations. But this existence condition does not imply that we are not allowed to refer to other elements in an explanation of some feature of the *a–b* bond. Suppose the *a–b* bond occurs in some SW and there are other elements near *a* and *b* in this structure. Then the *strength* of the *a–b* bond may be affected by these other elements of the SW and their positions relative to *a* and *b*, so it may be necessary to refer to these other elements in an explanation of the strength of this *a–b* bond. It is important to distinguish between these two kinds of cases: We may have three elements *a*, *b*, and *c* all bonded together by a set of binary bonds, and the strength of the *a–b* bond, say, may be affected by the presence of *c*. On the other hand, we may have a genuine tertiary bond between *a* and *b* and *c*. In the former case the relative constraints exist between the pairs alone, although probably in different strengths. In the latter case constrained pairs alone would not produce the relative constraints found in the tertiary bond; indeed, the tertiary bond stability is not a result of a combination of binary bonds.

4. Structured Wholes

Once again, let $P = \{p_1, \ldots , p_m\}$ be a finite set of elements of *Bas*. If the environmental conditions are such that the degree of freedom of any p_i is directly restricted by these conditions, then I say that p_i and P are *externally constrained.* Also, if the environmental conditions are such that the relative configuration positions of the p_i in any subset of P are directly restricted by these conditions, then I say that P is *externally constrained.* Now most things in the world are subject to some external constraints by the environment, but many of these constraints are so remote that they have no practical significance. Consider an amoeba in the middle of a large pond. This organism is perhaps constrained to stay within the pond, but there may be nothing in its local environment which is constraining it or its parts. I will count as the local environment of a thing that part of its environment which has significant effects on it, where "significant" is relative to the context under consideration. This is not a precise characterization, but I believe that it will be seen to be adequate for the purposes at hand. We can now say that a thing (or set of things) is *locally externally constrained* (is subject to a *local external constraint*) if it is externally constrained by its local environment.

Now let a and b be any two distinct elements of P. I say that a and b are *linked by a path of bonds* if and only if there is a set of elements, $\{a, p_1, p_2, \ldots, p_k, b\}$, in which the p_i are distinct from a and b, and also pairwise distinct, such that a is bonded to p_1, p_1 is bonded to p_2, …, and p_k is bonded to b. We allow that there may be no p_i, so that when a and b are directly bonded together, this bond also counts as a path. Also, when a and b are distinct, and are linked by a path (of bonds), they may also be directly bonded together. In other words, they may be directly connected by a bond and also related by a path. The concept of "path" used here is familiar from graph theory, except that we require that a and b be distinct, which is not always required in the graph theory literature.

Using the terminology which has been defined, and referring to the type of theory **T** previously introduced, I shall now present the existence condition for a structured whole. Actually, it is more convenient to break this task into two cases, according as the SW is not, or is, subject to local external constraints. For the sake of brevity, the present article is limited to the scope of SW's that are not externally constrained. An SW of this kind will be called an *unstressed structured whole*.

> **USW:** Existence Condition for an Unstressed Structured Whole. Let P be a finite set containing at least two elements of *Bas*, let B be a set of types of bonds definable in the language of **T**, and let E be a description of the local

environmental conditions external to *P*. Let *W* be an object described as follows: A list of pairs of elements of *P* bonded by binary bonds in *B*, a list of triples of elements of *P* bonded by tertiary bonds in *B*, etc. This list may contain zero or more bonds of any particular arity (but it must contain some bonds; see USW3). The elements of *P* are called the parts of *W*, and *W* is an Unstressed Structured Whole (USW) during a time interval if and only if all of the following conditions hold:

USW1. There are no local external constraints on *W* during the time interval.

USW2. *W* has a stable configuration during the time interval.

USW3. During the time interval, for any two parts of *W* there is a path of bonds that links these parts.

USW4. During the time interval, the particular bonding relations holding between particular parts of *W* remain the same.

USW5. The stable configuration of *W* is causally explainable in terms of the laws of **T**, attributes of the elements of *W*, and the description of the bonding relations between the elements of *W*.

Of course, the individual bonding relations are further explainable in the manner stated in **BB** in the previous section. The basic motivation for this condition should by now be clear. A USW consists of parts (the elements of *P*) that are bonded together in such a way that the entire set of parts is connected, i.e., each pair of parts is linked by some path of bonds. Moreover, the entire object has a stable configuration that results from the bonding relations between the parts (rather than having a configuration which results from external constraints). It is very important to note that **USW** is an existence condition for an unstressed SW, but not a uniqueness condition, and **USW** does not provide a criterion for the type identity of unstressed SW's. If W_1 and W_2 are two USW's, then they will certainly be of different types if they contain different types of parts. They will also be of different types if their sets of bonding relations are not the same. Yet, as described in **USW**, W_1 and W_2 may have the same kinds of parts and same bonding relations, but be different *types* of USW's. This is possible because a description of parts plus bonding relations *may not* be sufficient to determine a unique stable configuration. Whether or not this is the case will depend on exactly how bonding relations are characterized. For example, consider the compound bromochlorofluoromethane, which has the traditional "structural" formula:

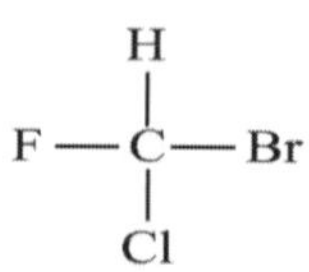

Although this type of formula was believed for some time to represent the "structure" of the molecule, it was eventually realized that the molecule actually is not two dimensional, but rather three dimensional. Moreover, the four bonds around the carbon atom, C, point in space towards the vertices of a tetrahedron with the C-atom in the center of this tetrahedron. Since there are four different atoms bonded to this C-atom, this molecule has two distinct three-dimensional stable configurations, called "enantiomorphs," (Parker 1982, p. 657). These two distinct configurations are mirror images of each other. They are distinct USW's. Thus, the existence condition **USW** does not provide a criterion of identity for types of USW's, just as the existence condition **BB** does not provide a criterion of identity for types of bonds. At the present time it appears to me that criteria for type identity of bonds and SW's are likely to be highly contextually dependent on substantive features of the relevant theory **T**, so I shall not attempt to state such criteria in this article.

Now consider a specially constructed accordion with an internal spring device that keeps the bellows expanded when no external force is applied. The unstressed configuration of this structure is its expanded-bellows rest position. But if strong, persisting, squeezing forces are applied to the ends of this accordion, it can be kept in a squeezed-up configuration. When the accordion is at rest, with its bellows expanded, it is unstressed and has an unstrained configuration. When it is squeezed up, it is subject to local external constraints (stresses) and it has a strained configuration. The existence condition stated above is clearly intended to apply to unstressed structured wholes. When there are no local external constraints on *W* we say that *W* is unstressed. When *W* is unstressed (condition USW1), then the stable configuration of the structured whole is explainable without *explicit* reference to the external environmental conditions *E*. Of course, *E* may be invoked in the explanations of the bonds in *W*. The reason *W* is characterized as unstressed is because of USW1; the local environment does not *by itself* directly constrain the parts of *W*.

Condition USW5 is actually rather deceptive, for the required explanations can be much more complicated than USW5 seems to suggest. The basic pattern of such explanations is this: Since each part in *W* is bonded, its degree of freedom will be restricted relative to the other parts to which it is bonded. In a structured whole all of these restrictions on individual parts must combine together somehow to produce a stable configuration of all of the parts in *W*. However, these restrictions will not usually be additive, for the bonds on one element may be affected by the other elements and their bonding relations. Thus, in general, the entire stable configuration (when it exists) will be the result of complex interactions between all of the parts and their bonding relations. Indeed, it is not obvious that a stable configuration must result from the fact that all of the parts in *W* are linked as stated in USW3. It is for this reason that USW2 is stated as an independent condition in the existence condition.

In the case of a USW the environmental conditions *E* do not produce local constraints, so the configuration is explainable without explicit reference to *E*. However, attributes of the parts of *W* may be affected by *E*, and these effects may indirectly affect the bonding relations. More generally, by affecting attributes of the parts, *E* may indirectly affect bond strengths and perhaps other features of bonds. Thus, the stable configuration of a USW may be indirectly affected by *E*. Yet, if we know the effects that *E* has on the parts, then we can use this information as part of a separate explanation of the states of the parts. We can then use the information about the states of the parts in the explanation mentioned in USW5. The basic idea is this: In a USW there are no local external constraints on *P*, so *E* does not directly contribute, through stresses, to the stable configuration of the whole. Clearly, many actual SW's are subject to stresses, i.e., local external constraints on *W*. If these stresses grow very strong, they may cause a breakdown of the structure, i.e., the complete destruction of a part or of a bond. This may produce a new kind of SW, or it may result in no SW at all. If the stresses are extremely weak, they may cause no significant change in the SW at all. If the stresses are significant, but moderate, they may leave the basic bonding pattern of the SW unchanged while producing some change in the stable configuration. In this latter case, I say that the structured whole is *strained* and that it has a *strained configuration*. In a sense, *strained* is a stronger notion than *stressed*, since *strain* involves stress together with a resulting change in the configuration. I use these terms in the way that is customary in mechanical and structural engineering. For example, a bridge is stressed when a load is applied to it, but if it also bends or is deformed in some way under this load, then it is strained. The term *structured whole*, SW, has been, and will continue to be, used for either the unstressed or the stressed cases.

To save space, I shall not explicate stressed structured wholes here. It is fairly straightforward to modify and extend the USW conditions to such an explication, but the explication is complex. Fortunately, the intuitive idea behind this extended explication is actually fairly simple. In order for an object to be a stressed structured whole (SSW), there must exist, at least theoretically, a corresponding unstressed structured whole (USW) in an environment that is like that of the SSW except for lacking the local external constraints on the SSW. An SSW and its corresponding USW must differ at most in their stable configurations and bond strengths. If there is such a difference, then the SSW is strained. The character and degree of this strain will be describable in terms of the differences in the stable configurations and bond strengths of the SSW and its corresponding USW. In addition, this strain should be explainable in terms of the USW5 explanation combined with relevant information about the nature of the local stresses caused by the constraints on the SSW.

5. A Model Social Structure

The preceding analysis of bonds and structured wholes was presented in a general and abstract form in order to include a broad range of cases. In the social science literature one finds a wide variety of uses of terms like 'collective action', 'aggregation', 'social structure', and the like (for example, see Kuipers 1984, SiS, and Bates and Harvey 1975). Straightforward aggregation of phenomena usually is not problematic, if done carefully. Unfortunately, there seems to be no consensus on the meaning of the term 'social structure', in spite of its frequency of use. A recent book on the subject, Crothers (1996), discusses at length many different conceptions of social structure. I propose that we can use the analysis of SW presented here as a semi-formal model for the use of 'social structure'. It is not practical to go into details here, even with any proposed example of an actual social structure. Instead, I shall present a simple model of social structure in robot actions. This model is entirely hypothetical and not intended to describe any existing or realizable robotic system. In many respects it is unrealistic and over-simplified. Yet, it can serve as a model for possible future robotic social structures.[3]

Suppose that there are three robots, NOD, ROD, and TOD, which can move about on a flat plane. These robots are each powered by little internal electric motors that receive power from an on-board battery. The robots also have television cameras, which are parts of their sensory apparatus. The lenses of these cameras need cleaning from time to time. The robots have on-board washers that clean these cameras. These washers require washing fluid called, "eyewash," which is stored in an on-board bottle. In addition, the robots have a small tank that stores light oil, which is used for lubricating some of their mechanical parts. In order to keep operating, from time to time the robots need to have their batteries recharged, their eyewash bottles refilled, and their oil tanks refilled. I shall call electricity for the battery, eyewash fluid, and oil the robots' "nutrients." In addition, these robots have on-board computers which can be programmed to give the robots various dispositions towards various types of behavior. The primary behavior of the robots is to roam around the

[3] Causey (1980) and Causey (1983) present an early sketch of some of the general ideas presented in the current article, but without detailed formulation of the **BB** and **USW** conditions. Since 1983 I have been largely occupied with administrative work, and with research on logic and artificial intelligence, and have only recently returned to the investigation of social structure.

flat plane with no preplanned route, observing and recording surface features, and developing a map of the plane. Incidentally, it is currently a major problem in artificial intelligence to program a robot to observe and build a readable map of territory. Therefore, the model system described here is currently an item of science fiction, although something like it may be feasible in the near future.

Now suppose that on this plane there are, at some distance from one another, three filling stations, named CHARGE, EYEWASH, and OIL. Station CHARGE is a battery charging station, EYEWASH is an eyewash filling station, and OIL is an oil filling station. When we first encounter these robots, they are free and independent. Each robot roams the plane with no interference or special pattern except one: Periodically it must replenish its supply of nutrients. When an on-board sensor detects that the eyewash level, say, is low, the robot heads, because of its internal program, towards EYEWASH, where it refills its eyewash bottle. The behavior is analogous for low levels of battery charge and of oil. Fortunately, the robots have a large battery and ample containers, so they do not need to replenish themselves often. Most of the time they roam freely, and independently of one another, about the plane,

Now suppose that the robots are reprogrammed at some time when all three robots have all of their batteries charged and their storage reservoirs full. According to their new programs, they are disposed to behave as follows: Robot NOD recharges its battery at station CHARGE, but it never uses EYEWASH or OIL. Robot ROD refills its oil tank at OIL, but it never uses EYEWASH or CHARGE, Robot TOD refills its eyewash bottle at EYEWASH, but it never uses CHARGE or OIL. When robot NOD is low on eyewash, its new program directs it to get eyewash from TOD, which is also programmed to give some of its extra eyewash to its companion robots. When robot NOD is low on oil, it then gets oil from ROD in a similar manner. Likewise, robots ROD and TOD get some of their nutrients from each other and from NOD in a similar manner. Thus, each robot refills one nutrient at one station and gets its other two nutrients from the other robots. Since each of them has a large battery and ample reservoirs, they are able to accomplish this satisfactorily.

It may be thought that spatial relations are essentially involved in the description of this robotic system because the previous description requires them periodically to move towards one another. But these movements can be avoided by assuming that the robots are equipped with arbitrarily long umbilical cords and with transmitting and receiving devices. Then, when one needs a nutrient from another, he merely signals the other, they extend their umbilicals, and transfer some nutrient. We may also assume that they get their nutrients from their respective filling stations in the same manner. Then they are free to move around the plane spatially any way they like. Now let us consider the configuration space of all possible kinds of behavior that these robots can perform. They can perform many kinds of actions, such as getting recharged at CHARGE, making observations of the landscape, moving in various directions, etc. Some of these actions, such as moving in a particular direction, involve spatial concepts, but not all of the actions do involve space. We can describe the action of NOD getting eyewash from TOD, using its umbilical cord, without referring to specific distances or locations. Thus, when the robots x and y are programmed to share nutrients through umbilicals, the program imposes a constraint on their relative degree of freedom with respect to each other. In plain language, we could say that the robots are dependent upon each other for supplies of nutrients. In principle, with suitable detailed information, x and y could satisfy condition **BB** for the existence of a binary bond. The combined effect of such behavioral bonds between NOD, TOD, and ROD could result in the satisfaction of condition **USW** for the existence of an unstressed structured whole. If this happened the robotic system would be an SW composed of objects in a configuration space of behavior (or actions). A *social structure* of this kind would be more than a mere aggregate, and it would be susceptible to reductive explanations of its attributes, including behavioral dispositions.

This robotic social structure illustrates that SW's need not involve spatial relations. In particular, social structures, described in suitably abstract language, need not involve spatial relations. This point can be disputed. For instance, one might argue that the individual robots are material objects existing in spatial locations, so therefore any SW of which they are parts must be spatially locatable. This, and related ontological issues, are discussed in detail in Ruben (1983). Ruben's article distinguishes a person's being a *part* of a social entity from being a *member* of that entity, and he does correctly point out some confusions that easily arise when discussing social structures. I believe that these confusions can be avoided if we use careful descriptions of wholes and their parts. In the present example, the parts are really not the material robots, but rather a more abstract kind of entity existing in a configuration space of behavior. Of course, this raises further questions about the relation between the material robots and these abstractions. That is a problem for future investigation and analysis.

6. Concluding Remarks

As Kuipers correctly shows in Chapter 3 of SiS, a scientific explanation often makes use of identification and aggregation. In a typical microreduction we identify some kinds of things with structured wholes composed of simpler kinds of things. When we say that one particular thing *a* is an SW composed of parts, we mean that *a* has integrity as a unit, and this implies that it has some degree of stability as an object in the world. It is therefore not adequate in a microreduction merely to describe a configuration of parts. A microreduction asserts that a type of whole, *W*, in the reduced theory is identical with a type of SW, say, *C*, in the reducing theory. *C* is a kind of compound thing in the reducing theory, and the thing-identity is $W = C$. A basic requirement of an adequate microreduction is that all identities of this kind that are used must be empirically justified, and this implies that *C* must be a kind of entity, not a mere aggregation. Thus, I require that *C* be a kind of thing composed of parts that are bonded together. The bonds must result from "forces of nature," where this term is to be very broadly understood, including social bonds of the type described in the preceding section. Furthermore, if these bonds are genuine empirical phenomena, rather than figments of our imagination, they should be causally responsible for the structure exhibited by SW's of type *C*.

For the reasons just stated, I believe that an adequate analysis of microreductive explanations requires an analysis of bonding relations and structured wholes. Moreover, in order for the concept of "bond" to have empirical significance, bonds must determine the possible SW's that can exist in an environment. The underlying theory must also be able to explain the existence conditions for SW's, and also explain the relative stability of SW's under various stressful environmental conditions. It is these very general considerations that lead to **BB**, **USW**, and their extensions (not presented here).

I cannot prove that the particular analyses presented here provide completely adequate explications of the concepts of bond and of structured whole. Yet, I do believe that my overall analysis provides an advance in the direction of an adequate explication. In fact, I believe that these conditions will turn out to be applicable to many, if not most, significant scientific investigations involving bonds or SW's in both the natural and the social sciences. Indeed, they may apply to some situations that are often considered to be nothing more than cases of aggregation. For instance, the classical kinetic theory of gases is often described as an example of aggregation together with identification; Kuipers does this in Chapter 3 of SiS. He is correct that statistical aggregation is used together with certain identifying assumptions. But I believe that we can also consider an ideal gas in the kinetic theory, together with its container, to be a kind of SW. Recall the example of *Marb* ⊕ *Bag* presented at the end of Section 2 of this article. I suggested there that the combined system consisting of the set of marbles together with the plastic bag enclosing them is an SW. Similarly, consider a swarm *S* of ideal, nearly point-sized molecules trapped within an enclosing box *B*. In the kinetic theory, it is assumed that these ideal molecules are in random motion and collide elastically with the walls of *B*. The system, *S* ⊕ *B*, consisting of the molecules together with the container appears to be an SW just as *Marb* ⊕ *Bag* does. The relative degrees of freedom of individual molecules are constrained with respect to each other and with respect to the container *B*. Thus, there are bonds between individual molecules and between molecules and *B*. In the simple derivation of the ideal gas law the exact nature of these bonds is not very interesting since the principal calculations make use of statistical aggregation. Furthermore, in the simplest form of the kinetic theory, it is assumed that there are no physical interactions between molecules, other than perhaps an occasional elastic collision. Thus, all mutual constraints result solely from the walls of *B*. However, if we elaborate our statistical theory of gases by introducing interactions between molecules, such as van der Waals forces, then *S* ⊕ *B* with these additional interactions becomes a more convincing example of an SW. The original *S* ⊕ *B*, with no intermolecular interactions, can be viewed as a limiting case of an SW, just as the simplest form of kinetic theory can be viewed as a limiting case of kinetic theories of gases; Kuipers (1982) shows the limiting assumptions that are used.

The analyses presented here depend on important concepts such as: configuration space, degrees of freedom, stable configuration, and stress, among others. I have not attempted to analyze these latter concepts in detail at this time. Their exact meanings will often depend on the domain of investigation and therefore be context dependent. I hope that I have shown how the concepts of bond and structured whole are intimately related, and that they are essential in an analysis of microreductive explanation. If this has been accomplished, it should be a useful addition to Kuipers' excellent treatment of structures in science in his SiS.[4]

University of Texas
Department of Philosophy
C 3500
Austin, TX 78712
USA

[4] I wish to thank Atocha Aliseda and Melinda B. Fagan for helpful comments on the first draft of this article.

REFERENCES

Bates, F.L. and C.C. Harvey (1975). *The Structure of Social Systems*. New York: Gardner Press.

Causey, R.L. (1977). *Unity of Science. Synthese Library*, vol. 109. Dordrecht and Boston: D. Reidel.

Causey, R.L. (1980). Structural Explanations in Social Science. In: T. Nickles (ed.), *Scientific Discovery, Logic, and Rationality*, pp. 355-373. Dordrecht and Boston: D. Reidel.

Causey, R.L. (1983). Philosophy and Behavioral Science. In J.L. Capps (ed.), *Philosophy and Human Enterprise* (U.S. Military Academy Class of 1951 Lecture Series, 1982-1983), pp. 57-80. West Point, NY: English Department, U.S. Military Academy.

Crothers, C. (1996). *Social Structure*. London and New York: Routledge.

Goldstein, H. (1950). *Classical Mechanics*. Reading, MA: Addison-Wesley.

Kuipers, T.A.F. (1982). The Reduction of Phenomenological to Kinetic Thermostatics. *Philosophy of Science* **49,** 107–119.

Kuipers, T.A.F. (1984). Utilistic Reduction in Sociology: The Case of Collective Goods. In: W. Balzer, D.A. Pearce, H.-J. Schmidt (eds.), *Reduction in Science*, pp. 239-267. Dordrecht and Boston: D. Reidel.

Kuipers, T.A.F. (2001/SiS). *Structures in Science. Synthese Library*, vol. 301. Dordrecht: Kluwer Academic Publishers.

Parker, S.P., ed. (1982). *McGraw-Hill Concise Encyclopedia of Science and Technology*. New York: McGraw-Hill.

Ruben, D.-H. (1983). Social Wholes and Parts. *Mind* (New Series) **92**, 219–238.

Theo A. F. Kuipers

CAUSAL COMPOSITION AND STRUCTURED WHOLES

REPLY TO ROBERT CAUSEY

Robert Causey's contribution reminds me of at least two preliminary points. First, as I also state in the Foreword to SiS, his work, notably his *Unity of Science*, has played an important role in my work, witness in particular Ch. 5, but also Ch. 3 and 6. It is an honor for me that he now presents new ideas in the context of my analysis of reduction of laws and concepts. Second, 'structures' in the title SiS can refer to at least three main uses: the primarily intended meta-sense of patterns in scientific knowledge and knowledge acquisition, the also intended mathematical sense of structures as used to formally represent objects of scientific interest, and finally the ontological-cum-epistemological sense of the nature of certain kinds of objects in the real world, the sense intended by Causey. He develops the notion of a "structured whole" in terms of bonding relations between elements of a (macro-) object (and perhaps its boundary), also simply called bonds, a stable configuration and a theory causally explaining the bonds and the stable configuration. In this way, Causey builds a notion that is at least characteristically, if not fundamentally, presupposed in cases of successful microreduction. In this reply I restrict myself to situating the idealized character of many examples of microreduction and to questioning whether a structured whole is a prerequisite for a genuine reduction.

Causal Composition

Robert Causey is quite right in suggesting that in typical cases of microreduction of a law – the crucial aggregation step together with one or more identification steps – the relevant macro-system or -object is a "structured whole" of one kind or another. As he also rightly notes at the end of his paper, the microreduction of the ideal gas law is an extreme case, since the bonds between the molecules are neglected. The same extreme character holds for my second favorite example of microreduction, that of Olson's quasi-

In: R. Festa, A. Aliseda and J. Peijnenburg (eds.), *Cognitive Structures in Scientific Inquiry* (*Poznań Studies in the Philosophy of the Sciences and the Humanities*, vol. 84), pp. 463-465. Amsterdam/New York, NY: Rodopi, 2005.

law about collective goods. Like Causey, I do not see this highly idealized character of paradigmatic examples as a reason to view more realistic putative cases of reduction as completely different in some qualitative sense or as no reduction at all. Instead, as I have shown in detail in the case of Van der Waals (Kuipers 1985), the reductive explanation of a concretized law is itself a concretization of the reductive explanation of the corresponding idealized law. However, in this case the term 'aggregation' remains adequate, but in other realistic cases it is not. See, for example, point (1) of my reply to Weber and De Preester.

As I suggest in SiS (p. 87), in cases where more than one type of element is involved, 'synthesis' or 'composition' can better replace the term 'aggregation'. The last term or, still more specifically, the term 'causal composition' seems particularly adequate to characterize the causal explanation of (some aspect of) the stable configuration characteristic of a structured whole *W*, that is, an explanation "in terms of the laws of [some theory] *T*, attributes of the elements of *W*, and the description of the bonding relations between the elements of *W*" (USW5 in Causey's paper).

Are Structured Wholes Presupposed in Microreduction?

Causey also links his notion of a structured whole to my notion of a "structure representation function" (SiS, Ch. 5). Apart from a minor terminological point, this suggests an interesting question. The minor point is that I wanted to use the term 'structure representation function' primarily to refer to the type of values the representation function assigns to certain objects, viz. the function assigns mathematical structures to what I call "macro-objects" or, more generally, "aggregates." These aggregates correspond to Causey's structured wholes or they are at least candidates for them, that is, they form the kind of objects that may be qualified as structured wholes.

Now the interesting question is whether being such a structured whole is a necessary condition for a successful microreduction. In Ch. 5 I distinguish between the reduction of laws and concepts, and I distinguish a singular, a multiple and a quasi-form of each. Let us concentrate on the singular forms. Recall that in Causey's notion of a structured whole the notion of a "stable configuration" which can be causally explained (USW5) is crucial. I certainly believe that obeying a macro-law requires a configuration that is in some sense stable, and hence, if it can be causally explained in terms of bonds between the elements themselves or between the elements and the boundary of the system, the configuration is a structured whole. However, this does not imply that every conceivable (singular) micro-reduction of a law governing an aggregate requires that this aggregate is a structured whole, for the relevant explanation may be of a different nature. The situation is similar for the case of microreduction of macro-properties, that is, properties of macro-objects. In SiS (p. 138) I claim the following: "Concept reduction only requires concepts at the side to be reduced, which is, of course, supposed to imply that these concepts are relatively stable and intersubjectively applicable." Hence, it seems that (singular) concept (micro-)reduction already requires a stable configuration. But again this need not imply that the relevant explanation is of the kind required for a structured whole.

For example, although in the case (see Causey's Section 2) of the balloons that are maintained in a certain configuration, say a sheeplike cloud, only by external forces, the notion of a structured whole does certainly not apply, the sheeplike cloud of balloons is nevertheless the aggregate effect of the external forces operating on the individual balloons, which can hence be microreduced in that sense. To be sure, such aggregates are not very typical, and Causey's other examples, including those of the "social structure" of robots, are more interesting. I should add that I have no doubt that detailed analysis would show that circuit examples such as the very instructive example of Weber and De Preester, presented in this volume to illustrate the microreduction of laws of artificial systems, and my own favorite example for introducing the idea of (actual and nomic) truth approximation (ICR, Ch. 7), are also typical cases of structured wholes.

REFERENCE

Kuipers, T. (1985). The Paradigm of Concretization: the Law of Van der Waals. *Poznań Studies in The Philosophy of the Sciences and the Humanities*, vol. 8, pp. 185-199. Amsterdam/Atlanta: Rodopi.

SCIENCE AND ETHICS

Henk Zandvoort

KNOWLEDGE, RISK, AND LIABILITY. ANALYSIS OF A DISCUSSION CONTINUING WITHIN SCIENCE AND TECHNOLOGY

ABSTRACT. In this paper I present my reflections on the ethics of science as described by Merton and as actually practiced by scientists and technologists. This ethics was the subject of Kuipers' paper "'Default norms' in Research Ethics" (Kuipers 2001). There is an implicit assumption in this ethics, notably in Merton's norm of communism, that knowledge is always, or unconditionally good, and hence that scientific research, and the dissemination of its results, is unconditionally good. I will give here reasons why scientists are not permitted to proceed, as they actually do, on the basis of this assumption. There is no factual or other binding justification for this assumption, and the activities it gives rise to frequently conflict with the broadly accepted ethical principle of restricted liberty. A recent discussion on the risks and hazards of science and on the issue of relinquishment is presented. What is shown in this paper is that the scientists and technologists participating in this discussion frequently violate core values of science relating to logical and empirical scrutiny and systematic criticism, as mentioned in Merton's norms of universalism, organized skepticism, and disinterestedness. It is concluded that, in order to live up to these values and in order to operate in agreement with broader ethical principles, science should stimulate open and critical discussion on the hazards and negative effects of science and technology, and on the present failure on the part of law and politics to control those hazards and negative effects. Science should also take the possibility of relinquishing certain themes of research seriously as long as such flaws in the systems of law and political decision-making persist.

1. Introduction and Overview

In "'Default norms' in research ethics" Kuipers discusses the ethical aspects of the activities of scientists, using Merton's description of the ethos of science as his starting point. As this is the only chapter in Kuipers' two books that deals with ethical rather than methodological and epistemological aspects of science, it takes a special place in Kuipers' work. As my own interests and activities have shifted from epistemology and methodology to ethics, I very much welcome Kuipers' interest in the ethical aspects of science, and I am grateful

In: R. Festa, A. Aliseda and J. Peijnenburg (eds.), *Cognitive Structures in Scientific Inquiry* (*Poznań Studies in the Philosophy of the Sciences and the Humanities,* vol. 84), pp. 469-498. Amsterdam/New York, NY: Rodopi, 2005.

for the opportunity to add my reflections on the ethics of science in general and Merton's description of it in particular.

The view expressed in Kuipers' paper is that the norms that make up this ethics – universalism, communism, disinterestedness, and organized skepticism – should function as default norms of scientific research: they should be respected, unless there are compelling reasons for deviating from the norms. Kuipers also asserts that there are many "grey areas" relating to situations where Merton's norms do not provide clear prescriptions for behavior. The possibilities for reducing these "grey areas" by formulating alternative or more elaborate prescriptive codes are, in Kuipers' opinion, very limited. In Kuipers' view, in scientific research there will necessarily remain many decision problems with ethical aspects for which an individual researcher will have "to find his own way."

My approach point is somewhat different from that of Kuipers. I will not focus primarily, as Kuipers does, on the precision with which the ethics of science has been or can be stated. Instead, my claim will be that a certain aspect of this ethics – incorporated in Merton's norm for communism – conflicts with broader ethical principles such as restricted liberty and reciprocity, whereas other elements – embodied by Merton's norms of universalism, disinterestedness, and organized skepticism – which *are* consistent with and at least partially related to such broader ethical norms, are not sufficiently respected. More specifically I will claim that it cannot be taken for granted, contrary to what Merton's norm of communism presupposes, that scientific knowledge is always, that is unconditionally good, and hence that scientific research, and the dissemination of its results, is unconditionally good. Rather, it is a value judgment that cannot be considered as a factual truth or an unassailable dogma, and its uncritical acceptance conflicts both with scientific norms such as (in Merton's terms) universalism and skepticism, and with broader ethical norms such as restricted liberty. More specifically, this value judgment cannot be based on the assumption that knowledge always has good consequences, since this assumption is false. I will give reasons why scientists are not allowed to proceed on the basis of the dogma that "knowledge is good," and why they should address the issue of which directions in research are desirable, and which parts of research should better be abandoned as long as the social institutions that are intended to control the application of results are not equal to the task. In addition, I will explain why scientists and technologists should critically consider the mechanisms for collective decision-making and the principles and practices of the current legal systems, in the light of empirical and theoretical evidence showing that these institutions in their present form are inadequate for controlling the use and effects of the results of science.

After reviewing Merton's norms for science in section 2 I will go on to explain, in section 3, why scientists are not permitted to work on the assumption that knowledge is good. In relation to this, I will argue in section 4 that in the ethos of science adequate norms for responsibility and liability are lacking. Section 5 presents an overview of liability in positive law and its development in the last 200 years, and explains the relevance of this to the ethics of science. Sections 6 and 7 present a recent discussion on whether science should relinquish (abandon) certain areas of research in view of the risks and hazards associated with the outcomes. This discussion serves, in part, as an illustration of the issues addressed in sections 3 and 4. It exemplifies the role of the dogma "knowledge is good" in discussions on the role of science in society. In addition and related to this, the discussion demonstrates that present-day science and technology do not consistently live up to Merton's norms of universalism and organized skepticism, whereas disinterestedness has become dubious (8). Section 9 draws together the conclusions obtained of this paper.

2. Merton's Norms for Science

The essay in which Merton describes the norms for science was originally published in 1942 under the title "Science and Technology in a Democratic Social Structure." It was later republished as "Science and Democratic Social Structure," and finally as "The Normative Structure of Science" in Merton (1973). The references made in what follows are to the latter publication. Merton starts from what he calls the institutional goal of science, which he takes to be the extension of certified knowledge (p. 270). Both the technical methods deployed in scientific research, and the ethos of science, which is "that affectively toned complex of values and norms which is held to be binding on the man of science" (p. 268), are considered as functional or necessary for achieving this goal:

> The institutional goal of science is the extension of certified knowledge. The technical methods employed toward this end provide the relevant definition of knowledge: empirically confirmed and logically consistent statements of regularities (which are, in effect, predictions). The institutional imperatives (mores) derive from the goal and the methods. The entire structure of technical and moral norms implements the final objective. The technical norm of empirical evidence, adequate and reliable, is a prerequisite for sustained true prediction; the technical norm of logical consistency, a prerequisite for systematic and valid prediction. The mores of science possess a methodological rationale but they are binding, not only because they are procedurally efficient, but because they are believed right and good. They are moral as well as technical prescriptions. (Merton 1973, p. 270)

According to Merton, the ethos of modern science consists of four "sets of institutional imperatives," namely universalism, communism, disinterestedness, and organized skepticism. These institutional norms form the starting point of Kuipers' chapter. I will summarize them below, as much as possible from the basis of Merton's original wording, and adding my own comments.

Universalism: "truth-claims, whatever their source, are to be subjected *to preestablished impersonal criteria*: consonant with observation and with previously confirmed knowledge. The acceptance or rejection of claims entering the lists of science is not to depend on the personal or social attributes of their protagonists; his race, nationality, religion, class, and personal qualities are as such irrelevant." (p. 270) "Universalism finds further expression in the demand that careers be open to talents," and hence that scientific careers may not be restricted on grounds other than those of lack of talent. (p. 272)

Communism, "in the nontechnical and extended sense of common ownership of goods": "the substantive findings of science are a product of social collaboration and are assigned to the community." (p. 273) "The institutional conception of science as part of the public domain is linked with the imperative for [full and open – HZ] communication of findings." (p. 274)

Comment. From the rest of Merton's paper it is clear that common ownership should extend to (members of) society at large, not merely to the community of science, i.e. those who have actually contributed. Merton does not give an explanation for this generosity of science toward society, but it would be understandable if intended as a return for society's (financial and other) support of science. In actual fact, it does serve as the sole argument of science for its claim for support from society.

Disinterestedness: personal and group interests should be subordinated to the interests of research (=the extension of certified knowledge). Disinterestedness according to Merton's definition does not refer to the individual motives of scientists, but rather to "a distinctive pattern of institutional control of a wide range of motives which characterizes the behavior of scientists. For once the institution enjoins disinterested activity, it is in the interest of scientists to conform on pain of sanctions and, in so far as the norm has been internalized, on pain of psychological conflict" (p. 276). The success of disinterestedness is witnessed by "[t]he virtual absence of fraud in the annals of science, which appears exceptional when compared with the record of other spheres of activity" (p. 276) and eventually by the successes of science in its technological applications.[1]

Comment. If the ultimate evidence of the success of disinterestedness is considered to be the successes of science in its technological applications, then apparently the unwanted or negative consequences are not ascribed to science. I will return to this point in section 3.

Organized skepticism: at one point described as "the temporary suspension of judgment and the detached scrutiny of beliefs in terms of empirical and logical criteria" (p. 277). Organized skepticism is both a methodological and an institutional mandate (in view of the institutional goal of science, the extension of certified knowledge).[2]

Merton points out that the ethos of science may conflict and actually has conflicted with the norms of society at large of which the institution of science is a part. Thus, universalism conflicts with nationalism, and with any system of castes within nations. (On the other hand, "The ethos of democracy includes universalism as a dominant guiding principle." (p. 273)) The norm of communism is incompatible with the definition of technology as "private property" in a capitalist economy; and organized skepticism has periodically involved science in conflict with other institutions, such as organized religion: "Science which asks questions of fact, including potentialities, concerning every aspect of nature and society may come into conflict with other attitudes toward these same data which have been crystallized and often ritualized by other institutions. The scientific investigator does not preserve the cleavage between the sacred and the profane, between that which requires uncritical respect and that which can be objectively analyzed." (pp. 277-8)

---

[1] "Every new technology bears witness to the integrity of the scientist. Science realizes its claims." (Merton 1973, p. 277) These technological successes exemplify Francis Bacon's utilitarian defense of science as a theoretical activity, expressed by Bacon in the following remark: "Now these two directions – the one active, the other contemplative – are one and the same thing; and what in operation is most useful, that in knowledge is most true." In the same vein Merton states that "[i]t is probable that the reputability of science and its lofty ethical status in the estimate of the layman is in no small measure due to technological achievements." (Merton 1973, p. 277)

[2] The importance of organized skepticism, or systematic criticism, for obtaining reliable knowledge was elaborated by Karl Popper in his writings on the methodology of science. See for instance his books *The Logic of Scientific Discovery*, first published in English in 1959 and in German as *Logik der Forschung* in 1934, and *Conjectures and Refutations: The Growth of Scientific Knowledge*, first published in1963.

3. Is Knowledge Good?

An important assumption which is part of, or presupposed by, the ethos of science as expressed by Merton (and others) is that knowledge and its dissemination is good. That is, in an absolute sense, unconditional. Hence the situation that scientific research is considered to be an unconditionally good activity, with moreover its public funding justified, provided that the results are disseminated to others. There may have been times and places where the unqualified assumption that scientific knowledge is good was quite tenable, or at least not objected to. However, under the present circumstances this assumption is not warranted.[3]

Especially during the last 50 years it has become more and more evident that scientific knowledge, through its application to technology, has resulted in and continues to result in serious negative consequences (such as death and illnesses; pollution; depletion of vital natural resources; etc.), and that the hazards that science and technology give rise to are increasingly unbounded and uncontrolled. Hazardous areas include the atomic, biological and chemical developments in the science and technology of the second half of the 20th century, and present-day developments in biotechnology, computing, and nanotechnology. The hazards and the actual negative effects and abuses of science and technology seem to be increasing as society proceeds into the 21st century. Among the factors that contribute to this increase are cheaper communication and transportation, and the fact that the hazards of for example biological science and technology affect increasingly fundamental aspects of all life on earth. By all practical standards these hazards are unlimited; it is not possible to indicate meaningful boundaries and to claim that negative effects will certainly not exceed them.

Anyone can become a victim of the known and unknown hazards of modern technological activities, including those who have not consented to, and/or who are opposed to such activities. Genetically engineered agricultural crops may serve as an example. For individuals or groups it is virtually impossible to find protection from the potential harm of such activities. Even if someone is not directly affected by a certain danger, he or she may still be forced through the tax system to contribute to the restoration or repair of the relevant damage. Examples that illustrate this mechanism of forced

---

[3] It should be noted that Merton worked on the ethos of science (that was published in 1942) before the time of the atomic bomb. But he was aware of opposition, both from within and from outside science, to the adverse social (war time and peace time) consequences of science, as he wrote about this opposition, and about the related discussion of the responsibility of scientists, in a paper entitled "Science and the social order" that came out in 1938 (Merton 1973, pp. 254-266, see especially pp. 261-3).

contribution to restoration are BSE (mad cow disease) and accidents such as the fireworks explosion in the town of Enschede.[4]

Science, conceived as an institution producing technological feasibilities, does not control the implementation of these feasibilities or their conditions of use. That has implicitly been delegated to institutions outside science, notably the political and legal systems of states. But law and politics have proven to be incapable of preventing or controlling the negative side-effects and hazards of technology. This even holds true for the states that are seen as the most democratic, the most developed, or the most reliable. It is even more true of the political and legal systems of less democratic, etc. states. When it comes to preventing or controlling negative side effects or abuses of modern scientific and technological knowledge, it is the weakest existing political or legal system that matters most.

The pattern displayed by history so far is that whatever has become technologically feasible, has also been put into practice. There is no reason to assume that this historical trend will soon disappear.

The unqualified prescription that newly acquired scientific knowledge should be made public and available to all would cause no problems if there were no dangers or negative effects at all, or if it could be asserted in an objective way – according to "preestablished impersonal criteria: consonant with observation and with previously confirmed knowledge," see Merton's norm of universalism – that the positive effects outweigh the negative effects. For very large portions of scientific and technological knowledge and know-how, the assumption that there are no hazards or negative effects at all is certainly false. As I will explain below, the only approach to the second issue – of asserting, "consonant with observation and with previously confirmed knowledge" that positive effects outweigh negative effects – that is consistent with generally held ethical principles is to obtain the informed consent of all those who are subjected to the possible effects. Essentially, there is no ethical basis for weighing the positive effects for some against the negative effects for others, if there is no prior consent on the part of all concerned to such a procedure.

---

[4] In addition to 22 fatalities, the costs of the fireworks explosion of May 13, 2000 in Enschede were estimated at 1100 million Guilders or 500 million Euros. ("Vuurwerkramp: het cijferwerk is nu begonnen," *NRC Handelsblad*, 12 oktober 2000) Of this, at least 350 million will be drawn from the national tax system. (The government provided 80 million for uninsured damage of local businesses, and made a 270 million contribution to the costs, estimated at about 500 million, of rebuilding. (Information from http://www.nu.nl, 25-8-2000, 11-11-2000.)) Another substantial part of the costs is covered by the insurances of the victims. The amount for which the actor, SE Fireworks, was insured was in one source estimated as being "between 1 and 10 million" (http://www.nu.nl 16-5-2000). The costs may be compared with annual sales of firework in the Netherlands to the tune of 100 million Guilders.

Because of the practical difficulty or impossibility of preventing the proliferation and dissemination of scientific and technological knowledge and know-how, and because of the irreversibility of the effects of proliferation and dissemination, one may moreover question whether it is justifiable to *perform* research in certain areas.

The assertion "knowledge is good" does not satisfy scientific norms of reliability and criticism as expressed e.g. by Merton's norms. According to the norm universalism, "truth-claims, whatever their source, are to be subjected *to preestablished impersonal criteria*: consonant with observation and with previously confirmed knowledge" whereas organized skepticism involves "the temporary suspension of judgment and the detached scrutiny of beliefs in terms of empirical and logical criteria."

4. Restricted Liberty, Responsibility and Liability

The ethical principle of *restricted liberty* asserts that everyone is free to act as he/she pleases, provided that he/she does not harm others. This ethical principle has a considerable history in both western moral thinking and that of other cultures. It was defended by, for instance, J.S. Mill in his essay "On Liberty," published in 1859. The principle is also consistent with, and at least partially related to core values of science as expressed in Merton's norms of universalism and organized skepticism.[5] If one accepts this principle, and if one also accepts that persons differ in what and how they value, then activities with potentially harmful and irreversible effects can only be justified by obtaining the informed consent of all who will be subjected to those risks (Van Velsen 2000). No one has shown that there are alternative ways to justify such activities. Alternatives, that is, that meet "preestablished impersonal criteria: consonant with observation and with previously confirmed knowledge" as required by the norm of universalism.

At present this informed consent has not been obtained for many developments in science and technology. On the contrary, many people actively oppose some of these developments and their applications, often for the reasons presented above. The case of genetically engineered agricultural

[5] See e.g. Merton's above-quoted remark to the effect that "The ethos of democracy includes universalism as a dominant guiding principle." (Merton 1973, p. 273) See also Popper's *The Open Society and Its Enemies*. One may also consider here the theory of argumentation, which includes norms that on the one hand are similar to Merton's universalism and organized skepticism, and that on the other hand are closely related to the broad ethical principles of equality and autonomy which together lead to restricted liberty as defined in the text.

crops provides an example of such opposition. Given the abundance of historical cases of actual harm from (applications of) science and technology, it is impossible to defend the claim that fears of further harm are unfounded. One need only think here of: pesticides and herbicides; Chernobyl; asbestos; CFKs and ozone depletion; CO_2 and climate change; harm caused by medicines such as DES and Softenon; etc.

Another time-honored ethical principle is the principle that everyone is responsible for (the consequences of) his/her own actions. In view of the principle of restricted liberty mentioned above, responsibility should be related to liability for damage. Speaking generally, the counterpart to restricted liberty is *reciprocity*. According to the latter principle, anyone who violates a certain right of another, loses this right him/herself to the extent needed for restoring the situation to what it was preceding the original violation. For activities for which there was no informed consent, reciprocity implies having a duty to repair or compensate for any damage done to others (Van Velsen 2000).

This responsibility for the hazards and negative effects of contested scientific research often literally cannot be borne, either by individual scientists or by science as an institution. This is not merely because of the limited financial capacity of science and scientists, but also because many actual and possible effects of science and technology such as deaths and many environmental consequences are irreversible. This circumstance adds to the importance of obtaining the informed consent of all who may be hurt by the activities concerned. (If all potential damage were repairable, and if the means for repair were secured, then the preceding consent requirement would be much less pressing.)

It is sometimes remarked, and often implicitly assumed, that scientific research and the dissemination and application of its results are ethically permissible, because they are legally allowed; hence, that the actors have been discharged of the responsibility for possible negative consequences. This is not a valid inference, since it presupposes that the procedures of collective decision-making that govern legislation are sound. This assumption is contradicted by the results of the science of public choice.[6] In democratic states, the procedures of collective decision-making are at best based on majority rule. Why should a minority be bound by the opinions or desires of a majority? As long as some preceding unanimous consent with this procedure of collective decision-making is lacking, it is altogether unclear why its results should be binding. Hence, just because of the fact that something is allowed by positive law, it cannot be concluded that it is ethically allowed, and that the

[6] For an overview of these results, see e.g. Mueller (1989).

actors do not bear responsibility for any consequences.[7] According to Merton's norms, any element of responsibility and liability for consequences is lacking. This would be acceptable if all knowledge was good, also in the sense of having (always or only) good effects. Indeed, the lack of the element of responsibility and liability might be explained by the belief that all knowledge is good; but as explained above, this belief is untenable. More to the point, the assertion that all knowledge is good does not satisfy the requirements for reliability formulated in Merton's norms for scientific claims. The absence of this element of responsibility brings the ethics of science into conflict with the ethical principles of restricted liberty and reciprocity mentioned above.

Another remark that is sometimes made in response to the above is that since the requirement of obtaining the informed consent of all relevant people is virtually impossible, it would mark the end of all scientific research. In response the following can be said. Firstly, a lot of interesting, and potentially useful research in the area of science and technology can be done that is not surrounded by large-scale and unbounded risks and hazards such as those associated with a number of research areas that are now actually being pursued. There are enormous differences in this respect between different, but equally interesting and potentially useful themes of research. Besides, there is also much relevant and very important work to be done in areas of the social sciences and humanities, such as ethics/law; and the empirical and theoretical study of individual and collective decision-making. See 7.7 below. Secondly, if the legal liability regulations were in better shape than they are at present, the present difficulties associated with obtaining informed consent, and with any remaining lack of consent, would greatly diminish. This second point will be explained in the next section and will recur in section 7.4.

5. Liability in Positive Law[8]

It is relevant to our discussion to consider the nature and development of liability for technological activities in positive law. The most relevant part of liability law is known in the Anglo-Saxon legal systems as tort law.[9] Largely in

---

[7] It was noticed by Rousseau that every majority decision, in order to be binding for the voters, should be preceded by at least one consensus decision, namely, to take future collective decisions with majority rule. Since then, many have questioned and in fact denied the binding force of political decisions based on (at best) majority vote, and hence the legitimacy of their enforcement by the state. For an example in the field of political philosophy, see Simmons (1993). For the relevant discussion in the field of public choice, see Mueller (1989).

[8] This section is based in part on Zandvoort (2000a).

[9] Tort law is that part of the law which deals with wrongful acts – 'tort' meaning 'wrong' – for which (financial) compensation can be obtained in a civil court by the person wronged, unlike

agreement with the ethical principles outlined in the previous section, the reigning principle of liability in tort law has long been that any unlawful damage or harm must be repaired or compensated by the actor, irrespective of whether the actor has or has not been careless or negligent. This is called strict liability. Strict liability was the dominant principle of liability in Roman law, as well as in European and Anglo-Saxon law until the 18th century. During the 19th century this principle was abandoned by making the duty to repair or to compensate subject to conditions and limits of various sorts, notably through the introduction of the principle of "no liability without fault," and of limited corporate liability.[10] The effect was that many legal possibilities for recovering damages due to technological development (industrial and traffic accidents; nuisance from water mills, roads, rail- and waterways, etc.) diminished or disappeared.

This transition from strict to conditional forms of liability was motivated, at least in England and the USA, by the desire to promote technological, and hence economic development (Zweigert and Kötz 1987, p 688; Van Dunné 1993). Judges and legislators saw this as sufficient justification for systematically reducing the possibilities to obtain redress for harm or nuisance caused by industrial activities because, as was sometimes explicitly stated, everyone would profit from the economic development resulting from these activities (Horwitz 1977). As was remarked earlier, these arguments do not seem tenable in the light of the experiences of the 20th century.

The 20th century saw some moves back to stricter forms of liability. Product liability is often quoted as an example. In spite of such moves contemporary liability law remains, in many important respects, conditional. For instance, the Dutch product liability law, in compliance with the directive of the European Community on product liability, excludes liability for the so-called risk of development. This means that a producer is not liable for damage caused by a faulty product if "it was not possible to discover the existence of the fault, given the state of scientific and technological knowledge at the time the product was brought into circulation".[11] This liability condition, together with similar other ones, has huge implications for the controlling of technology. It releases the producers of, for instance, genetically modified

---

wrongs that are breaches of contract. (The latter are dealt with in contract law.) See e.g. Zweigert and Kötz (1987, esp. chapter 47) for an overview of tort law in the various legal systems.

[10] The following focuses on developments concerning "no liability without fault." For a brief historical overview concerning the limited liability of corporations, see Zandvoort (2000a), section 3.

[11] *Burgerlijk Wetboek*, Book 6, Section 3: Product Liability, Art. 185.1.e. (Translation by the author, HZ.) For an analysis of the contents of the European Union Directive and Dutch legislation concerning product liability as well as a description of the historical background, see Van Empel and Ritsema (1987).

crops from liability for much possible future harm, and hence removes an important motive for being cautious and prudent. (Agricultural products are actually excluded from European and Dutch product liability law, but this is irrelevant to the present example. More relevant, in the present context, is the fact that illnesses like BSE/Creutzfeld Jacob (mad cow disease) have an incubation time of some 10 years.)

In an innovative technological society governed by conditional and limited liability, more and more activities come into being that have risks or side-effects for which the actors cannot be held liable. Usually, the advantages of new technological activities and possibilities are clear from the outset, whereas important harmful effects become manifest only later. In addition, such activities are usually legally allowed as long as their harmfulness has not been proven. If damage does occur, it mainly affects non-actors, who cannot influence the development, production and dissemination of the technologies in question, even if some may actually have tried to stop the activities.[12] Strict liability would promote prudence. It would stimulate research into adverse effects, and foster a more adequate control of technological risks. Conditional liability, on the other hand, comes down to an explicit refusal to control the adverse effects of new technologies.[13]

The above shows that the stipulations on liability in contemporary positive law do not compensate for the missing element of liability in the norms for science that was identified in section 4. This is notably because liability in contemporary positive law is conditional rather than strict. The stricter form that liability law once had was much more in agreement with the ethical principles of restricted liberty and reciprocity than is the case at present.[14] The historical transformation of liability law from strict to conditional also shows that law is amenable to change. If (re)transformed toward strict liability, liability law would be an important instrument for controlling the hazards and negative effects of technology. This would not be a panacea for all problems relating to the hazards and negative effects of science and technology. This is so, if only because much possible and actual damage from technology cannot

---

[12] It should be clear, at least intuitively, that these circumstances are likely to result in decisions at the level of individuals – natural persons or legal persons such as corporations – that are not optimal from the collective point of view. More particularly, there is no guarantee whatever that the resulting development would represent progress in a non-arbitrary sense. See for this point 7.7 below.

[13] In terms of the preceding note, it is likely that conditional liability promotes individual decisions that are sub-optimal or even negative from the collective point of view.

[14] This not only refers to that element of former liability law which required full and unconditional repair of or compensation for unlawful damage, but also to the circumstance that the relevant legal stipulations of what was and was not lawful were generally less contested than they are now.

be repaired or adequately compensated for; but strict liability would surely help enormously to diminish some of these problems.

6. Bill Joy on Risks and Relinquishment

The discussion on the hazards and the uncontrolled nature of scientific and technological development and on the ethical aspects involves a considerable history. The next two sections present and discuss a recent contribution.

In the spring of 2000, Bill Joy, co-founder of and chief scientist at SUN Microsystems, published an essay entitled “Why the Future Doesn’t Need Us” (2000) in the magazine *Wired*.[15] Referring to research and technology areas such as genetics, nanotechnology, computers, and robotics, Joy states that present day society is not prepared for the effective management and control of the consequences of these technologies. In his words: “We are being propelled into this new century with no plan, no control, no brakes.”

Joy argues that science should relinquish doing research into potentially dangerous areas. He points to the unilateral US abandonment, without preconditions, of the development of biological weapons as a hopeful historical precedent. According to Joy, this decision stemmed from the realization that while it would take enormous effort to create those weapons, they could from then on easily be duplicated and fall into the hands of rogue nations or terrorist groups. Hence, Joy proceeds, this decision to abandon further development was based on the consideration that the people of the USA would be safer without, than with the possession of these biological weapons.

Joy also thinks that scientists and technologists carry personal responsibility:

> The experiences of the atomic scientists clearly show the need to take personal responsibility, the danger that things will move too fast, and the way in which a process can take on a life of its own. We can, as they did, create insurmountable problems in almost no time flat. We must do more thinking up front if we are not to be similarly surprised and shocked by the consequences of our inventions.
>
> My continuing professional work is on improving the reliability of software. Software is a tool, and as a tool builder I must struggle with the uses to which the tools I make are put. I have always believed that making software more reliable, given its many uses, will make the world a safer and better place; if I were to come to believe the opposite, then I would be morally obligated to stop this work. I can now imagine such a day may come.

---

[15] Bill Joy, “Why the Future Doesn’t Need Us,” *Wired*, 8/4/2000, http://www.wired.com/wired/archive/8.04/joy.html

Reflecting on his discussions with other people, Joy says he sees "cause for hope in the voices for caution and relinquishment and in those people [he has] discovered who are as concerned as [he is] about our current predicament." But he also states that

> … many other people who know about the dangers still seem strangely silent. When pressed, they trot out the 'this is nothing new riposte – as if awareness of what could happen is response enough. They tell me, There are universities filled with bio ethicists who study this stuff all day long. They say, All this has been written about before, and by experts. They complain, Your worries and your arguments are already old hat.
>
> I don't know where these people hide their fear. As an architect of complex systems I enter this arena as a generalist. But should this diminish my concerns? I am aware of how much has been written about, talked about, and lectured about so authoritatively. But does this mean it has reached people? Does this mean we can discount the dangers before us?

Joy expresses the hope of participating in a much larger discussion on the issues raised, "with people from many different backgrounds, in settings not predisposed to fear or favor technology for its own sake." He reports having proposed to the American Academy of Arts and Sciences to take these issues up as an extension of its work with the Pugwash Conferences.

7. Reactions to Joy's Paper

Joy's essay evoked many reactions. I will focus on reactions that have become available through the internet.[16] I have found no single reaction that places the potentially far-reaching effects of science and technology as described by Joy and others in doubt. But there is less unanimity on whether something should or can be done to control these hazards and if so, what should be done. Many respondents share Joy's views on the uncontrolled nature of science and technology and the need for relinquishment. But usually they have no clear ideas on how improved control or relinquishment might be accomplished. There are also many respondents who fiercely reject Joy's call for relinquishment. They essentially claim that *the development of science and technology should evolve as it actually does, under (political, legal, etc.) circumstances as they actually are.*

Below, I will present and discuss the arguments brought forward for this latter claim in the discussion triggered by Joy's paper. I purport to show that

---

[16] Some of these reactions have been collected by The Center for the Study of Technology and Society, Inc., which presents itself as a non-profit making think tank. See http://www.tecsoc.org/innovate/focusbilljoy.htm. A sample of reactions was also collected by the editors of *Wired*. See *Wired*, section Rants & Raves, on the topic "Why the future doesn't need us": http://www.wired.com/wired/archive/8.07/rants.html. I will refer below to this source as Rants & Raves.

none of these arguments can support the claim, and that the claim must be viewed as an expression of personal belief, that has no objective or otherwise binding foundation. In the first six subsections below I try to group the arguments into different categories, without wanting to make claims as to whether categories do or do not overlap, etc. The section ends with a number of general comments (7.7).

7.1. *"Science and Technology Are Unconditionally/Absolutely Good, That Is Intrinsically, Irrespective of Consequences"*

Much of the verbal and nonverbal behavior of many scientists and technologists is based on this assumption.[17] Occasionally the assumption is made explicit. Robotics expert Moravec reportedly claimed that science and technology should proceed to create robots, even if they were to supplant humans as Earth's superior species.[18] The following statement is another example:

> The not-very-joyous Bill Joy makes me think of a dinosaur whining because it's not going to be the final point on the evolutionary scale. If the universe has evolved humans because our intervention is necessary to produce the next step up on the developmental ladder, so be it. I trust the universe knows best where it's going and what the point of it all is. Joy fears that if we simply move beyond Earth in order to survive, our nanotechnological time bomb will follow us. On the other hand, perhaps the coming "superior robot species" would see fit to terraform a planet or two that could be kept as a human reserve – like a galactic Jurassic Park. (Stephen H. Miller, editor in chief, *Competitive Intelligence Magazine,* quoted in Rants & Raves)

Discussion. Strictly speaking, this quotation contains no argument or conclusion. I assume that its author wants to express that the unhampered development of science and technology and the results thereof are good, irrespective of what the results may be. This is a normative statement, expressing a value judgment, for which the author does not give any arguments or foundation. A normative statement cannot be derived from factual statements alone, and so can be denied without conflict with whatever factual statement, however well-founded its truth may be. This is a well-known but often ignored truism. The implication is that no one can be logically forced to accept a value judgment on the basis of his acceptance of factual statements

---

[17] Merton apparently asserts that this assumption is part of the ethos of science when he says, in a passage quoted above in section 2, that "The mores of science possess a methodological rationale but they are binding, not only because they are procedurally efficient, but because they are believed right and good." (Merton 1973, p. 270)

[18] See http://www.tecsoc.org/innovate/focusbilljoy.htm; see also Damien Cave, "Killjoy," interview with Bill Joy in the magazine *Salon*, April 10, 2000, http://www.salon.com/tech/view/2000/04/10/joy/index.html

whatever they might be.[19] Neither does the statement under scrutiny follow from other normative statements which are unanimously accepted. To require from others that they accept this value judgment, and to demand their tolerance for the activities associated with it (unrestricted development of science and technology), contradicts the principle of restricted liberty. The author does not explain why others would be forbidden to assert and execute similar but opposing opinions, although the latter would inevitably lead to mutual violence.

The author violates other rules for rational argumentation or discussion as well. Qualifications such as "the not-very-joyous Bill Joy" and the comparison of Joy to a dinosaur are tendentious, personal, and irrelevant. Attacks such as these do not serve the purpose of rational discussion, which is to obtain consistent agreement on stated assertions. Such violations of elementary rules for rational discussion occur frequently in reactions to Joy, although there are no similar offences in Joy's essay.

7.2. *Fatalism*

The term fatalism refers here to people who profess that the course of scientific and technological development cannot be altered, and that we (that is, all of us) must live with the consequences, come what may. Where the previous argument was based on a value judgment, fatalists apparently base their conclusion on a factual claim concerning the inevitability or necessity of the course of events. However, as will become clear below, both types of argument are not as distinct from each other as this characterization might suggest.

---

[19] Violation of this is known as the naturalistic fallacy, or the is-ought fallacy. It seems that David Hume was the first to pinpoint this fallacy. After having claimed, and illustrated by examples, that there cannot be any difficulty "in proving, that vice and virtue are not matters of fact," he made the following "...observation, which may, perhaps, be found of some importance. In every system of morality, which I have hitherto met with, I have always remark'd, that the author proceeds for some time in the ordinary way of reasoning, and establishes the being of a God, or makes observations concerning human affairs; when of a sudden I am surpriz'd to find, that instead of the usual copulations of propositions, is, and is not, I meet with no proposition that is not connected with an ought, or an ought not. This change is imperceptible; but is, however, of the last consequence. For as this ought, or ought not, expresses some new relation or affirmation, 'tis necessary that it shou'd be observ'd and explain'd; and at the same time that a reason should be given, for what seems altogether inconceivable, how this new relation can be a deduction from others, which are entirely different from it. But as authors do not commonly use this precaution, I shall presume to recommend it to the readers; and am persuaded, that this small attention wou'd subvert all the vulgar systems of morality, and let us see, that the distinction of vice and virtue is not founded merely on the relations of objects, nor is perceiv'd by reason." (David Hume, *A Treatise of Human Nature* (1740), Book III, Of Morals, Part I, Of Virtue and Vice in General, Sect. I, *Moral Distinctions not deriv'd from Reason.)* As the quotations in the text show, Hume's observations and recommendations are still highly relevant today.

The following reaction of Michael Dertouzos, director of MIT's Laboratory for Computer Science, in the *MIT Technology Review*, may serve as an example of what I call here fatalism.[20]

> What troubles me with this argument [i.e. Joy's argument leading to the conclusion that science and technology should relinquish certain areas – HZ] is the arrogant notion that human logic can anticipate the effects of intended or unintended acts, and the more arrogant notion that human reasoning can determine the course of the universe.
> … We shouldn't forget that what we do as human beings is part of nature. I am not advocating that we do as we please, on the grounds that it is natural, but rather that we hold nature—including our actions—in awe. As we fashion grand strategies to "regulate the ozone problem," or any other complex aspect of our world, we should be respectful of the unpredictable ways nature may react. And we should approach with equal respect the presumption that the natural human urge to probe our universe should be restricted.
>
> I suggest we broaden our perspective to the fullness of our humanity, which besides reason includes feelings and beliefs. Sometimes, as we drive the car of scientific and technological progress, we'll veer because our reason says so. At other times we'll follow our feelings, or we'll be guided by faith. Most of the time, we'll steer with all three of these human forces guiding us in concert, as they have guided human actions for thousands of years. As we do so, we should stay vigilant, ready to stop, when danger is imminent, using our full humanity to make that determination. If we do so, our turning point will be very different from where it may seem today, based on early rational assessments...that have failed us so often. Let us have faith in ourselves, our fellow human beings and our universe. And let's keep in mind that our car is not the only moving thing out there.

Discussion. This quotation is not simply an illustration of fatalism. Dertouzos both asserts and denies that what will happen, also should happen, and he both asserts and denies that the actual course of events cannot be altered. He says that what happens is good, except when it is not good, in which case it must be corrected by "using our full humanity." Furthermore, he suggests that the course of events cannot be altered (and that it is arrogant to think it can) while asserting that sometimes the course of events must be corrected. A consistent fatalist would remain silent, rather than to try to influence the course of events by influencing the opinions and behavior of others, as Dertouzos is in fact doing. Perhaps he is not a fatalist after all, but rather someone who is claiming that the present unfettered development of science and technology *should* continue and *should* be tolerated. But for this claim he gives no objective or otherwise binding reasons.

---

[20] Michael Dertouzos, 'Not by Reason Alone', *MIT Technology Review*, September/October 2000, http://www.techreview.com/articles/oct00/dertouzos.htm. See also the reaction of Ray Kurtzweil to this opinion, and the rejoinder of Dertouzos, found in http://www.lcs.mit.edu/about/director.html

Dertouzos claims that "we" (must) have faith in ourselves and our fellow human beings. But experience amply shows that such faith is unwarranted when it comes to science and technology and the institutions that are supposed to control this. Dertouzos's faith in human beings is in conflict with experience. Demanding such faith from others conflicts with basic logical and scientific norms, and demanding the tolerance of others for the (potentially harmful) activities resulting from this faith conflicts with restricted liberty, as explained above in 7.1.

Dertouzos shows a completely uncritical attitude towards completely unfounded dogmas expressed in unclear terms, that are apparently chosen primarily for their capacity to resonate the disjointed feelings and emotions of the reader. He asserts and denies statements in a logically arbitrary way. To summarize, Dertouzos does not respect the basic norms of science. In Merton's terms, he violates universalism and skepticism, whereas his disinterestedness is suspect to say the least. If Dertouzos was be bound by norms such as universalism and skepticism, one would expect him to be much more modest and restrained with respect to the issues at stake than he actually is.

A more straightforward example of fatalism is this:

> For some problems, there are no solutions. This may be one of them. If knowledge and technology are the independent entities that I think they are, they will propagate. (Jim Gray, senior researcher, Microsoft Research, quoted in Rants & Raves)

Discussion. The assumption that science and technology develop independently or autonomously is false. Science and technology are the deliberate work of human agents, who have the power to decide to reorient their activities. Furthermore, the systems of political decision-making and of law, which largely determine the funding of science as well as the implementation and (conditions of) the use of technology, are made by human beings, and are amenable to change.

7.3. *"Positive Effects Outweigh Negative Effects"*

Many people who claim that the development of science and technology should not be restricted try to justify their claim by stating that the positive effects outweigh the negative effects and risks. The following is an example of this:

> Forgo the possibilities? After working all of my life to make precisely such possibilities a reality, and much of it quite consciously? No way. And I will fight with every means at my disposal not to be stopped. Not only because of my own drive and desires, but because I honestly believe that only in transforming itself does humanity have a chance of a long-term future. Yes, it will be a changed humanity. But at least we and our descendants will have a future – and one that doesn't cycle back to the Dark Ages. (Samantha Atkins, software architect, quoted in Rants & Raves.)

Discussion. It is exactly this claim that the positive effects of science and technology outweigh the negative effects has been questioned, for certain technologies at least. The claim neglects the question of the acceptability of certain costs such as deaths and incurable diseases. Why, for instance, are a number of people allowed to suffer or die in order to let a number of other people live happier lives? Is happiness at all measurable? Why are some sacrifices allowed, and others not? No objective or generally accepted or otherwise well founded answers to such questions are available. Atkins's suggestion that the only alternative to the unrestrained development of science and technology is "cycling back to the Dark Ages" is rhetorical nonsense. It would make more sense to claim instead that the unrestrained development of science and technology does not lead to a long-term future.

7.4. *"Science and Technology Are Actually Under Control"*

Some people defend the claim that technology is and will be kept under control by (other) social mechanisms. Thus, John Seely Brown, chief scientist at Xerox and director of Xerox PARC, and Paul Duguid, a researcher at the University of California in Berkeley, have argued that social pressure and discussion can (and will) have an effective control on evolving technology, and that there are critical social mechanisms active that keep technology under control and that "allow society to shape its future".[21] A historical example that purportedly shows the presence of these critical social mechanisms is nuclear technology. Another example these authors provide has to do with genetic engineering:

> Barely a year ago, the technology [of genetic engineering – HZ] seemed to be an unstoppable force. Major chemical and agricultural interests were barrelling down an open highway. In the past year, however, road conditions changed dramatically for the worse: Cargill faced Third World protests against its patents; Monsanto (PHA) suspended research on sterile seeds; and champions of genetically modified foods, who once saw an unproblematic and lucrative future, are scurrying to counter consumer boycotts of their products.

Discussion. Examples do not prove general claims. If the examples given are evidence of some slow-down in specific cases, they do not show the least control, not even in these specific cases. So far history has witnessed major technological accidents of many types. I would remind the reader of the

---

[21] "Ideas to Feed Your Business: Re-Engineering the Future," *The Standard, Intelligence for the Internet Economy*, April 13, 2000, http://www.thestandard.com/article/display/0,1151,14013,00.html. The authors have expressed similar views in their contribution "Don't Count Society Out – a Response to Bill Joy" to the National Science Foundation report *Societal Implications of Nanoscience and Nanotechnology*. (Section 6. Statements on Societal Implications, 6.1. Overviews, pp. 30-36. See http://itri.loyola.edu/nano/NSET.Societal.Implications/nanosi.pdf for the text of this report.)

examples mentioned in Section 4: Chernobyl; asbestos; ozone depletion; CO_2; DES/Softenon; etc. These examples substantiate the considerable hazards and risks of science and technology. As was stressed earlier, there is reason to believe that the scope and severity of the hazards are increasing as science and technology develop. The authors do not draw conclusions from this. Their expectations as to the absence of accidents in the future display unrestrained wishful thinking.[22]

At this point I would like to refer to what was said about liability laws in section 5. I claimed there that stricter forms of liability are a viable mechanism for controlling the risks and hazards deriving from technology. Joy refers in his article to the possibility of strict liability as an alternative to the regulation of research and development. Generally, Joy and others are uncomfortable with the idea of regulation (that is, preventive government restrictions being imposed on research and development activities) because it requires government surveillance, which they fear will give rise to privacy issues. Joy quotes a paper written by David Forrest who dealt with the prospects for regulating nanotechnology. Forrest noticed that

> ...if we used strict liability as an alternative to regulation it would be impossible for any developer to internalize the cost of the risk (destruction of the biosphere), so theoretically the activity of developing nanotechnology should never be undertaken.[23]

---

[22] More wishful thinking is delivered in the following examples:

> I always worry that formulations about the future fail to account for the rise of new economies and the natural positive biases that humans have (i.e., we assume that human behavior will not change in the presence of accurately projected threats). I can imagine a number of positive ways that humanity in the future could and, in my view, will handle the technological threats Joy cites.
>
> For example, you can imagine in an increasingly interconnected and educated world, with world population declining by 2050, the very real need for governments to become more peaceful and more people-centered as a natural result of their own self-interests in domestic issues. There is a chance that this could create a world where the spread of things Joy talks about are effectively banned. (Eric Schmidt, chief executive officer, Novell, quoted in Rants & Raves)

Comment. Schmidt's rosy outlook on the world in 2050 does not comply with experience to date. Schmidt gives no explanation for why his wishes should come true.

> It is hard for me to see how any group of technologists or scientists can be large enough to be effective in halting some type of research that would ultimately be harmful to humanity. It could be argued that the ultimate things of potential harm would best be discovered or invented by a more enlightened group rather than someone with bad intentions. For example, Einstein was worried that if we didn't develop the bomb, the Germans would. I have a fundamental belief that the positive forces of human nature are more dominant than the negative ones. The world is becoming increasingly enlightened and part of the reason is that people like us have invented or otherwise enabled technologies that increase the dissemination of information across cultures. Still, I'd be happy to help Bill in his efforts, because he's got such a good mind and I respect his concerns. (Jim Clark, founder of Silicon Graphics, Netscape, Healtheon, and myCFO, quoted in Rants & Raves)

Forrest added "Besides, if civilization is destroyed there won't be anyone around to collect damages." Both Joy and Forrest apparently conclude from this that, in the case of the hazards under consideration, strict liability is no viable alternative to regulation.[24] They seem to see the issue as a dilemma, i.e. a matter of either-or, but they are mistaken, since these options do not exclude each other. Forrest and Joy also seem to conclude, from the fact that the potential liability of the risks and hazard under consideration are literally unbearable (a fact that was noticed above in section 4), that there is no preventive effect arising from strict liability. However, also in the case of irreparable damage, whether actors are liable for compensation or not makes a difference. In addition, requirements relating to "financial evidence of responsibility" may be introduced for specific activities, as has actually been

---

Comment. Clark's claim that the world is becoming increasingly enlightened is a dogma (and a vague one) for which he provides no arguments.

> It is now obvious that the real dangers to human existence come from biotechnology and not from nanotechnology. If and when a self-reproducing robot is built, it will be using the far more powerful and flexible designs that biologists are now discovering.
>
> There is a long and impressive history of biologists taking seriously the dangers to which their work is leading. The famous self-imposed 1975-1976 moratorium on DNA research showed biologists behaving far more responsibly than physicists did 30 years earlier. In addition, there is a strong and well-enforced code of laws regulating experiments on human subjects. The problems of regulating technology are human and not technical. The way to deal with these deep human problems is to build trust between people holding opposing views. Joy's article seems more likely to build distrust. (Freeman Dyson, physicist and author of *The Sun, the Genome, and the Internet*: Quoted in Rants & Raves)

Comment. Dyson's statements have an authoritarian tone but provide no proof. He wants to build trust between people holding opposing views, but he ignores the fact that the opposing views of different people often cannot be jointly effectuated, and that it is impossible, for instance, to both build and not build a nuclear plant.

[23] Forrest, D. R., "Regulating Nanotechnology Development," paper written for an MIT course TPP32 on Law, Technology, and Public Policy 23 March 1989). http://www.foresight.org/NanoRev/Forrest1989.html.

[24] Joy's conclusion is that "Forrest's analysis leaves us with only government regulation to protect us – not a comforting thought." Forrest's own account is as follows:

> Baram [reference to Michael S. Baram, *Alternatives to Regulation*, D.C. Heath and Company, Lexington, MA, p. 56 (1982)] points out that, historically, success with using non governmental standards as an alternative to regulation depended on two conditions: (1) the technologies and risks were well-understood, and (2) potential liability was significant enough to force responsible industry behavior. The potential liability of a runaway replicating assembler is the worth of our biosphere, price enough to insure significant caution. But nanotechnology may not be sufficiently well-understood to merit this voluntary approach. Furthermore, most sources agree that if the potential effects of the substance or product in question are clearly irreversible or hazardous to human health or the environment, that item should be subjected to standards enforcement [references]. Some products of nanotechnology could fall into that category. This is the primary argument for regulatory control of nanotechnology development efforts, and why alternatives to regulation would be inappropriate.

done in certain areas of environmental liability law[25] but which is completely absent in many other areas of technological activity, such as genetic engineering or, for that matter, nanotechnology.

7.5. *"Relinquishment May Be Worse than Unrestricted Continuation of Scientific and Technological Development"*

The following is a statement of this argument:

> If we outlaw nanotech, it'll just go underground. We won't ever get a consensus of everyone on earth to not do it. And then the rest of us will be unprepared when one of the secret laboratories makes a breakthrough and uses it against us (whether in commerce or in war). We could build a worldwide police state to find and track and monitor and imprison people who investigate these "forbidden" technologies. That would work about as well as the drug war, and throw the "right to be left alone" basis of our civilization out the window besides.
>
> My guess is that the answer is sort of like what Silicon Valley has been doing already - agility and speed. If you learn to move ahead faster than the problems arise, then they won't catch up with you." (John Gilmore, cofounder, Electronic Frontier Foundation, quoted in Rants & Raves)

A more detailed reaction of this type was given by Glenn H. Reynolds, law professor, University of Tennessee, and Dave Kopel, research director of the Independence Institute. They argue against relinquishment, not because scientific and technological development will not have negative effects, but because the effects of relinquishment will be worse. As evidence they provide the history of the British and American biological warfare program, which started in 1940 and ended with its abandonment in 1972 when the Biological Weapons Convention was signed.[26] According to Reynolds and Kopel, the Biological Weapons Convention ". had exactly the opposite result that its sponsors intended. Before the United States, the Soviet Union, and other nations agreed to a ban on biological warfare, both the U.S. and Soviet programs proceeded more or less in tandem, with both giving biowar a low priority. But after the ban, the Soviet Union drastically increased its efforts. (So did quite a few smaller countries, most of them signatories of the Convention.)" From this they conclude that

> … "relinquishment" would probably *accelerate* the progress of destructive nanotechnology. In a world where nanotechnology is outlawed, outlaws would have an

---

[25] As in the case of the Oil Protection Act which was enacted in 1990 in the USA in response to the environmental accident with the Exxon Valdez oil vessel. See Zandvoort 2000.

[26] Glenn H. Reynolds and Dave Kopel, "Wait a Nano-Second…Crushing Nanotechnology would be a Terrible Thing," guest comment on the website of *National Review Online. America's Conservative Magazine* posted 7/5/2000; url: http://www.nationalreview.com/comment/comment070500c.html For the details of this history these authors refer to "Ed Regis's excellent history of biological warfare, *The Biology of Doom*."

> additional incentive to develop nanotechnology. And given that research into nanotechnology – like the cruder forms of biological and chemical warfare – can be conducted clandestinely on small budgets and in difficult-to-spot facilities, the likelihood of such research going on is rather high. Terrorists would have the greatest incentive possible to develop nanotechnologies far more deadly than old-fashioned biological warfare.

Discussion. The authors suggest that at present all relevant developments are in the open. This is not plausible at all. It is difficult to see that the statement that relinquishment would accelerate rather than stop developments follows from the facts stated in evidence. As with other examples discussed above, there is a strong impression of wishful thinking. At best, the authors show that relinquishment is not simple to accomplish, but this in itself is not an argument against relinquishment. In addition, Reynolds and Kopel incorrectly present the issue as a dichotomous one: relinquishment or not. But alternatively, or additionally, legal conditions of liability for consequences may be made more strict, as was discussed in 7.4 and 5 above. Apart from this, there is the question of who should determine which one of two perceived risks (relinquishment versus uncontrolled development) should receive the largest weight. This cannot be (entirely) established in an objective way. Hence the need for informed consent to justify possibly harmful (research and development) activities remains.

7.6. *"The Public Would Consent If Properly Educated and Informed"*

In 7.4 above Brown and Duguid were quoted on genetic engineering. In the sequel to that quotation, they suggest that people will consent to, for instance, genetically engineered crops, once the costs and benefits have been explained to them properly:

> Almost certainly, those who support genetic modification will have to look beyond the technology if they want to advance it. They need to address society directly – not just by putting labels on modified foods, but by educating people about the costs and the benefits of these new agricultural products. Having ignored social concerns, however, proponents have made the people they need to educate profoundly suspicious and hostile.[27]

Discussion. Brown and Duguid apparently recognize the importance of informed consent.[28] However, their assumption that people would consent if

---

[27] John Seely Brown, Paul Duguid, "Ideas to Feed Your Business: Re-Engineering the Future," *The Standard, Intelligence for the Internet Economy*, April 13, 2000, http://www.thestandard.com/article/display/0,1151,14013,00.html

[28] Another reaction expressing this is the following:

> At the very least, let's bring in people from all walks of life on discussions of this nanotechnology, or the projected integration of humans and robots. Is this what people *really* want? (Diane Wills, software engineer, Hewlett-Packard, quoted in Rants & Raves)

properly informed begs the question. To begin with, it is a fact that at present many people do *not* consent to certain developments. There are neither empirical nor theoretical grounds for the assumption that everyone, if properly informed, would consent to the form that scientific and technological development presently takes, and to the (legal, political) conditions under which this development occurs. Regarding the issues that are relevant I will mention here the following.

There are no empirical or theoretical reasons to assume that different persons value the same things or situations in the same way, neither are there reasons why they should. This holds for things or situations that are given or can be realized with certainty, but the possibilities for differing evaluations increase when there is an element of probability or risk at stake. Different persons may differ in their attitude towards risk. For instance, someone who is willing to take part in a lottery the expected utility of which is lower than the stake is called risk prone, and someone who is willing to take part in a lottery the expected utility of which is $\geq 0,1$ times the stake is more risk prone than one who is willing to do lotteries with expected utilities of $\geq 0,5$ times the stake. Neither of them need to be irrational, in the sense of being incoherent, or being inconsistent with objective facts or knowledge. Even a risk averse person, who does not want to take part in a lottery even if the expected utility exceeds the stake, need not be irrational.[29] The above holds for cases, such as a lottery or a Russian roulette game, where there is reliable knowledge about the possible outcomes, including probabilities. For the activities discussed in the present paper, knowledge about possible outcomes and their probabilities is largely absent and/or unreliable, which once more widens the margins for differing valuations between different people.[30]

7.7. General Comments. Topics Neglected in the Discussion

In section 3 it was noticed that science as a social institution relies upon other social institutions for the implementation and control of the technological feasibilities it generates, the most important of these institutions being law and politics. In all the reactions discussed above, a critical attitude toward these institutions is absolutely lacking. This is remarkable given the historical evidence that in their current form these institutions are incapable of

[29] This has been made clear in the science of decision theory. See e.g. Lindley (1971) for an introduction.

[30] Considerable empirical knowledge has been obtained on how people make decisions in situations of chance and uncertainty, and which factors may influence the choices made. This knowledge is highly relevant to understanding how people value risk, and for understanding some of the sources for interpersonal differences that do occur. See Hogarth (1987) for an overview of results.

preventing and controlling the negative effects of science and technology. In addition, there are theoretical reasons for questioning the soundness of these institutions in their present form.

To begin with, the decisions about collective issues such as the public funding of science, the development of new technologies, and the (legal) conditions of their use, are made on the basis of majority decision making at best. This elementary fact remains completely unnoticed, but is highly relevant to the discussion on the ethical aspects of scientific and technological activity. This way of making collective decisions is characterized by serious ethical and other flaws. These flaws are well known in the scientific field of public choice which studies political collective decision-making. Thus, even if all voters are properly skilled and informed, majority decision-making need not lead to optimal outcomes, and may even lead to negative results. The flaws can be aggravated if decision-making is not direct but "staggered" in one way or another. An example is representational government in combination with block (e.g., partisan) voting. These and other problems attached to majority decision-making (such as the fact that majority it leads to unstable results because of the phenomenon known as "cycling") have been amply documented in the relevant literature.[31] Given the relevance of these problems and of the various proposals in the literature that are aimed at solving or diminishing them, it is a grave omission not to take them into consideration in the present discussion.

A second relevant element that is not taken into account (with the exception of a remark from Joy discussed in 7.4) is the actual and possible role of liability law. As was discussed in 5 above, the actual conditional forms of liability are inconsistent with the restricted liberty principle presented in 4, and inconsistent with the aim of controlling the adverse effects of new technologies. The (re)introduction of strict liability would be more consistent with this restricted liberty principle, and would at least partially compensate for the flaws of majority decision making mentioned above.

8. Conclusions on Universalism, Organized Skepticism and Disinterestedness

The reactions to Joy's call for relinquishment discussed in the previous section exemplify the dogma "knowledge is good" and show that and how individual scientists and technologists violate the norms of universalism and skepticism. The authors quoted impose their subjective beliefs and value judgments upon others, while failing to show how these beliefs and value judgments follow

[31] See especially Mueller (1989) for an overview of results obtained in the area of public choice.

from well-founded empirical or theoretical knowledge and/or shared normative principles. In the light of the criteria of empirical and logical scrutiny which have such a central position in the ethos of science, many of these beliefs and value judgments emerge as unfounded dogmas. It would be expected that persons committed to the principles of science would display a much more skeptical and reserved attitude towards theses that defy vindication in terms of logic and empirical fact. The impression cannot be avoided that the people quoted are defending the interests of scientists, rather than the interests of science in the sense specified by Merton. It is dubious, in other words, whether the norm of disinterestedness is being adhered to.

If Merton's norms were adopted, one would also expect that the institutions of science would stimulate open and critical discussion on issues such as the one brought forward by Joy and others, but this is not the case. For example, Joy's proposal that the AAAS start a broad discussion on the subject of relinquishment has, to my knowledge, gone unanswered.[32]

9. Overview of Conclusions

The conclusions of this paper can be summarized as follows. The assumption that scientific knowledge and its dissemination is unconditionally good is part of, or presupposed by, the ethos of science as described by Merton, notably in Merton's norm of communism. The assumption is implicit or explicit in many of the utterances of the scientists and technologists who have been quoted in

---

[32] In December 2001 the AAAS website (www.aaas.org) showed no signs of such a broader discussion taking place. There is a *Scientific freedom, responsibility and law* program with activities covering subjects such as: the use of scientific evidence in the court; misconduct in scientific research; and certification of electronic publications. Closest to Joy's topic is a report filed on this page entitled *Stem Cell Research and Application: Monitoring the Frontiers of Biomedical Research,* produced by the American Association for the Advancement of Science and the Institute for Civil Society, November 1999 (http://www.aaas.org/spp/dspp/SFRL/projects /stem/report.pdf). After having noted that "This research raises ethical and policy concerns, but these are not unique to stem cell research," the report concludes that "Federal funding for stem cell research is necessary in order to promote investment in this promising line of research, to encourage sound public policy, and to foster public confidence in the conduct of such research." It is recommended that "Public and private research on human stem cells derived from all sources (embryonic, fetal, and adult) should be conducted in order to contribute to the rapidly advancing and changing scientific understanding of the potential of human stem cells from these various sources." The report does not address the broader issues raised by Joy and in the present paper.

In the magazine *Fortune* of November 26 2001, Bill Joy among others was asked about his reaction to the terrorist attacks of September 11 2001 on Washington and New York. He was quoted as saying that "I felt after I wrote my article ["Why the Future Doesn't Need Us" – HZ] that there was no political will to address these problems [i.e. the problems discussed in that article and illustrated in the September 11 events]. That's changed. We're closer to the discussion we need to have. We're not quite there yet." (p. 58)

this paper. However, the assumption does not live up to core values of science regarding the systematic criticism and logical and empirical scrutiny featuring in Merton's norms of universalism and organized skepticism. The assumption is unjustified with regard to the actual and potential negative effects quoted above. To assume that scientific knowledge is unconditionally good and to proceed on that basis not only conflicts with the core values and principles of science, but it also brings scientists and technologists into conflict with broadly held ethical norms such as the restricted liberty principle. In addition, the spokesmen for science and technology display widespread uncritical and unreflective attitudes towards politics and law, which determine the implementation of technological feasibilities and the conditions of their use, while completely ignoring the relevant knowledge from research areas such as decision theory and public choice.

Increasing parts of scientific research and technological development can be seen as potentially harmful if not disastrous activities. In view of broadly held ethical principles of restricted liberty and reciprocity, such activities can only be justified by obtaining the informed consent of all who are subjected to the possible consequences, and in the case of any damage caused by activities for which there was no informed consent, the actors should be liable for restoration or compensation. The ethics of science, as represented by Merton's norms and as exemplified by the utterances and behavior of many scientists and technologists, does not recognize these principles. Of course, if scientific knowledge is good then these ethical principles are irrelevant, but if science is not to be a religion with dogmas, it should be critical about this assumption.

The fact that activities (such as scientific research and technological development) are legally permitted does not imply that they are also ethically permitted, given the procedures actually in use for collective (political) decision-making. It also does not follow, from the fact that there is no legal liability for the consequences of certain activities, that there should be no liability.

As witnessed by the discussion triggered by Bill Joy's essay on the hazards of science and technology and on relinquishment, spokesmen from the fields of science and technology frequently violate core elements of the ethos of science when issues concerning science and society, such as the ones addressed by Joy, are at stake. When claiming that scientific and technological development should proceed unhampered and unconditionally, core principles of scientific thinking and scientific attitude are being violated. The arguments put forward in favor of this claim do not live up to elementary criteria of logical and empirical adequacy, as demanded by the principle that "truth-claims, whatever their source, are to be subjected *to preestablished impersonal criteria*: consonant with observation and with previously confirmed knowledge" (universalism), whereas signs of "temporary suspension of judgment and the detached scrutiny of beliefs in terms of empirical and logical criteria" (organized skepticism) are absent. The "methodological and institutional mandate" of organized skepticism is not being respected, and the disinterestedness (in the sense specified by Merton) of these spokesmen for science and technology is dubious.

The proponents of the claim that the unconstrained pursuit of science and technology is good, either in itself or because of the consequences, do not succeed in showing, on the basis of logic, empirical truths and/or shared ethical values or norms, the correctness of their claim. They violate the ethical principle of restricted liberty while trying to force their unfounded claim upon others, and they neglect the fact that the similar but opposing views and behavior of others can only lead to mutual violence.

In the discussion concerning the risks and hazards of science and technology presented in this paper, the reigning procedures of collective decision-making and the reigning principles of liability in positive law are given virtually no attention. This is an omission in the light of the ethos of science, for the following reasons: (1) science as a social institution relies upon politics and law for the implementation of its results and the control of negative effects; (2) history shows that politics and law, at least in their present form, are not equal to these tasks; (3) this inadequacy of actual political procedures and actual legal principles and practices can be understood on theoretical grounds which are documented in the relevant literature. If knowledge is good, one should not contradict it. This means, among other things, that the empirical and theoretical knowledge of public choice should not be contradicted, and that it should be admitted that nobody can be ethically bound by the decisions of others.

Merton pointed out that science, because of its core values of systematic criticism and logical and empirical scrutiny, has often clashed with other areas of society, such as organized religion. With respect to the claim that knowledge is (always) good and that science should proceed undisturbed, there may again be a conflict between science and the rest of society, this time because science, by neglecting its core values, is transforming itself into a religion, based upon unfounded dogmas and with an offensive and intolerant stance toward others.

If science is to adhere to its core values, its institutions and supporters should initiate and stimulate open and critical discussion on the issues addressed above, both within science (and technology), and with society at large. There are notably two social institutions beyond science itself that should receive critical attention in these discussions, namely, the actual legal systems, and the actual procedures for collective decision-making. As long as the inadequacies of these institutions for controlling the social effects of science and technology persist, relinquishment from certain areas of research should be taken very seriously.

ACKNOWLEDGMENTS

This paper has benefited a great deal from an unpublished paper by J.F.C. van Velsen (1998), which was submitted to Science as a contribution to a discussion in Science regarding science and society. I am also indebted to him for comments on a draft version. I furthermore acknowledge comments from T.A.F. Kuipers and the editors of this volume, and from the members of the Department of Philosophy of Delft University of Technology.

Department of Philosophy
Faculty of Technology, Policy and Management
Delft University of Technology
P.O. Box 5015
2600 GA Delft
The Netherlands

REFERENCES

Dunné, J.M., van (1993). *Verbintenissenrecht, deel 2*. Tweede herziene druk. Deventer: Kluwer.

Empel, M., van and H.A. Ritsema (1987). *Aansprakelijkheid voor Produkten*. Deventer: Kluwer.

Hogarth, R.M. (1987). *Judgement and Choice. The Psychology of Decision*. Second revised edition. Chichester/New York/Brisbane/Toronto: John Wiley & Sons.

Horwitz, M.J. (1977). *The Transformation of American Law 1780-1860*. Cambridge, Mass.: Harvard University Press.

Kuipers, T.A.F. (2001). 'Default Norms' in Research Ethics. In: *Structures in Science*, pp. 343-356. Dordrecht: Kluwer.

Lindley, D.V. (1971). *Making Decisions*. Chichester, UK: Wiley.

Merton, R.K. (1973). The Normative Structure of Science. In: *The Sociology of Science*, pp. 267-278. Chicago/London: The University of Chicago Press.

Mueller, D.C. (1989). *Public Choice II*. Cambridge, UK: Cambridge University Press.

Simmons, A.J. (1993). *On the Edge of Anarchy: Locke, Consent, and the Limits of Society*. Princeton, N.J.: Princeton University Press.

Velsen, J.F.C., van (forthcoming). Science and Its Search for Support.

Velsen, J.F.C., van (2000). Relativity, Universality and Peaceful Coexistence. *Archiv für Rechts- und Sozialphilosophie* **86**, 88-108.

Zandvoort, H. (2000a). Controlling Technology Through Law: The Role of Legal Liability. In: D.Brandt, J. Cernetic (eds.), *Preprints of 7th IFAC Symposium on Automated Systems Based on Human Skill.* Joint Design of Technology and Organisation. June 15-1 2000, Aachen, Germany, pp. 247-250. Duesseldorf: VDI/VDE-Gesellschaft Mess- und Automatisierungs-technik (GMA).

Zandvoort H. (2000b). Self-Determination, Strict Liability, and Ethical Problems in Engineering. In: P.A. Kroes, A.W.M. Meijers (eds.), *The Empirical Turn in the Philosophy of Technology.* (*Research in Philosophy and Technology*, vol. 20, pp. 219-243). Amsterdam: JAI (Elsevier Science).

Zweigert, K. and H. Kötz (1987). *An Introduction to Comparative Law.* Second revised edition. Oxford: Clarendon Press.

Theo A. F. Kuipers

SELF-APPLICATION OF MERTON'S NORMS
REPLY TO HENK ZANDVOORT

As one might expect, Henk Zandvoort delivered a very interesting and sound contribution. Moreover, it is very provocative. Before making some critical remarks, I shall first try to summarize Zandvoort's main argument. On the basis of a very informative characterization of Merton's CUDOS norms, using literal quotations, he questions the main presupposition in one of the norms on the basis of (a meta-application of) two other ones. Specifically, he argues that the underlying assumption of the "communism" norm is that scientific knowledge and its dissemination are unconditionally good and that this assumption has not been evaluated in accordance with the "universalism" – and the "organised-scepticism" norm, notably as a consequence of violating the "disinterestedness" norm. Serious evaluation of the goodness assumption easily leads to the conclusion that it has many known and hence, probably, also many as yet unknown irreparable exceptions.

Zandvoort also argues that, in contrast to Merton's norms, research ethics should take account of generally recognized ethical principles, notably those of restricted liberty and responsibility. They support the classical legal principle of strict liability, rather than the modern legal principle of conditional liability, that is, liability only if the actor was "careless" or "negligent." Combining the restricted validity of the goodness assumption with strict liability, Zandvoort's far-reaching conclusion for scientific research is that "preceding informed consent" is needed "of all who may be hurt by the activities concerned." Since a sound realization of such a consent is as yet almost impossible, he finally supports the recent claim of Bill Joy that "science should relinquish from doing research into potentially dangerous areas", where Joy sees "the unilateral US abandonment, without preconditions, of the development of biological weapons" as a hopeful historical precedent. Zandvoort, quite convincingly, shows, on the basis of the reactions to Joy's plea that scientists do not evidently exemplify the disinterestedness norm in this discussion. In the following I first very briefly comment on these reactions or, as the case may

In: R. Festa, A. Aliseda and J. Peijnenburg (eds.), *Cognitive Structures in Scientific Inquiry* (*Poznań Studies in the Philosophy of the Sciences and the Humanities,* vol. 84), pp. 499-501. Amsterdam/New York, NY: Rodopi, 2005.

be, on Zandvoort's discussion of them and then discuss a point about problematic political systems.

Arguments Against Relinquishment of Certain Directions of Research

In his Section 7 Zandvoort reviews six arguments against relinquishment used by scientists in the discussion with Bill Joy. In all cases the risk of bias due to self-interest is evident. I first quote Zandvoort's characterization of them and then give my comment, with or without a substantial argument.

(1) "Science and technology are unconditionally/absolutely good, that is, in themselves, irrespective of consequences." This clear example of a deontological principle illustrates how naïve such principles can be in a pure form.

(2) "Fatalism," that is, the view that "the course of scientific and technological development cannot be altered, and that we should live with the consequences, come what may." The fatalistic position is, strictly speaking, just false, since developments can be blocked, for it is logically possible to reach effective agreement among politicians and scientists. The case of "reproductive cloning," as opposed to "therapeutic cloning," may become an example.

(3) "Positive effects outweigh negative effects." Although Zandvoort is quite right in claiming that it is difficult to evaluate this claim, in particular for future developments, I would like to suggest that public opinion on overall "past performance" of science and technology should here be taken as the crucial criterion, followed by two inductive leaps. First, a fair sample of well-established developments can be evaluated by carefully interviewing people all over the world, which may support, first, the conclusion that the general claim is true in the eyes of almost all people and for almost all well-established developments and if so, second, that this will be the case for developments in the near future, hence leaving room for a later break in the public opinion. Of course, the second inductive jump is in itself already more problematic than the first, but even more so because it should be compared with the price of missing possible positive developments due to blockades, which is also very difficult to estimate. Be this as it may, in my opinion public opinion on overall past performance is crucial. Among other things, it circumvents the problem of bias due to self-interest when scientists would have to judge past performance.

(4) "Science and technology are actually under control." I can easily agree with Zandvoort that this is again a rather naïve contribution to the debate.

(5) "Relinquishment may be worse than unrestricted continuation of scientific and technological development." Here I would like to quote the very last sentences of SiS in which I compare the risks of a general code of incorruptible research conduct with the risks of ethical review procedures for

research proposals: "Pettit (1992) argues that such procedures endanger valuable research on human beings. Without precautionary measures, 'it is likely to carry us along a degenerating trajectory', avoiding all kinds of important research which might lead to ethical blockades. Hence, the question is whether a general code is possible that is not the start of a degenerating trajectory but a useful new point of reference in the interest of science and society." Indeed, relinquishment may block valuable research even more than ethical review procedures and general codes. However, I should concede that in some cases the continuation of research may be worse.

(6) "The public would consent if properly educated and informed." This is indeed also a case of unprovable wishful thinking, but, referring to (3), I would suggest that the public opinion should primarily be investigated with respect to overall past performance of science and technology. Although education and (neutral) information remain important in this case, many lay people roughly know already what they are talking about. Precisely such people should inform the rest of the public. The merits and problems with IVF may be a typical modern case in point.

Problematic Political Systems

In Section 3, Zandvoort writes: "when it comes to preventing or controlling negative side effects or abuses of modern scientific and technological knowledge, it is the weakest existing political or legal system that matters most." Here I think that a distinction should be drawn between negative side effects and abuses that can be prevented or controlled within a country and effects and abuses that are likely to become worldwide. In the first case it seems perfectly legitimate to me that a country allows the relevant research. It cannot be held responsible for the fact that other countries may not be able to maintain control of the negative side effects or abuses of applying the openly published results. For example, it may be that a new apartment building technology can only be applied safely under very strict conditions that require government prescription and control of a kind that some countries are not yet able to install and maintain. However, in the second case, when effects and abuses cannot be controlled within countries, the situation is different. The unilateral USA relinquishment of biological weapon research is of course at least partially inspired by the risk that technological information, although attempts are made to keep it secret, nevertheless falls into the hands of enemies of the USA.

REFERENCE

Pettit, Ph. (1992). Instituting a Research Ethics. Chilling and Cautionary Tales. *Bioethics* **6** (2), 89-112.

BIBLIOGRAPHY OF THEO A.F. KUIPERS

Biographical Notes

Theo A.F. Kuipers (b. Horst, Limburg, NL, 1947) studied mathematics at the Technical University of Eindhoven (1964-7) and philosophy at the University of Amsterdam (1967-71). In 1978 he received his Ph.D. degree from the University of Groningen, defending a thesis on inductive logic (*Studies in Inductive Probability and Rational Expectation, Synthese Library*, vol. 123, 1978). The supervisors were J.J.A. Mooij and A.J. Stam. From 1971 to 1975 he was deputy secretary of the Faculty of Philosophy of the University of Amsterdam. In 1975 he was appointed Assistant Professor of the philosophy of science in the Faculty of Philosophy of the University of Groningen; in 1985 he became associate professor and full professor since 1988. He married Inge E. de Wilde in 1971.

A synthesis of his work on confirmation, empirical progress and truth approximation, entitled *From Instrumentalism to Constructive Realism*, appeared in 2000 (*Synthese Library*, vol. 287). A companion synthesis of his work on the structure of theories, research programs, explanation, reduction, and computational discovery and evaluation, entitled *Structures in Science*, appeared in 2001 (*Synthese Library*, vol. 301).

The works he has edited include *What is Closer-to-the-Truth? A Parade of Approaches to Truthlikeness* (*Poznań Studies in the Philosophy of the Sciences and the Humanities*, vol. 10, 1987). He also edited, with Anne Ruth Mackor, *Cognitive Patterns in Science and Common Sense. Groningen Studies in Philosophy of Science, Logic, and Epistemology* (*Poznań Studies in the Philosophy of the Sciences and the Humanities*, vol. 45, 1995).

He was one of the main supervisors of the Ph.D. theses of Henk Zandvoort (1985), Rein Vos (1988), Maarten Janssen (1990), Gerben Stavenga (1991), Roberto Festa (1992), Frank Berndsen (1995), Jeanne Peijnenburg (1996), Anne Ruth Mackor (1997), Rick Looijen (1998), Sjoerd Zwart (1998), Eite Veening (1998), Alexander van den Bosch (2001), and Esther Stiekema (2002). In one way or another, he was also involved in several other Ph.D. theses in Groningen, Amsterdam (VU and UvA), Rotterdam, Nijmegen, Utrecht, Ghent, Leuven, Lublin and Helsinki.

During the academic years 1982/3 and 1996/7 he was a fellow of the Netherlands Institute of Advanced Study (NIAS) at Wassenaar.

Besides working in the Faculty of Philosophy, being Dean for a number of periods, he is an active member of the Graduate School for Behavioral and Cognitive Neurosciences (BCN) of which he chaired the research committee for a number of years.

On the national level he was one of the initiators of the section of philosophy of science as well as of the Foundation for Philosophical Research (SWON) of the National Science Foundation (ZWO/NWO). During 1997-2003 he was 'the philosopher member' of the Board of the Humanities of NWO. Since 2000 he has chaired the Dutch Society for Philosophy of Science.

He is a member of the Coordination Committee of the Scientific Network on Historical and Contemporary Perspectives of Philosophy of Science in Europe of the European Science Foundation (ESF).

His research group, which is working on the program *Cognitive Structures in Knowledge and Knowledge Development*, received the highest possible scores from the international assessment committee of Dutch philosophical research in the periods 1989-93 and 1994-8.

Publications

1971

0. *Inductieve Logica en Haar Beperkingen* (unpublished masters thesis). University of Amsterdam. 1971, 64 pp.

1972

1. De Wetenschapsfilosofie van Karl Popper. *Amersfoortse Stemmen* **53** (4), 1972, 122-6.
2. Inductieve Waarschijnlijkheid, de Basis van Inductieve Logica. *Algemeen Nederlands Tijdschrift voor Wijsbegeerte* **64** (4), 1972, 291-6.
3. A Note on Confirmation. *Philosophica Gandensia* **10**, 1972, 76-7.
4. Inductieve Logica. *Intermediair* **49**, 1972, 29-33.

1973

5. A Generalization of Carnap's Inductive Logic. *Synthese* **25**, 1973, 334-6. Reprinted in: J. Hintikka (ed.), *Rudolf Carnap* (*Synthese Library*, vol. 73). Dordrecht: Reidel, 1977.

1976

6. Inductive Probability and the Paradox of Ideal Evidence. *Philosophica* **17** (1), 1976, 197-205.

1977

7. Het Verschijnsel Wetenschapsfilosofie, Bespreking van Herman Koningsveld, het Verschijnsel Wetenschap. *Kennis en Methode* **I** (3), 1977, 271-9.
8. A Two-Dimensional Continuum of a Priori Probability Distributions on Constituents. In: M. Przełęcki, K. Szaniawski, R. Wójcicki (eds.), *Formal Methods in the Methodology of Empirical Sciences* (*Synthese Library*, vol. 103), pp. 82-92. Dordrecht: Reidel, 1977.

1978

9. On the Generalization of the Continuum of Inductive Methods to Universal Hypotheses. *Synthese* **37**, 1978, 255-84.
10. *Studies in Inductive Probability and Rational Expectation.* Ph.D. thesis University of Groningen, 1978. Also published as: *Synthese Library*, vol. 123, Dordrecht: Reidel, 1978, 145 pp.
11. Replicaties, een Reactie op een Artikel van Louis Boon. *Kennis en Methode* **II** (3), 1978, 278-9.

1979

12. Diminishing Returns from Repeated Tests. *Abstracts 6-th LMPS-Congress*, Section 6, Hannover, 1979, 118-22.
13. Boekaankondiging: G. de Brock e.a., De Natuur: Filosofische Variaties. *Algemeen Nederlands Tijdschrift Voor Wijsbegeerte* **71**.3, 1979, 200-1.

1980

14. A Survey of Inductive Systems. In: R. Jeffrey (ed.), *Studies in Inductive Logic and Probability*, pp. 183-92. Berkeley: University of California Press, 1980.
15. Nogmaals: Diminishing Returns from Repeated Tests. *Kennis en Methode* **IV** (3), 1980, 297-300.
16. a.Comment on D. Miller's "Can Science Do Without Induction?"
    b.Comment on I. Niiniluoto's "Analogy, Transitivity and the Confirmation of Theories." In: L.J. Cohen, M. Hesse (eds.), *Applications of Inductive Logic*, (1978), pp.151-2/244-5. Oxford: Clarendon Press, 1980.

1981

17. (Ed.) *Hoofdfiguren in de Hedendaagse Filosofie van de Natuurwetenschappen* (redactie, voorwoord (89) en inleiding (90-3)). *Wijsgerig Perspectief* **21** (4), (1980-) 1981. 26 pp.

1982

18. The Reduction of Phenomenological to Kinetic Thermostatics. *Philosophy of Science* **49** (1), 1982, 107-19.
19. Approaching Descriptive and Theoretical Truth. *Erkenntnis* **18** (3), 1982, 343-78.

1983

20. Methodological Rules and Truth. *Abstracts 7-th LMPS-Congress,* vol. **3** (Section 6), Salzburg, 1983, 122-5.
21. Non-Inductive Explication of Two Inductive Intuitions. *The British Journal for the Philosophy of Science* **34** (3), 1983, 209-23.

1984

22. Olson, Lindenberg en Reductie in de Sociologie. *Mens en Maatschappij* **59** (1), 1984, 45-67.
23. Two Types of Inductive Analogy by Similarity. *Erkenntnis* **21** (1), 1984, 63-87.
24. Oriëntatie: Filosofie in Polen (samenstelling, inleiding en vertaling). *Wijsgerig Perspectief* **24** (6), (1983-)1984, 216-21.
25. Empirische Mogelijkheden: Sleutelbegrip van de Wetenschapsfilosofie. *Kennis en Methode* **VIII** (3), 1984, 240-63.
26. Inductive Analogy in Carnapian Spirit. In: P.D. Asquith, Ph. Kitcher (eds.), *PSA 1984, Volume One* (Biennial Meeting Philosophy of Science Association in Chicago), pp. 157-67. East Lansing: PSA, 1984.

27. Utilistic Reduction in Sociology: The Case of Collective Goods. In: W. Balzer, D.A. Pearce, H.-J. Schmidt (eds.), *Reduction in Science. Structure, Examples, Philosophical Problems* (*Synthese Library*, vol. 175, Proc. Conf. Bielefeld, 1983), pp.239-67. Dordrecht: Reidel, 1984.

28. What Remains of Carnap's Program Today? In: E. Agazzi, D. Costantini (eds.), *Probability, Statistics, and Inductive Logic*, *Epistemologia* **7**, 1984, 121-52; Proc. Int. Conf. 1981 at Luino, Italy. With discussions with D. Costantini (149-51) and W. Essler (151-2) about this paper and with E. Jaynes (71-2) and D.Costantini (166-7) about theirs.

29. An Approximation of Carnap's Optimum Estimation Method. *Synthese* **61**, 1984, 361-2.

30. Approaching the Truth with the Rule of Success. In: P. Weingartner, Chr. Pühringer (eds.), *Philosophy of Science – History of Science*, Selection 7th LMPS Salzburg 1983, *Philosophia Naturalis* **21** (2/4), 1984, 244-53.

1985

31. The Paradigm of Concretization: The Law of Van der Waals. *Poznań Studies in the Philosophy of the Sciences and the Humanities*, vol. 8 (ed. J. Brzeziński), Amsterdam: Rodopi, 1985, pp. 185-99.

32. (met Henk Zandvoort), Empirische Wetten en Theorieën. *Kennis en Methode* **9** (I), 1985. 49-63.

33. The Logic of Intentional Explanation. In: J. Hintikka, F.Vandamme (Eds.), *The Logic of Discourse and the Logic of Scientific Discovery* (Proc. Conf. Gent, 1982), *Communication and Cognition* **18** (1/2), 1985, 177-98. Translated as: Logika wyjaśniania intencjonalnego. *Poznańskie Studia z Filozofii Nauki* **10**, 1986, 189-218.

34. Een Beurs voor de Verdeling van Arbeidsplaatsen. *Filosofie & Praktijk* **6** (4), 1985, 205-11.

1986

35. Some Estimates of the Optimum Inductive Method. *Erkenntnis* **24**, 1986, 37-46.

36. The Logic of Functional Explanation in Biology. In: W. Leinfellner, F. Wuketits (eds.), *The Tasks of Contemporary Philosophy* (Proc. 10th Wittgenstein Symp. 1985), pp. 110-4. Wenen: Hölder-Pichler-Temsky, 1986.

37. Intentioneel Verklaren van Handelingen. In: Proc. Conf. Handelingspsychologie, ISvW- Amersfoort 1985. *Handelingen*. O-nr, 1986, 12-18.

38. Explanation by Specification. *Logique et Analyse* **29** (116), 1986, 509-21.

1987

39. (Ed.) *What is Closer-To-The-Truth? A Parade of Approaches to Truthlikeness* (*Poznań Studies in the Philosophy of the Sciences and the Humanities*, vol. 10). Amsterdam: Rodopi, 1987, 254 pp. Introduction: 1-7.

40. A Structuralist Approach to Truthlikeness, in 39: 79-99.

41. Truthlikeness of Stratified Theories, in 39: 177-86.

42. (Ed.) *Holisme en Reductionisme in de Empirische Wetenschappen, Kennis en Methode* **11** (I), 1987. 136 pp., Voorwoord: 4-5.

43. Reductie van Wetten: een Decompositiemodel, in 42: 125-35.

44. Fascinaties: Wetenschappelijk Plausibel en Toch Taboe. *VTI* (contactblad Ver. tot Instandhouding Int. School v. Wijsbegeerte), nr.13, juli 1987, 5-8; discussie met J. Hilgevoord in nr. 14, 1987, 6-9.

45. A Decomposition Model for Explanation and Reduction. *Abstracts LMPS-VIII*, Moscow, 1987, vol. 4, 328-31.

46. Truthlikeness and the Correspondence Theory of Truth. In: P. Weingartner, G. Schurz (eds.), *Logic, Philosophy of Science and Epistemology*, Proc. 11th Wittgenstein Symp.1986, pp. 171-6. Wenen: Hölder-Pichler-Temsky, 1987.

47. Reductie van Begrippen: Stappenschema's. *Kennis en Methode* **11** (4), 1987, 330-42.

1988

48. Voorbeelden van Cognitief Wetenschapsonderzoek. *WO-NieuwsNet* **I** (I), 1988, 13-29.

49. Structuralistische Explicatie van Dialectische Begrippen. *Congresbundel Filosofiedag Maastricht 1987*, pp. 191-7. Delft: Eburon, 1988.

50. Inductive Analogy by Similarity and Proximity. In: D.H. Helman (ed.), *Analogical Reasoning*, pp. 299-313. Dordrecht: Kluwer Academic Publishers, 1988.

51. (with Hinne Hettema), The Periodic Table – its Formalization, Status, and Relation to Atomic Theory. *Erkenntnis* **28**, 1988, 387-408.

52. Cognitive Patterns in the Empirical Sciences: Examples of Cognitive Studies of Science. *Communication and Cognition* **21** (3/4), 1988, 319-41. Translated as: Modele kognitywistyczne w naukach empirycznych: przykłady badań nad nauką, *Poznańskie Studia z Filozofii Humanistyki* **14** (1), 1994, 15-41.

1989

53. (Ed.) *Arbeid en Werkloosheid.* Redactie, inleiding, discussie thema-nummer *Wijsgerig Perspectief* **29** (4), (1988-) 1989.

54. (with Maarten Janssen), Stratification of General Equilibrium Theory: A Synthesis of Reconstructions. *Erkenntnis* **30**, 1989, 183-205.

55. *Onderzoeksprogramma's Gebaseerd op een Idee. Impressies van een Wetenschapsfilosofische Praktijk*, inaugural address University of Groningen. Assen: Van Gorcum, 1989. 32 pp.

56. How to Explain the Success of the Natural Sciences. In: P. Weingartner, G. Schurz (eds.), *Philosophy of the Natural Sciences* Proc. 13th Int. Wittgenstein Symp. 1988, pp. 318-22. Wenen: Hölder-Pichler-Temsky, 1989.

1990

57. (Ed. with J. Brzeziński, F. Coniglione, and L. Nowak) *Idealization I: General Problems, Idealization II: Forms and Applications (Poznań Studies in the Philosophy of the Sciences and the Humanities,* vol. 16+17), Rodopi, Amsterdam-Atlanta, 1990.

58. Reduction of Laws and Concepts. In 57 I: 241-76.

59. Het Objectieve Waarheidsbegrip in Waarder. *Kennis en Methode* **XIV** (2), 1990, 198-211. (Met een reactie van Hans Radder: 212-15).

60. (met Hauke Sie), Industrieel en Academisch Onderzoek. *De Ingenieur*, nr. 6 (juni), 1990, 15-8.

61. Interdisciplinariteit en Gerontologie. In: D. Ringoir en C. Tempelman (ed.), *Gerontologie en Wetenschap*, pp. 143-9. Nijmegen: Netherlands Institute of Gerontology, 1990.

62. Het Belang van Onware Principes. *Wijsgerig Perspectief* **31** (1), 1990, 27-9.

1991

63. Economie in de Spiegel van de Natuurwetenschappen: Overeenkomsten, Plausibele Verschillen en Specifieke Rariteiten. *Kennis en Methode* **XV** (2), 1991, 182-97.

64. Realisme en Convergentie, of Hoe het Succes van de Natuurwetenschappen Verklaard Moet Worden. In: J. van Brakel en D. Raven (ed.), *Realisme en Waarheid*, pp. 61-83. Assen: Van Gorcum, 1991.

65. On the Advantages of the Possibility-Approach. In: A. Ingegno (ed.), *Da Democrito a Collingwood*, pp. 189-202. Firenze: Olschki, 1991.

66. Structuralist Explications of Dialectics. In: G. Schurz and G. Dorn (Eds.), *Advances in Scientific Philosophy. Essays in honour of Paul Weingartner on the occasion of the 60-th anniversary of his birthday* (*Poznań Studies in the Philosophy of the Sciences and the Humanities,* vol. 24), pp.295-312. Amsterdam-Atlanta: Rodopi, 1991.

67. Dat Vind Ik Nou Mooi. In: R.Segers (ed.), *Visies op Cultuur en Literatuur. Opstellen naar aanleiding van het werk van J.J.A. Mooij*, pp. 69-75. Amsterdam: Rodopi, 1991.

1992

68. (Ed.) *Filosofen in Actie*. Delft: Eburon, 1992. 255 pp.

69. Methodologische Grondslagen voor Kritisch Dogmatisme. In: J.W. Nienhuys (ed.), *Het Vooroordeel van de Wetenschap*, ISvW-conferentie 23/24 februari 1991, pp. 43-51. Utrecht: Stichting SKEPSIS, 1992.

70. (with Rein Vos and Hauke Sie), Design Research Programs and the Logic of Their Development. *Erkenntnis* **37** (1), 1992, 37-63. Translated as: Projektowanie programów badawczych i logika ich rozwoju. *Projektowanie i Systemy* **15**, 1995, pp. 29-48.

71. Truth Approximation by Concretization. In: J. Brzeziński and L. Nowak (eds.), *Idealization III: Approximation and Truth (Poznań Studies in the Philosophy of the Sciences and the Humanities*, vol. **25**), pp. 159-79. Amsterdam-Atlanta: Rodopi, 1992.

72. Naive and Refined Truth Approximation. *Synthese* **93**, 1992, 299-341.

73. Wetenschappelijk Onderwijs. In: *ABC van Minder Docentafhankelijk Onderwijs*, 25 jarig jubileum uitgave, pp. 133-7. Groningen: COWOG, 1992.

1993

74. On the Architecture of Computational Theory Selection. In: R. Casati & G. White (eds.), *Philosophy and the Cognitive Sciences*, pp. 271-78. Kirchberg: Austrian Ludwig Wittgenstein Society, 1993.

75. Computationele Wetenschapsfilosofie. *Algemeen Nederlands Tijdschrift voor Wijsbegeerte* **85** (4), 1993, 346-61.

76. De Pavarotti's van de Analytische Filosofie. *Filosofie Magazine* **2** (8), 1993, 36-9. Bewerking in: D. Pels en G. de Vries, *Burgers en Vreemdelingen*, t.g.v. afscheid L.W. Nauta, pp. 99-107. Amsterdam: Van Gennep, 1994. Reacties van Menno Lievers, Anthonie Meijers, Filip Buekens en Stefaan Cuypers, gevolgd door repliek TK: *Filosofie Magazine* **3** (1), 1994, 37-40.

77. Wetenschappelijk Onderwijs en Wijsbegeerte van een Wetenschapsgebied. *Universiteit en Hogeschool* **40** (1), 1993, 9-18.

1994

78. (with Andrzej Wiśniewski) An Erotetic Approach to Explanation by Specification. *Erkenntnis* **40** (3), 1994, 377-402.

79. (with Kees Cools and Bert Hamminga), Truth Approximation by Concretization in Capital Structure Theory. In: B. Hamminga and N.B. De Marchi (eds.), *Idealization VI: Idealization in Economics* (*Poznań Studies in the Philosophy of the Sciences and the Humanities,* vol. 38), pp. 205-28. Amsterdam-Atlanta: Rodopi, 1994.

80. Falsificationisme Versus Efficiënte Waarheidsbenadering. Of de Ironie van de List der Rede. *Algemeen Nederlands Tijdschrift voor Wijsbegeerte* **86** (4), 1994, 270-90.

81. The Refined Structure of Theories. In: M. Kuokkanen (ed.), *Idealization VII: Structuralism, Idealization, Approximation* (*Poznań Studies in the Philosophy of the Sciences and the Humanities*, vol. 42), pp. 3-24. Amsterdam-Atlanta: Rodopi, 1994.

1995

82. Observationele, Referentiële en Theoretische waarheidsbenadering (Reactie op Ton Derksen). *Algemeen Nederlands Tijdschrift voor Wijsbegeerte* **87** (1), 1995, 33-42.

83. Falsificationism Versus Efficient Truth Approximation. In: W. Herfel, W. Krajewski, I. Niiniluoto and R. Wojcicki (eds.), *Theories and Models in Scientific Processes* (*Poznań Studies in the Philosophy of the Sciences and the Humanities*, vol. 44), pp. 359-86. Amsterdam-Atlanta: Rodopi, 1995. (Extended and translated version of 80).

84. Ironie van de List der Rede. *Wijsgerig Perspectief* **35** (6), (1994-)1995, 189-90.

85. (Ed. with Anne Ruth Mackor), *Cognitive Patterns in Science and Common Sense. Groningen Studies in Philosophy of Science, Logic, and Epistemology*. With a foreword by Leszek Nowak. *Poznań Studies in the Philosophy of the Sciences and the*

Humanities, vol. 45. Amsterdam-Atlanta: Rodopi, 1995. With a general introduction ("Cognitive Studies of Science and Common Sense", pp. 23-34) and special introductions to the four parts.

86. Explicating the Falsificationist and the Instrumentalist Methodology by Decomposing the Hypothetico-Deductive Method. In 85: 165-86.

87. (with Hinne Hettema), Sommerfeld's *Atombau*: A Case Study in Potential Truth Approximation. In 85: 273-97.

88. Verborgen en Manifeste Psychologie in de Wetenschapsfilosofie. *Nederlands Tijdschrift voor Psychologie* **50** (6), 1995, 252.

1996

89. Truth Approximation by the Hypothetico-Deductive Method. In: W. Balzer, C.U. Moulines and J.D. Sneed (eds), *Structuralist Theory of Science: Focal Issues, New Results*, pp.83-113. Berlin: Walter de Gruyter, 1996.

90. Wetenschappelijk en Pseudowetenschappelijk Dogmatisch Gedrag. *Wijsgerig Perspectief* **36** (4), (1995-)1996, 92-7.

91. Het Softe Paradigma. Thomas Kuhn Overleden. *Filosofie Magazine* **5** (7), 1996, 28-31.

92. Explanation by Intentional, Functional, and Causal Specification. In: A. Zeidler-Janiszewska (ed.), *Epistemology and History. Humanities as a Philosophical Problem and Jerzy Kmita's Approach to It* (*Poznań Studies in the Philosophy of the Sciences and the Humanities*, vol. 47), pp.209-36. Amsterdam-Atlanta: Rodopi, 1996.

93. Efficient Truth Approximation by the Instrumentalist, Rather Than the Falsificationist Method. In: I. Douven and L. Horsten (eds.), *Realism in the Sciences* (*Louvain Philosophical Studies*, vol. 10, pp. 115-30. Leuven: Leuven University Press, 1996.

1997

94. Logic and Philosophy of Science: Current Interfaces. (Introduction to the proceedings of a special symposium with the same name). In: M.L. Dalla Chiara, K. Doets, D. Mundici and J. van Benthem (eds.), *Logic and Scientific Methods*, vol. 1, (*10th LMPS International Congress*, Florence, August, 1995), pp.379-81. Dordrecht: Kluwer Academic Publishers, 1997.

95. The Carnap-Hintikka Programme in Inductive Logic. In: Matti Sintonen (Ed.), *Knowledge and Inquiry: Essays on Jaakko Hintikka's Epistemology and Philosophy of Science.* (*Poznań Studies in the Philosophy of the Sciences and the Humanities*, vol. 51), pp. 87-99. Amsterdam-Atlanta: Rodopi, 1997. With a comment by Hintikka, pp. 317-18.

96. Boekaankondiging: A. Derksen (ed.), *The Scientific Realism of Rom Harré*. Tilburg: Tilburg University Press, 1994, *Algemeen Nederlands Tijdschrift voor Wijsbegeerte* **89** (2), 1997, 174.

97. The Dual Foundation of Qualitative Truth Approximation. *Erkenntnis* **47** (2), 1997, 145-79.

98. Comparative Versus Quantitative Truthlikeness Definitions: Reply to Thomas Mormann. *Erkenntnis* **47** (2), 1997, 187-92.

1998

99. Confirmation Theory. *The Routledge Encyclopedia of Philosophy*, vol. 2, 1998, 532-36.

100. Pragmatic Aspects of Truth Approximation. In: P. Weingartner, G. Schurz and G. Dorn (eds.), *The Role of Pragmatics in Contemporary Philosophy*, pp.288-300. Proceedings of the 20th International Wittgenstein-Symposium, August 1997. Vienna: Hölder-Pichler-Temsky, 1998.

1999

101. Kan Schoonheid de Weg Wijzen naar de Waarheid? *Algemeen Nederlands Tijdschrift voor Wijsbegeerte* **91** (3), 1999, 174-93.

102. The Logic of Progress in Nomological, Design and Explicative Research. In: J. Gerbrandy, M. Marx, M. de Rijke, and Y. Venema (eds.), *JFAK. Essays Dedicated to Johan van Benthem on the Occasion of his 50th Birthday*, CD-ROM, Amsterdam University Press, Series Vossiuspers, Amsterdam, ISBN 90 5629 104 1, 1999. (Unique) Book edition vol. 3, 1999, pp. 37-46.

103. Zeker Lezen: Wetenschapsfilosofie. *Wijsgerig Perspectief* **39** (6), 1999, 170-1.

104. De Integriteit van de Wetenschapper. In: E. Kimman, A. Schilder, en F. Jacobs (ed.), *Drieluijk: Godsdienst, Samenleving, Bedrijfsethiek, Liber Amicorum voor Henk van Luijk*, pp. 99-109. Amsterdam: Thela-Thesis, 1999.

105. Abduction Aiming at Empirical Progress or Even at Truth Approximation, Leading to a Challenge for Computational Modelling. In: J. Meheus, T. Nickles (eds.), *Scientific Discovery and Creativity*, special issue of *Foundations of Science* **4** (3), 1999, 307-23.

2000

106. *From Instrumentalism to Constructive Realism. On Some Relations Between Confirmation, Empirical Progress, and Truth Approximation* (*Synthese Library*, vol. 287). Dordrecht: Kluwer Academic Publishers, 2000.

107. Filosofen als Luis in de Pels. Over Kritiek, Dogma's en het Moderne Turven van Publicaties en Citaties. In: J. Bremmer (ed.), *Eric Bleumink op de Huid Gezeten. Opstellen aangeboden door het College van Decanen ter gelegenheid van zijn afscheid als Voorzitter van het College van Bestuur van de Rijksuniversiteit Groningen op 24 mei 2000*, pp.89-103. Groningen: Uitgave RUG, 2000.

108. (with Hinne Hettema), The Formalisation of the Periodic Table. In: W. Balzer, J. Sneed, U. Moulines (eds), *Structuralist Knowledge Representation. Paradigmatic Examples* (*Poznań Studies in the Philosophy of the Sciences and the Humanities*, vol. 75), pp. 285-305. Amsterdam-Atlanta: Rodopi, 2000. (Revised version of 51.)

2001

109. Epistemological Positions in the Light of Truth Approximation. In: T.Y. Cao (ed.), *Philosophy of Science* (Proceedings of the 20th World Congress of Philosophy, Boston, 1998, vol. 10), pp. 79-88. Bowling Green: Philosophy Documentation Center, Bowling Green State University, 2001.

110 Naar een Alternatieve Impactanalyse. *De Academische Boekengids*, 26. Amsterdam: AUP, 2001, p. 16.

111 *Structures in Science. Heuristic Patterns Based on Cognitive Structures. An Advanced Textbook in Neo-Classical Philosophy of Science.* (*Synthese Library*, vol. 301). Dordrecht: Kluwer Academic Publishers, 2001.

112 Qualitative Confirmation by the HD-Method. *Logique et Analyse* **41** (164), 1998 (in fact 2001), 271-99.

2002

113 Beauty, a Road to The Truth. *Synthese* **131** (3), 291-328.

114 Poppers Filosofie van de Natuurwetenschappen. *Wijsgerig Perspectief* **42** (2), 2002, 17-31.

115 Quantitative Confirmation, and its Qualitative Consequences. *Logique et Analyse* **42** (167/8), 1999 (in fact 2002), 447-82.

116 Aesthetic Induction, Exposure Effects, Empirical Progress, and Truth Approximation. In: R. Bartsch e.a (ed.), *Filosofie en Empirie. Handelingen 24e NV-Filosofiedag*, 2-11-2002, pp.194-204, Amsterdam: UvA-Wijsbegeerte, 2000.

117 O dwóch rodzajach idealizacji i konkretyzacji. Przypadek aproksymacji prawdy. In: J. Brzeziński, A. Klawiter, T. Kuipers, K Łastowski, K. Paprzycka and P. Przybysz (eds), *Odwaga Filozofowania. Leszkowi Nowakowi w darze*, pp. 117-139. Poznań: Wydawnictwo Fundacji Humaniora, 2002.

2003

2004

118 Inference to the Best Theory, Rather Than Inference to the Best Explanation. Kinds of Abduction and Induction. In: F. Stadler (ed.), *Induction and Deduction in the Sciences*, Proceedings of te ESF-workshop *Induction and Deduction in the Sciences*, Vienna, July, 2002, pp. 25-52, followed by a commentary of Adam Grobler, pp. 53-36, Dordrecht: Kluwer Academic Publishers, 2004.

119 De Logica van de G-Hypothese. Hoe Theologisch Onderzoek Wetenschappelijk Kan Zijn. In: K. Hilberdink (red.), *Van God Los? Theologie tussen Godsdienst en Wetenschap* 59-74, Amsterdam: KNAW, 2004.

2005

120 The Threefold Evaluation of Theories: A Synopsis of *From Instrumentalism to Constructive Realism* (2000) + replies to 17 contributions. In: Roberto Festa, Atocha Aliseda, and Jeanne Peijnenburg (eds.), *Confirmation, Empirical Progress, and Truth Approximation, Essays in Debate with Theo Kuipers*, Volume 1. *Poznań Studies in the Philosophy of the Sciences and the Humanities.* This volume.

121 Structures in Scientific Cognition: A Synopsis of *Structures in Science. Heuristic Patterns Based on Cognitive Structures* (2001) + replies to 17 contributions. In: Roberto Festa, Atocha Aliseda, and Jeanne Peijnenburg (Eds.), *Cognitive Structures in Scientific Inquiry, Essays in Debate with Theo Kuipers*, Volume 2. *Poznań Studies in the Philosophy of the Sciences and the Humanities.* The companion volume.

To appear

- Inductive Aspects of Confirmation, Information, and Content. To appear in the volume of *The Library of Livings Philosophers* (Schilpp) dedicated to Jaakko Hintikka.
- Empirical and Conceptual Idealization and Concretization. The Case of Truth Approximation. To appear in English edition Liber Amicorum for Leszek Nowak. It appeared already in the Polish edition: 117.

INDEX OF NAMES

MONOGRAPHS-IN-DEBATE

CONTENTS OF BACK ISSUES

VOLUME 81 (2004)

Evandro Agazzi

RIGHT, WRONG AND SCIENCE
THE ETHICAL DIMENSIONS OF THE TECHNO-SCIENTIFIC ENTERPRISE
(Edited by Craig Dilworth)

VOLUME 83 (2005)

CONFIRMATION, EMPIRICAL PROGRESS AND TRUTH APPROXIMATION

ESSAYS IN DEBATE WITH THEO KUIPERS, VOLUME 1

(Edited by Roberto Festa, Atocha Aliseda and Jeanne Peijnenburg)